Digital Signal Processing

T0177947

K. Deergha Rao · M. N. S. Swamy

Digital Signal Processing

Theory and Practice

 Springer

K. Deergha Rao
Department of Electronics
and Communication Engineering
Vasavi College of Engineering
(affiliated to Osmania University)
Hyderabad, Telangana
India

M. N. S. Swamy
Department of Electrical and Computer
Engineering
Concordia University
Montreal, QC
Canada

ISBN 978-981-13-4058-1   ISBN 978-981-10-8081-4   (eBook)
https://doi.org/10.1007/978-981-10-8081-4

This Springer imprint is published by the registered company Springer Nature Singapore Pte Ltd.
part of Springer Nature
The registered company address is: 152 Beach Road, #21-01/04 Gateway East, Singapore 189721, Singapore

असतोमा सदगमय
तमसोमा ज्योतिर्गमय
मृत्योर्मा अमृतं गमय

Lead me from the unreal to the real

Lead me from darkness to enlightenment

Lead me from death to immortality

To

My parents Dalamma and Boddu,

My beloved wife Sarojini,

and my beloved teacher Prof. D. C. Reddy
                                K. Deergha Rao

To

My parents, teachers

and my beloved wife Leela.
                                M. N. S. Swamy

Preface

Digital signal processing (DSP) is now a core subject in electronics, communications, and computer engineering curricula. The motivation in writing this book is to include modern topics of increasing importance not included in the textbooks available on the subject of digital signal processing and also to provide a comprehensive exposition of all aspects of digital signal processing. The text is integrated with MATLAB-based programs to enhance the understanding of the underlying theories of the subject.

This book is written at a level suitable for undergraduate and master students as well as for self-study by researchers, practicing engineers, and scientists. Depending on the chapters chosen, this text can be used for teaching a one- or two-semester course, for example, introduction to digital signal processing, multirate digital signal processing, multirate and wavelet signal processing, digital filters design and implementation.

In this book, many illustrative examples are included in each chapter for easy understanding of the DSP concepts. An attractive feature of this book is the inclusion of MATLAB-based examples with codes to encourage readers to implement on their personal computers to become confident of the fundamentals and to gain more insight into digital signal processing. In addition to the problems that require analytical solutions, problems that require solutions using MATLAB are introduced to the reader at the end of each chapter. Another attractive feature of this book is that many real-life signal processing design problems are introduced to the reader by the use of MATLAB and programmable DSP processors. This book also introduces three chapters of growing interest not normally found in an upper division text on digital signal processing. In less than 20 years, wavelets have emerged as a powerful mathematical tool for signal and image processing. In this textbook, we have introduced a chapter on wavelets, wherein we have tried to make it easy for readers to understand the wavelets from basics to applications. Another chapter is introduced on adaptive digital filters used in the signal processing problems for faster and acceptable results in the presence of changing environments and changing system requirements. The last chapter included in this book is on DSP processors, which is a growing topic of interest in digital signal processing.

This book is divided into 13 chapters. Chapter 1 presents an introduction to digital signal processing with typical examples of digital signal processing applications. Chapter 2 discusses the time-domain representation of discrete-time signals and systems, linear time-invariant (LTI) discrete-time systems and their properties, characterization of discrete-time systems, representation of discrete-time signals and systems in frequency domain, representation of sampling in frequency domain, reconstruction of a bandlimited signal from its samples, correlation of discrete-time signals, and discrete-time random signals. Chapter 3 deals with z-transform and analysis of LTI discrete-time systems. In Chap. 4, discrete Fourier transform (DFT), its properties, and fast Fourier transform (FFT) are discussed. Chapter 5 deals with analog filter approximations and IIR filter design methodologies. Chapter 6 discusses FIR filter design methodologies. Chapter 7 covers various structures such as direct form I & II, cascade, parallel, and lattice structures for the realization of FIR and IIR digital filters. The finite word length effects on these structures are also analyzed. Chapters 8 and 9 provide an in-depth study of the multirate signal processing concepts and design of multirate filter banks. A deeper understanding of Chaps. 8 and 9 is required for a thorough understanding of the discrete wavelet transforms discussed in Chap. 10. The principle of adaptive digital filters and their applications are presented in Chap. 11. Chapter 12 deals with the estimation of spectra from finite duration observations of the signal using both parametric and nonparametric methods. Programmable DSP processors are discussed in Chap. 13.

The salient features of this book are as follows.

- Provides comprehensive exposure to all aspects of DSP with clarity and in an easy way to understand.
- Provides an understanding of the fundamentals, design, implementation, and applications of DSP.
- DSP techniques and concepts are illustrated with several fully worked numerical examples.
- Provides complete design examples and practical implementation details such as assembly language and C language programs for DSP processors.
- Provides MATLAB implementation of many concepts:

  - Digital FIR and IIR filter design
  - Finite word length effects analysis
  - Discrete Fourier transform
  - Fast Fourier transform
  - z-Transform
  - Multirate analysis
  - Filter banks
  - Discrete wavelet transform
  - Adaptive filters

- – Power spectral estimation
- – Design of digital filters using MATLAB graphical user interface (GUI) filter designer SPTOOL

- Provides examples of important concepts and to reinforce the knowledge gained.

Hyderabad, India	K. Deergha Rao
Montreal, Canada	M. N. S. Swamy

Contents

About the Authors

K. Deergha Rao is currently a Professor in the Department of Electronics and Communication Engineering, Vasavi College of Engineering, affiliated to Osmania University, Hyderabad, India, and is Former Director and Professor at the Research and Training Unit for Navigational Electronics (NERTU), Osmania University. He was a Postdoctoral Fellow and Part-Time Professor in the Department of Electrical and Computer Engineering, Concordia University, Montreal, Canada, for 4 years. He has conducted several research projects for leading Indian organizations. His teaching areas include digital signal processing, channel coding, signals and systems, and MIMO wireless communications, and his current research focuses on wireless channel coding, OFDM wireless communication channel estimation, compressive sensing, and VLSI for communications. He has presented papers at several IEEE international conferences in the USA, Switzerland, Thailand, and Russia. He has more than 100 publications to his credit, including more than 60 in IEEE journals and conference proceedings. He is a senior member of IEEE. He was awarded the IETE K. S. Krishnan Memorial Award in 2013 for the best system-oriented paper. He has authored two books, *Channel Coding Techniques for Wireless Communications* and *Signals and Systems* (both with Springer), and co-authored the book *Digital Signal Processing*.

M. N. S. Swamy is a Research Professor and holds the Concordia Chair in Signal Processing in the Department of Electrical and Computer Engineering at Concordia University, Montreal, Canada, where he served as the Founding chair of the Department of Electrical Engineering from 1970 to 1977 and the Dean of Engineering and Computer Science from 1977 to 1993. He received his M.Sc. and Ph.D. degrees in Electrical Engineering from the University of Saskatchewan, Saskatoon, Canada. He was made an Honorary Professor by the National Chiao Tung University, Taiwan, in 2009, which is equivalent to an honorary doctorate at the institute. He has also been associated with the Department of Electrical Engineering, Technical University of Nova Scotia, Halifax, NS, Canada; the University of Calgary, Calgary, AB, Canada; and the Department of Mathematics, University of Saskatchewan. He has published extensively in the areas of circuits, systems, and signal processing and holds six patents. He is a coauthor of eight books and several book chapters. He is a recipient of numerous IEEE Circuits and Systems Society Awards, including the Education Award in 2000, the Golden Jubilee Medal in 2000, and Guillemin–Cauer Best Paper Award. He is a Fellow of several professional societies including IEEE. The Indian Institute of Science (IISc) in Bengaluru, a premier research institution of India, has established a gold medal and a scholarship in his name. In 1993, he was awarded the commemorative medal for the 125th Anniversary of Canada, issued by the Governor General of Canada, in recognition of his significant contributions to Canada and the community. Recently, the journal *Circuits, Systems and Signal Processing* (CSSP) has instituted a Best Paper Award in his name.

Chapter 1
Introduction

1.1 What Is Digital Signal Processing?

A *signal* is defined as any physical quantity that varies with time, space, or any other independent variable or variables. The world of science and engineering is filled with signals: images from remote space probes, voltages generated by the heart and brain, radar and sonar echoes, seismic vibrations, speech signals, signals from GPS satellites, signals from human genes, and countless others. Signal processing is concerned with theory and methods for extraction of information from signals or alteration of signals with a purpose. The method of extraction of the information depends on the type of signal and the nature of information carried by the signal. Thus, the concern of signal processing is to represent the signal mathematically and to use an appropriate algorithm to extract information present in the signal. The information extraction can be carried out in the original domain of the signal or in a transformed domain. Most signals in nature are in analog form being continuous in time with continuous amplitude. A speech signal is an example of an analog signal. In most cases, these signals originate as sensory data from the real world: seismic vibrations, visual images, sound waves, etc. Digital signal processing (DSP) includes the mathematics, the algorithms, and the techniques used to manipulate these signals after they have been converted into a digital form.

1.2 Why Digital Signal Processing?

The block diagram of a typical real-time digital signal processing system is shown in Fig. 1.1. Basically, digital processing of an analog signal consists of three steps: conversion of the analog signal into digital form, processing of the digital signal so obtained, and finally, conversion of the processed output into analog signal.

© Springer Nature Singapore Pte Ltd. 2018
K. D. Rao and M. N. S. Swamy, *Digital Signal Processing*,
https://doi.org/10.1007/978-981-10-8081-4_1

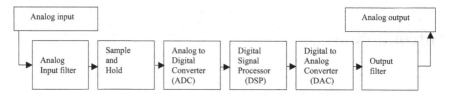

Fig. 1.1 Block diagram of a real-time digital signal processing system

The analog input signal is applied to the input filter. The input filter is a lowpass analog anti-aliasing filter. The analog input filter limits the bandwidth of the analog input signal. The analog-to-digital converter (ADC) converts the analog input signal into digital form; for wideband signals, ADC is preceded by a sample and hold circuit. The output of the ADC is an N-bit binary number depending on the value of the analog signal at its input. The sample and hold device provides the input to the ADC and will be required if the input signal must remain relatively constant during the conversion of the analog signal to digital format. Once converted to digital form, the signal can be processed using digital techniques. The digital signal processor is the heart of the system; it implements various DSP algorithms. The digital signal processor may be a large programmable digital computer or a microprocessor programmed to perform the desired operations on the input signal. The architectures of standard microprocessors are not suited to the DSP characteristics, and this has led to the development of new kinds of processors with very fast speed; for example, ADSP2100, Motorola DSP56000, and TMS320C50 fixed-point processors, and analog devices SHARC, TigerSHARC and Texas Instruments TMS320C67xx (floating-point processors) are configured to perform a specified set of operations on the input signal. The digital-to-analog converter (DAC) converts the processed digital data into analog form, followed by an analog filter to give the final output. The output of the DAC is continuous, but contains high-frequency components that are unwanted. To eliminate the high-frequency components, the output of the DAC is passed through a lowpass output filter.

There are several advantages of digital signal processing over analog signal processing. The most important among them are the following:

- **Flexibility**—Digital implementation allows flexibility to reconfigure the DSP operations by simply changing the program.
- **Accuracy**—DSP provides any desirable accuracy by simply increasing the number of bits (word length), while tolerance limits have to be met in the analog counterpart.
- **Easy Storage**—Digital signals can be easily saved on storing media, such as magnetic tape, disk, and optical disk without loss of information. They can also be easily transported and processed off-line in remote laboratories.
- **Processing**—DSP allows for the implementation of more sophisticated signal processors than its analog counterparts do.

- **Cost Effectiveness**—With the recent advances in very large-scale integrated (VLSI) circuit technology, the digital implementation of the signal processing system is cheaper.
- **Perfect Reproducibility**—No variations due to component tolerances.
- **No Drift**—No drifts in performance with temperature and age.
- **Immunity to Noise**—DSP is immune to noise.
- **Easy Processing of VLF Signals**—DSP is applicable to easy processing of the very low-frequency (VLF) signals such as seismic signals, whereas an analog processing system requires very large-size inductors and capacitors.

Digital signal processing has also some disadvantages over analog signal processing. They are:

- **Finite Word Length**—Cost considerations limit the DSP implementation with less number of bits which may create degradation in system performance.
- **System Complexity**—Increased complexity in the digital processing of an analog signal because of the need for devices such as ADC, DAC, and the associated filters.
- **Speed Limitation**—Signals having extremely wide bandwidths require fast sampling rate ADC and fast digital signal processors. But the speed of operation of ADC and digital signal processors has a practical limitation.

In several real-world applications, the advantages of DSP overweigh the disadvantages, and DSP applications are increasing tremendously in view of the decreasing hardware cost of digital processors.

1.3 Typical Signal Processing Operations

Various types of signal processing operations are employed in practice. Some typical signal processing operations are given below.

1.3.1 Elementary Time-Domain Operations

The basic time-domain signal processing operations are scaling, shifting, addition, and multiplication. Scaling is the multiplication of a signal by a positive or a negative constant. Shifting operation is a shift replica of the original signal. The addition operation consists of adding two or more signals to form a new signal. Multiplication operation is to perform the product of two or more signals to generate a new signal.

1.3.2   Correlation

Correlation of signals is necessary to compare one reference signal with one or more signals to determine the similarity between them and to determine additional information based on the similarity. Applications of cross-correlation include cross-spectral analysis, detection of signals buried in noise, pattern matching, and delay measurements.

1.3.3   Digital Filtering

Digital filtering is one of the most important operations in DSP. Filtering is basically a frequency-domain operation. Filter is used to pass certain band of frequency components without any distortion and to block other frequency components. The range of frequencies that is allowed to pass through the filter is called the passband, and the range of frequencies that is blocked by the filter is called the stopband.

1.3.4   Modulation and Demodulation

Transmission media, such as cables and optical fibers, are used for transmission of signals over long distances; each such medium has a bandwidth that is more suitable for the efficient transmission of signals in the high-frequency range. Hence, for transmission over such channels, it is necessary to transform the low-frequency signal to a high-frequency signal by means of a modulation operation. The desired low-frequency signal is extracted by demodulating the modulated high-frequency signal at the receiver end.

1.3.5   Discrete Transformation

Discrete transform is the representation of discrete-time signals in the frequency domain, and inverse discrete transform converts the signals from the frequency domain back to the time domain. The discrete transform provides the spectrum of a signal. From the knowledge of the spectrum of a signal, the bandwidth required to transmit the signal can be determined. The transform domain representations provide additional insight into the behavior of the signal and make it easy to design and implement DSP algorithms, such as those for digital filtering, convolution, and correlation.

1.3.6 Quadrature Amplitude Modulation

Quadrature amplitude modulation (QAM) uses double-sideband (DSB) modulation to modulate two different signals so that they both occupy the same bandwidth. Thus, QAM achieves as much efficiency as that of single-sideband (SSB), since it takes up as much bandwidth as the SSB modulation method does. In QAM, the two band-limited low-frequency signals are modulated by two carrier signals (in-phase and quadrature components) and are summed, resulting in a composite signal. Multiplication of the composite signal by both the in-phase and quadrature components of the carrier separately results in two signals, and lowpass filtering of these two signals yields the original modulating signals.

1.3.7 Multiplexing and Demultiplexing

Multiplexing is a process where multiple analog message signals or digital data streams are combined into one signal for transmission over a shared medium. The reverse process to extract the original message signals at the receiver end is known as demultiplexing. Frequency-division multiplexing (FDM) is used to combine different voice signals in a telephone communication system [1, 2] resulting in a high-bandwidth composite signal, which is modulated and transmitted.

The composite baseband signal can be obtained by demodulating the FDM signal at the receiver side. The individual signals can be separated from the composite signal by demultiplexing which is accomplished by passing the composite signal through a bandpass filter with center frequency equal to the carrier frequency of the amplitude modulation. Then, the original low-frequency narrow-bandwidth individual user signals are recovered by demodulating the bandpass filter output.

Code-division multiplexing (CDM) is a communication networking technique in which multiple data signals are combined for simultaneous transmission over a common frequency band.

1.4 Application Areas of DSP

Digital signal processing is a very rapidly growing field that is being used in many areas of modern electronics, where the information is to be handled in a digital format or controlled by a digital processor. Some typical application areas of DSP are as follows:

- **Speech Processing**—Speech compression and decompression for voice storage system and for transmission and reception of voice signals; speech synthesis in message warning systems.

- **Communication**—Elimination of noise by filtering and echo cancellation, adaptive filtering in transmission channels.
- **Biomedical**—Spectrum analysis of ECG signals to identify various disorders in the heart and spectrum analysis of EEG signals to study the malfunctions or disorders in the brain.
- **Consumer Electronics**—Music synthesis and digital audio and video.
- **Seismology**—Spectrum analysis of seismic signals (i.e., signals generated by movement of rocks) can be used to predict earthquakes, volcanic eruptions, nuclear explosions, and earth movement.
- **Image Processing**—Two-dimensional filtering on images for image enhancement, fingerprint matching, image compression, medical imaging, identifying hidden images in the signals received by radars, etc.
- **Navigation**—Global positioning system (GPS) satellite signal processing for air, sea, and land navigation.
- **Genomic Signal Processing**—Processing of sequences of a human genome to explore the mysteries.

1.5 Some Application Examples of DSP

1.5.1 Telecommunications

DSP has revolutionized the telecommunication industry in many areas: signaling tone generation and detection, frequency band shifting, filtering to remove power line hum, etc. Two specific examples from the telephone network, namely compression and echo control, are briefly summarized below.

Compression

Most of the digital information is *redundant*. When a voice signal is digitized at 8 kHz sampling rate and if 8-bit quantization is used, then it results in a 64-Kbps-data-rate signal. Several DSP algorithms called data compression algorithms have been developed to convert digitized voice signals into data streams that require fewer bits/sec. Decompression algorithms are used to restore the signal to its original form. In general, reducing the data rate from 64 to 32 Kbps results in no loss of voice quality, but a reduced data rate of 8 Kbps causes noticeable distortion in voice quality. However, it is still usable for long-distance telephone networks. The highest reduced data rate of 2 Kbps results in highly distorted quality, but usable in military and undersea communications [3].

Echo Control

Echo is a serious problem in long-distance telephone communications. When a person speaks into a telephone, his/her voice signal travels to the connecting receiver and a portion of it returns as an echo. The echo becomes very irritating as the distance between the speaker and the receiver becomes large. As such, it is highly objectionable in intercontinental communications. Digital signal processing attacks this type of problems by measuring the returned signal and generating an appropriate *anti-signal* to cancel the echo.

1.5.2 Noise Reduction in Speech Signals

Most of the energy in speech signals lies in the frequency band of 0–3 kHz. Utilizing this fact, we can design a digital lowpass filter to remove all the high-frequency components beyond 3 kHz in it, thus saving bandwidth without loss of intelligibility of the speech signal. A voice signal 'Don't fail me again' [4] is considered to illustrate noise reduction in speech signals. The noisy signal corresponding to the voice signal is shown in Fig. 1.2. When a lowpass filter, designed to preserve frequency components in the frequency band 0–2.5 kHz, is applied on the noisy voice signal, then the voice signal after filtering is as shown in Fig. 1.3. When the filtered voice signal is connected to a loud speaker, audio quality of the reconstructed signal is observed to be almost the same as the original voice signal.

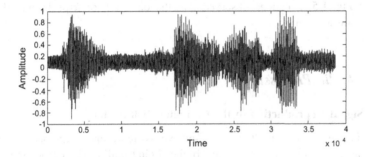

Fig. 1.2 Noisy voice signal 'Don't fail me again'

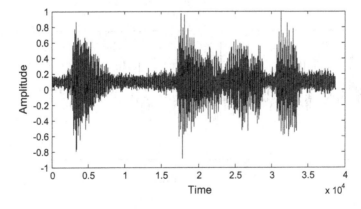

Fig. 1.3 Reconstructed voice signal 'Don't fail me again'

1.5.3 ECG Signal Processing

The electrocardiogram (ECG) signal recorded from human heart represents the electrical activity of the heart. The processing of ECG signal yields information, such as amplitude and timing, required for a physician to analyze a patient's heart condition [5]. Detection of R-peaks and computation of R-R interval of an ECG record is an important requirement of comprehensive arrhythmia analysis systems.

In practice, various types of externally produced interferences appear in an ECG signal [6]. Unless these interferences are removed, it is difficult for a physician to make a correct diagnosis. A common source of noise is the 60- or 50-Hz power lines. This can be removed by using a notch filter with a notch at 60 or 50 Hz. The other interferences can be removed with careful shielding and signal processing techniques. Data compression finds use in the storage and transmission of the ECG signals. Due to their efficiency for processing non-stationary signals and robustness to noise, wavelet transforms have emerged as powerful tools for processing ECG signals. A noisy ECG signal is shown in Fig. 1.4.

An approach [7] using the Daubechies discrete wavelet transform (DWT) can be applied on the noisy ECG signal for the detection of R-peaks and compression. The reconstructed ECG signal with 82% data reduction is shown in Fig. 1.5.

From Fig. 1.5, it can be observed that the detection of R-peaks and desired compression are achieved.

1.5.4 Audio Processing

Audio Signal Reproduction in the Compact Disk System

The digital signal from a CD is in 16-bit words, representing the acoustic information at a 44.1-kHz sampling rate. If the digital audio signal from the CD is directly converted into analog signal, images with frequency bands centered at

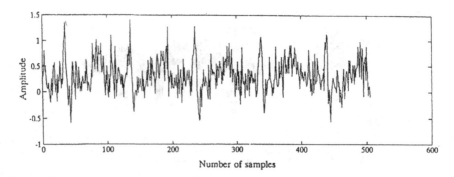

Fig. 1.4 Noisy ECG signal

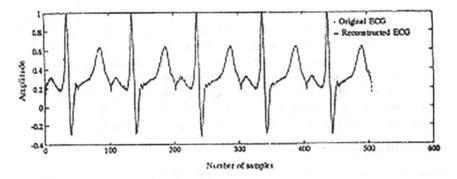

Fig. 1.5 Reconstructed ECG signal

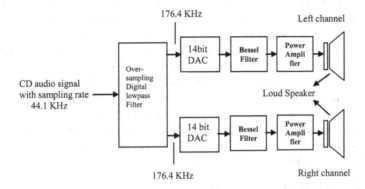

Fig. 1.6 Reproduction of audio signal in compact disc

multiples of 44.1 kHz would be produced. They could cause overloading if passed on to the amplifier and loudspeaker of the CD player. To avoid this, the digital signals are processed further by passing them through a digital filter operating at four times the audio sampling rate of 44.1 kHz before being applied to the 14-bit DAC. The effect of raising the sampling frequency is to push the image frequencies to higher frequencies. Then, they can be filtered using a simple filter like a Bessel filter. Raising the sampling frequency also helps to achieve a 16-bit signal-to-noise ratio performance with a 14-bit DAC [8]. The schematic block diagram for the reproduction of audio from a compact disk is shown in Fig. 1.6.

1.5.5 Image Processing

Medical Imaging

Computed tomography (CT) is a medical imaging example of digital signal processing. X-rays from many directions are passed through the patient's body part being examined. The images can be formed with the detected X-rays. Instead of forming images this way, the signals are converted into digital data and stored in a computer and then the information is used to obtain images that appear to be *slices through the body*. These images provide more details for a better diagnosis and treatment than the conventional techniques. Magnetic resonance imaging (MRI) is another imaging example of DSP. MRI discriminates between different types of soft tissues in an excellent manner and also provides information about physiology, such as blood flow through arteries. MRI implementation depends completely on DSP techniques.

Image Compression

A reasonable size digital image in its original form requires large memory space for storage. Large memories and high data transfer rates are bottlenecks for cheaper commercial systems. Similar to voice signals, images contain a tremendous amount of redundant information and can be represented with reduced number of bits. Television and other moving pictures are especially suitable for compression, since most of the images remain the same from frame to frame. Commercial imaging products such as video telephones, computer programs that display moving pictures, and digital television take advantage of this technology. The JPEG2000 is the new standard [9] for still image compression. It is a discrete wavelet transform (DWT)-based standard. For example, Lenna image shown in Fig. 1.7a is of size 512×512 pixels and contains 2,097,152 bits with 8 bits per pixel. The image coding method such as JPEG2000 is used to represent the image with 0.1 bit per pixel requiring only 26,214 bits. The reconstructed image with 0.1 bit per pixel is shown in Fig. 1.7b and is without distortion.

Image Restoration and Enhancement

Image restoration and enhancement algorithms are used to improve the quality of images taken under extremely unfavorable conditions, and from unmanned satellites and space exploration vehicles. These include DSP techniques for brightness and contrast enhancement, edge detection, noise reduction, motion blur reduction, etc.

(a)   **(b)**

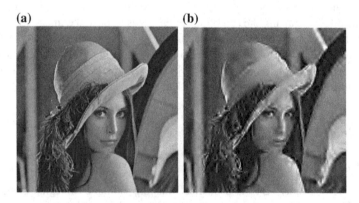

Fig. 1.7 **a** Original image. **b** Reconstructed image compressed with 0.1 bpp

1.5.6 GPS Signal Processing

GPS Positioning

Navigation systems are used to provide moving objects with information about their positioning. An example is the satellite-based global positioning system (GPS), which consists of a constellation of 24 satellites at high altitudes above the earth. Figure 1.8 shows an example of the GPS used in air, sea, and land navigation. It requires signals at least from four satellites to find the user position (X, Y, and Z) and clock bias from the user receiver. The measurements required in a GPS receiver for position finding are the ranges, i.e., the distances from GPS satellites to the user. The ranges are deduced from measured time or phase differences based on a comparison between the received and receiver-generated signals. To measure the time, the replica sequence generated in the receiver is to be compared to the satellite sequence.

The job of the correlator in the user GPS receiver is to determine as to which codes are being received, as well as their exact timing. When the received and receiver-generated sequences are in phase, the correlator supplies the time delay. Now, the range can be obtained by multiplying the time delay by the velocity of light. For example, assuming the time delay as 3 ms (equivalent to 3 blocks of the C/A code of satellite 12), the correlation of satellite 12 producing a peak after 3 ms [10] is shown in Fig. 1.9.

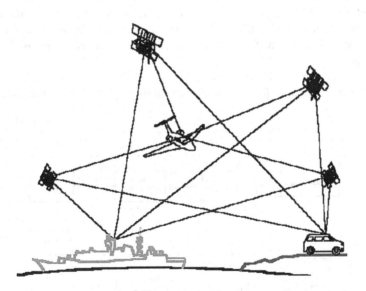

Fig. 1.8 A pictorial representation of GPS positioning for air, sea, and land navigation

Fig. 1.9 Correlation of
satellite 12 producing a peak

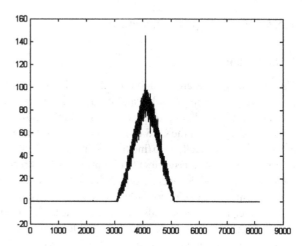

GPS Location-Based Mobile Emergency Services

The knowledge of a mobile user's location by the service provider can enhance the class of services and applications that can be offered to the mobile user. This class of applications and services is termed as location-based services. A location-based service is a service that makes use of position or location information. Based on the way the information is utilized, a variety of services may be developed. Wireless emergency services are a type of 'Location-Based Services (LBS)' which is useful for emergency service requests such as ambulance, fire, and police. The 4G mobile phones are equipped with GPS receiver within them for finding the position of the mobile user. The block diagram of a mobile emergency service system is shown in Fig. 1.10.

The service consists of the following messages: *emergency location immediate request* and *emergency location immediate answer*. When user has dialed the

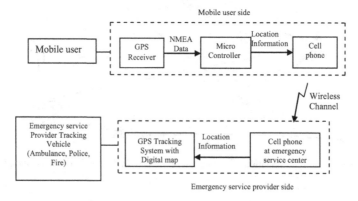

Fig. 1.10 Block diagram of GPS location-based mobile emergency services

emergency number, *emergency location immediate request* is sent to the service provider. This request consists of the user's precise location information in the form of latitude and longitude. The GPS receiver within the mobile phone provides the position information very accurately. After receiving the emergency immediate request from the user, the service provider identifies the service (like ambulance, police, and fire services) and sends *emergency location immediate answer* to the mobile user. The service provider has the digital maps of all the geographical positions. Whenever an emergency service request is received, a mark will appear on the corresponding digital map. This mark will indicate the user's location. By using the GPS tracking system with digital map from his side, the service provider can reach the spot easily and in time.

1.5.7 Genomic Signal Processing

Lee Hood has observed [11], 'The sequence of the human genome would be perhaps the most useful tool ever developed to explore the mysteries of human development and disease.' The genetic code describes only the relation between the sequences of bases, also called nucleotides **a**denine (A), **t**hymine (T), **c**ytosine (C), and **g**uanine (G), in deoxyribonucleic acid (DNA) that encode the proteins and the sequence of the amino acids ACDEFGHIKLMNPQRSTVWY in those proteins. The flow of genetic information from DNA to function is shown below

$$\textbf{DNA} \rightarrow \textbf{RNA(ribonucleic acid)} \rightarrow \textbf{Protein} \rightarrow \textbf{Function}$$

Gene Prediction Using Short-Time Fourier Transform (STFT)

A gene is a sequence made up of the four bases and can be divided into two subregions called the exons and introns [12] as shown in Fig. 1.11. Only the exons are involved in protein-coding. The gene prediction is based on the period-3 property [13, 14]. Using the electron–ion interaction-potential (EIIP) values for nucleotides, one defines the numerical sequence of a DNA stretch and computes its STFT [15]. The magnitude of the STFT is evaluated for window size 120 for a DNA stretch of *C. elegans* (GenBank accession number AF099922), containing 8000 nucleotides starting from location 7021. The five exons of the gene F56F11.4 in the C-elegans chromosome III are clearly seen in the STFT magnitude plot, shown in Fig. 1.12.

Identification of the Resonant Recognition Model (RRM) Characteristic Frequency of Protein Sequences

For example, the tumor suppression genes contribute to cancer when they are inactivated or lost as a result of DNA damage (mutations). The interaction of JC virus T-antigen with tumor suppression blastoma leads to brain cancer. Consider the

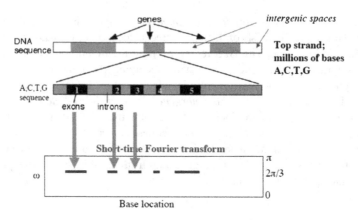

Fig. 1.11 Regions in a DNA molecule

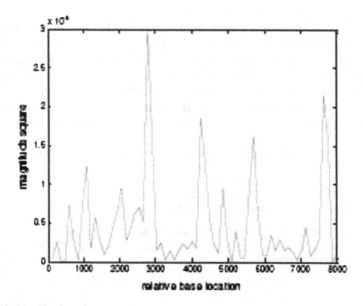

Fig. 1.12 Identification of gene F56F11.4 with five exons (@ Adopted from IEEE 2008)

following two proteins called retinoblastoma human protein and retinoblastoma mouse protein with amino acid chains (represented by single protein retino letter symbols) with lengths 1257 and 461, respectively. Using EIIP values for the amino acids in the protein sequence in converting the protein sequence into a numerical sequence, the Fourier transforms of the above two protein sequences and the consensus spectra are shown in Fig. 1.13 (these plots show the squared magnitudes).

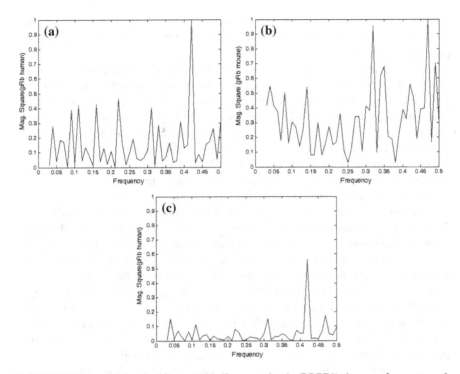

Fig. 1.13 a Spectra of retinoblastoma binding protein 1 (RBBP1) human, **b** spectra of retinoblastoma binding protein 1 (RBBP1) mouse, and **c** cross-spectral function of the spectra presented in (**a**) and (**b**)

In Fig. 1.13c, the sharp peak at the frequency 0.4316 is the characteristic frequency of the tumor suppression proteins retinoblastoma. It is concluded in [15] that one particular biological interaction is characterized by one RRM characteristic frequency.

1.6   Scope and Outline

The motivation in writing this book is to modernize the digital signal processing teaching by including additional topics of increasing importance on the subject, and to provide a comprehensive exposition of all aspects of digital signal processing. The text is integrated with MATLAB-based programs to enhance the understanding of the underlying theories of the subject.

One of the most fundamental concepts of digital signal processing is the idea of sampling an analog signal to provide a set of numbers which, in some sense, is representative of the analog signal being sampled. The most common form of sampling that we refer throughout this book is periodic sampling. That is, the samples are uniformly spaced in the dimension of time t, occurring T (sampling period) seconds apart.

Chapter 2 introduces basic classes of discrete-time signals and systems, basic system properties such as linearity, time-invariance, causality, and stability, time-domain representation of linear time-invariant (LTI) systems through convolution sum and the class of LTI systems represented by linear constant coefficient difference equations, frequency-domain representation of signals and systems through the Fourier transform, effect of sampling in the frequency domain, Nyquist sampling theorem, reconstruction of a band-limited signal from its samples, correlation of discrete-time signals, and discrete-time random signals.

The z-transform is the mathematical tool used for analysis and design of discrete-time systems like Laplace transform for continuous systems. In Chap. 3, we develop the z-transform as a generalization of the Fourier transform. The basic theorems and properties of the z-transform and the methods for inverse z-transform are also presented. The extensive use of the z-transform in the representation and analysis of linear time-invariant systems is also described in this chapter.

The discrete Fourier transform (DFT) and an efficient algorithm for its computation, known as the fast Fourier transform (FFT), have been responsible for a major shift to digital signal processing. The FFT has reduced the computation time drastically, thus enabling the implementation of sophisticated signal processing algorithms. Chapter 4 discusses the DFT and the FFT in detail.

Digital filter design is one of the most important topics in DSP, it being at the core of most of the DSP systems. From specification to implementation, techniques for designing infinite impulse response (IIR) digital filters are treated in detail in Chap. 5. Several solved design examples using MATLAB programs as well as GUI MATLAB SPTOOL are provided throughout the chapter to help the reader to design IIR filters.

Chapter 6 describes the characteristics of finite impulse response (FIR) digital filters, various types of linear phase FIR transfer functions, and their frequency response. Various techniques used for designing FIR filters are also detailed in Chap. 6. Several solved design examples, using MATLAB programs as well as GUI MATLAB SPTOOL, to design FIR filters are included for a better understanding of the concepts.

Chapter 7 focuses on the development of various structures for the realization of digital FIR and IIR filters. In practical implementation of digital filters, the effect of coefficient inaccuracies and arithmetic errors due to finite precision is dependent on the specific structure used. Hence, this chapter analyzes the effects of coefficient quantization and arithmetic errors due to round-off errors in the context of implementation of digital filters.

The process of digitally converting the sampling rate of a signal from a given rate (1/T) to a different rate (1/T') is called sampling rate conversion. This is also known as multirate digital signal processing. It is especially an important part of modern digital communications in which digital transmission systems such as teletype, facsimile, low-bit-rate speech, and video are required to handle data at several rates. The multirate digital signal processing became practically attractive with the invention of polyphase decomposition, since it often results in computational efficiency. The theory of perfect reconstruction filter banks is well established enabling us to design and implement the same easily. The multirate filter banks have immense potential for applications such as in sub-band coding, voice privacy systems, image processing, multiresolution, and wavelet analysis. Chapters 8 and 9 of this text are devoted to the area of multirate digital signal processing.

It has been realized that there is a close relation between multirate filter banks and wavelet transforms. It can be observed that wavelet analysis is closely related to octave-band filter banks. The wavelet transform has grown increasingly popular for a variety of applications in view of the fact that it can be applied to non-stationary signals and resolving signals both in time and frequency. An important application of discrete wavelet transform is the JPEG2000 for still image compression. All the above concepts are presented in Chap. 10.

An adaptive filter is a digital filter that adapts automatically to changes in its input signals. The adaptive filters are useful in many practical applications where fixed coefficient filters are not appropriate. The principle of adaptive digital filters and their applications are presented in Chap. 11. Spectral analysis of signals has several practical applications such as communication engineering and study of biological signals in medical diagnosis. Chapter 12 deals with the estimation of spectra from finite duration observations of signal using both parametric and nonparametric methods.

The implementation of DSP uses a variety of hardware approaches, ranging from the use of off-the-shelf microprocessors to field-programmable gate arrays (FPGAs) to custom integrated circuits (ICs). Programmable 'DSP processors,' a class of microprocessors optimized for DSP, are a popular solution for several reasons. Chapter 13 deals with an introduction to DSP processors, key features of various DSP processors, internal architectures, addressing modes, important instruction sets, and implementation examples.

References

1. L.W. Couch II, *Digital and Analog Communication Systems* (Macmillan, New York, NY, 1983)
2. A.V. Oppenheim, A.S. Willsky, *Signals and Systems* (Prentice-Hall, EngleWood Cliffs, NJ, 1983)
3. S.W. Smith, *The Scientist and Engineer's Guide to Digital Signal Processing* (California Technical Publishing, San Diego, USA, 1999)
4. http://www.jedisaber.com/SW/ESB.asp. Accessed 31 October 2008
5. A.F. Shackil, Microprocessors and MD. IEEE Spectr. **18**, 45–49 (1981)
6. W.J. Tompkins, J.G. Webster (eds.), *Design of Microcomputer-Based Medical Instrumentation* (Prentice Hall, EngleWood Cliffs, NJ, 1981)
7. K. Deergha Rao, DWT based detection of R-peaks and data compression of ECG signal. IETE J. Res. **43**, 345–349 (1997)

8. E.C. Ifeachor, B.W. Jervis, *Digital Signal Processing—A Practical Approach* (Prentice Hall, EngleWood Cliffs, NJ, 2002)

9. Ch. Christopoulos, A. Skodras, T. Ebrahimi, The JPEG2000 still image coding system: an overview. IEEE Trans. Consum. Electron. **46**(4), 1103–1127 (2000)

10. K. Deergha Rao, M.N.S. Swamy, New approach for suppression of FM jamming in GPS receivers. IEEE Trans. Aerosp. Electron. Syst. **42**(4), 1464–1474 (2006)

11. D.L. Brutlag, Understanding the human genome, in *Scientific American: Introduction to Molecular Medicine*, ed. by P. Leder, D.A. Clayton, E. Rubenstein (Scientific American, New York, 1994), pp. 153–168

12. B. Alberts, D. Bray, A. Johnson, J. Lewis, M. Raff, K. Roberts, P. Walter, *Essential Cell Biology* (Garland Publishing Inc., New York, 1998)

13. D. Anastassiou, Genomic signal processing. IEEE Signal Process. Mag. **18**(4), 8–20 (2001)

14. P.P. Vaidyanatha, B.-J. Yoon, The role of signal-processing concepts in genomics and proteomics. J. Franklin Inst. **341**, 111–135 (2004)

15. K. Deergha Rao, M.N.S. Swamy, Analysis of genomics and proteomics using DSP techniques. IEEE Trans. Circuits Syst. I Regul. Pap. **08**, 370–378

16. I. Cosic, Macromolecular bioactivity: is it resonant interaction between macromolecules? Theory and applications. IEEE Trans. Biomed. Eng. **41**(12), 1101–1114 (1994)

Chapter 2
Discrete-Time Signals and Systems

Digital signal processing deals basically with discrete-time signals, which are processed by discrete-time systems. The characterization of discrete-time signals as well as discrete-time systems in time domain is required to understand the theory of digital signal processing. The discrete-time signals and discrete-time systems are often characterized conveniently in a transform domain. In this chapter, the fundamental concepts of discrete-time signals as well as discrete-time systems are considered. First, the basic sequences of discrete-time systems and their classification are emphasized. The input–output characterization of linear time-invariant (LTI) systems by means of convolution sum is described. Next, we discuss the transform domain representation of discrete-time sequences by discrete-time Fourier transform (DTFT) in which a discrete-time sequence is mapped into a continuous function of frequency. The Fourier transform domain representation of discrete-time sequences is described along with the conditions for the existence of DTFT and its properties. Later, the frequency response of discrete-time systems, frequency-domain representation of sampling process, reconstruction of band-limited signals from its samples are discussed. Finally, cross-correlation of discrete-time signals and time-domain representation of discrete-time random signals are reviewed.

2.1 Discrete-Time Signals

As defined in Chap. 1, a signal is a physical quantity that varies with one or more independent variables. If the independent variable is discrete in time, the signal defined at discrete instants of time is called discrete-time signal. Hence, it is represented as a sequence of numbers called samples. Thus, a continuous signal is continuous both in time and amplitude, while a discrete-time signal is continuous in amplitude but discrete in time. A digital signal is one that is discrete in both time and amplitude. Thus, it is a finely quantized discrete-time signal with amplitudes

© Springer Nature Singapore Pte Ltd. 2018
K. D. Rao and M. N. S. Swamy, *Digital Signal Processing*,
https://doi.org/10.1007/978-981-10-8081-4_2

represented by a finite number of bits. Figure 2.1 illustrates continuous-time, discrete-time, and digital signals.

An nth number or sample value of a discrete-time sequence is denoted as $x(n)$, n being an integer varying from $-\infty$ to ∞. Here, $x(n)$ is defined only for integer values of n. The following is an example of a discrete-time signal with real-valued samples for positive values of n.

$$x(n) = \{1, 0.4, 0.6, 367, 5, 7, 34, 98, 0, 1, 45, 7, 0\}, \quad \text{for } n = 0, 1, 2, \ldots \quad (2.1)$$

For the above signal, $x(0) = 1$, $x(1) = 0.4$, $x(2) = 0.6$, and so on.

A complex sequence $x(n)$ can be written as $x(n) = x_R(n) + jx_I(n)$, where $x_R(n)$ and $x_I(n)$ are real sequences corresponding to the real and imaginary parts of $x(n)$, respectively.

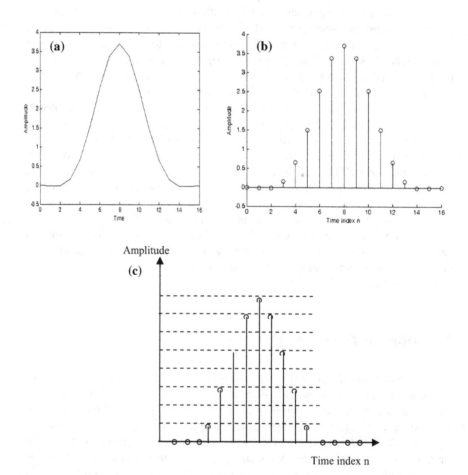

Fig. 2.1 **a** A continuous-time signal, **b** a discrete-time signal, and **c** a digital signal

2.1.1 Elementary Operations on Sequences

The elementary operations on sequences are multiplication, addition, scalar multiplication, time shifting, and time reversal.

Multiplication operation consists of multiplying two or more sequences to generate a new sequence. The schematic representation of the multiplication operation for two sequences $x_1(n)$ and $x_2(n)$ is shown in Fig. 2.2.

For example, if $x_1(n) = \{1, 5, 6, 7\}$ and $x_2(n) = \{4, 3, 1, 2\}$, then

$$y(n) = x_1(n)x_2(n) = \{4, 15, 6, 14\}$$

Addition operation consists of adding two or more sequences to form a new sequence. This operation can be performed by using an adder. The schematic representation of an adder for the addition of two sequences $x_1(n)$ and $x_2(n)$ is shown in Fig. 2.3.

For example, if $x_1(n) = \{1, 2, 3, 1\}$ and $x_2(n) = \{2, 5, 3, 4\}$, then

$$y(n) = x_1(n) + x_2(n) = \{1+2,\ 2+5,\ 3+3,\ 1+4\} = \{3,\ 7,\ 6, 5\}$$

Scalar multiplication is the multiplication of a sequence by a positive or a negative constant. If $x(n)$ is a discrete-time sequence, the scalar multiplication operation generates a sequence, $y(n) = Kx(n)$ where K is a constant. Its schematic representation is shown in Fig. 2.4.

For example, if $x(n) = \{2, 5, 1, 4\}$ and $K = 2$, then

$$y(n) = Kx(n) = \{4, 10, 2, 8\}$$

Fig. 2.2 Schematic representation of a multiplication operation

Fig. 2.3 Schematic representation of an addition operation

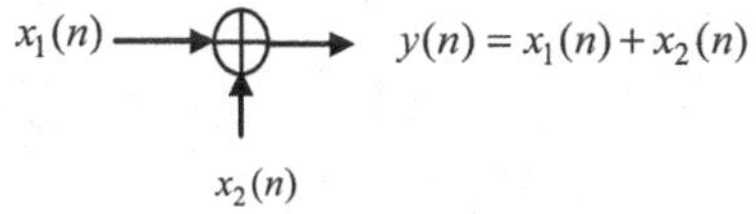

Fig. 2.4 Schematic representation of scalar multiplication

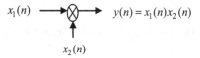

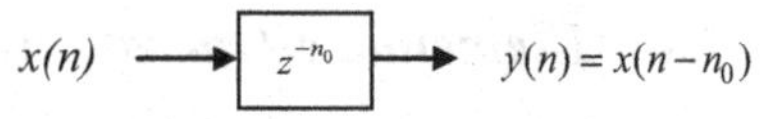

Fig. 2.5 Schematic
representation of shifting
operation

$$x(n) \longrightarrow \boxed{z^{-n_0}} \longrightarrow y(n) = x(n-n_0)$$

Shifting operation consists of shifting the original sequence by a certain number of samples. A discrete-time sequence $x(n)$ shifted by n_0 samples produces the sequence $y(n) = x(n - n_0)$.

Figure 2.5 gives a schematic representation of the shift operation, where the symbol z^{-n_0} is used to denote the shift by n_0 samples. It is seen that if $n_0 > 0$, $y(n)$ is nothing but $x(n)$ delayed by n_0 samples. If $n_0 < 0$, then the signal $x(n)$ is shifted to the left by n_0 samples; that is, the input signal is advanced by n_0 samples. Such an advance is not possible to be realized in real time, but is possible to do so in non-real-time application by storing the signal in memory and recalling it any time.

For example, if $x(n) = \{2, 4, 3, 1\}$, the shifted sequence $x(n - 2)$ for $n_0 = 2$ and the shifted sequence $x(n + 1)$ for $n_0 = 1$ are shown in Fig. 2.6.

The time-reversal operation generates time-reversed version of a sequence. For example, the sequence $x(-n)$ is time-reversed version of sequence $x(n)$.

The sequence $x(n) = \{1, 3, 3, 1\}$ and its time-reversed version are shown in Fig. 2.7a, b respectively.

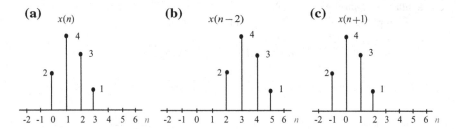

Fig. 2.6 Illustration of shifting operation

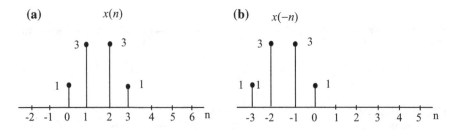

Fig. 2.7 a Sequence $x(n)$ and **b** its time-reversed version

2.1.2 Basic Sequences

The unit sample sequence, the unit step sequence, and the exponential and sinusoidal sequences are the most common sequences.

Unit sample sequence
The unit sample sequence is defined by (Fig. 2.8).

$$\delta(n) = \begin{cases} 0 & n \neq 0 \\ 1 & n = 0 \end{cases} \tag{2.2}$$

The unit sample sequence is often referred to as a discrete-time impulse or simply an impulse sequence. More generally, any sequence can be expressed as

$$x(n) = \sum_{k=0}^{\infty} x(k)\delta(n-k) \tag{2.3}$$

Unit step sequence
The unit step sequence is given by

$$u(n) = \begin{cases} 1 & n \geq 0 \\ 0 & n < 0 \end{cases} \tag{2.4}$$

The unit step sequence shifted by k samples is given by

$$u(n) = \begin{cases} 1 & n \geq k \\ 0 & n < k \end{cases} \tag{2.5}$$

The unit step sequence in terms of a sum of delayed impulses may be written as

$$u(n) = \delta(n) + \delta(n-1) + \delta(n-2) + \cdots$$

$$u(n) = \sum_{k=0}^{\infty} \delta(n-k) \tag{2.6}$$

The unit step sequence is shown in Fig. 2.9.
Conversely, the impulse sequence can be expressed as the backward difference of the unit step sequence and the unit step delayed by one sample:

Fig. 2.8 Unit sample
sequence

Fig. 2.9 Unit step sequence

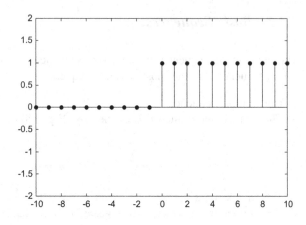

$$\delta(n) = u(n) - u(n-1) \tag{2.7}$$

Exponential and Sinusoidal sequences

Exponential sequences are extremely important in representing and analyzing linear and time-invariant systems. The general form of an exponential sequence is

$$x(n) = Aa^n \tag{2.8}$$

If A and a are real numbers, then the sequence is real. If $0 < a < 1$ and A is positive, then the sequence values are positive and decrease with increasing n. An example of an exponential is shown in Fig. 2.10.

Let $a = e^{j\theta}$, then Eq. (2.8) can be rewritten as

$$x(n) = Ae^{j\theta n} = A \cos(\theta n) + jA \sin(\theta n) \tag{2.9}$$

Fig. 2.10 An exponential sequence

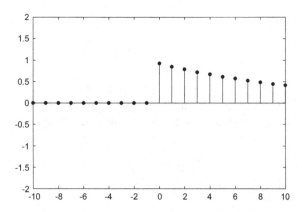

From the above equation, we get the real and imaginary parts of $x(n)$ as

$$x_{re}(n) = A \cos(\theta n), \tag{2.10a}$$

$$x_{im}(n) = A \sin(\theta n). \tag{2.10b}$$

Thus, for $n > 0$, the real and imaginary parts of complex exponential sequence are real sinusoidal sequences.

2.1.3   Arbitrary Sequence

An arbitrary sequence can be represented as a weighted sum of some of the basic sequences and its delayed versions. For example, an arbitrary sequence as weighted sum of unit sample sequence and its delayed versions is shown in Fig. 2.11.

2.2   Classification of Discrete-Time Signals

2.2.1   Symmetric and AntiSymmetric Signals

A real-valued signal $x(n)$ is said to be symmetric if it satisfies the condition

$$x(-n) = x(n) \tag{2.11a}$$

Example of a symmetric sequence is shown in Fig. 2.12a.
On the other hand, a signal $x(n)$ is called antisymmetric if it follows the condition

$$x(-n) = -x(n) \tag{2.11b}$$

An example of antisymmetric sequence is shown in Fig. 2.12b.

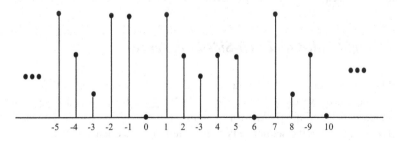

Fig. 2.11 An example of an arbitrary sequence

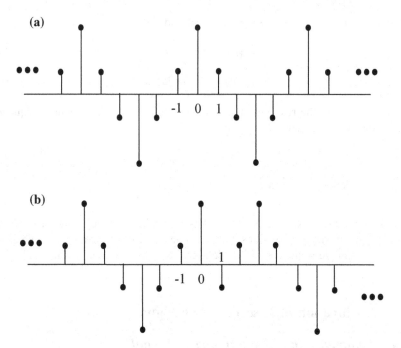

Fig. 2.12 **a** An example of symmetric sequence and **b** an example of antisymmetric sequence

2.2.2 Finite and Infinite Length Sequences

A signal is said to be of finite length or duration if it is defined only for a finite time interval:

$$-\infty < N_1 \leq n \leq N_2 < \infty \qquad (2.12)$$

The length of the sequence is $N = N_2 - N_1 + 1$. Thus, a finite sequence of length N has N samples. A discrete-time sequence consisting of N samples is called a N-point sequence. Any finite sequence can be viewed as an infinite length sequence by adding zero-valued samples outside the range (N_1, N_2). Also, an infinite length sequence can be truncated to produce a finite length sequence.

2.2.3 Right-Sided and Left-Sided Sequences

A right-sided sequence is an infinite sequence $x(n)$ for which $x(n) = 0$ for $n < N_1$, where N_1 is a positive or negative integer. If $N_1 \geq 0$, the right-sided sequence is said to be causal. Similarly, if $x(n) = 0$ for $n > N_2$, where N_2 is a positive or negative integer, then the sequence is called a left-sided sequence. Also, if $N_2 \leq 0$, then the sequence is said to be anti-causal.

2.2.4 *Periodic and Aperiodic Signals*

A sequence $x(n) = x(n + N)$ *for* all n is periodic with a period N, where N is a positive integer. The smallest value of N for which $x(n) = x(n + N)$ is referred as the fundamental period. A sequence is called aperiodic, if it is not periodic. An example of a periodic sequence is shown in Fig. 2.13.

For example, consider $x(n) = \cos\left(\frac{\pi n}{4}\right)$. The relation $x(n) = x(n+N)$ is satisfied if ω_0 is an integer multiple of 2π, i.e., $\omega_0 N = 2\pi m$; $N = 2\pi \frac{m}{\omega_0}$. For this case, $\omega_0 = \frac{\pi}{4}$; for $m = 1$, $N = 2\pi \frac{4}{\pi} = 8$. Hence, $x(n) = \cos\left(\frac{\pi n}{4}\right)$ is periodic with fundamental period $N = 8$, whereas $x(n) = \sin 2n$ is aperiodic because $\omega_0 N = 2N = 2\pi m$ is not satisfied for any integer value of m in making N to be an integer. As another example, consider $x(n) = \sin\left(\frac{\pi n}{4}\right) + \cos 2n$. In this case, $\sin\left(\frac{\pi n}{4}\right)$ is periodic and $\cos 2n$ is aperiodic. Since the sum of periodic and aperiodic signals is aperiodic, the signal $x(n) = \sin\left(\frac{\pi n}{4}\right) + \cos 2n$ is aperiodic.

2.2.5 *Energy and Power Signals*

The total energy of a signal $x(n)$, real or complex, is defined as

$$E = \sum_{n=-\infty}^{\infty} |x(n)|^2 \tag{2.13}$$

By definition, the average power of an aperiodic signal $x(n)$ is given by

$$P = \underset{N \to \infty}{Lt} \frac{1}{2N+1} \sum_{n=-N}^{N} |x(n)|^2 \tag{2.14a}$$

The signal is referred to as an energy signal if the total energy of the signal satisfies the condition $0 < E < \infty$. It is clear that for a finite energy signal, the average power P is zero. Hence, an energy signal has zero average power. On the

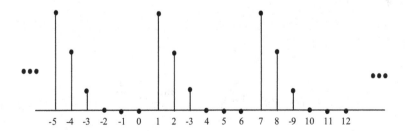

Fig. 2.13 An example of a periodic sequence

other hand, if E is infinite, then P may be finite or infinite. If P is finite and nonzero, then the signal is called a power signal. Thus, a power signal is an infinite energy signal with finite average power.

The average power of a periodic sequence $x(n)$ with a period I is given by

$$P = \frac{1}{I} \sum_{n=0}^{I-1} |x(n)|^2 \qquad (2.14b)$$

Hence, periodic signals are power signals.

Example 2.1 Determine whether the sequence $x(n) = a^n u(n)$ is an energy signal or a power signal or neither for the following cases:

$$|a| < 1, \ (b)|a| = 1, \ (c)|a| > 1.$$

Solution For $x(n) = a^n u(n)$, E is given by

$$E = \sum_{-\infty}^{\infty} |x(n)|^2 = \sum_{0}^{\infty} |a^n|^2$$

$$P = \lim_{N\to\infty} \frac{1}{2N+1} \sum_{-\infty}^{\infty} |x(n)|^2 = \lim_{N\to\infty} \frac{1}{2N+1} \sum_{0}^{N} |a^{2n}|$$

(a) For $|a| < 1$,

$$E = \sum_{-\infty}^{\infty} |x(n)|^2 = \sum_{0}^{\infty} |a^n|^2 = \frac{1}{1-|a|^2} \text{ is finite}$$

$$P = \lim_{N\to\infty} \frac{1}{2N+1} \sum_{n=0}^{N} |a^{2n}| = \lim_{N\to\infty} \frac{1}{2N+1} \frac{1-|a|^{2(N+1)}}{1-|a|^2} = 0$$

The energy E is finite, and the average power P is zero. Hence, the signal $x(n) = a^n u(n)$ is an energy signal for $|a| < 1$.

(b) For $|a| = 1$,

$$E = \sum_{0}^{\infty} |a^n|^2 \to \infty$$

$$P = \lim_{N\to\infty} \frac{1}{2N+1} \sum_{n=0}^{N} |a^{2n}| = \lim_{N\to\infty} \frac{N+1}{2N+1} = \frac{1}{2}$$

The energy E is infinite, and the average power P is finite. Hence, the signal $x(n) = a^n u(n)$ is a power signal for $|a| = 1$.

(a) For $|a| > 1$,

$$E = \sum_0^\infty |a^n|^2 \to \infty$$

$$P = \lim_{N \to \infty} \frac{1}{2N+1} \sum_{n=0}^N |a^{2n}| = \lim_{N \to \infty} \frac{1}{2N+1} \frac{|a|^{2(N+1)}-1}{|a|^2-1} \to \infty$$

The energy E is infinite, and also the average power P is infinite. Hence, the signal $x(n) = a^n u(n)$ is neither an energy signal nor a power signal, for $|a| > 1$.

Example 2.2 Determine whether the following sequences

  (i) $x(n) = e^{-n} u(n)$ (ii) $x(n) = e^n u(n)$ (iii) $x(n) = nu(n)$ and
(iv) $x(n) = \cos \pi n\, u(n)$ are energy or power signal, or neither energy nor power signal.

Solution (i) $x(n) = e^{-n} u(n)$. Hence, E and P are given by

$$E = \sum_{-\infty}^\infty |x(n)|^2 = \sum_0^\infty e^{-2n} = \frac{1}{1 - e^{-2}} \quad \text{is finite}$$

$$P = \lim_{N \to \infty} \frac{1}{2N+1} \sum_{n=0}^N |x(n)|^2 = \lim_{N \to \infty} \frac{1}{2N+1} \sum_{n=0}^N e^{-2n}$$

$$= \lim_{N \to \infty} \frac{1}{2N+1} \frac{1 - e^{-2(N+1)}}{1 - e^{-2}} = 0$$

The energy E is finite, and the average power P is zero. Hence, the signal $x(n) = e^{-n} u(n)$ is an energy signal.

  (ii) $x(n) = e^n u(n)$. Therefore, E and P are given by

$$E = \sum_{-\infty}^\infty |x(n)|^2 = \sum_0^\infty e^{2n} \to \infty$$

$$P = \lim_{N \to \infty} \frac{1}{2N+1} \sum_{n=0}^N |x(n)|^2 = \lim_{N \to \infty} \frac{1}{2N+1} \sum_{n=0}^N e^{2n}$$

$$= \lim_{n \to \infty} \frac{1}{2N+1} \frac{e^{2(N+1)} - 1}{e^2 - 1} \to \infty$$

The energy E is infinite, and also the average power P is infinite. Hence, the signal $x(n) = e^n u(n)$ is neither an energy signal nor a power signal.

(iii) $x(n) = nu(n)$. Hence, E and P are given by

$$E = \sum_{-\infty}^{\infty} |x(n)|^2 = \sum_{0}^{\infty} n^2 \to \infty$$

$$P = \lim_{N\to\infty} \frac{1}{2N+1} \sum_{-\infty}^{\infty} |x(n)|^2$$

$$= \lim_{N\to\infty} \frac{1}{2N+1} \sum_{n=0}^{N} n^2 = \lim_{N\to\infty} \frac{N(N+1)(2N+1)}{6(2N+1)} \to \infty$$

The energy E is infinite, and also the average power P is infinite. Hence, the signal $x(n) = nu(n)$ is neither an energy signal nor a power signal.

(iv) $x(n) = \cos\pi n\, u(n)$. Since $\cos \pi n = (-1)^n$, E and P are given by

$$E = \sum_{-\infty}^{\infty} |x(n)|^2 = \sum_{0}^{\infty} |\cos\pi n|^2 = \sum_{0}^{\infty} (-1)^{2n} \to \infty$$

$$P = \lim_{N\to\infty} \frac{1}{2N+1} \sum_{-\infty}^{\infty} |x(n)|^2$$

$$= \lim_{N\to\infty} \frac{1}{2N+1} \sum_{n=0}^{N} (-1)^{2n} = \lim_{N\to\infty} \frac{N+1}{2N+1} = \frac{1}{2}$$

The energy E is not finite, and the average power P is finite. Hence, the signal $x(n) = \cos \pi n u(n)$ is a power signal.

2.3   The Sampling Process of Analog Signals

2.3.1   Impulse-Train Sampling

The acquisition of an analog signal at discrete time intervals is called sampling. The sampling process mathematically can be treated as a multiplication of a continuous-time signal $x(t)$ by a periodic impulse train $p(t)$ of unit amplitude with period T. For example, consider an analog signal $x_a(t)$ as shown in Fig. 2.14a and a periodic impulse train $p(t)$ of unit amplitude with period T as shown in Fig. 2.14b is referred to as the sampling function, the period T as the sampling period, the fundamental frequency $\omega_T = (2\pi/T)$ as the sampling frequency in radians. Then, the sampled version $x_p(t)$ is shown in Fig. 2.14c.

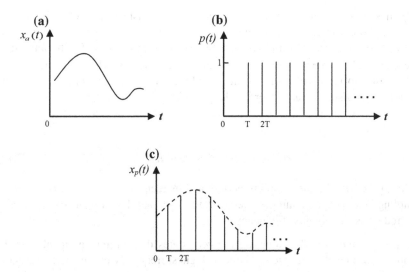

Fig. 2.14 **a** Continuous-time signal, **b** pulse train, and **c** sampled version of (**b**)

In the time domain, we have

$$x_p(t) = x_a(t)p(t) \tag{2.15}$$

where

$$p(t) = \sum_{n=-\infty}^{\infty} \delta(t - nT) \tag{2.15a}$$

$x_p(t)$ is the impulse train with the amplitudes of the impulses equal to the samples of $x_a(t)$ *at intervals T, 2T, 3T,....*

Therefore, the sampled version of signal $x_p(t)$ mathematically can be represented as

$$x_p(t) = \sum_{n=-\infty}^{\infty} x_a(nT)\delta(t - nT) \tag{2.16}$$

2.3.2 Sampling with a Zero-Order Hold

In Sect. 2.3.1, the sampling process establishes a fact that the band-limited signal can be uniquely represented by its samples. In a practical setting, it is difficult to generate and transmit narrow, large-amplitude pulses that approximate impulses. Hence, it is more convenient to implement the sampling process using a zero-order hold. It samples analog signal at a given sampling instant and holds the sample value until the succeeding sampling instant. A block diagram representation of the analog-to-digital conversion (ADC) process is shown in Fig. 2.15. The amplitude of each signal sample is quantized into one of the 2^b levels, where b is the number of bits used to represent a sample in the ADC. The discrete amplitude levels are encoded into distinct binary word of length b bits.

A sequence of samples $x(n)$ is obtained from an analog signal $x_a(t)$ according to the relation,

$$x(n) = x_a(nT) \quad -\infty < n < \infty. \tag{2.17}$$

In Eq. (2.16), T is the sampling period, and its reciprocal, $F_T = 1/T$, is called the sampling frequency, in samples per second. The sampling frequency F_T is also referred to as the Nyquist frequency.

Sampling Theorem The sampling theorem states that an analog signal must be sampled at a rate at least twice as large as highest frequency of the analog signal to be sampled. This means that

$$F_T \geq 2f_{max} \tag{2.18}$$

where f_{max} is maximum frequency component of the analog signal. The frequency $2f_{max}$ is called the Nyquist rate.

For example, to sample a speech signal containing up to 3 kHz frequencies, the required minimum sampling rate is 6 kHz, that is 6000 sample per second. To sample an audio signal having frequencies up to 22 kHz, the required minimum sampling rate is 44 kHz, that is 44,000 samples per second.

A signal whose energy is concentrated in a frequency band range $f_L < |f| < f_H$ is often referred to as a *bandpass signal*. The sampling process of such signals is generally referred to as *bandpass sampling*. In the bandpass sampling process, to prevent aliasing effect, the bandpass continuous-time signal can be sampled at sampling rate greater than twice the highest frequency (f_H)

$$F_T \geq 2f_H \tag{2.19}$$

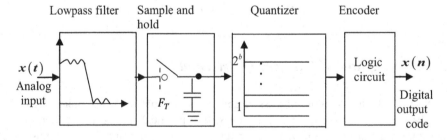

Fig. 2.15 A block diagram representation of an analog-to-digital conversion process

The bandwidth of the bandpass signal is defined as

$$\Delta f = f_H - f_L \qquad (2.20)$$

Consider that the highest frequency contained in the signal is an integer multiple of the bandwidth that is given as

$$f_H = c(\Delta f) \qquad (2.21)$$

The sampling frequency is to be selected to satisfy the condition as

$$F_T = 2(\Delta f) = \frac{f_H}{c} \qquad (2.22)$$

2.3.3 Quantization and Coding

Quantization and coding are two primary steps involved in the process of A/D conversion. Quantization is a nonlinear and non-invertible process that rounds the given amplitude $x(n) = x(nT)$ to an amplitude $x\_\{k\}$ that taken from the finite set of values at time $t = nT$. Mathematically, the output of the quantizer is defined as

$$x_q(n) = Q[x(n)] = \hat{x}_k \qquad (2.23)$$

The procedure of the quantization process is depicted as

$$x_1 \ \hat{x}_1 \ x_2 \quad \hat{x}_2 \ x_3 \ \hat{x}_3 \ x_4 \quad \hat{x}_4 \ x_5 \quad \hat{x}_5 \ \cdots \cdots \cdots \cdots \cdots \cdots$$

The possible outputs of the quantizer (i.e., the quantization levels) are indicated by $\hat{x}_1 \ \hat{x}_2 \ \hat{x}_3 \ \hat{x}_4 \ \cdots \ \hat{x}_L$ where L stands for number of intervals into which the signal amplitude is divided. For uniform quantization,

$$\hat{x}_{k+1} - \hat{x}_k = \Delta \quad k = 1, 2, \cdots, L.$$

$$x_{k+1} - x_k = \Delta \quad \text{for finite } x_k, x_{k+1}. \qquad (2.24)$$

where Δ is the quantizer step size.

The *coding* process in an A/D converter assigns a unique binary number to each quantization level. For L levels, at least L different binary numbers are needed. With word length of n bits, 2^n distinct binary numbers can be represented. Then, the step size or the resolution of the A/D converter is given by

$$\Delta = \frac{A}{2^n} \qquad (2.25)$$

where A is the range of the quantizer.

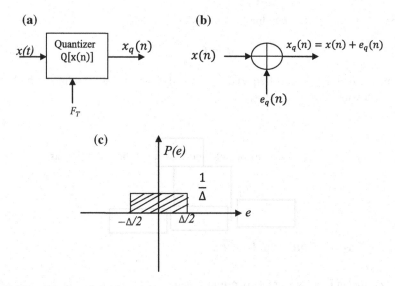

Fig. 2.16 **a** Quantizer, **b** mathematical model, and **c** power spectral density of quantization noise

Quantization Error

Consider an n-bit ADC sampling analog signal $x(t)$ at sampling frequency of F_T
as shown in Fig. 2.16a. The mathematical model of the quantizer is shown in
Fig. 2.16b. The power spectral density of the quantization noise with an assumption
of uniform probability distribution is shown in Fig. 2.16c.

If the quantization error is uniformly distributed in the range $(-\Delta/2, \Delta/2)$ as
shown in Fig. 2.16, the mean value of the error is zero and the variance (the
quantization noise power) σ_e^2 is given by

$$P_{qn} = \sigma_e^2 = \int\limits_{-\Delta/2}^{\Delta/2} q_e^2(n)P(e)\mathrm{d}e = \frac{\Delta^2}{12} \qquad (2.26)$$

The quantization noise power can be expressed by

$$\sigma_e^2 = \frac{\text{quantization step}^2}{12} = \frac{A^2}{12} \times \frac{1}{2^{2n}} = \frac{A^2}{12}2^{-2n} \qquad (2.27)$$

The effect of the additive quantization noise on the desired signal can be
quantified by evaluating the signal-to-quantization noise (power) ratio (SQNR) that
is defined as

$$\text{SQNR} = 10\log_{10}\frac{P_x}{P_{qn}} \qquad (2.28)$$

where $P_x = \sigma_x^2 = E[x^2(n)]$ is the signal power and $P_{qn} = \sigma_e^2 = E\left[e_q^2(n)\right]$ is the quantization noise power.

2.4 Discrete-Time Systems

A discrete-time system is defined mathematically as a transformation that maps an input sequence x(n) into an output sequence y(n). This can be denoted as

$$y(n) = \Re[x(n)] \tag{2.29}$$

where $\Re$ is an operator.

2.4.1 Classification of Discrete-Time Systems

Linear systems

A system is said to be linear if and only if it satisfies the following conditions:

$$\Re[ax(n)] = a\Re[x(n)] \tag{2.30}$$

$$\Re[x_1(n) + x_2(n)] = \Re[x_1(n)] + \Re[x_2(n)] = y_1(n) + y_2(n) \tag{2.31}$$

where a is an arbitrary constant, and $y_1(n)$ and $y_2(n)$ are the responses of the system when $x_1(n)$ and $x_2(n)$ are the respective inputs. Equations (2.30) and (2.31) represent the homogeneity and additivity properties, respectively.

The above two conditions can be combined into one representing the principle of superposition as

$$\Re[ax_1(n) + bx_2(n)] = a\Re[x_1(n)] + b\Re[x_2(n)] \tag{2.32}$$

where a and b are arbitrary constants.

Example 2.3 Check for linearity of the following systems described by the following input–output relationships

(i) $y(n) = \sum_{k=-\infty}^{n} x(k)$

(ii) $y(n) = x^2(n)$

(iii) $y(n) = x(n - n_0)$, where n_0 is an integer constant

Solution (i) The outputs $y_1(n)$ and $y_2(n)$ for inputs $x_1(n)$ and $x_2(n)$ are, respectively, given by

$$y_1(n) = \sum_{k=-\infty}^{n} x_1(k)$$

$$y_2(n) = \sum_{k=-\infty}^{n} x_2(k)$$

The output $y(n)$ due to an input $x(n) = ax_1(n) + bx_2(n)$ is then given by

$$y(n) = \sum_{k=-\infty}^{n} ax_1(k) + bx_2(k) = a\sum_{k=-\infty}^{n} x_1(k) + b\sum_{k=-\infty}^{n} x_2(k) = ay_1(n) + by_2(n)$$

Hence the system described by $y(n) = \sum_{k=-\infty}^{n} x(k)$ is a linear system.
(ii) The outputs $y_1(n)$ and $y_2(n)$ for inputs $x_1(n)$ and $x_2(n)$ are given by

$$y_1(n) = x_1^2(n)$$
$$y_2(n) = x_2^2(n)$$

The ouput $y(n)$ due to an input $x(n) = ax_1(n) + bx_2(n)$ is then given by

$$y(n) = (ax_1(n) + bx_2(n))^2 = a^2 x_1^2(n) + 2abx_1(n)x_2(n) + b^2 x_2^2(n)$$

$$ay_1(n) + by_2(n) = ax_1^2(n) + bx_2^2(n) \neq y(n)$$

Therefore, the system $y(n) = x^2(n)$ is not linear.
(iii) The outputs $y_1(n)$ and $y_2(n)$ for inputs $x_1(n)$ and $x_2(n)$, respectively, are given by

$$y_1(n) = x_1(n - n_0)$$
$$y_2(n) = x_2(n - n_0)$$

The output $y(n)$ due to an input $x(n) = ax_1(n) + bx_2(n)$ is then given by

$$y(n) = ax_1(n - n_0) + bx_2(n - n_0)$$
$$= ay_1(n) + by_2(n)$$

Hence, the system $y(n) = x(n - n_0)$ is linear.
Time-Invariant systems
A time-invariant system (shift invariant system) is one in which the internal parameters do not vary with time. If $y_1(n)$ is output to an input $x_1(n)$, then the system is said to be time invariant if, for all n_0, the input sequence $x_1(n) = x(n - n_0)$ produces the output sequence $y_1(n) = y(n - n_0)$ i.e.

$$\Re[x(n - n_0)] = y(n - n_0) \tag{2.33}$$

where n_0 is a positive or negative integer.

Example 2.4 Check for time-invariance of the system defined by

$$y(n) = \sum_{k=-\infty}^{n} x(k)$$

Solution The output $y(n)$ of the system delayed by n_0 can be written as

$$y(n - n_0) = \sum_{k=-\infty}^{n-n_0} x(k)$$

For example, for an input $x_1(n) = x(n - n_0)$, the output $y_1(n)$ can be written as

$$y_1(n) = \sum_{k=-\infty}^{n} x(k - n_0)$$

Substitution of the change of variables $k_1 = k - n_0$ in the above summation yields

$$y_1(n) = \sum_{k_1=-\infty}^{n-n_0} x(k_1) = y(n - n_0)$$

Hence, it is a time-invariant system.

Example 2.5 Check for time-invariance of the down-sampling system with a factor of 2, defined by the relation

$$y(n) = x(2n) \quad -\infty < n < \infty$$

Solution For an input $x_1(n) = x(n - n_0)$, the output $y_1(n)$ of the compressor system can be written as

$$y_1(n) = x(2n - n_0)$$
$$y(n - n_0) = x(2(n - n_0))$$

Comparing the above equations, it can be observed that $y_1(n) \neq y(n - n_0)$. Thus, the down-sampling system is not time invariant.

Causal System

A system is said to be causal, if its output at time instant n depends only on the present and past input values, but not on the future input values.

It implies that for every choice of n_0, if $x_1(n) = x_2(n)$ for $n \leq n_0$, then $y_1(n) = y_2(n)$ for $n \leq n_0$.

For example, a system defined by

$$y(n) = x(n+2) - x(n+1)$$

is not causal, as the output at time instant n depends on future values of the input. But, the system defined by

$$y(n) = x(n) - x(n-1)$$

is causal, since its output at time instant n depends only on the present and past values of the input.

Stable System

A system is said to be stable, if and only if every bounded input sequence produces a bounded output sequence. The input $x(n)$ is bounded if there exists a fixed positive finite value β_x such that

$$|x(n)| \leq \beta_x < \infty \quad \text{for all} \ n \tag{2.34a}$$

Similarly, the output $y(n)$ is bounded if there exists a fixed positive finite value β_y such that

$$|y(n)| \leq \beta_y < \infty \quad \text{for all} \ n \tag{2.34b}$$

and this type of stability is called *bounded-input bounded-output* (BIBO) stability.

Example 2.6 Check for stability of the system described by the following input–output relation

$$y(n) = x^2(n)$$

Solution Assume that the input $x(n)$ is bounded such that $|x(n)| \leq \beta_x < \infty$ for all n

Then, $|y(n)| = |x(n)|^2 \leq \beta_x^2 < \infty$

Hence, $y(n)$ is bounded and the system is stable.

Example 2.7 Check for stability, causality, linearity, and time-invariance of the system described by

$$\Re[x(n)] = (-1)^n x(n)$$

This transformation outputs the current value of $x(n)$ multiplied by either ± 1.

It is stable, since it does not change the magnitude of $x(n)$ and hence satisfies the conditions for bounded-input bounded-output stability.

It is causal, because each output depends only on the current value of $x(n)$.

Let

$$y_1(n) = \Re[x_1(n)] = (-1)^n x_1(n)$$
$$y_2(n) = \Re[x_2(n)] = (-1)^n x_2(n)$$

Then, $\Re[ax_1(n) + bx_2(n)] = (-1)^n ax_1(n) + (-1)^n bx_2(n) = ay_1(n) + by_2(n)$
Hence, it is linear.

$$y(n) = \Re[x(n)] = (-1)^n x(n) \quad \Re[x(n-1)] = (-1)^n x(n-1)$$
$$\Re[x(n-1)] \neq y(n-1)$$

Therefore, it is not time invariant.

Example 2.8 Check for stability, causality, linearity, and time-invariance of the system described by

$$\Re[x(n)] = x(n^2)$$

Solution Stable, since if $x(n)$ is bounded, $x(n^2)$ is also bounded.

It is not causal, since, for example, if $n = 4$, then the output $y(n)$ depends upon the future input because $y(4) = \Re[x(4)] = x(16)$

$$y_1(n) = \Re[x_1(n)] = x_1(n^2); y_2(n) = \Re[x_2(n)] = x_2(n^2);$$

$$\Re[ax_1(n) + bx_2(n)] = ax_1(n^2) + bx_2(n^2)$$
$$= ay_1(n) + by_2(n)$$

Therefore, it is linear.

$$y(n) = \Re[x(n)] = x(n^2)$$
$$\Re[x(n-1)] \neq y(n-1)$$

Hence, it is not time invariant.

2.4.2   Impulse and Step Responses

Let the input signal $x(n)$ be transformed by the system to generate the output signal $y(n)$. This transformation operation is given by

$$y(n) = \Re[x(n)]$$

If the input to the system is a unit sample sequence (i.e., impulse input $\delta(n)$), then the system output is called as impulse response and denoted by $h(n)$. If the input to the system is a unit step sequence $u(n)$, then the system output is called as step response. In the next section, we show that a linear time-invariant discrete-time system is characterized by its impulse response or step response.

2.5 Linear Time-Invariant Discrete-Time Systems

Linear time-invariant systems have significant signal processing applications, and hence, it is of interest to study the properties of such systems.

2.5.1 Input–Output Relationship

An arbitrary sequence $x(n)$ can be expressed as a weighted linear combination of unit sample sequences given by

$$x(n) = \sum_{k=-\infty}^{\infty} x(k)\,\delta(n-k) \tag{2.35}$$

Now, the discrete-time system response $y(n)$ is given by

$$y(n) = \Re[x(n)] = \Re\left[\sum_{k=-\infty}^{\infty} x(k)\delta(n-k)\right] \tag{2.36}$$

From the principle of superposition, the above equation can be written as

$$y(n) = \sum_{k=-\infty}^{\infty} x(k)\Re[\delta(n-k)] \tag{2.37}$$

Let the response of the system due to input $\delta(n-k)$ be $h_k(n)$, that is,

$$h_k(n) = \Re[\delta(n-k)]$$

Then, the system response $y(n)$ for an arbitrary input $x(n)$ is given by

$$y(n) = \sum_{k=-\infty}^{\infty} x(k)h_k(n)$$

Since $\delta(n-k)$ is a time-shifted version of $\delta(n)$, the response $h_k(n)$ is the time-shifted version of the impulse response $h(n)$, since the operator is time invariant. Hence, $h_k(n) = h(n-k)$. Thus,

$$y(n) = \sum_{k=-\infty}^{\infty} x(k)h(n-k) \tag{2.38}$$

The above equation for $y(n)$ is commonly called the convolution sum and represented by

$$y(n) = x(n) * h(n) \tag{2.39}$$

where the symbol * stands for convolution. The discrete-time convolution operates on the two sequences $x(n)$ and $h(n)$ to produce the third sequence $y(n)$.

Example 2.9 Determine discrete convolution of the following sequences for large value of n.

$$h(n) = \left(\frac{1}{5}\right)^n u(n)$$

$$x(n) = (-1)^n u(n)$$

Solution

$$y(n) = x(n) * h(n)$$

$$= \sum_{k=-\infty}^{\infty} x(k)h(n-k)$$

$$= \sum_{k=-\infty}^{\infty} \left(\frac{1}{5}\right)^k u(k)(-1)^{n-k} u(n-k)$$

$$= (-1)^n \sum_{k=0}^{n} \left(\frac{1}{5}\right)^k (-1)^{-k}$$

$$= (-1)^n \sum_{k=0}^{n} \left(-\frac{1}{5}\right)^k = (-1)^n \frac{\left(1 - \left(-\frac{1}{5}\right)^{n+1}\right)}{1 - \left(-\frac{1}{5}\right)} = (-1)^n \frac{\left(1 - \left(-\frac{1}{5}\right)^{n+1}\right)}{1 + \frac{1}{5}}$$

For large n, $\left(-\frac{1}{5}\right)^{n+1}$ tends to zero and hence,

$$y(n) = (-1)^n \frac{1}{1.2}$$

Example 2.10 Determine discrete convolution of the following two finite duration sequences

$$h(n) = \left(\frac{1}{3}\right)^n u(n)$$

$$x(n) = \left(\frac{1}{5}\right)^n u(n)$$

Solution The impulse response $h(n) = 0$ *for* $n < 0$; hence, the given system is causal, and $x(n) = 0$ for $n < 0$; therefore, the sequence $x(n)$ is causal sequence.

$$
y(n) = x(n) * h(n) = \sum_{k=0}^{n} \left(\frac{1}{5}\right)^{k} \left(\frac{1}{3}\right)^{n-k} = \left(\frac{1}{3}\right)^{n} \sum_{k=0}^{n} \left(\frac{3}{5}\right)^{k}
$$
$$
= \left(\frac{1}{3}\right)^{n} \frac{1 - (3/5)^{n+1}}{1 - (3/5)}
$$

2.5.2   Computation of Linear Convolution

Matrix Method

If the input $x(n)$ is of length N_1 and the impulse sequence $h(n)$ is of length N_2, then the convolution sequence is of length $N_1 + N_2 - 1$. Thus, the linear convolution given by Eq. (2.38) can be written in matrix form as

$$
\begin{bmatrix} y(0) \\ y(1) \\ y(2) \\ y(3) \\ \vdots \\ y(N_1 - 1) \\ y(N_1) \\ \vdots \\ y(N_1 + N_2 - 2) \end{bmatrix}_{(N_1 + N_2 - 1) \times 1}
$$

$$
= \begin{bmatrix}
x(0) & 0 & 0 & \cdots & 0 \\
x(1) & x(0) & 0 & \cdots & 0 \\
x(2) & x(1) & x(0) & \cdots & 0 \\
\vdots & x(2) & x(1) & \cdots & \vdots \\
x(N_1 - 1) & \vdots & x(2) & \cdots & \vdots \\
0 & x(N_1 - 1) & \vdots & \cdots & \vdots \\
0 & 0 & x(N_1 - 1) & \cdots & \vdots \\
\vdots & \vdots & \vdots & \vdots & \vdots \\
0 & 0 & 0 & \cdots & x(0)
\end{bmatrix}_{(N_1 + N_2 - 1) \times (N_1 + N_2 - 1)}
\begin{bmatrix} h(0) \\ h(1) \\ h(2) \\ h(3) \\ \vdots \\ h(N_2 - 1) \\ 0 \\ \vdots \\ 0 \end{bmatrix}_{(N_1 + N_2 - 1) \times 1}
$$

$$(2.40)$$

The following example illustrates the above procedure for computation of linear convolution

Example 2.11 Find the convolution of the sequences $x(n) = \{6, -3\}$ and $h(n) = \{-3, 6, 3\}$.

Solution Using Eq. (2.40), the linear convolution of $x(n)$ and $h(n)$ is given by

$$\begin{bmatrix} y(0) \\ y(1) \\ y(2) \\ y(3) \end{bmatrix} = \begin{bmatrix} 6 & 0 & 0 & 0 \\ -3 & 6 & 0 & 0 \\ 0 & -3 & 6 & 0 \\ 0 & 0 & -3 & 6 \end{bmatrix} \begin{bmatrix} -3 \\ 6 \\ 3 \\ 0 \end{bmatrix} = \begin{bmatrix} -18 \\ 45 \\ 0 \\ -9 \end{bmatrix}$$

Thus,

$$y(n) = x(n) * h(n) = \{-18, 45, 0, -9\}$$

Graphical Method for Computation of Linear Convolution

Evaluation of sum at any sample n consists of the following four important operations.

(i)   Time reversing or reflecting of the sequence $h(k)$ about $k = 0$ sample to give $h(-k)$
(ii) Shifting the sequence $h(-k)$ to the right by n samples to obtain $h(n - k)$
(iii) Forming the product $x(k)h(n - k)$ sample by sample for the desired value of n
(iv) Summing the product over the index k in $y(n)$ for the desired value of n

The length of the convolution sum sequence $y(n)$ is given by $n = N_1 + N_2 - 1$, where N_1 is length of the sequence $x(n)$ and N_2 is length of the sequence $h(n)$.

Example 2.12 Compute the convolution of the sequences of Example 2.11 using the graphical method.

Solution The sequences $x(n)$ and $h(n)$ are as shown as in Fig. 2.17

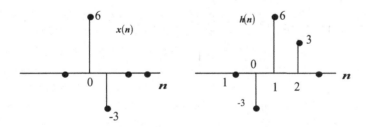

Fig. 2.17 Sequences $x(n)$ and $h(n)$

2.5.3   Computation of Convolution Sum Using MATLAB

The MATLAB function conv(a,b) can be used to compute convolution sum of two sequences a and b as illustrated in the following example (Fig. 2.18).

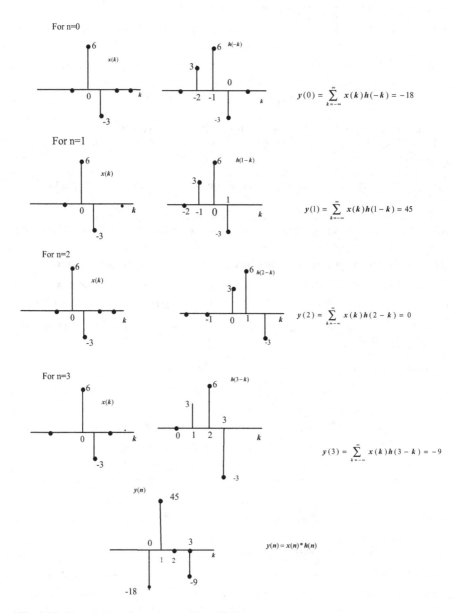

Fig. 2.18 Convolution of sequences $x(n)$ and $h(n)$

Example 2.13 Compute convolution sum of the sequences $x(n) = \{2, -1, 0, 0\}$ and $h(n) = \{-1.2.1\}$, using MATLAB.

Program 2.1 Illustration of convolution

```
a = [ 2 -1 0 0];% first sequence
b = [-1 2 1];% second sequence
c = conv(a,b);% convolution of first sequence and second sequence
len = length(c)-1;
n = 0:1:len;
stem(n,c)
xlabel('Time index n'); ylabel('Amplitude');
axis([0 5 -3 5])
```

2.5.4   Some Properties of the Convolution Sum

Starting with the convolution sum given by (2.39), namely $y(n) = x(n) * h(n)$, we can establish the following properties (Fig. 2.19):

(1) The convolution sum obeys the commutative law

$$x(n) * h(n) = h(n) * x(n) \qquad (2.41a)$$

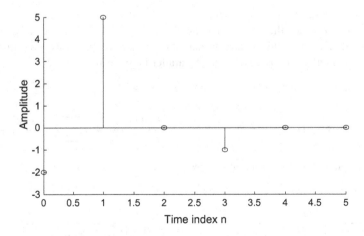

Fig. 2.19 Sequence generated by the convolution

(2) The convolution sum obeys the associative law

$$(x(n) * h_1(n)) * h_2(n) = x(n) * (h_1(n) * h_2(n)) \qquad (2.41b)$$

(3) The convolution sum obeys the distributive law

$$x(n) * (h_1(n) + h_2(n)) = x(n) * h_1(n) + x(n) * h_2(n) \qquad (2.41c)$$

Let us now interpret the above relations physically.

(1) The commutative law shows that the output is the same if we interchange the roles of the input and the impulse response. This is illustrated in Fig. 2.20.
(2) To interpret the associative law, we consider a cascade of two systems whose impulse responses are $h_1(n)$ and $h_2(n)$. Then $y_1(n) = x(n) * h(n)$ if $x(n)$ is the input to the system with the impulse response $h_1(n)$. If $y_1(n)$ is now fed as the input to the system with impulse response $h_2(n)$, then the overall system output is given by

$$\begin{aligned} y(n) = y_1(n) * h_2(n) &= [x(n) * h_1(n)] * h_2(n) \\ &= x(n) * [h_1(n) * h_2(n)], \text{by associative law} \\ &= x(n) * h(n) \end{aligned}$$

This equivalence is shown in Fig. 2.21. Hence, if two systems with impulse responses $h_1(n)$ and $h_2(n)$ are cascaded, then the overall system response is given by

$$h(n) = h_1(n) * h_2(n) \qquad (2.42)$$

This can be generalized to a number of LTI systems in cascade.

(3) We now consider the distributive law given by (2.41c). This can be easily interpreted as two LTI systems in parallel and that the overall system impulse response $h(n)$ of the two systems in parallel is given by

Fig. 2.20 Interpretation of the commutative law

Fig. 2.21 Interpretation of the associative law

$$h(n) = h_1(n) + h_2(n) \tag{2.43}$$

This is illustrated in Fig. 2.22.

Example 2.14 Consider the system shown in Fig. 2.23 with $h(n)$ being real. If $y_2(n) = y_1(-n)$, find the overall impulse response $h_1(n)$ that relates $y_2(n)$ to $x(n)$.

Solution From Fig. 2.23, we have the following relations:

$$
\begin{aligned}
y(n) &= x(n) * h(n) \\
y_1(n) &= y(-n) * h(n) \\
y_2(n) &= y_1(-n) = y(n) * h(-n) \\
&= (x(n) * h(n)) * h(-n) \\
&= x(n) * (h(n) * h(-n)) = x(n) * h_1(n)
\end{aligned}
$$

Hence, the overall impulse response $= h_1(n) = h(n) * h(-n)$.

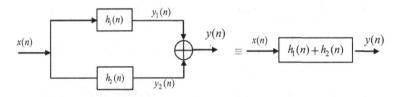

Fig. 2.22 Interpretation of distributive law

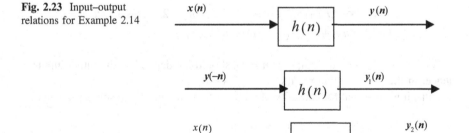

Fig. 2.23 Input–output relations for Example 2.14

2.5.5 Stability and Causality of LTI Systems in Terms of the Impulse Response

The output of a LTI system can be expressed as

$$|y(n)| = \left| \sum_{k=-\infty}^{\infty} h(k)x(n-k) \right| \le \sum_{k=-\infty}^{\infty} |h(k)||x(n-k)|$$

For bounded input $x(n)$

$$|x(n)| \le \beta_x < \infty$$

we have

$$|y(n)| \le \beta_x \sum_{k=-\infty}^{\infty} |h(k)| \tag{2.44}$$

It is seen from (2.44) that $y(n)$ is bounded if and only if $\sum_{k=-\infty}^{\infty} |h(k)|$ is bounded. Hence, the necessary and sufficient condition for stability is that

$$S = \sum_{k=-\infty}^{\infty} |h(k)| < \infty. \tag{2.45}$$

The output $y(n_0)$ of a LTI causal system can be expressed as

$$y(n_0) = \sum_{k=-\infty}^{\infty} h(k)x(n_0-k)$$

$$= h(-\infty)x(n_0+\infty) + \cdots + h(-2)x(n_0+2) + h(-1)x(n_0+1)$$
$$+ h(0)x(n_0) + h(1)x(n_0-1) + h(2)x(n_0-2) + \cdots$$

For a causal system, the output at $n = n_0$ should not depend on the future inputs. Hence, in the above equation, $h(k) = 0$ for $k < 0$.

Thus, it is clear that for causality of a LTI system, its impulse response sequence

$$h(n) = 0 \quad \text{for } n < 0. \tag{2.46}$$

Example 2.15 Check for the stability of the systems with the following impulse responses:

(i) Ideal delay: $h(n) = \delta(n - n_d)$, (ii) forward difference: $h(n) = \delta(n+1) - \delta(n)$,

(iii) Backward difference: $h(n) = \delta(n) - \delta(n-1)$, (iv) $h(n) = u(n)$,

(v) $h(n) = a^n u(n)$, where $|a| < 1$, and (vi) $h(n) = a^n u(n)$, where $|a| \ge 1$.

Solution Given impulse responses of the systems, stability of each system can be tested by computing the sum

$$S = \sum_{k=-\infty}^{\infty} |h(k)|$$

In case of (i), (ii), and (iii), it is clear that $S<\infty$. As such, the systems corresponding to (i), (ii), and (iii) are stable.

For the impulse response given in (iv), the system is unstable since

$$S = \sum_{n=0}^{\infty} u(n) = \infty.$$

This is an example of an infinite duration impulse response (IIR) system.

In case of (v), $S = \sum_{n=0}^{\infty} |a|^n$. For $|a|<1$, $S<\infty$, and hence the system is stable. This is an example of a stable IIR system.

Finally, in case of (vi), $|a| \geq 1$, and the sum is infinite, making the system unstable.

Example 2.16 Check the following systems for causality:

(i) $h(n) = \left(\frac{3}{4}\right)^n u(n)$, (ii) $h(n) = \left(\frac{1}{2}\right)^n u(n+2) + \left(\frac{3}{4}\right)^n u(n)$,

(iii) $h(n) = \left(\frac{1}{2}\right)^n u(-n-1)$, (iv) $h(n) = \left(\frac{3}{4}\right)^{|n|}$, and

(v) $h(n) = u(n+1) - u(n)$

Solution

(i) $h(n) = 0$ for $n<0$; hence, the system is causal
(ii) $h(n) \neq 0$ for $n<0$; hence, the system is not causal
(iii) $h(n) \neq 0$ for $n<0$; thus, the system is not causal
(iv) $h(n) = \left(\frac{3}{4}\right)^{|n|}$, hence $h(n) \neq 0$ for $n<0$; so, the sytem is not causal
(v) $h(n) = u(n+1) - u(n)$, $h(n) \neq 0$ for $n<0$; so, the system is not causal.

Example 2.17 Check the following systems for stability:

(i) $h(n) = \left(\frac{1}{3}\right)^n u(n-1)$,   (ii) $h(n) = u(n+2) - u(n-5)$,

(iii) $h(n) = 5^n u(-n-3)$,

(iv) $h(n) = \sin\left(\frac{n\pi}{4}\right) u(n)$, and (v) $h(n) = \left(\frac{1}{2}\right)^{|n|} \cos\left(\frac{\pi n}{4}\right)$

Solution

(i) The system is stable, since $S = \sum_{k=-\infty}^{\infty} |h(k)| < \infty$.
(ii) $h(n) = u(n+2) - u(n-5)$. The system is stable, since S is finite.

(iii) $h(n) = 5^n u(-n - 3)$. Hence, $\sum_n |h(n)| = \sum_{n=-\infty}^{-3} 5^n = \sum_{n=3}^{\infty} \left(\frac{1}{5}\right)^n < \infty$
Therefore, the system is stable.

(iv) $h(n) = \sin\left(\frac{n\pi}{4}\right) u(n)$
Summing $|h(n)|$ over all positive n, we see that S tends to infinity. Hence, the system is not stable.

(v) $h(n) = \left(\frac{1}{2}\right)^{|n|} \cos\left(\frac{\pi n}{4}\right)$
$|h(n)|$ is upper bounded by $\left(\frac{1}{2}\right)^{|n|}$. Thus, $S = \sum_{k=-\infty}^{\infty} |h(k)| < \infty$.
Hence, the system is stable.

2.6   Characterization of Discrete-Time Systems

Discrete-time systems are characterized in terms of difference equations. An important class of LTI discrete-time systems is one that is characterized by a linear difference equation with constant coefficients. Such a difference equation may be of two types, namely non-recursive and recursive.

2.6.1   Non-recursive Difference Equation

A non-recursive LTI discrete-time system is one that can be characterized by a linear constant coefficient difference equation of the form

$$y(n) = \sum_{m=-\infty}^{\infty} b_m x(n - m) \qquad (2.47)$$

where b_m's represent constants. By assuming causality, the above equation can be written as

$$y(n) = \sum_{m=0}^{\infty} b_m x(n - m) \qquad (2.48)$$

In addition, if $x(n) = 0$ for $n < 0$ and $b_m = 0$ for $m > N$, then Eq. (2.48) becomes

$$y(n) = \sum_{m=0}^{N} b_m x(n - m) \qquad (2.49)$$

Thus an LTI, causal, and non-recursive system can be characterized by an Nth-order linear non-recursive difference equation. The Nth-order non-recursive difference equation has a finite impulse response (FIR). Therefore, an FIR filter is characterized by a non-recursive difference equation.

2.6.2 Recursive Difference Equation

The response of a discrete-time system depends on the present and previous values of the input as well as the previous values of the output. Hence, a linear time-invariant, causal, and recursive discrete-time system can be represented by the following Nth-order linear recursive difference equation

$$y(n) = \sum_{m=0}^{N} b_m x(n - m) - \sum_{m=1}^{N} a_m y(n - m) \tag{2.50}$$

where a_m and b_m are constants. An Nth-order recursive difference equation has an infinite impulse response. Hence, an infinite impulse response (IIR) filter is characterized by a recursive difference equation.

Example 2.18 An initially relaxed LTI system was tested with an input signal $x(n) = u(n)$, and found to have a response as shown in Table 2.1.

(i) Obtain the impulse response of the system.
(ii) Deduce the difference equation of the system.

Solution

(i) From Table 2.1, it can be observed that the response $y(n)$ for an input $x(n) = u(n)$ is given by

$$y(n) = \{1, 2, 4, 6, 10, 10, 10, \ldots\}$$

Similarly, for an input $x(n) = u(n - 1)$, the response $y(n - 1)$ is given by

$$y(n - 1) = \{0, 1, 2, 4, 6, 10, 10, 10, \ldots\}$$

For an input $x(n) = u(n) - u(n - 1)$, the response of an LTI system is the impulse response $h(n)$ given by

Table 2.1 Response of an LTI system for an input $x(n) = u(n)$

yn	1	2	3	4	5	...	100	...
$y(n)$	1	2	4	6	10	...	10	...

$$h(n) = y(n) - y(n-1) = \{1, 1, 2, 2, 4\}$$

(ii) The difference equation is given by

$$y(n) = \sum_{m=0}^{4} h(m)x(n-m)$$

Hence, the difference equation of the system can be written as

$$y(n) = x(n) + 1x(n-1) + 2x(n-2) + 2x(n-4) + 4x(n-5)$$

2.6.3 Solution of Difference Equations

A general linear constant coefficient difference equation can be expressed as

$$y(n) = -\sum_{k=1}^{N} a_k y(n-k) + \sum_{k=0}^{M} b_k x(n-k) \qquad (2.51)$$

The solution of the difference equation is the output response $y(n)$. It is the sum of two components which can be computed independently as

$$y(n) = y_c(n) + y_p(n) \qquad (2.52)$$

where $y_c(n)$ is called the complementary solution and $y_p(n)$ is called the particular solution.

The complementary solution $y_c(n)$ is obtained by setting $x(n) = 0$ in Eq. (2.51). Thus $y_c(n)$ is the solution of the following homogeneous difference equation

$$\sum_{k=0}^{N} a_k y(n-k) = 0 \qquad (2.53a)$$

where $a_0 = 1$. To solve the above homogeneous difference equation, let us assume that

$$y_c(n) = \lambda^n \qquad (2.53b)$$

where the subscript c indicates the solution to the homogeneous difference equation. Substituting $y_c(n)$ in Eq. (2.53a), the following equation can be obtained:

$$\sum_{k=0}^{N} a_k \lambda^{n-k} = 0 = \lambda^{n-N} \left[\lambda^N + a_1 \lambda^{N-1} + \cdots + a_{N-1}\lambda + a_N \right] = 0 \qquad (2.54)$$

which takes the form

$$\lambda^N + a_1\lambda^{N-1} + \cdots + a_{N-1}\lambda + a_N = 0 \tag{2.55}$$

The above equation is called the characteristic equation, which consists of N roots represented by $\lambda_1, \lambda_2, \cdots, \lambda_N$. If the N roots are distinct, then the complementary solution can be expressed as

$$y_c(n) = \alpha_1\lambda_1^n + \alpha_2\lambda_2^n + \cdots + \alpha_N\lambda_N^n \tag{2.56a}$$

where $\alpha_1, \alpha_2, \ldots, \alpha_N$ are constants which can be obtained from the specified initial conditions of the discrete-time system. For multiple roots, the complementary solution $y_c(n)$ assumes a different form. In the case when the root λ_1 of the characteristic equation is repeated m times, but $\lambda_2, \ldots, \lambda_N$ are distinct, then the complementary solution $y_c(n)$ assumes the form

$$\lambda_1^n\left(\alpha_1 + \alpha_2 n + \cdots + \alpha_m n^{m-1}\right) + \beta_2\lambda_2^n + \cdots + \beta_{N-M}\lambda_{N-M}^n \tag{2.56b}$$

In case the characteristic equation consists of complex roots $\lambda_1, \lambda_2 = a \pm jb$, then the complementary solution results in $y_c(n) = (\alpha^2 + \beta^2)^{n/2}(C_1\cos n\theta + C_2\sin n\theta)$, where $\theta = \tan^{-1} b/a$ and C_1 and C_2 are constants.

We now look at the particular solution $y_p(n)$ of Eq. (2.51). The particular solution $y_p(n)$ is any solution that satisfies the difference equation for the specific input signal $x(n)$, for ≥ 0, i.e.,

$$y(n) + \sum_{k=1}^{N} a_k y(n-k) = \sum_{k=0}^{M} b_k x(n-k) \tag{2.57}$$

The procedure to find the particular solution $y_p(n)$ assumes that $y_p(n)$ depends on the form of $x(n)$. Thus, if $x(n)$ is a constant, then $y_p(n)$ is implicitly a constant. Similarly, if $x(n)$ is a sinusoidal sequence, then $y_p(n)$ is implicitly a sinusoidal sequence and so on.

In order to find out the overall solution, the complementary and particular solutions must be added. Hence,

$$y(n) = y_c(n) + y_p(n) \tag{2.58}$$

Example 2.19 Determine impulse response for the case of $x(n) = \delta(n)$ of a discrete-time system characterized by the following difference equation

$$y(n) + 2y(n-1) - 3y(n-2) = x(n) \tag{2.59}$$

Solution First, we determine the complementary solution by setting $x(n) = 0$ and $y(n) = \lambda^n$ in Eq. (2.59), which gives us

$$\lambda^n + 2\lambda^{n-1} - 3\lambda^{n-2} = \lambda^{n-2}(\lambda^2 + 2\lambda - 3)$$
$$= \lambda^{n-2}(\lambda - 1)(\lambda + 3) = 0$$

Hence, the zeros of the characteristic polynomial $\lambda^2 + 2\lambda - 3$ are $\lambda_1 = -3$ and $\lambda_2 = 1$.

Therefore, the complementary solution is of the form

$$y_c(n) = \alpha_1(-3)^n + \alpha_2(1)^n \tag{2.60}$$

For impulse $x(n) = \delta(n)$, $x(n) = 0$ for $n > 0$ and $x(0) = 1$. Substituting these relations in Eq. (2.59) and assuming that $y(-1) = 0$ and $y(-2) = 0$, we get

$$y(0) + 2y(-1) - 3y(-2) = x(0) = 1$$

i.e., $y(0) = 1$. Similarly $y(1) + 2y(0) - 3y(-1) = x(1) = 0$ yields $y(1) = -2$.

Thus, from Eq. (2.60), we get

$$\alpha_1 + \alpha_2 = 1 \text{ and } -3\alpha_1 + \alpha_2 = -2$$

Solving these two equations, we obtain $\alpha_1 = 3/4; \alpha_2 = 1/4$.

Since $x(n) = 0$ for $n > 0$, there is no particular solution. Hence, the impulse response is given by

$$h(n) = y_c(n) = 0.75(-3)^n + 0.25(1)^n \tag{2.61}$$

Example 2.20 A discrete-time system is characterized by the following difference equation

$$y(n) + 5y(n-1) + 6y(n-2) = x(n) \tag{2.62}$$

Determine the step response of the system, i.e., $x(n) = u(n)$.

Solution For the given difference equation, total solution is given by

$$y(n) = y_c(n) + y_p(n)$$

First, we determine the complementary solution by setting $x(n) = 0$ and $y(n) = \lambda^n$ in Eq. (2.62), which gives us

$$\lambda^n + 5\lambda^{n-1} + 6\lambda^{n-2} = \lambda^{n-2}(\lambda^2 + 5\lambda + 6) = 0$$

Hence, the zeros of the characteristic polynomial $\lambda^2 + 5\lambda + 6$ are $\lambda_1 = -3$ and $\lambda_2 = -2$.

Therefore, the complementary solution is of the form

$$y_c(n) = \alpha_1(-3)^n + \alpha_2(-2)^n$$

The particular solution for the step input is of the form

$$y_p(n) = K$$

For $n > 2$, substituting $h(n)$ and $x(n) = 1$ in Eq. (2.62), we get $K + 5K + 6K = 1; K = \frac{1}{12}$, and $y_p(n) = \frac{1}{12}$.

Therefore, the solution for given difference equation is

$$y(n) = \alpha_1(-3)^n + \alpha_2(-2)^n + \frac{1}{12} \tag{2.63}$$

For $n = 0$, Eq. (2.62) becomes

$$y(0) + 5y(-1) + 6y(-2) = x(0)$$

Assuming $y(-1) = y(-2) = 0$, from the above equation, we get $y(0) = x(0) = 1$
and for $n = 1$, $y(1) + 5y(0) + 6y(-1) = x(1) = 1$, i.e., $y(1) = -4$.
Then, we get from Eq. (2.63)

$$\alpha_1 + \alpha_2 + \frac{1}{12} = 1$$
$$-3\alpha_1 - 2\alpha_2 + \frac{1}{12} = -4$$

Solving these equations, we arrive at $\alpha_1 = \frac{27}{12}$ and $\alpha_2 = \frac{-16}{12}$. Then, the step response is given by

$$y(n) = \frac{27}{12}(-3)^n - \frac{16}{12}(-2)^n + \frac{1}{12} \tag{2.64}$$

Example 2.21 A discrete-time system is characterized by the following difference equation

$$y(n) - 2y(n-1) + y(n-2) = x(n) - x(n-1) \tag{2.65}$$

Determine the response y(n), $n \geq 0$ when the system input is $x(n) = (-1)^n u(n)$ and the initial conditions are $y(-1) = 1$ and $y(-2) = -1$.

Solution For the given difference equation, the total solution is given by

$$y(n) = y_c(n) + y_p(n)$$

First, determine the complementary solution by setting $x(n) = 0$ and $y(n) = \lambda^n$ in Eq. (2.65); this gives

$$\lambda^n - 2\lambda^{n-1} + \lambda^{n-2} = \lambda^{n-2}\left(\lambda^2 - 2\lambda + 1\right) = 0$$

Hence, the zeros of the characteristic polynomial $\lambda^2 - 2\lambda + 1$ are $\lambda_1 = \lambda_2 = 1$. It has repeated roots; thus, the complementary solution is of the form

$$y_c(n) = 1^n(\alpha_1 + n\,\alpha_2).$$

The particular solution for the step input is of the form

$$y_p(n) = K(-1)^n u(n)$$

Substituting $x(n) = (-1)^n u(n)$ and $y_p(n) = K(-1)^n u(n)$ in Eq. (2.65), we get

$$K(-1)^n u(n) - 2K(-1)^{n-1}u(n-1) + K(-1)^{n-2}u(n-2)$$
$$= (-1)^n u(n) - (-1)^{n-1}u(n-1)$$

For $n = 2$, the above equation becomes $K + 2K + K = 2$; $K = \frac{1}{2}$. Therefore, the particular solution is given by

$$y_p(n) = \frac{1}{2}(-1)^n u(n)$$

Then, the total solution for given difference equation is

$$y(n) = 1^n(\alpha_1 + n\alpha_2) + \frac{1}{2}(-1)^n u(n). \tag{2.66}$$

For $n = 0$, Eq. (2.65) becomes

$$y(0) - 2y(-1) + y(-2) = 1$$

Using the initial conditions $y(-1) = 1, y(-2) = -1$, we get $y(0) = 4$. Then, for $n = 1$, from Eq. (2.65), we get $y(1) = 5$. Thus, we get from Eq. (2.66)

$$\alpha_1 + (1/2) = 4$$
$$\alpha_1 + \alpha_2 - (1/2) = 5$$

Solving these two equations, we arrive at $\alpha_1 = (7/2)$ and $\alpha_2 = 2$. Thus, the response of the system for the given input is

$$y(n) = 1^n\left(\frac{7}{2} + 2n\right) + \frac{1}{2}(-1)^n u(n) \tag{2.67}$$

2.6.4 Computation of Impulse and Step Responses Using MATLAB

The impulse and step responses of LTI discrete-time systems can be computed using MATLAB function

$$y = \text{filter}(b, a, x)$$

where b and a are the coefficient vectors of difference equation describing the system, x is the input data vector, and y is the vector generated assuming zero initial conditions. The following example illustrates the computation of the impulse and step responses of an LTI system.

Example 2.22 Determine the impulse and step responses of a discrete-time system described by the following difference equation

$$y(n) - 2y(n - 1) = x(n) + 0.1x(n - 1) - 0.06x(n - 2) \qquad (2.68)$$

Solution Program 2.2 is used to compute and plot the impulse and step responses, which are shown in Fig. 2.24a, b, respectively.

Program 2.2 Illustration of impulse and step responses computation

```
clear;clc;
flag = input('enter 1 for impulse response, and 2 for step response');
len = input('enter desired response length = ');
b = [1-2];%b coefficients of the difference equation
a = [1 0.1 -0.06]; %a coefficients of the difference equation
if flag ==1;
x = [1,zeros(1,len-1)];
```

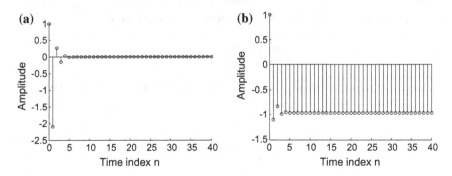

Fig. 2.24 a Impulse response and b step response for Example 2.22

```
end
if flag ==2;
x = [ones(1,len)];
end
y = filter(b,a,x);
n = 0:1:len-1;
stem(n,y)
xlabel('Time index n'); ylabel('Amplitude');
```

2.7 Representation of Discrete-Time Signals and Systems in Frequency Domain

2.7.1 Fourier Transform of Discrete-Time Signals

The discrete-time Fourier transform (DTFT) of a finite energy sequence $x(n)$ is defined as

$$\mathfrak{F}[x(n)] = X(e^{j\omega}) = \sum_{n=-\infty}^{\infty} x(n)e^{(-j\omega n)} \tag{2.69}$$

From $X(e^{j\omega})$, $x(n)$ can be computed as

$$x(n) = \frac{1}{2\pi} \int_{-\pi}^{\pi} X(e^{j\omega})e^{(j\omega n)}d\omega \tag{2.70}$$

Equation (2.70) is called the inverse Fourier transform.

Convergence of the DTFT

The existence of DTFT of $x(n)$ depends on the convergence of the series in Eq. (2.69). Now, we look at the condition for convergence.

Let $X_k(e^{j\omega}) = \sum_{k=-\infty}^{\infty} x(n)e^{(-j\omega n)}$ denote the partial sum of the weighted complex exponentials in Eq. (2.69). Then for uniform convergence of $X(e^{j\omega})$,

$$\lim_{k \to \infty} X_k(e^{j\omega}) = X(e^{j\omega}) \tag{2.71}$$

Hence, for uniform convergence of $X(e^{j\omega})$, $x(n)$ must be absolutely summable, i.e.,

$$\sum_{n=-\infty}^{\infty} |x(n)| < \infty, \tag{2.72}$$

Then,

$$|X(e^{j\omega})| = \left| \sum_{n=-\infty}^{\infty} x(n)e^{-j\omega n} \right| \leq \sum_{n=-\infty}^{\infty} |x(n)||e^{-j\omega n}| \leq \sum_{n=-\infty}^{\infty} |x(n)| < \infty \qquad (2.73)$$

guaranteeing the existence of $X(e^{j\omega})$, for all values of ω. Consequently, Eq. (2.72) is only a sufficient condition for the existence of the DTFT, but is not a necessary condition.

2.7.2   Theorems on DTFT

We will now consider some important theorems concerning DTFT that can be used in digital signal processing. All these properties can be proved using the definition of DTFT. The following notation is adopted for convenience:

$$X(e^{j\omega}) = \mathfrak{F}[x(n)] \qquad (2.74a)$$

$$x(n) = \mathfrak{F}^{-1}[X(e^{j\omega})] \qquad (2.74b)$$

Linearity: If $x_1(n)$ and $x_2(n)$ are two sequences with Fourier transforms $X_1(e^{j\omega})$ and $X_2(e^{j\omega})$, then the Fourier transform of a linear combination of $x_1(n)$ and $x_2(n)$ is given by

$$\mathfrak{F}[a_1x_1(n) + a_2x_2(n)] = a_1X_1(e^{j\omega}) + a_2X_2(e^{j\omega}) \qquad (2.75)$$

where a_1 and a_2 are arbitrary constants.

Time Reversal: If $x(n)$ is a sequence with Fourier transform $X(e^{j\omega})$, then the Fourier transform of time-reversed sequence $x(-n)$ is given by

$$\mathfrak{F}[x(-n)] = X(e^{-j\omega}) \qquad (2.76)$$

Time shifting: If $x(n)$ is a sequence with Fourier transform $X(e^{j\omega})$, then the Fourier transform of the delayed sequence $x(n - k)$, where k an integer, is given by

$$\mathfrak{F}[x(n - k)] = e^{-j\omega k}X(e^{j\omega}) \qquad (2.77)$$

Therefore, time shifting results in a phase shift in the frequency domain.

Frequency shifting: If $x(n)$ is a sequence with Fourier transform $X(e^{j\omega})$, then the Fourier transform of the sequence $e^{j\omega_0 n}x(n)$ is given by

$$\mathfrak{F}\left[e^{j\omega_0 n}x(n)\right] = X\left(e^{j(\omega-\omega_0)}\right) \qquad (2.78)$$

Thus, multiplying a sequence $x(n)$ by a complex exponential $e^{j\omega_0 n}$ in the time domain corresponds to a shift in the frequency domain.

Differentiation in Frequency: If $x(n)$ is a sequence with Fourier transform $X(e^{j\omega})$, then the Fourier transform of the sequence $nx(n)$ is given by

$$\mathfrak{F}[nx(n)] = j\frac{\mathrm{d}}{\mathrm{d}\omega}X(e^{j\omega}) \qquad (2.79)$$

Convolution Theorem If $x_1(n)$ and $x_2(n)$ are two sequences with Fourier transforms $X_1(e^{j\omega})$ and $X_2(e^{j\omega})$, then the Fourier transform of the convolution of $x_1(n)$ and $x_2(n)$ is given by

$$\mathfrak{F}[x_1(n) * x_2(n)] = X_1(e^{j\omega})X_2(e^{j\omega}) \qquad (2.80)$$

Hence, convolution of two sequences $x_1(n)$ and $x_2(n)$ in the time domain is equal to the product of their frequency spectra. In the above equation, since $X_1(e^{j\omega})$ and $X_2(e^{j\omega})$ are periodic in ω with period 2π, the convolution is a periodic convolution.

Windowing Theorem If $x(n)$ and $w(n)$ are two sequences with Fourier transforms $X(e^{j\omega})$ and $W(e^{j\omega})$, then the Fourier transform of the product of $x(n)$ and $w(n)$ is given by

$$\mathfrak{F}[x(n)w(n)] = X(e^{j\omega}) * W(e^{j\omega}) = \frac{1}{2\pi}\int_{-\pi}^{\pi}X(e^{j\theta})W(e^{j(\omega-\theta)})\mathrm{d}\theta \qquad (2.81)$$

The above result is called the windowing theorem.

Correlation Theorem If $x_1(n)$ and $x_2(n)$ are two sequences with Fourier transforms $X_1(e^{j\omega})$ and $X_2(e^{j\omega})$, then the Fourier transform of the correlation $r_{x_1 x_2}(l)$ of $x_1(n)$ and $x_2(n)$ defined by

$$r_{x_1 x_2}(l) = \sum_{n=-\infty}^{\infty} x_1(n)x_2(n-l) \qquad (2.82a)$$

is given by

$$\mathfrak{F}[r_{x_1 x_2}(l)] = \mathfrak{F}\left[\sum_{n=-\infty}^{\infty} x_1(n)x_2(n-l)\right] = X_1(e^{j\omega})X_2(e^{-j\omega}) \qquad (2.82b)$$

which is called the cross-energy density spectrum of the signals $x_1(n)$ and $x_2(n)$.

Parseval's theorem If $x(n)$ is a sequence with Fourier transform $X(e^{j\omega})$, then the energy E of $x(n)$ is given by

$$E = \sum_{-\infty}^{\infty} |x(n)|^2 = \frac{1}{2\pi} \int_{-\pi}^{\pi} |X(e^{j\omega})|^2 \, d\omega \qquad (2.83)$$

where $|X(e^{j\omega})|^2$ is called the energy density spectrum.

Proof The energy E of $x(n)$ is defined as

$$E = \sum_{-\infty}^{\infty} |x(n)|^2 = \sum_{-\infty}^{\infty} x(n)x^*(n) \qquad (2.84)$$

$$= \sum_{-\infty}^{\infty} x(n) \frac{1}{2\pi} \int_{-\pi}^{\pi} X^*(e^{j\omega}) e^{-j\omega n} d\omega,$$

using Eq. (2.70).

Interchanging the integration and summation signs, the above equation can be rewritten as

$$E = \frac{1}{2\pi} \int_{-\pi}^{\pi} X^*(e^{j\omega}) \sum_{-\infty}^{\infty} x(n) e^{-j\omega n} d\omega = \frac{1}{2\pi} \int_{-\pi}^{\pi} X^*(e^{j\omega}) X(e^{j\omega}) d\omega$$

$$= \frac{1}{2\pi} \int_{-\pi}^{\pi} |X(e^{j\omega})|^2 d\omega$$

Thus,

$$E = \sum_{-\infty}^{\infty} |x(n)|^2 = \frac{1}{2\pi} \int_{-\pi}^{\pi} |X(e^{j\omega})|^2 d\omega \qquad (2.85)$$

The above theorems concerning DTFT are summarized in Table 2.2.

Using the definitions of DTFT pair given by (2.69) and (2.70), we may establish the DTFT pairs for some useful functions. These are given in Table 2.3.

2.7.3   Some Properties of the DTFT of a Complex Sequence $x(n)$

From Eq. (2.69), the DTFT of a time-reversed sequence $x(-n)$ can be written as

Table 2.2 Some properties of discrete-time Fourier transforms

Property	Sequence	DTFT
Linearity	$a_1 x_1(n) + a_2 x_2(n)$	$a_1 X_1(e^{j\omega}) + a_2 X_2(e^{j\omega})$
Time shifting	$x(n-k)$	$e^{-j\omega k} X(e^{j\omega})$
Time reversal	$x(-n)$	$X(e^{-j\omega})$
Frequency shifting	$e^{j\omega_0 n} x(n)$	$X(e^{j(\omega - \omega_0)})$
Differentiation in the frequency domain	$nx(n)$	$j\dfrac{d}{d\omega} X(e^{j\omega})$
Convolution theorem	$x_1(n) * x_2(n)$	$X_1(e^{j\omega}) X_2(e^{j\omega})$
Windowing theorem	$x_1(n) x_2(n)$	$X_1(e^{j\omega}) * X_2(e^{j\omega})$
Correlation theorem	$\displaystyle\sum_{-\infty}^{\infty} x_1(n) x_2(n-l)$	$X_1(e^{j\omega}) X_2(e^{-j\omega})$
Parseval's theorem	$\displaystyle\sum_{-\infty}^{\infty} |x(n)|^2 = \frac{1}{2\pi} \int_{-\pi}^{\pi} |X(e^{j\omega})|^2 d\omega$	

Table 2.3 Some useful DTFT pairs

$x(n)$	DTFT
$\delta(n)$	1
$1 \ (-\infty < n < \infty)$	$\displaystyle\sum_{k=-\infty}^{\infty} 2\pi\delta(\omega + 2\pi k)$
$a^n u(n), |a| < 1$	$\dfrac{1}{1 - ae^{-j\omega}}$
$\dfrac{\sin(\omega_c n)}{\pi n}$	$\begin{cases} 1 & |\omega| < \omega_c \\ 0 & \omega_c < |\omega| < \pi \end{cases}$
$\begin{cases} 1 & 0 \leq n \leq L \\ 0 & \text{otherwise} \end{cases}$	$\dfrac{\sin \omega(L+1)/2}{\sin \omega/2} e^{-j\omega L/2}$
$e^{-j\omega_0 n}$	$\displaystyle\sum_{k=-\infty}^{\infty} 2\pi\delta(\omega - \omega_0 + 2\pi k)$

$$\mathfrak{F}[x(-n)] = \sum_{n=-\infty}^{\infty} x(-n)e^{-j\omega n} = \sum_{l=\infty}^{-\infty} x(l)e^{j\omega l} = X\left(e^{-j\omega}\right) \qquad (2.86a)$$

Similarly, the DTFT of the complex conjugate sequence $x*(n)$ can be expressed as

$$\mathfrak{F}[x^*(n)] = \sum_{n=-\infty}^{\infty} x^*(n)e^{-j\omega n} = \left(\sum_{n=-\infty}^{\infty} x(n)e^{j\omega n}\right)^* = X^*\left(e^{-j\omega}\right) \qquad (2.86b)$$

From the above two equations, it can be easily shown that

$$\mathfrak{F}[x^*(-n)] = X^*\left(e^{j\omega}\right) \qquad (2.87)$$

The sequence $x(n)$ can be represented as a sum of conjugate symmetric sequence $x_e(n)$ and a conjugate antisymmetric sequence $x_o(n)$ as

$$x(n) = x_e(n) + x_o(n) \tag{2.88}$$

where

$$x_e(n) = \frac{1}{2}[x(n) + x^*(-n)] \tag{2.89}$$

and

$$x_o(n) = \frac{1}{2}[x(n) - x^*(-n)]. \tag{2.90}$$

The DTFT $X(e^{j\omega})$ can be split into

$$X(e^{j\omega}) = X_e(e^{j\omega}) + X_o(e^{j\omega}) \tag{2.91}$$

where $X_e(e^{j\omega})$ and $X_o(e^{j\omega})$ are the DTFTs of $x_e(n)$ and $x_o(n)$, respectively. Using Eqs. (2.69), (2.87), and (2.89), $x_e(e^{j\omega})$ can be expressed as

$$\begin{aligned} X_e(e^{j\omega}) &= \mathfrak{F}[x_e(n)] = \frac{1}{2}(\mathfrak{F}[x(n)] + \mathfrak{F}[x^*(-n)]) \\ &= \frac{1}{2}[X(e^{j\omega}) + X^*(e^{j\omega})] = \operatorname{Re}[X(e^{j\omega})] \end{aligned} \tag{2.92}$$

In a similar way, using Eqs. (2.69), (2.76), and (2.90), $x_o(e^{j\omega})$ can be written as

$$\begin{aligned} X_o(e^{j\omega}) &= \mathfrak{F}[x_o(n)] = \frac{1}{2}(\mathfrak{F}[x(n)] - \mathfrak{F}[x^*(-n)]) \\ &= \frac{1}{2}[X(e^{j\omega}) - X^*(e^{j\omega})] = j\operatorname{Im}[X(e^{j\omega})] \end{aligned} \tag{2.93}$$

A complex sequence $x(n)$ can be decomposed into a sum of its real and imaginary parts as

$$x(n) = x_R(n) + jx_I(n) \tag{2.94}$$

where

$$x_R(n) = \frac{1}{2}[x(n) + x^*(n)] \tag{2.95}$$

and

$$jx_I(n) = \frac{1}{2}[x(n) - x^*(n)] \tag{2.96}$$

The DTFT of $x_R(n)$ can be written as

$$\mathfrak{F}[\mathrm{Re}(x(n))] = \mathfrak{F}\left[\frac{1}{2}(x(n) + x^*(n))\right]$$
$$= \frac{1}{2}\left[X(e^{j\omega}) + X^*(e^{-j\omega})\right] \tag{2.97}$$

Similarly, the DTFT of $jx_I(n)$ can be expressed as

$$\mathfrak{F}[j\mathrm{Im}(x(n))] = \mathfrak{F}\left[\frac{1}{2}(x(n) - x^*(n))\right]$$
$$= \frac{1}{2}\left[X(e^{j\omega}) - X^*(e^{-j\omega})\right] \tag{2.98}$$

The above properties of the DTFT of a complex sequence are summarized in Table 2.4.

2.7.4   Some Properties of the DTFT of a Real Sequence x(n)

Since $e^{-j\omega n} = \cos\omega n - j\sin\omega n$, the DTFT $X(e^{j\omega})$ given by Eq. (2.69) can be expressed as

$$X(e^{j\omega}) = \sum_{n=-\infty}^{\infty} x(n)\cos\omega n - j\sum_{n=-\infty}^{\infty} x(n)\sin\omega n \tag{2.99}$$

The Fourier transform $X(e^{j\omega})$ is a complex function of ω and can be written as the sum of the real and imaginary parts as

$$X(e^{j\omega}) = X_R(e^{j\omega}) + jX_I(e^{j\omega}) \tag{2.100}$$

Table 2.4 Some properties of DTFT of a complex sequence

Sequence	DTFT
$x^*(n)$	$X^*(e^{-j\omega})$
$x^*(-n)$	$X^*(e^{j\omega})$
$x_R(n) = \mathrm{Re}[x(n)]$	$\frac{1}{2}[X(e^{j\omega}) + X^*(e^{-j\omega})]$
$jx_I(n) = j\mathrm{Im}[x(n)]$	$\frac{1}{2}[X(e^{j\omega}) - X^*(e^{-j\omega})]$
$x_e(n) = \frac{1}{2}[x(n) + x^*(-n)]$	$\mathrm{Re}[X(e^{j\omega})]$
$x_0(n) = \frac{1}{2}[x(n) + x^*(-n)]$	$j\mathrm{Im}[X(e^{j\omega})]$

From Eq. (2.99), the real and imaginary parts of $X(e^{j\omega})$ are given by

$$X_R(e^{jw}) = \sum_{n=-\infty}^{\infty} x(n) \cos \omega n \tag{2.101}$$

and

$$X_I(e^{j\omega}) = -\sum_{n=-\infty}^{\infty} x(n) \sin \omega n \tag{2.102}$$

Since $\cos(-\omega n) = \cos \omega n$ and $\sin(-\omega n) = -\sin \omega n$, we can obtain the following relations from Eqs. (2.101) and (2.102).

$$X_R(e^{-j\omega}) = \sum_{n=-\infty}^{\infty} x(n) \cos \omega n = X_R(e^{j\omega}) \tag{2.103a}$$

$$X_I(e^{-j\omega}) = \sum_{n=-\infty}^{\infty} x(n) \sin \omega n = -X_I(e^{j\omega}) \tag{2.103b}$$

indicating that the real part of DTFT is an even function of ω, while the imaginary part is an odd function of ω. Thus,

$$X(e^{j\omega}) = X^*(e^{-j\omega}) \tag{2.104}$$

In polar form, $X(e^{j\omega})$ can be written as

$$X(e^{j\omega}) = |X(e^{j\omega})| e^{j\theta\omega} \tag{2.105}$$

where

$$|X(e^{j\omega})| = \sqrt{[X_R(e^{j\omega})]^2 + [X_I(e^{j\omega})]^2} \tag{2.106a}$$

and

$$\theta(\omega) = \angle X(e^{j\omega}) = \text{phase of } X(e^{j\omega}) = \tan^{-1}\frac{X_I(e^{j\omega})}{X_R(e^{j\omega})} \tag{2.106b}$$

Using the above relations, it can easily be seen that $|X(e^{j\omega})|$ is an even function of ω, whereas the function $\theta(\omega)$ is an odd function of ω.

Now, the DTFT of $x_e(n)$, the even part of the real sequence $x(n)$ is given by

$$\begin{aligned}\mathfrak{F}[x_e(n)] &= \frac{1}{2}(\mathfrak{F}[x(n)] + \mathfrak{F}[x(-n)]) \\ &= \frac{1}{2}[X(e^{j\omega}) + X(e^{-j\omega})] = X_R(e^{j\omega})\end{aligned} \tag{2.107}$$

Table 2.5 Some properties of DTFT of a real sequence $x(n)$

$\mathcal{F}[x(n)] = X(e^{j\omega}) = X_R(e^{j\omega}) + jX_I(e^{j\omega})$
$\mathcal{F}[x_e(n)] = X_R(e^{j\omega})$
$\mathcal{F}[x_o(n)] = jX_I(e^{j\omega})$
$X_R(e^{j\omega}) = X_R(e^{-j\omega})$
$X_I(e^{j\omega}) = -X_I(e^{-j\omega})$
$X(e^{j\omega}) = X^*(e^{-j\omega})$
$|X(e^{j\omega})| = |X(e^{-j\omega})|$

$$\angle X(e^{j\omega}) = -\angle X(e^{-j\omega})$$

Thus, the DTFT of even part of a real sequence is the real part of $X(e^{j\omega})$.
Similarly, the DTFT of $x_o(n)$, the odd part of the real sequence $x(n)$ is given by

$$\mathcal{F}[x_o(n)] = \frac{1}{2}\left[X(e^{j\omega}) - X(e^{-j\omega})\right] = jX_I(e^{j\omega}) \tag{2.108}$$

Hence, the DTFT of the odd part of a real sequence is $jX_I(e^{j\omega})$.

The above properties of the DTFT of a real sequence are summarized in Table 2.5.

Example 2.24 A causal LTI system is represented by the following difference equation:

$$y(n) - ay(n-1) = x(n-1) \tag{2.109}$$

(i) Find the impulse response of the system $h(n)$, as a function of parameter a.
(ii) For what range of values would the system be stable?

Solutions

(i) Given

$$y(n) - ay(n-1) = x(n-1)$$

Taking Fourier transform on both sides of Eq. (2.109), we get

$$Y(e^{j\omega}) - ae^{-j\omega}Y(e^{j\omega}) = e^{-j\omega}X(e^{j\omega}) \tag{2.110}$$

From Eq. (2.110), we arrive at

$$H(e^{j\omega}) = \frac{Y(e^{j\omega})}{X(e^{j\omega})} = \frac{e^{-j\omega}}{1 - ae^{-j\omega}}$$

$$\mathfrak{F}[a^n u(n)] = \sum_{n=-\infty}^{\infty} a^n e^{-j\omega n} = \sum_{n=-\infty}^{\infty} (ae^{-j\omega})^n$$

$$= \frac{1}{1 - ae^{-j\omega}}$$

From the above equation and time shifting property, the impulse response is given by

$$h(n) = \mathfrak{F}^{-1}(H(e^{j\omega})) = \mathfrak{F}^{-1}\left(\frac{e^{-j\omega}}{1 - ae^{-j\omega}}\right) = a^{n-1}u(n-1)$$

(ii) Now,

$$\sum_{n=1}^{\infty} |h(n)| = \sum_{n=1}^{\infty} |a|^{n-1} < \infty \quad \text{for } |a| < 1. \tag{2.111}$$

Thus, the system is stable for $|a| < 1$.

Example 2.25 Find the impulse response of a system described by the following difference equation

$$y(n) - \frac{5}{6}y(n-1) + \frac{1}{6}y(n-2) = \frac{1}{3}x(n-1) \tag{2.112}$$

Solution   Taking Fourier transformation on both sides of Eq. (2.112), we get

$$Y(e^{j\omega}) - \frac{5}{6}e^{-j\omega}Y(e^{j\omega}) + \frac{1}{6}e^{-2j\omega}Y(e^{j\omega}) = \frac{1}{3}e^{-j\omega}X(e^{j\omega}) \tag{2.113}$$

From Eq. (2.113), we arrive at

$$H(e^{j\omega}) = \frac{Y(e^{j\omega})}{X(e^{j\omega})} = \frac{(1/3)e^{-j\omega}}{1 - (5/6)e^{-j\omega} + (1/6)e^{-2j\omega}}$$

$$= \frac{2}{1 - (1/2)e^{-j\omega}} - \frac{2}{1 - (1/3)e^{-j\omega}}$$

The impulse response $h(n)$ is given by

$$h(n) = \mathfrak{F}^{-1}\left(\frac{2}{1 - (1/2)e^{-j\omega}}\right) - \mathfrak{F}^{-1}\left(\frac{2}{1 - (1/3)e^{-j\omega}}\right)$$

$$= 2\left[\left(\frac{1}{2}\right)^n - \left(\frac{1}{3}\right)^n\right]u(n)$$

Example 2.26 Find the DTFT of $x(n) = \frac{(n+m-1)!}{n!(m-1)!} a^n u(n)$, $|a| < 1$

Solution   Let $x_1(n) = a^n u(n)$

The Fourier transform of $x_1(n)$ is given by

$$X_1(e^{j\omega}) = \sum_{n=0}^{\infty} (a)^n e^{-j\omega n} = \sum_{n=0}^{\infty} (ae^{-j\omega})^n = \frac{1}{1 - ae^{-j\omega}}$$

For $m = 2$,

$$x(n) = (n+1)a^n u(n)$$

Using the differentiation property of DTFT, the Fourier transform of $na^n u(n)$ is given by

$$j\frac{dX_1(e^{j\omega})}{d\omega} = j\frac{d}{d\omega}\left(\frac{1}{1 - ae^{-j\omega}}\right) = \frac{ae^{-j\omega}}{(1 - ae^{-j\omega})^2}$$

Using linearity property of the DTFT, the Fourier transform of $x(n)$ is denoted by

$$X(e^{j\omega}) = \frac{ae^{-j\omega}}{(1 - ae^{-j\omega})^2} + \frac{1}{(1 - ae^{-j\omega})} = \frac{1}{(1 - ae^{-j\omega})^2}$$

For $m = 3$,

$$x(n) = \left(\frac{(n+2)(n+1)}{2}\right)a^n u(n) = \frac{n^2 + 3n + 2}{2}a^n u(n)$$

$$= \frac{1}{2}\left[n^2 a^n u(n) + 3na^n u(n) + 2a^n u(n)\right]$$

Using the differentiation and linearity properties of DTFT, the Fourier transform of $x(n)$ is given by

$$X(e^{j\omega}) = \frac{1}{2}\left[j\frac{d}{d\omega}\left(\frac{ae^{-j\omega}}{(1 - ae^{-j\omega})^2}\right) + \frac{3ae^{-j\omega}}{(1 - ae^{-j\omega})^2} + \frac{2}{(1 - ae^{-j\omega})}\right]$$

$$= \frac{1}{2}\left[\frac{ae^{-j\omega}(1 + ae^{-j\omega})}{(1 - ae^{-j\omega})^3} + \frac{3ae^{-j\omega}}{(1 - ae^{-j\omega})^2} + \frac{2}{(1 - ae^{-j\omega})}\right]$$

$$= \frac{1}{2}\left[\frac{2}{(1 - ae^{-j\omega})^3}\right] = \frac{1}{(1 - ae^{-j\omega})^3}$$

In general, for $m = k$, the Fourier transform of $x(n)$ is given by

$$X(e^{j\omega}) = \frac{1}{(1 - ae^{-j\omega})^k}, \qquad (2.114)$$

where k is any integer value.

Example 2.27 Let $G_1(e^{j\omega})$ denote the DTFT of the sequence $g_1(n)$ shown in Fig. 2.25a. Express the DTFT of the sequence $g_2(n)$ in Fig. 2.25b in terms of $G_1(e^{j\omega})$. Do not evaluate $G_1(e^{j\omega})$.

Solution From Fig. 2.25b, $g_2(n)$ can be expressed in terms of $g_1(n)$ as

$$g_2(n) = g_1(n) + g_1(n - 4)$$

Applying DTFT on both sides, we obtain

$$G_2(e^{j\omega}) = G_1(e^{j\omega}) + e^{-j4\omega}G_1(e^{j\omega}) = (1 + e^{-j4\omega})G_1(e^{j\omega})$$

Example 2.28 Evaluate the inverse DTFT of each of the following DTFTs:

(a) $X_1(e^{jw}) = \sum\limits_{k=-\infty}^{\infty} \delta(\omega + 2\pi k)$

(b) $X_2(e^{j\omega}) = \frac{-\alpha e^{-j\omega}}{(1 - \alpha e^{-j\omega})^2} |\alpha| < 1$

Solution

(a) $X_1(e^{j\omega}) = \sum\limits_{k=\infty}^{\infty} \delta(\omega + 2\pi k)$

From Table 2.5,

$$\mathfrak{F}(1)\ (-\infty < n < \infty) = \sum\limits_{k=-\infty}^{\infty} 2\pi\delta(\omega + 2\pi k)$$

Hence,

$$\mathfrak{F}^{-1}[\delta(\omega + 2\pi k)] = \frac{1}{2\pi}, \qquad (-\infty < n < \infty)$$

(b) $X_2(e^{j\omega}) = \frac{-\alpha e^{-j\omega}}{(1 - \alpha e^{-j\omega})^2}, \qquad |\alpha| < 1$

From Example 2.26,

$$\frac{1}{(1 - \alpha e^{-j\omega})^m} \leftrightarrow \frac{(n + m - 1)!}{n!(m - 1)!}\alpha^n u(n)$$

For $m = 2$,

$$\frac{1}{(1 - \alpha e^{-j\omega})^2} \leftrightarrow \frac{(n+1)!}{n!(1)!} \alpha^n u(n)$$

$$\frac{1}{(1 - \alpha e^{-j\omega})^2} \leftrightarrow (n+1)\alpha^n u(n)$$

Then

$$\frac{-\alpha}{(1 - \alpha e^{-j\omega})^2} \leftrightarrow -(n+1)\alpha^{n+1}u(n)$$

$$\frac{-\alpha e^{-j\omega}}{(1 - \alpha e^{-j\omega})^2} \leftrightarrow -n\alpha^n u(n-1)$$

Example 2.29 A length-9 sequence $x(n)$ is shown in Fig. 2.26

If the DTFT of $x(n)$ is $X(e^{j\omega})$, calculate the following functions without computing $X(e^{j\omega})$.

(a) $X(e^{j0})$ (b) $X(e^{j\pi})$ (c) $\int_{-\pi}^{\pi} X(e^{j\omega})d\omega$ (d) $\int_{-\pi}^{\pi} |X(e^{j\omega})|^2 d\omega$

(e) $\int_{-\pi}^{\pi} \left|\frac{dX(e^{j\omega})}{d\omega}\right|^2 d\omega$

Solution From the given data,

$$x(-3) = 3, \ x(-2) = 0, \ x(-1) = 1, \ x(0) = -2, \ x(1) = -3, \ x(2) = 4, \ x(3)$$
$$= 1, \ x(4) = 0, \ x(5) = -1$$

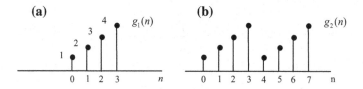

Fig. 2.25 a Sequence $g_1(n)$ and **b** sequence $g_2(n)$

(a) $X(e^{j0})$

From the definition of Fourier transform,

$$X(e^{j\omega}) = \sum_{n=-\infty}^{\infty} x(n)e^{-j\omega n}$$

$$X(e^{j0}) = \sum_{n=-\infty}^{\infty} x(n)$$

$$= [3+0+1-2-3+4+1+0-1] = 3$$

(b) $X(e^{j\pi})$

From the definition of Fourier transform,

$$X(e^{j\pi}) = \sum_{n=-\infty}^{\infty} x(n)e^{-j\pi}$$

$$X(e^{j\pi}) = -\sum_{n=-\infty}^{\infty} x(n) = -3$$

(c) $\int_{-\pi}^{\pi} X(e^{j\omega})\, d\omega$

From the definition of inverse Fourier transform,

$$x(n) = \frac{1}{2\pi} \int_{-\pi}^{\pi} X(e^{j\omega})e^{j\omega n}\, d\omega$$

Fig. 2.26 A length-9
sequence $x(n)$

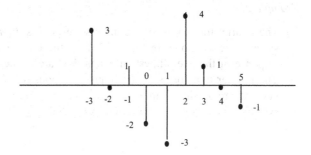

Hence,

$$\int_{-\pi}^{\pi} X(e^{j\omega})e^{j\omega n}d\omega = 2\pi x(0) = -4\pi$$

(d)
$$\int_{-\pi}^{\pi} |X(e^{j\omega})|^2 d\omega$$

From the definition of Parseval's theorem,

$$\sum_{n=-\infty}^{\infty} |x(n)|^2 = \frac{1}{2\pi}\int_{-\pi}^{\pi} |X(e^{j\omega})|^2 d\omega$$

Hence,

$$\int_{-\pi}^{\pi} |X(e^{j\omega})|^2 d\omega = 2\pi \sum_{n=-\infty}^{\infty} |x(n)|^2$$

$$= 2\pi(9+0+1+4+9+16+1+0+1) = 82\pi$$

(e)
$$\int_{-\pi}^{\pi} \left|\frac{dX(e^{j\omega})}{d\omega}\right|^2 d\omega$$

From differentiation property and Parseval's theorem of DTFT,

$$\int_{-\pi}^{\pi} \left|\frac{dX(e^{j\omega})}{d\omega}\right|^2 d\omega = 2\pi \sum_{n=-\infty}^{\infty} |nx(n)|^2$$

$$= 2\pi[81+0+1+0+9+64+9+0+25] = 189\pi$$

Example 2.30

(a) The Fourier transforms of the impulse responses, $h_1(n)$ and $h_2(n)$, of two LTI systems are as shown in Fig. 2.27. Find the Fourier transform of the impulse response of the overall system, when they are connected in cascade.

(b) The Fourier transforms of the impulse responses $h_1(n)$ and $h_2(n)$ of two LTI systems are as shown in Fig. 2.28. Find the Fourier transform of the overall system, when they are connected in parallel.

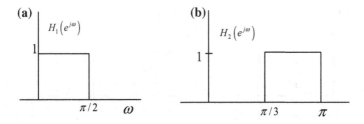

Fig. 2.27 **a** Fourier transform of $h_1(n)$ and **b** Fourier transform of $h_2(n)$

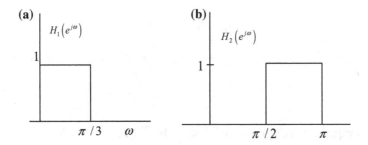

Fig. 2.28 **a** Fourier transform of $h_1(n)$ and **b** Fourier transform of $h_2(n)$

Solution

(a) The impulse response $h(n)$ of the overall system is given by

$$h(n) = h_1(n) * h_2(n)$$

Then, by the convolution property of the Fourier transform, the Fourier transform of the impulse response of the cascade system is given by

$$H_1(e^{j\omega})H_2(e^{j\omega})$$

The Fourier transform of impulse response of the cascade system is shown in Fig. 2.29a.

(b) The impulse response $h(n)$ of the overall system is given by

$$h(n) = h_1(n) + h_2(n)$$

Hence, the Fourier transform of impulse response of the cascade system is given by

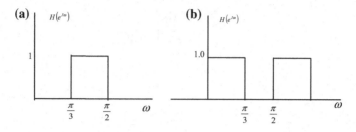

Fig. 2.29 **a** Fourier transform of the impulse response of the cascade system and **b** Fourier transform of the impulse response of the parallel system

$$H_1\left(e^{j\omega}\right) + H_2\left(e^{j\omega}\right)$$

The Fourier transform of the impulse response of the parallel system is shown in Fig. 2.29b.

2.8 Frequency Response of Discrete-Time Systems

For an LTI discrete-time system with impulse response $h(n)$ and input sequence $x(n)$, the output $y(n)$ is the convolution sum of $x(n)$ and $h(n)$ given by

$$y(n) = \sum_{k=-\infty}^{\infty} h(k)x(n-k) \tag{2.115}$$

To demonstrate the eigen function property of complex exponential for discrete-time systems, consider the input $x(n)$ of the form

$$x(n) = e^{j\omega n}, \quad -\infty < n < \infty \tag{2.116}$$

Then from Eq. (2.115), the output is given by

$$y(n) = \sum_{k=-\infty}^{\infty} h(k)e^{j\omega(n-k)} = \left(\sum_{k=-\infty}^{\infty} h(k)e^{-j\omega k}\right) e^{j\omega n} \tag{2.117}$$

The above equation can be rewritten as

$$y(n) = H(e^{j\omega})e^{j\omega n}, \tag{2.118a}$$

where

$$H(e^{j\omega}) = \sum_{n=-\infty}^{\infty} h(n)e^{-j\omega n}. \tag{2.118b}$$

$H(e^{j\omega})$ is called the frequency response of the LTI system whose impulse response is $h(n)$, $e^{j\omega n}$ is an eigen function of the system, and the associated eigenvalue is $H(e^{j\omega})$. In general $H(e^{j\omega})$ is complex and is expressed in terms of real and imaginary parts as

$$H(e^{j\omega}) = H_R(e^{j\omega}) + jH_I(e^{j\omega}) \tag{2.119}$$

where $H_R(e^{j\omega})$ and $H_I(e^{j\omega})$ are the real and imaginary parts of $H(e^{j\omega})$, respectively.

Furthermore, due to convolution, the Fourier transforms of the system input and output are related by

$$Y(e^{j\omega}) = H(e^{j\omega})X(e^{j\omega})$$

where $X(e^{j\omega})$ and $Y(e^{j\omega})$ are the Fourier transforms of the system input and output, respectively. Thus,

$$H(e^{j\omega}) = \frac{Y(e^{j\omega})}{X(e^{j\omega})} \tag{2.120}$$

The frequency response function $H(e^{j\omega})$ is also known as the transfer function of the system. The frequency response function provides valuable information on the behavior of LTI systems in the frequency domain. However, it is very difficult to realize a digital system since it is a complex function of the frequency variable ω.

In polar form, the frequency response can be written as

$$H(e^{j\omega}) = |H(e^{j\omega})|e^{j\theta(\omega)} \tag{2.121a}$$

where $|H(e^{j\omega})|$ the amplitude response term and $\theta(\omega)$ the phase-response term are given by

$$|H(e^{j\omega})|^2 = |H_R(e^{j\omega})|^2 + |H_I(e^{j\omega})|^2 \tag{2.121b}$$

$$\theta(\omega) = \tan^{-1}\left(\frac{H_I(e^{j\omega})}{H_R(e^{j\omega})}\right) \tag{2.121c}$$

Phase and Group Delays
If the input is a sinusoidal signal given by

$$x(n) = \cos(\omega n), \quad \text{for } -\infty < n < \infty, \tag{2.122a}$$

then from Eq. (2.121a) the output is

$$y[n] = \left|H(e^{j\omega_0})\right| \cos(\omega n + \theta(\omega)) \tag{2.122b}$$

The above equation can be rewritten as

$$y[n] = \left|H(e^{j\omega_0})\right| \cos\left(\omega\left(n + \frac{\theta(\omega)}{\omega}\right)\right), \tag{2.123a}$$
$$= \left|H(e^{j\omega_0})\right| \cos\left(\omega(n - \tau_p(\omega))\right)$$

It can be clearly seen that the above equation expresses the phase response as a time delay in seconds which is called as *phase delay* and is defined by

$$\tau_p(\omega) = -\frac{\theta(\omega)}{\omega} \tag{2.123b}$$

An input signal consisting of a group of sinusoidal components with frequencies within a narrow interval about ω, experiences different phase delays when processed by an LTI discrete-time system. As such, the signal delay is represented by another parameter called *group delay* defined as

$$\tau_g(\omega) = -\frac{d\theta(\omega)}{d\omega} \tag{2.123c}$$

Example 2.31 Determine the magnitude and phase responses of a system whose impulse response is given by $h(n) = \left(\frac{1}{2}\right)^n u(n)$.

Solution For $h(n) = \left(\frac{1}{2}\right)^n u(n)$, the frequency response is given by

$$H(e^{j\omega}) = \sum_{n=-\infty}^{\infty} \left(\frac{1}{2}\right)^n e^{-j\omega n} = \sum_{n=-\infty}^{\infty} \left(\frac{1}{2}e^{-j\omega}\right)^n$$
$$= \frac{1}{1 - 0.5e^{-j\omega}} = \frac{1}{1 - 0.5\cos\omega + j0.5\sin\omega}$$

The magnitude response is given by

$$\left|H(e^{j\omega})\right| = \frac{1}{\sqrt{(1 - 0.5\cos\omega)^2 + (0.5)^2 \sin^2\omega}} = \frac{1}{\sqrt{(1 + (0.5)^2 - 2(0.5)\cos\omega}}$$

The phase response is

$$\theta(\omega) = -\tan^{-1}\frac{0.5\sin\omega}{1 - 0.5\cos\omega}$$

Table 2.6   Magnitude and phase responses

ω	0	$\frac{\pi}{4}$	$\frac{\pi}{2}$	$\frac{3\pi}{4}$	π	$\frac{5\pi}{4}$	$\frac{3\pi}{2}$	$\frac{7\pi}{4}$	2π
$\lvert H(e^{j\omega})\rvert$	2	1.3572	0.8944	0.7148	0.67	0.715	0.8944	1.3572	2
$\theta(\omega)$	0°	$-28.675°$	$-26.565°$	$-14.64°$	0°	14.64°	26.565°	28.675°	0°

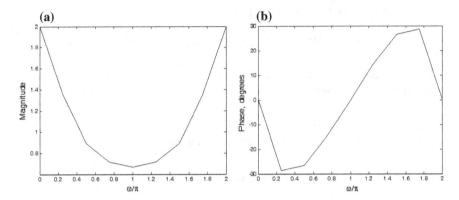

Fig. 2.30   a Magnitude and **b** phase responses of $h(n)$ of Example 2.31

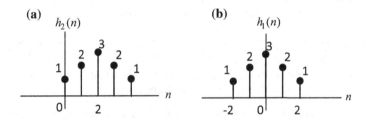

Fig. 2.31   a Impulse response of $h_1(n)$ and **b** *impulse* response of $h_2(n)$

The magnitude and phase values are given in Table 2.6 for various values of ω, and plotted in Fig. 2.30a, b, respectively.

Example 2.32 Compute the magnitude and phase responses of the impulse responses given in Fig. 2.31, and comment on the results.

Solution Since $h_1(n)$ is an even function of time, it has a real DTFT indicating that the phase is zero; that is, the phase is a horizontal line; $h_2(n)$ is the right-shifted version of $h_1(n)$. Hence, from time shifting property of DTFT, the transform of $h_2(n)$ is obtained by multiplying the transform of $h_1(n)$ by $e^{-j2\omega}$. This changes the slope of the phase linearly and can be verified as follows:

The frequency response of $h_1(n)$ is

$$H_1\left(e^{j\omega}\right) = e^{2j\omega} + 2e^{j\omega} + 3 + 2e^{-j\omega} + e^{-2j\omega}$$
$$= \left(e^{2j\omega} + e^{-2j\omega}\right) + 2(e^{j\omega} + e^{-j\omega}) + 3$$
$$= 2\cos2\omega + 4\cos\omega + 3$$

The magnitude response of $H_1(e^{j\omega})$ is

$$\left|H_1\left(e^{j\omega}\right)\right| = 2\cos2\omega + 4\cos\omega + 3$$

The phase response of $H_1(e^{j\omega})$ is zero.
The frequency response of $h_2(n)$ is

$$H_2\left(e^{j\omega}\right) = e^{-2j\omega}H_1\left(e^{j\omega}\right)$$
$$= e^{-2j\omega}(2\cos2\omega + 4\cos\omega + 3)$$

The magnitude response of $H_2(e^{j\omega})$ is

$$\left|H_2\left(e^{j\omega}\right)\right| = 2\cos2\omega + 4\cos\omega + 3$$

The phase response of $H_2(e^{j\omega})$ is given by
$$\angle H_2(e^{j\omega}) = \angle e^{-2j\omega} = -2\omega.$$

The magnitude and phase responses of $h_1(n)$ and $h_2(n)$ are shown in Fig. 2.32a, b, c, and d. From the magnitude and phase responses of $h_1(n)$ and $h_2(n)$, it is observed that $h_1(n)$ has zero phase and $h_2(n)$ has a linear phase response, whereas both $h_1(n)$ and $h_2(n)$ have the same magnitude responses.

Example 2.33 Trapezoidal integration formula is represented by a recursive difference equationas $y(n) - y(n-1) = 0.5x(n) + 0.5x(n-1)$. Determine $H(e^{j\omega})$ of the trapezoidal integration formula.

Solution Given

$$y(n) - y(n-1) = 0.5x(n) + 0.5x(n-1)$$

Taking Fourier transform on both sides of the above equation, we get

$$Y(e^{j\omega}) - e^{-j\omega}Y(e^{j\omega}) = 0.5X(e^{j\omega}) + 0.5e^{-j\omega}X(e^{j\omega})$$
$$Y(e^{j\omega})(1 - e^{-j\omega}) = 0.5X(e^{j\omega})(1 + e^{-j\omega})$$

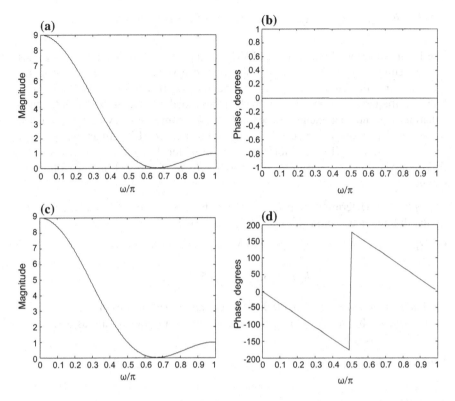

Fig. 2.32 **a** Magnitude response of $h_1(n)$, **b** phase response of $h_1(n)$, **c** magnitude response of $h_2(n)$, and **d** phase response of $h_2(n)$

$$H(e^{j\omega}) = \frac{Y(e^{j\omega})}{X(e^{j\omega})} = 0.5 \frac{(1 + e^{-j\omega})}{(1 - e^{-j\omega})}$$

$$= 0.5 \left[\frac{e^{-j\omega/2}(e^{j\omega/2} + e^{-j\omega/2})}{e^{-j\omega/2}(e^{j\omega/2} - e^{-j\omega/2})} \right] = -j0.5 \left[\frac{\cos(\omega/2)}{\sin(\omega/2)} \right]$$

The magnitude response is given by

$$\left| H(e^{j\omega}) \right| = 0.5 \left| \frac{\cos(\omega/2)}{\sin(\omega/2)} \right|$$

The phase response is given as follows:

If $0 < \omega < \pi$, then both $\cos \omega/2$ and $\sin \omega/2$ are positive, and hence, the phase is $\left(-\frac{\pi}{2} \right)$.

If $\pi < \omega < 2\pi$, then $\cos \omega/2$ is negative, but $\sin \omega/2$ are positive; hence, the phase is $\left(\frac{\pi}{2} \right)$.

2.8.1 *Frequency Response Computation Using MATLAB*

The M-file functions freqz(h,w) in MATLAB can be used to determine the values of the frequency response of an impulse response vector *h* at a set of given frequency points ω. Similarly, the M-file function freqz(b,a, ω) can also be used to find the frequency response of a system described by the recursive difference equation with the coefficients in vectors *b* and *a*. From frequency response values, the real and imaginary parts can be computed using MATLAB functions *real* and *imag*, respectively. The magnitude and phase of the frequency response can be determined using the functions *abs* and *angle* as illustrated in the following examples.

Example 2.34 Determine the magnitude and phase responses of a system described by the difference equation, $y(n) = 0.5x(n) + 0.5x(n-2)$.

Solution If $x(n) = \delta(n)$, then the impulse response $h(n)$ is given by

$$h(n) = 0.5\delta(n) + 0.5\delta(n-2)$$

Hence, $h(n)$ sequence is [0.5 0 0.5]. When this sequence is used in Program 2.3 given below, the resulting magnitude and phase responses are as shown in Fig. 2.33a, b, respectively.

Program 2.3

```
clear;clc;
w = 0:0.05:pi;
h = exp(j*w); %set h = exp(jw)
num = 0.5 + 0*h.^-1 + 0.5*h.^-2;
den = 1;
%Compute the frequency responses
```

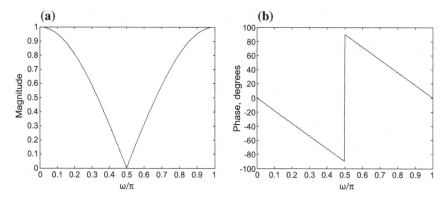

Fig. 2.33 a Magnitude response of $h(n)$ sequence and **b** phase response of $h(n)$ sequence

```
H = num/den;
%Compute and plot the magnitude response
mag = abs(H);
Figure (1),plot(w/pi,mag);
ylabel('Magnitude');xlabel('\omega/\pi');
%Compute and plot the phase responses
ph = angle(H)*180/pi;
Figure (2),plot(w/pi,ph);
ylabel('Phase, degrees');
xlabel('\omega/\pi');
```

Example 2.35 Determine the magnitude and phase responses of a system described by the following difference equation

$$y(n) - 2.1291y(n-1) + 1.7834y(n-2) - 0.5435y(n-3)$$
$$= 0.0534x(n) - 0.0009x(n-1) - 0.0009x(n-2) + 0.0534x(n-3)$$

Comment on the frequency response of the system.

Solution

Program 2.4

```
clear;close all;
num = [0.0534 -0.0009 -0.0009 0.0534];% numerator coefficients
den = [1-2.1291 1.7834 -0.5435];% denominator coefficients
w = 0:pi/255:pi;
%Compute the frequency responses
H = freqz(num,den,w);
%Compute and plot the magnitude response
mag = abs(H);
Figure (1),plot(w/pi,mag);
ylabel('Magnitude');xlabel('\omega/\pi');
%Compute and plot the phase responses
ph = angle(H)*180/pi;
Figure (2),plot(w/pi,ph);
ylabel('Phase, degrees');xlabel('\omega/\pi');
```

The frequency response shown in Fig. 2.34 characterizes a lowpass filter with nonlinear phase.

Example 2.36 Determine the magnitude and phase responses of a system described by the following difference equation

$$y(n) - 3.0538y(n-1) + 3.8281y(n-2) - 2.2921y(n-3) + 0.5507y(n-4)$$
$$= x(n) - 4x(n-1) + 6x(n-2) - 4x(n-3) + x(n-4).$$

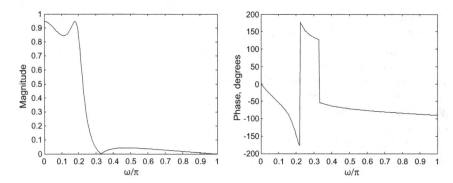

Fig. 2.34 a Magnitude response and **b** phase response

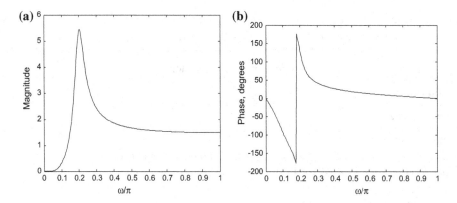

Fig. 2.35 a Magnitude response and **b** phase response

Comment on the frequency response of the system.

Solution Program 2.4 with variables num $= \begin{bmatrix} 1 & -4 & 6 & -4 & 1 \end{bmatrix}$ and den $=$ $\begin{bmatrix} 1 & -3.0538 & 3.8281 & -2.2921 & 0.5507 \end{bmatrix}$ is used, and the resultant magnitude and phase responses are shown in Fig. 2.35a, b, respectively. It is observed from this figure that the frequency response characterizes a narrowband bandpass filter.

Example 2.37 Consider the following difference equations, and verify whether any one of them has a linear phase.

$$\begin{aligned}
\text{(i)} \ y(n) = {}& -0.3x(n) + 0.11x(n-1) + 0.3x(n-2) \\
& + 1.22x(n-3) + 0.3x(n-4) + 0.11x(n-5) - 0.3x(n-6) \\
\text{(ii)} \ y(n) = {}& -0.5x(n) + 0.45x(n-1) + 0.58x(n-2) \\
& + 1.02x(n-3) + 0.1x(n-4) - 0.03x(n-5) - 0.18x(n-6)
\end{aligned}$$

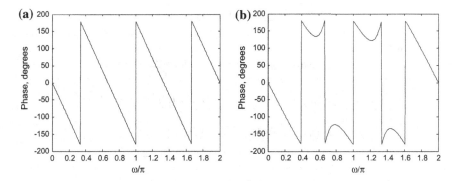

Fig. 2.36 a Phase response of (i) and **b** phase response of (ii)

Solution The phase responses of (i) and (ii) are shown in Fig. 2.36a, b, respectively. From these phase responses, it can be seen that (i) has a linear phase because the difference equation exhibits the symmetry property.

Example 2.38 An LTI system is described by the following difference equation

$$y(n) = x(n) + 2x(n-1) + x(n-2)$$

(a) Find the frequency response $H(e^{j\omega})$ and group delay $grd[H(e^{j\omega})]$ of the system.
(b) Determine the difference equation of a new system such that the frequency response $H_1(e^{j\omega})$ of the new system is related to $H(e^{j\omega})$ as $H_1(e^{j\omega}) = H(e^{j(\omega+\pi)})$.

Solution (a)

$$y(n) = x(n) + 2x(n-1) + x(n-2)$$
$$h(n) = \delta(n) + 2\delta(n-1) + \delta(n-2)$$

$$H(e^{j\omega}) = 1 + 2e^{-j\omega} + e^{-2j\omega}$$

$$= 2e^{-j\omega}\left[\left(\frac{1}{2}\right)(e^{j\omega}) + 1 + \left(\frac{1}{2}\right)(e^{-j\omega})\right] = 2e^{-j\omega}(\cos\omega + 1)$$

Hence,

$$\left|H(e^{j\omega})\right| = 2(\cos\omega + 1)$$

$$\angle H(e^{j\omega}) = -\omega$$

Therefore,

$$\text{group delay} = grad[H(e^{j\omega})] = -\frac{d\angle H(e^{j\omega})}{d\omega} = 1$$

(b) By frequency shifting property, $e^{-j\pi n}h(n) \leftrightarrow H(e^{j(\omega+\pi)})$. Therefore,

$$h_1(n) = e^{-j\pi n}h(n) = (-1)^n h(n)$$
$$= \delta(n) - 2\delta(n-1) + \delta(n-2)$$

Hence, the difference equation of the new system is

$$y(n) = x(n) - 2x(n-1) + x(n-2).$$

2.9   Representation of Sampling in Frequency Domain

As mentioned in Sect. 2.3, mathematically, the sampling process involves multiplying a continuous-time signal $x_a(t)$ by a periodic impulse train $p(t)$

$$p(t) = \sum_{n=-\infty}^{\infty} \delta(t - nT) \tag{2.124}$$

As a consequence, the multiplication process gives an impulse train $x_p(t)$, which can be expressed as

$$x_p(t) = x_a(t)p(t)$$
$$= \sum_{-\infty}^{\infty} x_a(t)\delta(t - nT) \tag{2.125}$$

Since $x_a(t)\delta(t - nT) = x_a(nT)\delta(t - nT)$, the above reduces to

$$x_p(t) = \sum_{-\infty}^{\infty} x_a(nT)\delta(t - nT) \tag{2.126}$$

If we now take the Fourier transform of (2.125) and use the multiplication property of the Fourier transform of a product, we get

$$X_p(j\Omega) = \frac{1}{2\pi}[X_a(j\Omega) * P(j\Omega)] \tag{2.127}$$

$$= \frac{1}{2\pi} \left[X_a(j\Omega) * \frac{2\pi}{T} \sum_{k-\infty}^{\infty} \delta(\Omega - k\Omega_T) \right] \tag{2.128}$$

where * denotes the convolution in the continuous-time domain, and $X_p(j\Omega), X_a(j\Omega)$, and $P(j\Omega)$ are the Fourier transforms of $x_p(t), x_a(t)$, and $p(t)$, respectively. Since $p(t)$ is periodic with a period T, it can be expressed as a Fourier series

$$p(t) = \frac{1}{T} \sum_{-\infty}^{\infty} e^{j\left(\frac{2\pi}{T}\right)kt}$$

Since the Fourier transform of $f(t) = e^{j\Omega_T t}$ is given by $F(j\Omega) = 2\pi\delta(\Omega - \Omega_T)$, we see that the Fourier transform of $p(t)$ is given by

$$P(j\Omega) = \frac{2\pi}{T} \sum_{k=-\infty}^{\infty} \delta(\Omega - k\Omega_T) \tag{2.128}$$

where $\Omega_T = 2\pi/T$. Substitution of (2.128) in (2.127) yields

$$X_p(j\Omega) = \frac{1}{T} \left[X_a(j\Omega) * \sum_{k=-\infty}^{\infty} \delta(\Omega - k\Omega_T) \right]$$

Since the convolution of $X_a(j\Omega)$ with a shifted impulse $\delta(\Omega - k\Omega_T)$ is the shifted function $X_a(j(\Omega - k\Omega_T))$, the above reduces to

$$X_p(j\Omega) = \frac{1}{T} \sum_{k-\infty}^{\infty} X_a(j\Omega - jk\Omega_T) \tag{2.129}$$

Equation (2.129) shows that the spectrum of $x_p(t)$ consists of an infinite number of shifted copies of the spectrum of $x_a(t)$, and the shifts in frequency are multiples of Ω_T; that is, $X_p(j\Omega)$ is a periodic function with a period of $\Omega_T = 2\pi/T$.

Since the continuous Fourier transform of $\delta(t - nT)$ is given by

$$\mathfrak{F}[\delta(t - nT)] = e^{-j\Omega T n}, \tag{2.130}$$

we have from Eq. (2.126) that

$$X_p(j\Omega) = \sum_{n=-\infty}^{\infty} x_a(nT)e^{-j\Omega T n} \tag{2.131}$$

If we now compare (2.131) and (2.125) and use the relation

$$x(n) = x_a(nT), \quad -\infty < n < \infty \tag{2.132}$$

and the fact that the DTFT of the sequence $x(n)$ is given by

$$X(e^{j\omega}) = \sum_{n=-\infty}^{\infty} x(n)e^{-j\omega n}, \tag{2.133}$$

then we obtain

$$X(e^{j\omega}) = X_p(j\Omega)\big|_{\Omega=\omega/T} \tag{2.134a}$$

or equivalently

$$X_p(j\Omega) = X(e^{j\omega})\big|_{\omega=\Omega T} \tag{2.134b}$$

Hence, we have from (2.134a) and (2.132) that

$$X(e^{j\omega}) = \frac{1}{T}\sum_{k-\infty}^{\infty} X_a(j\Omega - jk\Omega_T)\Bigg|_{\Omega=\omega/T} = \frac{1}{T}\sum_{k-\infty}^{\infty} X_a\left(j\frac{\omega}{T} - j\frac{2\pi k}{T}\right) \tag{2.135}$$

On the other hand, the above equation can also be expressed as

$$X(e^{j\Omega T}) = \frac{1}{T}\sum_{k-\infty}^{\infty} X_a(j\Omega - jk\Omega_T) \tag{2.136}$$

From Eq. (2.135) or (2.136), it can be observed that $X(e^{j\omega})$ is obtained by frequency scaling $X_p(j\Omega)$ using $\Omega = \omega/T$.

As mentioned earlier, the continuous-time Fourier transform $X_p(j\Omega)$ is periodic with respect to Ω having a period of $\Omega_T = (2\pi/T)$. In view of the frequency scaling, the DTFT $X(e^{j\omega})$ is also periodic with respect to ω with a period of 2π.

2.9.1 Sampling of Lowpass Signals

Sampling Theorem
If the highest component of frequency in analog signal $x_a(t)$ is Ω_m, then $x_a(t)$ is uniquely determined by its samples $x_a(nT)$, provided that

$$\Omega_T \geq 2\Omega_m \tag{2.137}$$

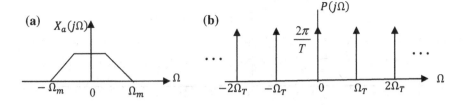

Fig. 2.37 a Spectrum of an analog signal and b spectrum of the pulse train

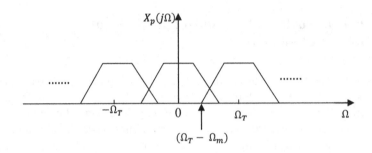

Fig. 2.38 Spectrum of an undersampled signal, showing aliasing (fold over region). Signals in the fold over region are not recoverable

where Ω_T is called the sampling frequency in radians. Equation (2.137) is often referred as the Nyquist condition.

The spectra of the analog signal $x_a(t)$ and the impulse train $p(t)$ with a sampling period $T = 2\pi/\Omega_T$ are shown in Fig. 2.37a, b, respectively.

Undersampling

If $\Omega_T < 2\Omega_m$, then the signal is undersampled and the corresponding spectrum $X_p(j\Omega)$ is as shown in Fig. 2.38. In this figure, the image frequencies centered at Ω_T will alias into the baseband frequencies and the information of the desired signal is indistinguishable from its image in the fold over region.

Oversampling

If $\Omega_T > 2\Omega_m$, then the signal is oversampled and its spectrum is shown in Fig. 2.39. Its spectrum is the same as that of the original analog signal, but repeats itself at every multiple of Ω_T. The higher-order components centered at multiples of Ω_T are called image frequencies.

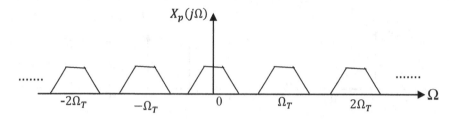

Fig. 2.39 Spectrum of an oversampled signal

2.10 Reconstruction of a Band-Limited Signal from Its Samples

According to the sampling theorem, samples of a continuous-time band-limited signal (i.e., its Fourier transform $X_a(j\Omega) = 0$ for $|\Omega| > |\Omega_m|$) taken frequently enough are sufficient to represent the signal exactly. The original continuous-time signal $x_a(t)$ can be fully recovered by passing the modulated impulse train $x_p(t)$ through an ideal lowpass filter, $H_{LP}(j\Omega)$, whose cutoff frequency satisfies $\Omega_m \leq \Omega_c \leq \Omega_T/2$. Consider a lowpass filter with a frequency response

$$H_{LP}(j\Omega) = \begin{cases} T & |\Omega| \leq \Omega_c \\ 0 & |\Omega| > \Omega_c \end{cases} \tag{2.138}$$

Applying the inverse continuous-time Fourier transform to $H_{LP}(j\Omega)$, we obtain the impulse response $h_{LP}(t)$ of the ideal lowpass filter given by

$$h_{LP}(t) = \frac{1}{2\pi} \int_{-\infty}^{\infty} H_{LP}(j\Omega) e^{j\Omega t} d\Omega = \frac{T}{2\pi} \int_{-\Omega_c}^{\Omega_c} e^{j\Omega t} d\Omega = \frac{\sin(\Omega_c t)}{(\pi t/T)}, \quad -\infty < t < \infty \tag{2.139}$$

For a given sequence of samples $x(n)$, we can form an impulse train $x_p(t)$ in which successive impulses are assigned an area equal to the successive sequence values, i.e.,

$$x_p(t) = \sum_{n=-\infty}^{\infty} x(n)\delta(t - nT) \tag{2.140}$$

The nth sample is associated with the impulse at $t = nT$, where T is the sampling period associated with the sequence $x(n)$. Therefore, the output $x_a(t)$ of the ideal lowpass filter is given by the convolution of $x_p(t)$ with the impulse response $h_{LP}(t)$ of the analog lowpass filter:

$$x_a(t) = \sum_{n=-\infty}^{\infty} x(n)h_{LP}(t - nT) \tag{2.141}$$

Substituting $h_{LP}(t)$ from Eq. (2.139) in Eq. (2.141) and assuming, for simplicity that $\Omega_c = \Omega_T/2 = \pi/T$, we get

$$x_a(t) = \sum_{n=-\infty}^{\infty} x(n) \frac{\sin[\pi(t - nT)/T]}{\pi(t - nT)/T} \tag{2.142}$$

The above expression indicates that the reconstructed continuous-time signal $x_a(t)$ is obtained by shifting in time the impulse response $h_{LP}(t)$ of the lowpass filter by an amount nT and scaling it in amplitude by the factor $x(n)$ for all integer values of n in the range $-\infty < n < \infty$ and then summing up all the shifted versions.

Example 2.39 A continuous-time signal $x_a(t)$ has its spectrum $X_a(j\Omega)$ as shown in Fig. 2.40. The signal $x_a(t)$ is input to the system shown in Fig. 2.41. $H(e^{j\omega})$ in Fig. 2.41 is an ideal LTI lowpass filter with a cutoff frequency of $(\pi/2)$. Sketch the spectrums of $x(n), y(n)$, and $y_r(t)$.

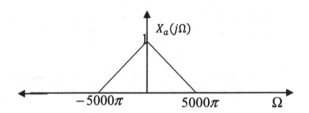

Fig. 2.40 Spectrum of signal $x_a(t)$

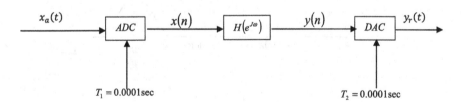

Fig. 2.41 Signal reconstruction system

Solution

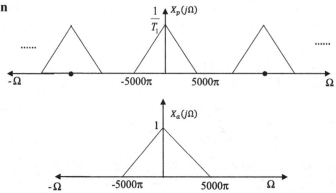

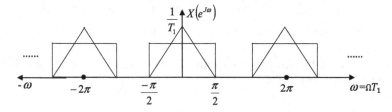

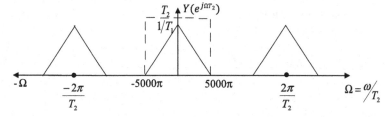

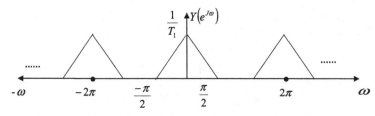

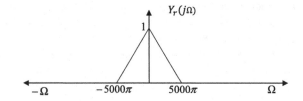

2.11 Discrete-Time Random Signals

A discrete-time random process $X(n)$ is an ensemble of the sample sequences $x(n)$. The statistical properties of $X(n)$ are similar to those of $X(t)$ of in the continuous time case, except that the index n is now an integer time variable. The time–frequency-domain statistical attributes of random signals, as well as the effect of filtering on such signals, can be studied by using the concept of random process.

2.11.1 Statistical Properties of Discrete-Time Random Signals

The *mean value* of a discrete-time random signal or process $X(n)$ at time index n is given by

$$\mu_{X(n)} = E[X(n)] \tag{2.143}$$

where $E[.]$ denotes the expected value. Without distinguishing between the random process $X(n)$ and the sequence $x(n)$, for simplification of mathematical notation, Eq. (2.143) can be written as

$$\mu_{x(n)} = E[x(n)] \tag{2.144}$$

The *variance* σ_x^2 of $x(n)$ at time index n can be expressed as

$$\sigma_x^2 = E\left[(x(n) - E[x(n)])^2\right] = E\left[x^2(n)\right] - E^2[x(n)] \tag{2.145}$$

Since the mean and variance of a discrete-time random signal are functions of the time index n, they can be represented as sequences.

The *autocorrelation* of a complex discrete-time random signal or process $x(n)$ at two different time indices m and n is defined by

$$r_{xx}(m,n) = E[x(m)x^*(n)] \tag{2.146}$$

where * denotes the complex conjugate.

The *cross-correlation* of two discrete-time random signals or processes $x(n)$ and $y(n)$ is defined by

$$r_{xy}(m,n) = E[x(m)y^*(n)] \tag{2.147}$$

If a random process at two time indices n and $n+k$ has the same statistics for any value of k, then the process is called a *stationary random process*. A process $X(n)$ is said to be *wide sense stationary* (WSS), if its mean is independent of the time index n, that is, has the same constant μ_X for all n, and its autocorrelation depends on the difference $(m-n)$ only, that is,

$$r_{xx}(k) = E[x(n+k)x^*(n)] \tag{2.148}$$

Two processes $x(n)$ and $y(n)$ are said to be *jointly stationary*, if each is stationary and their cross-correlation depends on $(m-n)$ only, that is,

$$r_{xy}(k) = E[x(n+k)y^*(n)] \tag{2.149}$$

2.11.2 Power of White Noise Input

The mean square value of a WSS random process is given by [from Eq. (2.148)]

$$E\left[|x(n)|^2\right] = r_{xx}(0) \tag{2.150}$$

and hence, using Eq. (2.145), its variance can be expressed as

$$\sigma_x^2 = r_{xx}(0) - |\mu_x|^2 \tag{2.151}$$

The power in a random process $X(n)$ is given by

$$P_x = E\left[\lim_{N\to\infty}\frac{1}{2N+1}\sum_{n=-N}^{N}|x(n)|^2\right] = \lim_{N\to\infty}\frac{1}{2N+1}\sum_{n=-N}^{N}E\left[|x(n)|^2\right] \tag{2.152}$$

Since a WSS random process has a constant mean square value for all values of n, the above equation becomes

$$P_x = E[|x(n)|^2]$$

Using Eqs. (2.150) and (2.151), the average power can be expressed as

$$P_x = r_{xx}(0) = \sigma_x^2 + |\mu_x|^2 \tag{2.153}$$

Since white noise is a WSS white random process with zero mean, the above equation can be written for white noise input as

$$r_{xx}(k) = \sigma_x^2 \delta(k) \tag{2.154}$$

The corresponding power spectrum is given by

$$P_{xx}(\omega) = \sigma_x^2 \tag{2.155}$$

Thus, the autocorrelation sequence of a white noise is an impulse sequence of area σ_x^2 and the power spectral density is of constant value σ_x^2 for all values of ω.

2.11.3   Statistical Properties of LTI System Output for White Noise Input

Consider an LTI system with an impulse response $h(n)$. If the input $x(n)$ to the system is a simple sequence of a WSS random process, then the output $y(n)$ is also a random process and related to its input process by

$$y(n) = \sum_{n=-\infty}^{\infty} h(l)x(n-l) \tag{2.156}$$

Now,

$$\mu_y = E[y(n)] = \sum_{l=-\infty}^{\infty} h(l)E[x(n-l)] = \mu_x \sum_{l=-\infty}^{\infty} h(l) \tag{2.157}$$

For real input $x(n)$, the autocorrelation of the output process $y(n)$ is defined by

$$r_{yy}(k) = E[y(n+k,n)] = E\left\{ \sum_{l=-\infty}^{\infty} \sum_{j=-\infty}^{\infty} h(l)h(j)x(n+k-j)x(n-l) \right\}$$
$$= \sum_{l=-\infty}^{\infty} h(l) \sum_{j=-\infty}^{\infty} h(j)E[x(n+k-j)x(n-l)] \tag{2.158}$$

For stationary input $x(n)$, $E[x(n+k-j)x(n-l)]$ depends on the time difference $k+l-j$. Hence, Eq. (2.158) can be rewritten as

$$r_{yy}(k) = \sum_{l=-\infty}^{\infty} h(l) \sum_{j=-\infty}^{\infty} h(j)r_{xx}(k+l-j) \tag{2.159}$$

Letting $j - k = m$, we can express the above equation as

$$r_{yy}(k) = \sum_{m=-\infty}^{\infty} r_{xx}(k-m) \sum_{k=-\infty}^{\infty} h(k)h(m+k)$$
$$= \sum_{m=-\infty}^{\infty} r_{xx}(k-m)\varnothing_{hh}(m) \tag{2.160}$$

where

$$\varnothing_{hh}(m) = \sum_{k=-\infty}^{\infty} h(k)h(m+k) \qquad (2.161)$$

stands for the autocorrelation of the deterministic impulse response $h(n)$

Taking the DTFT on both sides of Eq. (2.160), we obtain,

$$R_{yy}(e^{j\omega}) = \varnothing_{xx}(e^{j\omega})R_{xx}(e^{j\omega}) \qquad (2.162)$$

The DTFT of Eq. (2.161) yields

$$\varnothing_{xx}(e^{j\omega}) = |H(e^{j\omega})|^2 \qquad (2.163)$$

Hence, Eq. (2.162) becomes

$$R_{yy}(e^{j\omega}) = |H(e^{j\omega})|^2 R_{xx}(e^{j\omega}) \qquad (2.164)$$

Denoting the input and output power spectral densities $R_{xx}(e^{j\omega})$ and $R_{yy}(e^{j\omega})$ by $P_{xx}(e^{j\omega})$ and $P_{yy}(e^{j\omega})$, respectively, the above equation can be rewritten as

$$P_{yy}(e^{j\omega}) = |H(e^{j\omega})|^2 P_{xx}(e^{j\omega}) \qquad (2.165)$$

The inverse DTFT of $P_{yy}(e^{j\omega})$ yields the autocorrelation sequence $r_{yy}(k)$ as follows:

$$r_{yy}(k) = \frac{1}{2\pi} \int_{-\pi}^{\pi} P_{yy}(e^{j\omega})e^{j\omega k} d\omega \qquad (2.166)$$

The total average output power is given by

$$E[y^2(n)] = r_{yy}(0) = \frac{1}{2\pi} \int_{-\pi}^{\pi} P_{yy}(e^{j\omega}) d\omega \qquad (2.167)$$

Substituting Eq. (2.165) for $P_{yy}(e^{j\omega})$ in the above equation, we get

$$E[y^2(n)] = r_{yy}(0) = \frac{1}{2\pi} \int_{-\pi}^{\pi} |H(e^{j\omega})|^2 P_{xx}(e^{j\omega}) d\omega \qquad (2.168)$$

Based on Eq. (2.155), for white noise input, we obtain

$$r_{yy}(0) = \frac{\sigma_x^2}{2\pi} \int\limits_{-\pi}^{\pi} \left| H\left(e^{j\omega}\right) \right|^2 d\omega \qquad (2.169)$$

By making use of Parseval's theorem, Eq. (2.169) can be rewritten as

$$r_{yy}(0) = \sigma_x^2 \sum_{n=-\infty}^{\infty} |h(n)|^2 \qquad (2.170)$$

2.11.4 Correlation of Discrete-Time Signals

In many applications, it is often required to compare one reference sequence with one or more signals to determine the similarity between the pair and to acquire additional information based on the similarity. For example in GPS applications, the replica sequence generated in the user GPS receiver is the delayed version of the sequence transmitted by the GPS satellite, and by measuring the delay, one can determine the distance from the GPS satellite to the user which is used in determining the user position.

The similarity between a pair of finite energy signals $x(n)$ and $y(n)$ is given by the cross-correlation function $r_{xy}(l)$:

$$r_{xy}(l) = \sum_{n=-\infty}^{\infty} x(n)y(n-l) \quad l = 0, \pm1, \pm2, \ldots \qquad (2.171)$$

where l is called the lag time shift between the pair. When $y(n) = x(n)$, it reduces to the autocorrelation function $r_{xx}(l)$:

$$r_{xx}(l) = \sum_{n=-\infty}^{\infty} x(n)x(n-l) \quad l = 0, \pm1, \pm2, \ldots \qquad (2.172)$$

The autocorrelation and cross-correlation of sequences are easily computed using MATLAB as illustrated in the following example:

Example 2.40 Consider the sequence $x(n) = (4, 2, -1, 1, 3, 2, 1, 5)$. Compute its autocorrelation, the cross-correlation when $y(n) = x(n - 3)$, and the autocorrelation of $x(n)$ corrupted with random noise.

Solution The following Program 2.5 is used to compute the autocorrelation and cross-correlation of the sequences.

Program 2.5 Computation of autocorrelation (AC) and cross-correlation (CC)

```
clear;clc;
flag = input('Type in 0 for AC,1 for CC,2 for AC of noisy signal = ');
x = [ 4 2 −1 1 3 2 1 5 ];%sequence x(n)
y = [ 0 0 0 4 2 −1 1 3 2 1 5];% delayed version of sequence x(n), i.e., y(n)=x(n−3)
xn = x+randn(1,size(x));%noisy sequence x(n)
if flag ==0;
c = conv(x,fliplr(x));% autocorrelation of sequence x(n)
len1 = length(x)-1;
len2 = len1
end
if flag ==1;
c = conv(x,fliplr(y));% cross-correlation of x(n) and y(n) = x(n-3)
len1 = length(y)-1;
len2 = length(x)-1;
end
if flag ==2;
c = conv(xn,fliplr(xn));% autocorrelation of sequence x(n) corrupted with random
noise
len1 = length(xn)-1;
len2 = len1;
end
n = (-len1):len2;
stem(n,c)
xlabel('Lag index'); ylabel('Amplitude');
v = axis; axis([-len1 len2 v(3:end)])
```

The program starts running, when the input is entered. The autocorrelation of $x(n)$, cross-correlation of $x(n)$ and its delayed version $y(n) = x(n − 3)$, and the autocorrelation of the corrupted sequence $x(n)$ are shown in Fig. 2.42a, b, and c, respectively. $r_{xx}(l)$ is maximum at zero lag, as shown in Fig. 2.42a. From Fig. 2.42b, it can be observed that the peak of the cross-correlation sequence is exactly at a lag l equal to the delay used, indicating that the cross-correlation is useful to figure out the precise value of the delay. The noise corrupted sequence of $x(n)$ is generated by adding a random noise to it. The random noise is computed by using the MATLAB function randn, and as expected, the autocorrelation of the corrupted sequence of $x(n)$ still exhibits a pronounced peak at zero lag, as shown in Fig. 2.42c.

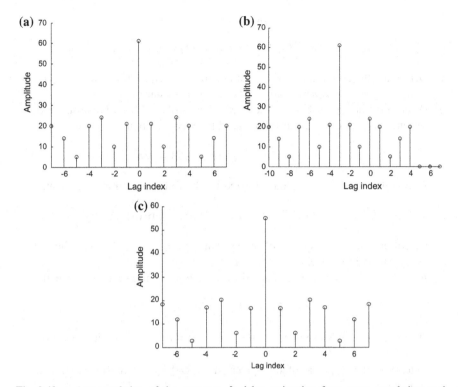

Fig. 2.42 **a** Autocorrelation of the sequence, **b** delay estimation from cross correlation, and **c** autocorrelation of sequence corrupted with random noise

2.12 Problems

1. Check the following for linearity, time-invariance, and causality.
   (i) $y(n) = 5nx^2(n)$. (ii) $y(n) = x(n)\sin 2n$. (iii) $y(n) = e^{-n}x(n+3)$

2. Given the input $x(n) = u(n)$ and the output $y(n) = \left(\frac{1}{2}\right)^{n-1}u(n-1)$ of a system.

   (i)   Determine the impulse response $h(n)$
   (ii) Is the system stable?
   (iii) Is the system causal?

3. Check for stability and causality of a system for the following impulse responses:
   (i) $h(n) = e^{2n}\sin\left(\frac{\pi n}{2}\right)u(n-1)$ (ii) $h(n) = \sin\left(\frac{\pi n}{2}\right)u(n)$

4. Determine if the following signals are periodic, and if periodic, find its period.
   (a) $\sin n$ (b) $e^{j\pi n/3}$ (c) $\sin\left(\frac{\pi n}{4}\right) + \sin\left(\frac{3\pi n}{4}\right)$

5. Determine the convolution of the sum of the two sequences

$$x_1(n) = (3, 2, 1, 2) \text{ and } x_2(n) = (1, 2, 1, 2).$$

6. Determine the convolution of the sum of the two sequences $x_1(n)$ and $x_2(n)$, if $x_1(n) = x_2(n) = c^n u(n)$ for all n, where c is a constant.

7. Determine the impulse response (i.e., when $x(n) = \delta(n)$ of a discrete-time system characterized by the following difference equation:

$$y(n) + y(n-1) - 6y(n-2) = x(n)$$

8. A discrete-time system is characterized by the following difference equation:

$$6\,y(n) - y(n-1) - y(n-2) = 6x(n)$$

Determine the step response of the system, i.e., $x(n) = u(n)$, given the initial conditions $y(-1) = 1$ and $y(-2) = -1$.

9. A discrete-time system is characterized by the following difference equation

$$y(n) - 5y(n-1) + 6y(n-2) = x(n)$$

Determine the response of the system for $x(n) = nu(n)$ and initial conditions y $(-1) = 1$, $y(-2) = 0$

10. Determine the response of the system described by the following difference equation

$$y(n) + y(n-1) = \sin 3n \; u(n)$$

11. Find the DTFT for the following sequences
    (a) $x_1(n) = u(n) - u(n-5)$ (b) $x_2(n) = \alpha^n(u(n) - u(n-8))$, $|\alpha| < 1$
    (c) $x_3(n) = n\left(\frac{1}{2}\right)^{|n|}$ (d) $x_4(n) = |\alpha|^n \sin \omega n$, $|\alpha| < 1$

12. Let $G_1(e^{j\omega})$ denote the DTFT of the sequence $g_1(n)$ shown in Fig. P2.1a. Express the DTFTs of the remaining sequences in Fig. P2.1 in terms of $G_1(e^{j\omega})$. Do not evaluate $G_1(e^{j\omega})$.

13. Determine the inverse DTFT of each of the following DTFTs:
    (a)$H_1(e^{j\omega}) = 1 + 4\cos\omega + 3\cos 2\omega$
    (b) $H_2(e^{j\omega}) = (3 + 2\cos\omega + 4\cos(2\omega))\cos(\omega/2)e^{-j\omega/2}$
    (c) $H_3(e^{j\omega}) = e^{-j\omega/4}$ (d) $H_4(e^{j\omega}) = e^{-j\omega}[1 + 4\cos\omega]$

14. Consider the system shown in Fig. P2.2a, where $\theta(\omega) = \arg(H(e^{j\omega}))$ is an ideal LTI lowpass filter with cutoff of $Y(e^{j\omega}) = H(e^{j\omega})X(e^{j\omega})$ rad/s and the spectrum of $X(e^{j\omega})$ is shown in Fig. P2.2b.

    (i) What is the maximum value of T to avoid aliasing in the ADC?
    (ii) If $1/T = 10$ kHz, then what will be the spectrum of $y_r(t)$.

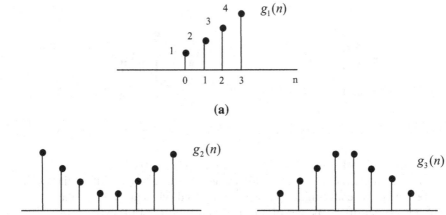

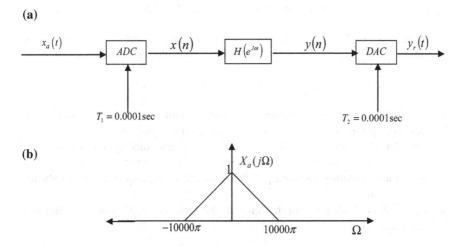

Fig. P2.1 Sequences $g_1(n)$, $g_2(n)$, and $g_3(n)$

Fig. P2.2

2.13 MATLAB Exercises

1. Using the function impz, write a MATLAB program to determine the impulse response of a discrete-time system represented by

$$y(n) - 5y(n-1) + 6y(n-2) = x(n) - 2x(n-1)$$

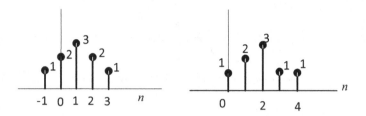

Fig. M2.1

2. Using MATLAB, verify the following properties of DTFT.
   (i) Linearity, (ii),time shifting, (iii) frequency shifting, (iv) differentiation in frequency, (v) convolution, (vi) windowing, and (vii) Parseval's relation.
3. Verify symmetry properties of a real sequence using MATLAB.
4. Verify symmetry properties of a complex sequence using MATLAB.
5. Determine the magnitude and phase responses of a system described by the following difference equation.

$$y(n) - y(n-1) + 0.24y(n-2) = 5x(n)$$

6. Using MATLAB Program 2, compute the magnitude and phase responses of Simpson's integration formula described by the following difference equation.

$$y(n) - y(n-2) = 0.333x(n) + 1.333x(n-1) + 0.333x(n-2)$$

7. Using the MATLAB function, generate a uniformly distributed random sequence in the range $[-1,1]$ and compute its mean and variance.
8. Consider the sequence $x(n) = (3,-2,0,1,4,5,2)$ and modify the program to determine the autocorrelation of the sequence $x(n)$ corrupted by a uniformly distributed random signal and plot the autocorrelation sequence and verify for peak at zero lag.
9. Write a MATLAB program to compute the cross-correlation of the following sequences.

$$x(n) = (3, -2, 0, 1, 4, 5, 2) \text{ and } y(n) = (-5, 4, 3, 6, -5, 0, 1).$$

   Plot the cross-correlation sequence.
10. Write a MATLAB program to determine the cross-correlation of the sequence $x(n) = (0,7,1,-3,4,9,-2)$ and its delayed version, $y(n) = x(n-3)$. Plot the cross-correlation sequence, and verify for peak at lag equal to the delay, i.e., 3.
11. Modify Program 2.3 to find the magnitude and phase responses of the following impulse sequences, and comment on the results (Fig. M2.1).

Chapter 3
The z-Transform and Analysis of LTI Systems in the Transform Domain

The DTFT may not exist for all sequences due to the convergence condition, whereas the z-transform exists for many sequences for which the DTFT does not exist. Also, the z-transform allows simple algebraic manipulations. As such, the z-transform has become a powerful tool in the analysis and design of digital systems. This chapter introduces the z-transform, its properties, the inverse z-transform, and methods for finding it. Also, in this chapter, the importance of the z-transform in the analysis of LTI systems is established.

3.1 Definition of the z-Transform

The z-transform of an arbitrary discrete time signal $x(n)$ is defined as

$$X(z) = Z[x(n)] = \sum_{n=-\infty}^{\infty} x(n)z^{-n} \tag{3.1}$$

where z is a complex variable. For the existence of the z-transform, Eq. (3.1) should converge. It is known from complex variables that if $\sum_{n=-\infty}^{\infty} x(n)z^{-n}$ is absolutely convergent, then Eq. (3.1) is convergent. Equation (3.1) can be rewritten as

$$X(z) = \sum_{n=0}^{\infty} x(n)z^{-n} + \sum_{n=-\infty}^{-1} x(n)z^{-n} \tag{3.2}$$

By ratio test, the first series is absolutely convergent if

$$\lim_{n\to\infty}\left|\frac{x(n+1)\,z^{-(n+1)}}{x(n)\quad z^{-n}}\right| = \lim_{n\to\infty}\left|\frac{x(n+1)}{x(n)}\right|\left|z^{-1}\right| < 1$$

© Springer Nature Singapore Pte Ltd. 2018
K. D. Rao and M. N. S. Swamy, *Digital Signal Processing*,
https://doi.org/10.1007/978-981-10-8081-4_3

or

$$|z| > \lim_{n \to \infty} \left| \frac{x(n+1)}{x(n)} \right| = r_1 \text{ (say)} \tag{3.3a}$$

Similarly, the second series in Eq. (3.2) is absolutely convergent if

$$\lim_{n \to -\infty} \left| \frac{x(n+1)}{x(n)} \right| \left| z^{-1} \right| < 1$$

or

$$|z| < \lim_{n \to -\infty} \left| \frac{x(n+1)}{x(n)} \right| = r_2 \text{ (say)} \tag{3.3b}$$

Thus, in general, Eq. (3.1) is convergent in some annulus

$$r_1 < |z| < r_2 \tag{3.4}$$

The set of values of z satisfying the above condition is called the region of convergence (ROC). It is noted that for some sequences $r_1 = 0$ or $r_2 = \infty$. In such cases, the ROC may not include $z = 0$ or $z = \infty$, respectively. Also, it is seen that no z-transform exists if $r_1 > r_2$.

The complex variable z in polar form may be written as

$$z = r e^{j\omega} \tag{3.5}$$

where r and ω are the magnitude and the angle of z, respectively. Then, Eq. (3.1) can be rewritten as

$$X(re^{j\omega}) = \sum_{n=-\infty}^{\infty} x(n)(re)^{-j\omega n} = \sum_{n=-\infty}^{\infty} x(n)e^{-j\omega n} r^{-n} \tag{3.6}$$

When $r = 1$, that is, when the contour $|z| = 1$, a unit circle in the z-plane, then Eq. (3.5) becomes the DTFT of $x(n)$.

Rational z-Transform

In LTI discrete-time systems, we often encounter with a z-transform which is a ratio of two polynomials in z:

$$X(z) = \frac{N(z)}{D(z)} = \frac{b_0 + b_1 z^{-1} + b_2 z^{-2} + \cdots + b_M z^{-M}}{1 + a_1 z^{-1} + a_2 z^{-2} + \cdots + a_N z^{-N}} \tag{3.7}$$

The zeros of the numerator polynomial $N(z)$ are called the zeros of $X(z)$, and those of the denominator polynomial $D(z)$ as the poles of $X(z)$. The number of finite zeros and poles in Eq. (3.7) are M and N, respectively. For example, the function $X(z) = \frac{z}{(z-1)(z-2)}$ has a zero at $z = 0$ and two poles at $z = 1$ and $z = 2$.

Example 3.1 Find the z-transform of the sequence $x(n) = \{1, 2, 3, 4, 5, 6, 7\}$.

Solution (i) For the given sequence $x(n) = \{1, 2, 3, 4, 5, 6, 7\}$, we can write $x(0) = 1$, $x(1) = 2$, $x(2) = 3$, $x(3) = 4$, $x(4) = 5$, $x(5) = 6$, and $x(6) = 7$. The z-transform of the sequence $x(n)$ is given by

$$X(z) = \sum_{n=0}^{6} x(n) z^{-n}$$

Hence,

$$X(z) = 1 + 2z^{-1} + 3z^{-2} + 4z^{-3} + 5z^{-4} + 6z^{-5} + 7z^{-6}$$

Therefore, $X(z)$ has finite values for all values of z except at $z = 0$. Therefore, the ROC is the entire z-plane except for $z = 0$.

Example 3.2 Find the z-transform of the sequence $x(n)$ tabulated below

n	-2	-1	0	1	2	3	4
$x(n)$	1	2	3	4	5	6	7

Solution

$$X(z) = \sum_{n=-2}^{4} x(n) z^{-n} = z^2 + 2z + 3 + 4z^{-1} + 5z^{-2} + 6z^{-3} + 7z^{-4}$$

Hence, $X(z)$ has finite values for all values of z except at $z = 0$ and $z = \infty$. Therefore, the ROC is the entire z-plane except for $z = 0$ and $z = \infty$.

Example 3.3 Determine the z-transform and the ROC for the following sequence:

$$x(n) = 2^n \quad \text{for } n \geq 0.$$

Solution From the definition of the z-transform,

$$X(z) = \sum_{n=-\infty}^{\infty} x(n) z^{-n} = \sum_{n=0}^{\infty} 2^n z^{-n} = \sum_{n=0}^{\infty} (2z^{-1})^n$$

$$= \frac{1}{1 - 2z^{-1}}, \quad |2z^{-1}| < 1$$

Thus, the ROC is $|z| > 2$.

Example 3.4 Determine the z-transform and the ROC for the following sequence:

$$x(n) = \begin{cases} \left(-\frac{1}{5}\right)^n & \text{for } n \geq 0 \\ -\left(\frac{1}{3}\right)^n & \text{for } n < 0 \end{cases}$$

Solution $X(z) = \sum_{n=-\infty}^{\infty} x(n)z^{-n} = \sum_{n=0}^{\infty} \left(-\frac{1}{5}\right)^n z^{-n} + \sum_{n=-\infty}^{-1} -\left(\frac{1}{3}\right)^n z^{-n}$

$= \frac{1}{1+(1/5)z^{-1}} + \frac{1}{1-(1/3)z^{-1}}$, for $\left|\frac{1}{5}\right| < |z|$ and $|z| < \left|\frac{1}{3}\right|$ respectively.

Thus, the ROC is $\left|\frac{1}{5}\right| < |z| < \left|\frac{1}{3}\right|$.

3.2 Properties of the Region of Convergence for the z-Transform

The properties of the ROC are related to the characteristics of the sequence $x(n)$. In this section, some of the basic properties of ROC are considered.

- **ROC should not contain poles**

  In the ROC, $X(z)$ should be finite for all z. If there is a pole p in the ROC, then $X(z)$ is not finite at this point, and the z-transform does not converge at $x = p$. Hence, ROC cannot contain any poles.

- **The ROC for a finite duration causal sequence is the entire z-plane except for $z = 0$**

  A causal finite duration sequence of length N is such that $x(n) = 0$ for $n < 0$ and for $n > N - 1$. Hence, $X(z)$ is of the form

  $$\begin{aligned} X(z) &= \sum_{n=0}^{N-1} x(n)z^{-n} \\ &= x(0) + x(1)z^{-1} + \cdots + x(N-1)z^{-N+1} \end{aligned} \tag{3.8}$$

  It is clear from the above expression that $X(z)$ is convergent for all values of z except for $z = 0$, assuming that $x(n)$ is finite. Hence, the ROC is the entire z-plane except for $z = 0$ and is shown as shaded region in Fig. 3.1.

- **The ROC for a non-causal finite duration sequence is the entire z-plane except for $z = \infty$.** A non-causal finite duration sequence of length N is such that $x(n) = 0$ for $n \geq 0$ and for $n \leq -N$. Hence, $X(z)$ is of the form

Fig. 3.1 ROC of a finite
duration causal sequence

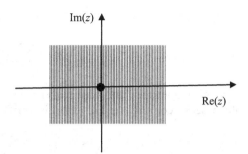

$$X(z) = \sum_{n=-N}^{-1} x(n)z^{-n}$$

$$= x(-N)z^N + \cdots + x(-2)z^2 + x(-1)z \tag{3.9}$$

It is clear from the above expression that $X(z)$ is convergent for all values of
except for $z = \infty$, assuming that $x(n)$ is finite. Hence, the ROC is the entire z-
plane except for $z = \infty$ and is shown as shaded region in Fig. 3.2.

- **The ROC for a finite duration two-sided sequence is the entire z-plane**
  except for $z = 0$ and $z = \infty$.

A finite duration of length $(N_2 + N_1 + 1)$ is such that $x(n) = 0$ for $n < -N_1$ and
for $n > N_2 - 1$, where N_1 and N_2 are positive. Hence, $x(z)$ is of the form

$$X(z) = \sum_{n=-N_1}^{N_2} x(n)z^{-n}$$

$$= x(-N_1)z^{N_1} + \cdots + x(-1)z + x(0) + x(1)z^{-1} + \cdots + x(N_2)z^{N_2} \tag{3.10}$$

It is seen that the above series is convergent for all values of z except for $z = 0$
and $z = \infty$.

- **The ROC for an infinite duration right-sided sequence is the exterior of a**
  circle which may or may not include $z = \infty$.

Fig. 3.2 ROC of a finite
duration non-causal sequence

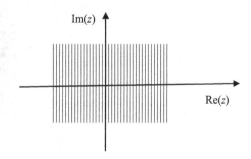

For such a sequence, $x(n) = 0$ for $n < N$. Hence, $X(z)$ is of the form

$$X(z) = \sum_{n=N}^{\infty} x(n) z^{-n} \qquad (3.11)$$

If $N \geq 0$, then the right-sided sequence corresponds to a causal sequence and the above series converges if Eq. (3.3a) is satisfied, that is,

$$|z| > \lim_{n \to \infty} \left| \frac{x(n+1)}{x(n)} \right| = r_1. \qquad (3.12)$$

Hence, in this case the ROC is the region exterior to the circle $|z| = r_1$, or the region $|z| > r_1$ including the point at $z = \infty$.

However, if N is a negative integer, say, $N = -N_1$, then the series (3.12) will contain a finite number of terms involving positive powers of z. In this case, the series is not convergent for $z = \infty$ and hence the ROC is the exterior of the circle $|z| = r_1$, but will not include the point at $z = \infty$.

As an example of an infinite duration causal sequence, consider

$$x(n) = \begin{cases} r_1^n & n \geq 0, \\ 0 & n < 0. \end{cases}$$

$$\text{Then} \quad X(z) = \sum_{n=0}^{\infty} r_1^n z^{-n} = \sum_{n=0}^{\infty} (r_1 z^{-1})^n = \frac{1}{1 - r_1 z^{-1}} \qquad (3.13)$$

Equation (3.13) holds only if $\left| r_1 z^{-1} \right| < 1$. Hence, the ROC is $|z| > r_1$. The ROC is indicated by the shaded region shown in Fig. 3.3 and includes the region $|z| > r_1$. It can be seen that $X(z)$ has a zero at $z = 0$ and pole at $z = r_1$. The zero is denoted by O and the pole by X.

- **The ROC for an infinite duration left-sided sequence is the interior of a circle which may or may not include $z = 0$.**

Fig. 3.3 ROC of an infinite duration causal sequence

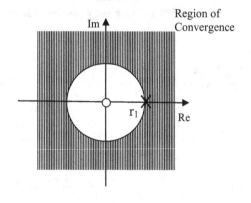

For such a sequence, $x(n) = 0$ for $n > N$. Hence, $X(z)$ is of the form

$$X(z) = \sum_{n=-\infty}^{N} x(n)z^{-n} \tag{3.14}$$

If $N < 0$, then the left-sided sequence corresponds to a non-causal sequence and the above series converges if Eq. (3.3b) is satisfied, that is,

$$|z| < \lim_{n \to -\infty} \left| \frac{x(n+1)}{x(n)} \right| = r_2 \tag{3.15}$$

Hence, in this case the ROC is the region interior to the circle $|z| = r_2$, or the region $|z| < r_2$ including the point at $z = 0$.

However, if N is a positive integer, then the series (3.14) will contain a finite number of terms involving negative powers of z. In this case, the series is not convergent for $z = 0$ and hence the ROC is the interior of the circle $|z| = r_2$, but will not include the point at $z = 0$.

As an example of an infinite duration non-causal sequence, consider

$$x(n) = \begin{cases} 0 & n \geq 0, \\ -r_2^n & n \leq -1. \end{cases} \tag{3.16}$$

Then,

$$X(z) = \sum_{n=-\infty}^{-1} -r_2^{-n}z^{-n} = -r_2^{-1}z \sum_{m=0}^{\infty} r_2^{-m}z^m$$

$$X(z) = \frac{1}{1 - r_2 z^{-1}} = \frac{z}{z - r_2} \quad \text{for } |z| < r_2 \tag{3.17}$$

Hence, the ROC is $|z| < r_2$, that is, the interior of the circle $|z| = r_2$. The ROC as well as the pole and zero of $X(z)$ are shown in Fig. 3.4.

- **The ROC of an infinite duration two-sided sequence is a ring in the z-Plane**

In this case, the z-transform $X(z)$ is of the form

$$X(z) = \sum_{n=-\infty}^{\infty} x(n)z^{-n} \tag{3.18}$$

and converges in the region $r_1 < |z| < r_2$, where r_1 and r_2 are given by (3.3a) and (3.3b), respectively. As mentioned before, the z-transform does not exist if $r_1 > r_2$.

Fig. 3.4 ROC of an infinite
duration non-causal sequence

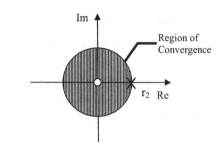

Fig. 3.5 ROC of an infinite
duration two-sided sequence

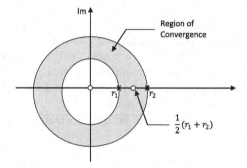

As an example, consider the sequence

$$x(n) = \begin{cases} r_1^n & n \geq 0 \\ -r_2^n & n \leq -1 \end{cases} \tag{3.19}$$

Then,

$$X(z) = \frac{z}{z - r_1} + \frac{z}{z - r_2} = \frac{z(2z - r_1 - r_2)}{(z - r_1)(z - r_2)} \tag{3.20}$$

where the region of convergence is $r_1 < |z| < r_2$. Thus, the ROC is a ring with a pole
on the interior boundary and a pole on the exterior boundary of the ring, without
any pole in the ROC. There are two zeros, one being located at the origin and the
other in the ROC. The poles and zeros as well as the ROC are shown in Fig. 3.5.

3.3   Properties of the z-Transform

Properties of the z-transform are very useful in digital signal processing. Some
important properties of the z-transform are stated and proved in this section. We
will denote in the following ROC of $X(z)$ by R $(r_1 < |z| < r_2)$ and those of $X_1(z)$ and
$X_2(z)$ by R_1 and R_2, respectively. Also, the region $(1/(r_2) < |z| < 1/(r_1))$ is denoted
by $(1/R)$.

Linearity: If $x_1(n)$ and $x_2(n)$ are two sequences with z-transforms $X_1(z)$ and $X_2(z)$, and ROCs R_1 and R_2, respectively, then the z-transform of a linear combination of $x_1(n)$ and $x_2(n)$ is given by

$$Z\{a_1x_1(n) + a_2x_2(n)\} = a_1X_1(z) + a_2X_2(z) \tag{3.21}$$

whose ROC is at least $(R_1 \cap R_2)$, a_1 and a_2 being arbitrary constants.

Proof

$$Z\{a_1x_1(n) + a_2x_2(n)\} = \sum_{n=-\infty}^{\infty} \{a_1x_1(n) + a_2x_2(n)\}z^{-n} \tag{3.22}$$

$$= a_1 \sum_{n=-\infty}^{\infty} x_1(n)z^{-n} + a_2 \sum_{n=-\infty}^{\infty} x_2(n)z^{-n} \tag{3.23}$$
$$= a_1X_1(z) + a_2X_2(z)$$

The result concerning the ROC follows directly from the theory of complex variables concerning the convergence of a sum of two convergent series.

Time Reversal: If $x(n)$ is a sequence with z-transform $X(z)$ and ROC R, then the z-transform of the time-reversed sequence $x(-n)$ is given by

$$Z\{x(-n)\} = X(z^{-1}) \tag{3.24}$$

whose ROC is $1/R$.

Proof From the definition of the z-transform, we have

$$Z[x(-n)] = \sum_{n=-\infty}^{\infty} x(-n)z^{-n} = \sum_{m=-\infty}^{\infty} x(m)z^{m}$$
$$= \sum_{m=-\infty}^{\infty} x(m)(z^{-1})^{-m} \tag{3.25}$$

Hence,

$$Z[x(-n)] = X(z^{-1}) \tag{3.26}$$

Since $(r_1 < |z| < r_2)$, we have $1/(r_2) < |z^{-1}| < 1/(r_1)$. Thus, the ROC of $Z[x(-n)]$ is $1/R$.

Time shifting: If $x(n)$ is a sequence with z-transform $X(z)$ and ROC R, then the z-transform of the delayed sequence $x(n-k)$, k being an integer, is given by

$$Z[x(n-k)] = z^{-k}X(z) \qquad (3.27)$$

whose ROC is the same as that of $X(z)$ except for $z = 0$ if $k > 0$, and $z = \infty$ if $k < 0$

Proof

$$Z\{x(n-k)\} = \sum_{n=-\infty}^{\infty} x(n-k)z^{-n} \qquad (3.28)$$

Substituting $m = n - k$,

$$Z[x(n-k)] = \sum_{m=-\infty}^{\infty} x(m)z^{-(m+k)} = z^{-k}\sum_{m=-\infty}^{\infty} x(m)z^{-m} \qquad (3.29)$$

$$= z^{-k}\sum_{m=-\infty}^{\infty} x(m)z^{-m}$$
$$= z^{-k}X(z) \qquad (3.30)$$

It is seen from Eq. (3.30) that, in view of the factor z^{-k}, the ROC of $Z[x(n-k)]$, is the same as that of $X(z)$ except for $z = 0$ if $k > 0$, and $z = \infty$ if $k < 0$. It is also observed that in particular, a unit delay in time translates into the multiplication of the z-transform by z^{-1}.

Scaling in the z-domain. If $x(n)$ is a sequence with z-transform $X(z)$, then $Z\{a^n x(n)\} = X(a^{-1}z)$ for any constant a, real or complex. Also, the ROC of $Z\{a^n x(n)\}$ is $|a|R$, i.e., $|a|r_1 < |z| < |a|r_2$.

Proof

$$Z\{a^n x(n)\} = \sum_{n=-\infty}^{\infty} a^n x(n)z^{-n} \qquad (3.31)$$

$$= \sum_{n=-\infty}^{\infty} x(n)\left(\frac{z}{a}\right)^{-n} = X\left(\frac{z}{a}\right) \qquad (3.32)$$

Since the ROC of $X(z)$ is $r_1 < |z| < r_2$, the ROC of $X(a^{-1}z)$ is given by $r_1 < |a^{-1}z| < r_2$, that is,

$$|a|r_1 < |z| < |a|r_2.$$

Differentiation in the z-domain. If $x(n)$ is a sequence with z-transform $X(z)$, then

$$Z\{nx(n)\} = -z\frac{\mathrm{d}X(z)}{\mathrm{d}z} \tag{3.33}$$

whose ROC is the same as that of $X(z)$.

Proof From the definition

$$Z[x(n)] = \sum_{n=-\infty}^{\infty} x(n)z^{-n}$$

Differentiating the above equation with respect to z, we get

$$\frac{\mathrm{d}X(z)}{\mathrm{d}z} = \sum_{n=-\infty}^{\infty} (-n)x(n)z^{-n-1} \tag{3.34}$$

Multiplying the above equation both sides by $-z$, we obtain

$$-z\frac{\mathrm{d}X(z)}{\mathrm{d}z} = -z\sum_{n=-\infty}^{\infty} (-n)x(n)z^{-n-1} \tag{3.35}$$

which can be rewritten as

$$-z\frac{\mathrm{d}X(z)}{\mathrm{d}z} = \sum_{n=-\infty}^{\infty} nx(n)z^{-n} = Z\{nx(n)\} \tag{3.36a}$$

Now, the region of convergence $r_a < |z| < r_b$ of the sequence $nx(n)$ can be found using Eqs. (3.3a) and (3.3b).

$$r_a = \lim_{n\to\infty} \left|\frac{(n+1)x(n+1)}{nx(n)}\right| = \lim_{n\to\infty}\left|\frac{x(n+1)}{x(n)}\right| = r_1$$

and

$$r_b = \lim_{n\to-\infty} \left|\frac{(n+1)x(n+1)}{nx(n)}\right| = n\lim_{n\to-\infty}\left|\frac{x(n+1)}{x(n)}\right| = r_2$$

Hence, the ROC of $Z[nx(n)]$ is the same as that of $X(z)$.

By repeated differentiation of Eq. (3.36a), we get the result

$$Z[n^k x(n)] = \left[-z\frac{\mathrm{d}\{X(z)\}}{\mathrm{d}z}\right]^k \tag{3.36b}$$

It is to be noted that the ROC of $Z[n^k x(n)]$ is also the same as that of $X(z)$.

Convolution of two sequences. If $x_1(n)$ and $x_2(n)$ are two sequences with z-transforms $X_1(z)$ and $X_2(z)$, and ROCs R_1 and R_2, respectively, then

$$Z[x_1(n) * x_2(n)] = X_1(z)X_2(z) \tag{3.37}$$

whose ROC is at least $R_1 \cap R_2$.

Proof

$$X(z) = \sum_{n=-\infty}^{\infty} x(n)z^{-n} \tag{3.38}$$

The discrete convolution of $x_1(n)$ and $x_2(n)$ is given by

$$x_1(n) * x_2(n) = \sum_{k=-\infty}^{\infty} x_1(k)x_2(n-k) = \sum_{k=-\infty}^{\infty} x_2(k)x_1(n-k) \tag{3.39}$$

Hence, the z-transform of the convolution is

$$Z[x_1(n) * x_2(n)] = \sum_{n=-\infty}^{\infty} \left[\sum_{k=-\infty}^{\infty} x_2(k)x_1(n-k) \right] z^{-n} \tag{3.40}$$

Interchanging the order of summation, the above equation can be rewritten as

$$
\begin{aligned}
Z[x_1(n) * x_2(n)] &= \sum_{k=-\infty}^{\infty} x_1(k) \sum_{n=-\infty}^{\infty} x_2(n-k)z^{-n} \\
&= \sum_{k=-\infty}^{\infty} x_1(k) \sum_{m=-\infty}^{\infty} x_2(m)z^{-(m+k)} \\
&= \sum_{k=-\infty}^{\infty} x_1(k)z^{-k} \sum_{m=-\infty}^{\infty} x_2(m)z^{-m}
\end{aligned}
\tag{3.41}
$$

Hence,

$$Z[x_1(n) * x_2(n)] = X_1(z)X_2(z) \tag{3.42}$$

Since the right side of Eq. (3.42) is a product of the two convergent sequences $X_1(z)$ and $X_2(z)$ with ROCs R_1 and R_2, it follows from the theory of complex variables that the product sequence is convergent at least in the region $R_1 \cap R_2$. Hence, the ROC of $Z[x_1(n) * x_2(n)]$ is at least $R_1 \cap R_2$.

Correlation of two sequences. If $x_1(n)$ and $x_2(n)$ are two sequences with z-transforms $X_1(z)$ and $X_2(z)$, and ROCs R_1 and R_2, respectively, then

$$Z[r_{x_1x_2}(l)] = X_1(z)X_2(z^{-1}) \tag{3.43}$$

whose ROC is at least $R_1 \cap (1/R_2)$.

Proof Since $r_{x_1x_2}(l) = x_1(l) * x_2(-l)$,

$$Z[r_{x_1x_2}(l)] = Z[x_1(l) * x_2(-l)] \tag{3.44}$$

$$
\begin{aligned}
&= Z[x_1(l)]Z[x_2(-l)], \text{ using Eq. (3.37)} \\
&= X_1(z)X_2(z^{-1}), \text{ using Eq. (3.24)}
\end{aligned}
\tag{3.45}
$$

Since the ROC of $X_2(z)$ is R_2, the ROC of $X_2(z^{-1})$ is $1/R_2$ from the property concerning time reversal. Also, since the ROC of $X_1(z)$ is R_1, it follows from Eq. (3.45) that the ROC of $Z[r_{x_1x_2}(l)]$ is at least $R_1 \cap (1/R_2)$.

Conjugate of a Complex Sequence. If $x(n)$ is a complex sequence with the z-transform $X(z)$, then

$$Z[x^*(n)] = [X(z^*)]^* \tag{3.46}$$

with the ROCs of both $X(z)$ and $Z[x^*(n)]$ being the same.

Proof The z-transform of $x^*(n)$ is given by

$$Z[x^*(n)] = \sum_{n=-\infty}^{\infty} x^*(n)z^{-n} \tag{3.47}$$

$$= \left[\sum_{n=-\infty}^{\infty} x(n)(z^*)^{-n} \right]^* \tag{3.48}$$

In the RHS of the above equation, the term in the brackets is equal to $X(z^*)$. Therefore, Eq. (3.48) can be written as

$$Z[x^*(n)] = [X(z^*)]^* = X^*(z^*) \tag{3.49}$$

It is seen from Eq. (3.49) that the ROC of the z-transform of conjugate sequence is identical to that of $X(z)$.

Real Part of a Sequence. If $x(n)$ is a complex sequence with the z-transform $X(z)$, then

$$Z\left[\text{Re}\{x(n)\}\right] = \frac{1}{2}\left[X(z) + X^*(z^*)\right] \tag{3.50}$$

whose ROC is the same as that of $X(z)$.

Proof

$$Z[\text{Re}\{x(n)\}] = Z\left[\frac{1}{2}\{x(n) + x^*(n)\}\right] \tag{3.51}$$

Since the z-transform satisfies the linearity property, we can write Eq. (3.51) as

$$Z[\text{Re}\{x(n)\}] = \frac{1}{2}Z[x(n)] + \frac{1}{2}Z[x^*(n)] \tag{3.52}$$

$$= \frac{1}{2}[X(z) + X^*(z^*)], \text{ using (3.49)} \tag{3.53}$$

It is clear that the ROC of $Z[\text{Re}\{x(n)\}]$ is the same as that of $X(z)$.

Imaginary Part of a Sequence. If $x(n)$ is a complex sequence with the z-transform $X(z)$, then

$$Z[\text{Im}\{x(n)\}] = \frac{1}{2j}[X(z) - X^*(z^*)] \tag{3.54}$$

whose ROC is the same as that of $X(z)$,

Proof Now

$$x(n) - x^*(n) = 2j\text{Im}\{x(n)\} \tag{3.55}$$

Thus,

$$\text{Im}\{x(n)\} = \frac{1}{2j}\{x(n) - x^*(n)\} \tag{3.56}$$

Hence,

$$Z[\text{Im}\{x(n)\}] = Z\left[\frac{1}{2j}\{x(n) - x^*(n)\}\right] \tag{3.57}$$

Again, since the z-transform satisfies the linearity property, we can write Eq. (3.57) as

$$Z[\text{Im}\{x(n)\}] = \frac{1}{2j}Z[x(n)] - \frac{1}{2j}Z[x^*(n)]$$

$$= \frac{1}{2j}[X(z) - X^*(z^*)], \text{ using (3.49)} \tag{3.58}$$

Again, it is evident that the ROC of the above is the same as that of $X(z)$.
The above properties of the z-transform are all summarized in Table 3.1.

Table 3.1 Some properties of the z-transform

Property	Sequence	z-transform	ROC
Linearity	$a_1 x_1(n) + a_2 x_2(n)$	$a_1 X_1(z) + a_2 X_2(z)$	At least $R_1 \cap R_2$
Time shifting	$x(n-k)$	$z^{-k} X(z)$	Same as R except for $z=0$ if $k>0$ and for $z=\infty$ if $k<0$
Time reversal	$x(-n)$	$X(z^{-1})$	$\frac{1}{R}$
Scaling in the z-domain	$a^n x(n)$	$X(a^{-1}z)$	$\lvert a \rvert R$
Differentiation in the z-domain	$nx(n)$	$-z\frac{dx(z)}{dz}$	R
Convolution theorem	$x_1(n) * x_2(n)$	$X_1(z)X_2(z)$	At least $R_1 \cap R_2$
Correlation theorem	$r_{x_1 x_2}(l) = \sum\limits_{n=-\infty}^{\infty} x_1(n)x_2(n-l)$	$X_1(z)X_2(z^{-1})$	At least $R_1 \cap \frac{1}{R_2}$
Conjugate complex sequence	$x^*(n)$	$[X(z^*)]^* = X^*(z^*)$	R
Real part of a complex sequence	$\mathrm{Re}[x(n)]$	$\frac{1}{2}[X(z) + X^*(z^*)]$	At least R
Imaginary part of a complex sequence	$\mathrm{Im}[x(n)]$	$\frac{1}{2j}[X(z) - X^*(z^*)]$	At least R
Time reversal of a complex conjugate sequence	$x^*(-n)$	$x^*(1/z^*)$	$\frac{1}{R}$

3.4   z-Transforms of Some Commonly Used Sequences

Unit Sample Sequence. The unit sample sequence is defined by

$$\delta(n) = \begin{cases} 1 & \text{for } n = 0 \\ 0 & \text{elsewhere} \end{cases} \tag{3.59}$$

By definition, the z-transform of $\delta(n)$ can be written as

$$X(z) = \sum_{n=-\infty}^{\infty} x(n)z^{-n} = 1z^0 = 1 \tag{3.60}$$

It is obvious from (3.60) that the ROC is the entire z-plane.

Unit Step Sequence. The unit step sequence is defined by

$$u(n) = \begin{cases} 1 & \text{for } n \geq 0 \\ 0 & \text{elsewhere} \end{cases} \tag{3.61}$$

The z-transform of $x(n)$ by definition can be written as

$$X(z) = \sum_{n=-\infty}^{\infty} x(n)z^{-n} = 1 + z^{-1} + z^{-2} + \cdots$$

$$= \frac{1}{1 - z^{-1}} = \frac{z}{z - 1} \quad \text{for } |z^{-1}| < 1 \tag{3.62}$$

Hence, the ROC for $X(z)$ is $|z| > 1$.

Example 3.5 Find the z-transform of $x(n) = \delta(n - k)$.

Solution By using the time shifting property, we get

$$Z[\delta(n - k)] = z^{-k}Z[\delta(n)] = z^{-k} \tag{3.63}$$

The ROC is the entire z-plane except for $z = 0$ if k is positive and for $z = \infty$ if k is negative.

Example 3.6 Find the z-transform of $x(n) = -u(-n - 1)$.

Solution We know that $Z[u(n)] = \frac{z}{z-1}$ for $|z| > 1$ from Eq. (3.62).
Hence, using the time shifting property

$$Z[u(n - 1)] = z^{-1}\frac{z}{z - 1} = \frac{1}{z - 1} \quad \text{for } |z| > 1, \tag{3.64}$$

Now, using the time reversal property (Table 3.1) we get

$$Z[u(-n - 1)] = \frac{1}{z^{-1} - 1} = \frac{z}{1 - z} \quad \text{for } |z| < 1$$

Hence,

$$Z[-u(-n - 1)] = \frac{z}{z - 1} \quad \text{for } |z| < 1 \tag{3.65}$$

Example 3.7 Find the z-transform of the sequence $x(n) = \{b^n u(n)\}$.

Solution Let $x_1(n) = u(n)$. From Eq. (3.62), $Z[u(n)] = X_1(z) = \frac{z}{z-1}$ for $|z| > 1$.
Using the scaling property, we get

$$Z[b^n u(n)] = X_1(b^{-1}z) = \frac{z}{z - b} \quad \text{for } |z| > |b|.$$

Example 3.8 Find the *z*-transform of $x(n) = nu(n)$.

Solution Let $x_1(n) = u(n)$. Again, using Eq. (3.62), we have $Z[u(n)] = X_1(z) = \frac{z}{z-1}$ for $|z| > 1$.

Using the differentiation property,

$$Z[nx(n)] = -z\frac{dX(z)}{dz}$$

we get

$$Z[nu(n)] = -z\frac{dX_1(z)}{dz} = -z\frac{d}{dz}\left(\frac{z}{z-1}\right) = \frac{z}{(z-1)^2} \quad \text{for } |z| > 1.$$

Example 3.9 Obtain the *z*-transform of the following sequence:

$$x(n) = \begin{cases} n^2 u(n) \\ 0 & \text{elsewhere} \end{cases}$$

Solution $X(z) = \sum\limits_{n=-\infty}^{\infty} x(n)z^{-n} = \sum\limits_{n=0}^{\infty} n^2 u(n)z^{-n}$.

Let $x(n) = n^2 x_1(n)$, where $x_1(n) = u(n)$. Then

$$X_1(z) = \frac{z}{z-1} \quad \text{for } |z| > 1$$

Using the differentiation property that

$$\text{if } x(n) \overset{ZT}{\leftrightarrow} X(z), \quad \text{then } n^2 x(n) \overset{ZT}{\leftrightarrow} \left(-z\frac{d}{dz}\right)^2 X(z)$$

we get

$$X(z) = -z\frac{d}{dz}\left(-z\frac{d}{dz}[X_1(z)]\right) = -z\frac{d}{dz}\left[\frac{z}{(z-1)^2}\right] = \frac{z(z+1)}{(z-1)^3}$$

The ROC of $X(z)$ is the same as that of $u(n)$, namely $|z| > 1$.

Example 3.10 Find the *z*-transform of $x(n) = \sin \omega n\, u(n)$.

Solution

$$Z\{\sin \omega n\, u(n)\} = Z\left\{\frac{e^{j\omega n} - e^{-j\omega n}}{2j}u(n)\right\} = \frac{1}{2j}\left[Z\{e^{j\omega n}u(n)\} - Z\{e^{-j\omega n}u(n)\}\right]$$

Using the scaling property, we get

$$\frac{1}{2j}[Z\{e^{j\omega n}u(n)\} - Z\{e^{-j\omega n}u(n)\}] = \frac{1}{2j}\left[\frac{z}{z - e^{j\omega}} - \frac{z}{z - e^{-j\omega}}\right]$$

$$= \frac{z\sin\omega}{z^2 - 2z\cos\omega + 1} \quad \text{for } |z| > 1$$

Therefore,

$$Z\{\sin\omega n\, u(n)\} = \frac{z\sin\omega}{z^2 - 2z\cos\omega + 1} \quad \text{for } |z| > 1.$$

Example 3.11 Find the z-transform of $x(n) = \cos\omega n\, u(n)$.

Solution $Z\{\cos\omega n u(n)\} = Z\left\{\frac{e^{j\omega n} + e^{-j\omega n}}{2} u(n)\right\} = \frac{1}{2}[Z\{e^{j\omega n}u(n)\} + Z\{e^{-j\omega n}u(n)\}].$

Using the scaling property, we get

$$\frac{1}{2j}[Z\{e^{j\omega n}u(n)\} + Z\{e^{-j\omega n}u(n)\}] = \frac{1}{2}\left[\frac{z}{z - e^{j\omega}} + \frac{z}{z - e^{-j\omega}}\right]$$

$$= \frac{z(z - \cos\omega)}{z^2 - 2z\cos\omega + 1} \quad \text{for } |z| > 1$$

Therefore,

$$Z\{\cos\omega n\, u(n)\} = \frac{z(z - \cos\omega)}{z^2 - 2z\cos\omega + 1} \quad \text{for } |z| > 1.$$

Example 3.12 Find the z-transform of the sequence $x(n) = [u(n) - u(n - 5)]$.

Solution $X(z) = \sum_{n=0}^{4} 1 + z^{-1} + z^{-2} + z^{-3} + z^{-4} = \frac{1 - z^{-5}}{1 - z^{-1}} = \frac{1}{z^4}\frac{z^5 - 1}{z - 1}.$
The ROC is the entire z-plane except for $z = 0$.

Example 3.13 Determine $X(z)$ for the function $x(n) = -\left[\frac{1}{2}\right]^n u(-n - 1)$.

Solution From Eq. (3.65), we have

$$Z[-u(-n - 1)] = \frac{z}{z - 1} \quad \text{for } |z| < 1$$

Now using the scaling property (Table 3.1).

$$Z\left\{-\left[\frac{1}{2}\right]^n u(-n - 1)\right\} = \frac{2z}{2z - 1} \quad \text{for } |z| < \frac{1}{2}$$

Thus, the ROC is $|z| < \frac{1}{2}$.

Example 3.14 Consider a system with input $x(n)$ and output $y(n)$. If its impulse response $h(n) = Ax(L - n)$, where L is an integer constant, and A is a known constant, find $Y(z)$ in terms of $X(z)$.

Solution

$$h(n) = Ax(L - n)$$
$$y(n) = x(n) * h(n)$$

By the convolution property of the z-transform, we have

$$Y(z) = H(z)X(z)$$

where

$$H(z) = Z\{Ax(L - n)\} = A \sum_{n=-\infty}^{\infty} x(-(n - L))z^{-n}$$

Letting $n - L = m$ in the above, we have

$$H(z) = A \sum_{m=-\infty}^{\infty} x(-m)z^{-(m+L)} = Az^{-L} \sum_{m=-\infty}^{\infty} x(-m)z^{-m}$$

$$= Az^{-L} \sum_{m=-\infty}^{\infty} x(-m)z^{-m}$$

$$= Az^{-L}X(z^{-1}) = Az^{-L}X(1/z)$$

Hence,

$$Y(z) = Az^{-L}X(1/z)X(z)$$

NOTE: It should be observed from Eqs. (3.64) to (3.65) that the z-transforms for both $u(n)$ and $-u(-n - 1)$ are the same. However, the ROC for the former is $|z| < 1$ while that for the latter is $|z| > 1$. Hence, it is very important to specify the ROC along with the z-transform of any sequence in order for us to uniquely determine $x(n)$ given an $X(z)$.

A list of some commonly used z-transform pairs are given in Table 3.2.

Initial Value Theorem

If a sequence $x(n)$ is causal, i.e., $x(n) = 0$ for $n < 0$, then

$$x(0) = \operatorname*{Lt}_{z \to \infty} X[z]. \tag{3.66}$$

Table 3.2 Some commonly used z-transform pairs

$x(n)$	$X(z)$	ROC
$\delta(n)$	1	Entire z-plane
$u(n)$	$\frac{1}{1-z^{-1}}$	$|z| > 1$
$nu(n)$	$\frac{z^{-1}}{(1-z^{-1})^2}$	$|z| > 1$
$-a^n u(-n-1)$	$\frac{1}{1-az^{-1}}$	$|z| < |a|$
$-na^n\{u(-n-1)\}$	$\frac{az^{-1}}{(1-az^{-1})^2}$	$|z| < |a|$
$\{\cos \omega n\}u(n)$	$\frac{1-z^{-1}\cos \omega}{1-2z^{-1}\cos \omega + z^{-2}}$	$|z| > 1$
$\{\sin \omega n\}u(n)$	$\frac{z^{-1}\sin \omega}{1-2z^{-1}\cos \omega + z^{-2}}$	$|z| > 1$

Proof Since $x(n)$ is causal, its z-transform $X[z]$ can be written as

$$X[z] = \sum_{n=0}^{\infty} x(n) \cdot z^{-n} = x(0) + x(1)z^{-1} + x(2)z^{-2} + \cdots. \qquad (3.67)$$

Now, taking the limits on both sides

$$\underset{z \to \infty}{\text{Lt}} \; X[z] = \underset{z \to \infty}{\text{Lt}} \; \{x(0) + x(1)z^{-1} + x(2)z^{-2} + \cdots\} = x(0) \qquad (3.68)$$

Hence, the theorem is proved.

Example 3.15 Find the initial value of a causal sequence $x(n)$ if its z-transform $X(z)$ is given by

$$X(z) = \frac{0.5z^2}{(z-1)(z^2 - 0.85z + 0.35)}.$$

Solution The initial value $x(0)$ is given by

$$x(0) = \lim_{z \to \infty} X(z) = \lim_{z \to \infty} \frac{0.5z^2}{(z-1)(z^2 - 0.85z + 0.35)} = \lim_{z \to \infty} \frac{0.5z^2}{z(z^2)} = 0.$$

3.5   The Inverse z-Transform

The z-transform of a sequence $x(n), Z[x(n)]$, defined by Eq. (3.1) is

$$X(z) = \sum_{m=-\infty}^{\infty} x(m)z^{-m} \qquad (3.69)$$

Multiplying the above equation both sides by z^{n-1} and integrating both sides on a closed contour C in the ROC of the z-transform $X(z)$ enclosing the origin, we get

$$\oint_C X(z)z^{n-1}\mathrm{d}z = \oint_C \sum_{m=-\infty}^{\infty} x(m)z^{-m}z^{n-1}\mathrm{d}z$$

$$= \oint_C \sum_{m=-\infty}^{\infty} x(m)z^{-m+n-1}\mathrm{d}z \tag{3.70}$$

Multiplying both sides of Eq. (3.70) by $\frac{1}{2\pi j}$, we arrive at

$$\frac{1}{2\pi j}\oint_C X(z)z^{n-1}\mathrm{d}z = \frac{1}{2\pi j}\oint_C \sum_{m=-\infty}^{\infty} x(n)z^{-m+n-1}\mathrm{d}z \tag{3.71}$$

By Cauchy integral theorem [1], we have

$$\frac{1}{2\pi j}\oint_C \sum_{m=-\infty}^{\infty} z^{-m+n-1}\mathrm{d}z = \begin{cases} 1 & \text{for } m=n \\ 0 & \text{for } m \neq n \end{cases} \tag{3.72}$$

Hence, the RHS of Eq. (3.71) becomes $x(n)$. That is,

$$\frac{1}{2\pi j}\oint_C X(z)z^{n-1}\mathrm{d}z = x(n).$$

Thus, the inverse z-transform of $X(z)$, denoted by $Z^{-1}[X(z)]$, is given by

$$Z^{-1}[X(z)] = x(n) = \frac{1}{2\pi j}\oint_C X(z)z^{n-1}\mathrm{d}z \tag{3.73}$$

It should be noted that given the ROC and the z-transform $X(z)$, the sequence $x(n)$ is unique. Table 3.2 can be used in most of the cases for obtaining the inverse transform. We will consider in Sect. 3.6 different methods of finding the inverse z-transform.

3.5.1   Modulation Theorem in the z-Domain

The z-transform of the product of two sequences (real or complex) $x_1(n)$ and $x_2(n)$ is given by

$$Z[x_1(n)x_2(n)] = \frac{1}{2\pi j}\oint_C X_1(v)X_2\left(\frac{z}{v}\right)v^{-1}\mathrm{d}v \tag{3.74}$$

where C is a closed contour which encloses the origin and lies in the ROC that is common to both $X_1(v)$ and $X_2\left(\frac{z}{v}\right)$.

Proof Let $x(n) = x_1(n)x_2(n)$.
  The inverse z-transform of $x_1(n)$ is given by

$$x_1(n) = \frac{1}{2\pi j} \oint_C X_1(v)v^{n-1}dv \tag{3.75}$$

Using Eq. (3.75), we get

$$x(n) = x_1(n)x_2(n) = \frac{1}{2\pi j} \oint_C X_1(v)v^{n-1}x_2(n)dv \tag{3.76}$$

Taking the z-transform of Eq. (3.76), we obtain

$$\begin{aligned} X(z) &= \sum_{n=-\infty}^{\infty} x(n)z^{-n} = \sum_{n=-\infty}^{\infty} \left[\frac{1}{2\pi j} \oint_C X_1(v)v^{n-1}x_2(n)dv \right] z^{-n} \\ &= \frac{1}{2\pi j} \oint_C X_1(v)dv \left[\sum_{n=-\infty}^{\infty} v^{-n}x_2(n)z^{-n} \right] v^{-1}dv \end{aligned} \tag{3.77}$$

Using the scaling property, we have that

$$\sum_{n=-\infty}^{\infty} v^{-n}x_2(n)z^{-n} = X_2\left(\frac{z}{v}\right)$$

Hence, Eq. (3.77) becomes

$$X(z) = \frac{1}{2\pi j} \oint_C X_1(v)X_2\left(\frac{z}{v}\right)v^{-1}dv$$

which is the required result.

3.5.2 Parseval's Relation in the z-Domain

If $x_1(n)$ and $x_2(n)$ are complex valued sequences, then

$$\sum_{n=-\infty}^{\infty} [x_1(n)x_2^*(n)] = \frac{1}{2\pi j} \oint_C X_1(v)X_2^*\left(\frac{1}{v^*}\right)v^{-1}dv \tag{3.78}$$

where C is a contour contained in the ROC common to the ROCs of $X_1(v)$ and $X_2^*\left(\frac{1}{v^*}\right)$.

Proof From Eq. (3.77), we have

$$Z[x_1(n)x_2(n)] = \frac{1}{2\pi j} \oint_C X_1(v)X_2\left(\frac{z}{v}\right)v^{-1}dv$$

Hence,

$$Z[x_1(n)x_2^*(n)] = \frac{1}{2\pi j} \oint_C X_1(v)X_2^*\left(\frac{z^*}{v^*}\right)v^{-1}dv \qquad (3.79)$$

where we have used the result concerning the z-transform of a complex conjugate (see Table 3.1). That is,

$$\sum_{n=-\infty}^{\infty} [x_1(n)x_2^*(n)]z^{-n} = \frac{1}{2\pi j} \oint_C X_1(v)X_2^*\left(\frac{z^*}{v^*}\right)v^{-1}dv \qquad (3.80)$$

Letting $z = 1$ in Eq. (3.80), we get

$$\sum_{n=-\infty}^{\infty} [x_1(n)x_2^*(n)] = \frac{1}{2\pi j} \oint_C X_1(v)X_2^*\left(\frac{1}{v^*}\right)v^{-1}dv$$

Hence, the theorem.

If $x_1(n) = x_2(n) = x(n)$ and the unit circle is included by the ROC of $X(z)$, then by letting $v = e^{j\omega}$ in (3.78), we get the energy of sequence in the z-domain to be

$$\sum_{n=-\infty}^{\infty} |x(n)|^2 = \frac{1}{2\pi j} \oint_C X(z)X^*\left(\frac{1}{z^*}\right)z^{-1}dz \qquad (3.81)$$

For the energy of real sequences in the z-domain, the above expression becomes

$$\sum_{n=-\infty}^{\infty} |x(n)|^2 = \frac{1}{2\pi j} \oint_C X(z)X\left(z^{-1}\right)z^{-1}dz \qquad (3.82)$$

This is the equivalent of the Parseval's relation in the frequency domain given by Eq. (2.85). Thus,

$$\sum_{n=-\infty}^{\infty} |x(n)|^2 = \frac{1}{2\pi j} \oint_C X(z)X\left(z^{-1}\right)z^{-1}dz = \frac{1}{2\pi} \int_{-\pi}^{\pi} |X(e^{j\omega})|^2 d\omega \qquad (3.83)$$

3.6 Methods for Computation of the Inverse z-Transform

The methods often used for computation of the inverse z-transform are:

1. Cauchy's Residue Theorem
2. Partial Fraction Expansion
3. Power Series Expansion.

3.6.1 Cauchy's Residue Theorem for Computation of the Inverse z-Transform

By Cauchy's residue theorem, the integral in Eq. (3.73) for rational z-transforms yields $Z^{-1}[X(z)] = x(n) =$ sum of the residues of the function $[X(z)z^{n-1}]$ at all the poles p_i enclosed by a contour C that lies in the ROC of $X(z)$ and encloses the origin. The residue at a simple pole p_i is given by

$$\operatorname*{res}_{z=p}[X(z)z^{n-1}] = \lim_{z \to p_i}[(z - p_i)X(z)z^{n-1}] \tag{3.84}$$

while for a pole p_i of multiplicity m, the residue is given by

$$\operatorname*{res}_{z=p}[X(z)z^{n-1}] = \frac{1}{(m-1)!}\lim_{z \to p_i}\frac{d^{m-1}}{dz^{m-1}}[(z - p_i)^m X(z)z^{n-1}] \tag{3.85}$$

We will now consider a few examples of finding the inverse z-transform using the residue method.

Example 3.16 Assuming the sequence $x(n)$ to be causal, find the inverse z-transform of

$$X(z) = \frac{z(z+1)}{(z-1)^3}.$$

Solution Since the sequence is causal, we have to consider the poles of $X(z)z^{n-1}$ for only $n \geq 0$. For $n \geq 0$, the function $X(z)z^{n-1}$ has only one pole at $z = 1$ of multiplicity 3. Thus, the inverse z-transform is given by

$$
\begin{aligned}
x(n) &= \frac{1}{(3-1)!}\lim_{z \to 1}\frac{d^2}{dz^2}\left[(z-1)^3 \frac{z(z+1)}{(z-1)^3}z^{n-1}\right] \\
&= \frac{1}{2!}\lim_{z \to 1}\frac{d^2}{dz^2}[(z+1)z^n] = \frac{1}{2}\lim_{z \to 1}[n(n+1)z^{n-1} + n(n-1)z^{n-2}] \\
&= n^2
\end{aligned}
$$

It should be mentioned that if $x(n)$ were not causal, then $X(z)z^{n-1}$ would have had a multiple pole of order n at the origin, and we would have to find the residue of $X(z)z^{n-1}$ at the origin to evaluate $x(n)$ for $n < 0$.

Example 3.17 If $x(n)$ is causal, find the inverse z-transform of

$$X(z) = \frac{1}{2(z - 0.8)(z + 0.4)}.$$

Solution Since the sequence is causal, we have to consider the poles of $X(z)z^{n-1}$ for only $n \geq 0$. Since $X(z)z^{n-1} = \frac{1}{2(z-0.8)(z+0.4)}z^{n-1}$, we see that for $n \geq 1, X(z)z^{n-1}$ has two simple poles at 0.8 and -0.4. However for $n = 0$, we have an additional pole at the origin. Hence, we evaluate $x(0)$ separately by evaluating the residues of $X(z)z^{-1} = \frac{1}{2z(z-0.8)(z+0.4)}$. Thus,

$$x(0) = \frac{1}{2(z - 0.8)(z + 0.4)}\bigg|_{z=0} + \frac{1}{2z(z - 0.8)}\bigg|_{z=-0.4} + \frac{1}{2z(z + 0.4)}\bigg|_{z=0.8}$$

$$= \frac{1}{2(-0.8)(0.4)} + \frac{1}{2(-0.4)(-1.2)} + \frac{1}{2(0.8)(1.2)} = 0$$

For $n > 0$

$$x(n) = \frac{z^{n-1}}{2(z - 0.8)}\bigg|_{z=-0.4} + \frac{z^{n-1}}{2(z + 0.4)}\bigg|_{z=0.8}$$

$$= \frac{(-0.4)^{n-1}}{2(-1.2)} + \frac{0.8^{n-1}}{2(1.2)} = \frac{1}{2.4} \cdot \left(0.8^{n-1} - (-0.4)^{n-1}\right)$$

Hence, for any $n \geq 0$

$$x(n) = \frac{1}{2.4} \cdot \left(0.8^{n-1} - (-0.4)^{n-1}\right)u(n - 1)$$

3.6.2 Computation of the Inverse z-Transform Using the Partial Fraction Expansion

Partial fractional expansion is another technique that is useful for evaluating the inverse z-transform of a rational function, and is a widely used method. To apply the partial fraction expansion method to obtain the inverse z-transform, we may consider the z-transform to be a ratio of two polynomials in either z or in z^{-1}. We now consider a rational function $X(z)$ as given in Eq. (3.7). It is called a proper rational function if $M > N$; otherwise, it is called an improper rational function. An improper rational function can be expressed as a proper rational function by dividing the numerator polynomial $N(z)$ by its denominator polynomial $D(z)$ and expressing $X(z)$ in the form

$$X(z) = \sum_{k=0}^{M-N} f_k z^{-k} + \frac{N_1(z)}{D(z)} \tag{3.86}$$

where the order of the polynomial $N_1(z)$ is less than that of the denominator polynomial. The partial fraction expansion can be now made on $N_1(z)/D(z)$. The inverse z-transform of the terms in the sum is obtained from the pair $\delta[n] \overset{z}{\leftrightarrow} 1$ (see Table 3.1) and the time shift property (see Table 3.2).

Let $X(z)$ be a proper rational function expressed as

$$X(z) = \frac{N(z)}{D(z)} = \frac{b_0 + b_1 z^{-1} + b_2 z^{-2} + \cdots + b_M z^{-M}}{1 + a_1 z^{-1} + a_2 z^{-2} + \cdots + a_N z^{-N}} \tag{3.87}$$

For simplification, eliminating negative powers Eq. (3.87) can be rewritten as

$$X(z) = \frac{N(z)}{D(z)} = \frac{b_0 z^N + b_1 z^{N-1} + b_2 z^{N-2} + \cdots + b_M z^{N-M}}{z^N + a_1 z^{N-1} + a_2 z^{-2} + \cdots + a_N} \tag{3.88}$$

Since $X(z)$ is a proper fraction, so will be $[X(z)/z]$. If all the poles p_i are simple, then, $[X(z)/z]$ can be expanded in terms of partial fractions as

$$\frac{X(z)}{z} = \sum_{i=1}^{N} \frac{c_i}{z - p_i} \tag{3.89}$$

where

$$c_i = (z - p_i) \frac{X(z)}{z} \bigg|_{z=v} \tag{3.90}$$

If $[X(z)/z]$ has a multiple pole, say at p_j, with a multiplicity of k, in addition to $(N - k)$ simple poles at p_i, then the partial fraction expansion given in Eq. (3.89) has to be modified as follows

$$\frac{X(z)}{z} = \frac{c_{j1}}{z - p_j} + \frac{c_{j2}}{(z - p_j)^2} + \cdots + \frac{c_{jk}}{(z - p_j)^k} + \sum_{i=1}^{N-k} \frac{c_i}{z - p_i} \tag{3.91}$$

where c_i is still given by (3.90) and c_{jk} by

$$c_{jk} = \frac{1}{(k-j)!} \frac{\mathrm{d}^{(k-j)}}{\mathrm{d}z^{k-j}} \left\{ (z-p_j)^k \frac{X(z)}{z} \right\} \Bigg|_{z=p_j} \tag{3.92}$$

Hence,

$$X(z) = \frac{c_{j1}z}{z-p_j} + \frac{c_{j2}z}{(z-p_j)^2} + \cdots + \frac{c_{jk}z}{(z-p_j)^k} + \sum_{i=1}^{N-k} \frac{c_i z}{z-p_i} \tag{3.93}$$

Then inverse z-transform is obtained for each of the terms on the right-hand side of (3.91) by the use of Tables 3.1 and 3.2. We will now illustrate the method by a few examples.

Example 3.18 Assuming the sequence $x(n)$ to be right sided, find the inverse z-transform of the following

$$X(z) = \frac{z}{(z-a)(z-b)}.$$

Solution The given function has poles at $z = a$ and $z = b$. Since $X(z)$ is a right-sided sequence, the ROC of $X(z)$ is the exterior of a circle around the origin that includes both the poles. Now $X(z)/z$ can be expressed in partial fraction expansion as

$$\frac{X(z)}{z} = \frac{a}{a-b} \frac{1}{z-a} - \frac{b}{a-b} \frac{1}{z-b}$$

Hence,

$$X(z) = \frac{a}{a-b} \frac{1}{1-az^{-1}} - \frac{b}{a-b} \frac{1}{1-bz^{-1}}$$

We can now find the inverse transform of each term using Table 3.2 as

$$x(n) = \frac{a}{a-b} a^n u(n) - \frac{b}{a-b} b^n u(n).$$

Example 3.19 Assuming the sequence $x(n)$ to be causal, find the inverse z-transform of the following

$$X(z) = \frac{10z^2 - 3z}{10z^2 - 9z + 2}.$$

Solution Dividing the numerator and denominator by z^2, we can rewrite $X(z)$ *as*

$$= \frac{10 - 3z^{-1}}{10 - 9z^{-1} + 2z^{-2}}$$

$$= \frac{4}{2 - z^{-1}} - \frac{5}{5 - 2z^{-1}}$$

$$= \frac{2}{1 - 0.5z^{-1}} - \frac{1}{1 - 0.4z^{-1}}$$

Each term in the above expansion is a first-order z-transform and can be recognized easily to evaluate the inverse transform as

$$Z^{-1}\{X(z)\} = x(n) = 2(0.5)^n u(n) - (0.4)^n u(n).$$

Example 3.20 Assuming the sequence $x(n)$ to be causal, determine the inverse z-transform of the following

$$X(z) = \frac{z(z+1)}{(z-1)^3}.$$

Solution $X(z)/z$ can be written in partial fraction expansion as

$$\frac{X(z)}{z} = \frac{z+1}{(z-1)^3} = \frac{A}{z-1} + \frac{B}{(z-1)^2} + \frac{C}{(z-1)^3}$$

Solving for A, B, and C, we get $A = 0, B = 1, C = 2$. Hence, $X(z)$ can be expanded as

$$X(z) = \frac{z}{(z-1)^2} + \frac{2z}{(z-1)^3}$$

Making use of Table 3.2, the inverse z-transform of $X(z)$ can be written as

$$Z^{-1}\{X(z)\} = x(n) = nu(n) + n(n-1)u(n) = n^2 u(n).$$

Example 3.21 If $x(n)$ is a right-handed sequence, determine the inverse z-transform for the function:

$$X(z) = \frac{1 + 2z^{-1} + z^{-3}}{(1 - z^{-1})(1 - 0.5z^{-1})}.$$

Solution $X(z) = \frac{1 + 2z^{-1} + z^{-3}}{(1 - z^{-1})(1 - 0.5z^{-1})} = \frac{z^3 + 2z^2 + 1}{z(z-1)(z-0.5)}.$

Now, $X(z)/z$ can be written in partial fraction expansion form as

$$\frac{X(z)}{z} = \frac{z^3 + 2z^2 + 1}{z^2(z-1)(z-0.5)} = \frac{A}{z} + \frac{B}{z^2} + \frac{C}{(z-1)} + \frac{D}{(z-0.5)}$$

Solving for A, B, C, and D, we get $A = 6, B = 2, C = 8, D = -13$. Hence,

$$X(z) = \frac{z^3 + 2z^2 + 1}{z(z-1)(z-0.5)} = 6 + \frac{2}{z} + \frac{8z}{(z-1)} - \frac{13z}{(z-0.5)}$$

Since the sequence is right handed, and the poles of $X(z)$ are located at $z = 0, 0.5$ and 1, the ROC of $X(z)$ is $|z| > 1$. Thus, from Table 3.2 we have

$$Z^{-1}\{X(z)\} = x(n) = 6\delta(n) + 2\delta(n-1) + 8u(n) - 13(0.5)^n u(n).$$

Example 3.22 Assuming $h(n)$ to be causal, find the inverse z-transform of

$$H(z) = \frac{(z-1)^2}{(z^2 - 0.1z - 0.56)}.$$

Solution Expanding $H(z)/z$ as

$$\frac{H(z)}{z} = \frac{(z-1)^2}{z(z-0.8)(z+0.7)} = \frac{A}{z} + \frac{B}{(z-0.8)} + \frac{C}{(z+0.7)}$$

Solving for A, B, and C, we get $A = -1.78$, $B = 0.033$, and $C = 2.75$. Therefore, $H(z)$ can be expanded as

$$H(z) = -1.7857 + \frac{0.0333z}{(z-0.8)} + \frac{2.7524z}{(z+0.7)}$$

Hence,

$$Z^{-1}\{H(z)\} = h(n) = -1.7857\delta(n) + 0.0333(0.8)^n u(n) + 2.7524(-0.7)^n u(n).$$

3.6.3 Inverse z-Transform by Partial Fraction Expansion Using MATLAB

The M-file residue z can be used to find the inverse z-transform using the power series expansion. The coefficients of the numerator and denominator polynomial for the above Example 3.22 can be written in descending powers of z as

$$\text{num} = [1 \quad -2 \quad 1];$$
$$\text{den} = [1 \quad -0.1 \quad -0.56];$$

The following MATLAB statement determines the residue (r), poles (p), and direct terms (k) of the partial fraction expansion of $H(z)$.

$$[r, p, k] = \text{residuez}(\text{num}, \text{den});$$

After execution of the above statements, the residues, poles, and constants obtained are:

$$\text{Residues}: \quad 0.0333 \quad 2.7524;$$
$$\text{Poles}: \quad 0.8000 \quad -0.7000;$$
$$\text{Constants}: \quad -1.7857$$

The desired expansion is

$$H(z) = -1.7857 + \frac{0.0333z}{(z - 0.8)} + \frac{2.7524z}{(z + 0.7)} \tag{3.94}$$

3.6.4 Computation of the Inverse z-Transform Using the Power Series Expansion

The z-transform of an arbitrary sequence defined by Eq. (3.1) implies that $X(z)$ can be expressed as power series in z^{-1} or z. In this expansion, the coefficient of the term z^{-n} indicates the value of the sequence $x(n)$. Long division is one way to express $X(z)$ in power series.

Example 3.23 Assuming $h(n)$ to be causal, fnd the inverse z-transform of the following

$$H(z) = \frac{z^2 + 2z + 1}{z^2 + 0.4z - 0.12}.$$

Solution We obtain the inverse z-transform by long division of the numerator by the denominator as follows

$$1+1.6z^{-1}+0.48z^{-2}+0z^{-3}+0.0576z^{-4}+\cdots$$

$$z^2+0.4z-0.12 \quad\Big|\quad z^2+2z+1$$

$$z^2+0.4z-0.12$$

$$1.6z+1.12$$

$$1.6z+0.64-0.192z^{-1}$$

$$0.48+0.192z^{-1}$$

$$0.48+0.192z^{-1}-0.0576z^{-2}$$

$$0.0576z^{-2}$$

$$0.0576z^{-2}+0.023o4z^{-3}-0.006912z^{-4}$$

$$-0.02304z^{-3}+0.006912z^{-4}$$

$$\cdots\cdots\cdots\cdots$$

Hence, $H(z)$ can be written as

$$H(z) = 1.0 + 1.6z^{-1} + 0.48z^{-2} + 0z^{-3} + 0.0576z^{-4} + \cdots$$

implying that

$$\{h[n]\} = \{1.0, \quad 1.6, \quad 0.48, \quad 0 \quad 0.0576, \quad \cdots\} \quad \text{for } n \geq 0.$$

Example 3.24 Find the inverse z-transform of the following

$$X(z) = \log(1 + bz^{-1}), \quad |b| < |z|.$$

Solution We know that power series expansion for $\log(1+u)$ is

$$\log(1+u) = u - \frac{u^2}{2} + \frac{u^3}{3} - \frac{u^4}{4} + \frac{u^5}{5} - \cdots$$

$$= \sum_{n=1}^{\infty} \frac{(-1)^{n+1}u^n}{n}, \quad |u| < 1$$

Letting $u = bz^{-1}$, $X(z)$ can be written as

$$X(z) = \log\left(1 + bz^{-1}\right) = \sum_{n=1}^{\infty} \frac{(-1)^{n+1} b^n z^{-n}}{n}, \quad |b| < |z|.$$

From the definition of z-transform of $x(n)$, we have

$$X(z) = \sum_{n=1}^{\infty} x(n) z^{-n}$$

Comparing the above two expressions, we get $x(n)$, i.e., the inverse z-transform of $X(z) = \log(1 + bz^{-1})$ to be

$$x(n) = \begin{cases} (-1)^{n+1} \cdot \frac{b^n}{n} & n > 0 \\ 0 & n \leq 0 \end{cases} \tag{3.95}$$

Example 3.25 Find the inverse z-transform of

$$X(z) = \frac{z}{z - b}, \quad \text{for} |z| > |b|.$$

Solution The sequence is a right-sided causal sequence as the region of convergence is $|z| > |b|$. We can use the long division as we did in Example 3.23 to express $z/(z - b)$ as a series in powers of z^{-1}. Instead, we will use binomial expansion.

$$\begin{aligned} X(z) &= \frac{z}{z - b} = \frac{1}{1 - bz^{-1}} \\ &= 1 + bz^{-1} + b^2 z^{-2} + \cdots \quad \text{for } \left| bz^{-1} \right| < 1 \\ &= \sum_{n=0}^{\infty} b^n z^{-n} \quad \text{for } |z| > |b| \end{aligned}$$

Hence,

$$Z^{-1}\{X(z)\} = x(n) = Z^{-1}\left\{ \frac{z}{z - b} \right\} = b^n u(n).$$

Example 3.26 Find the inverse z-transform of

$$X(z) = \frac{z}{z-b}, \quad \text{for } |z| < |b|.$$

Solution Since the region of convergence is $|z| < |b|$, the sequence is a left-sided sequence. We can use the long division to obtain $z/(z-b)$ as a power series in z. However, we will use the binomial expansion.

$$X(z) = \frac{z}{z-b} = -\frac{z}{b}\frac{1}{1-(z/b)}$$

$$= -\left(\frac{z}{b}\right)\left[1 + \left(\frac{z}{b}\right) + \left(\frac{z}{b}\right)^2 + \cdots\right] \quad \text{for } \left|\frac{z}{b}\right| < 1$$

$$= -\sum_{n=-1}^{\infty} b^n z^{-n} \quad \text{for } |z| < |b|$$

Hence,

$$Z^{-1}\{X(z)\} = x(n) = Z^{-1}\left\{\frac{z}{z-b}\right\} = -b^n u(-n-1).$$

Example 3.27 Using the z-transform, find the convolution of the sequences

$$x_1(n) = \{1, -3, 2\} \quad \text{and} \quad x_2(n) = \{1, 2, 1\}.$$

Solution Step 1: Determine z-transform of individual signal sequences

$$X_1(z) = Z[x_1(n)] = \sum_{n=0}^{2} x_1(n)z^{-1} = x_1(0) + x_1(1)z^{-1} + x_1(2)z^{-2}$$

$$= 1 - 3z^{-1} + 2z^{-2}$$

and

$$X_2(z) = Z[x_2(n)] = \sum_{n=0}^{2} x_2(n)z^{-1} = x_2(0) + x_2(1)z^{-1} + x_2(2)z^{-2}$$

$$= 1 + 2z^{-1} + z^{-2}$$

Step 2: Obtain $X(z) = X_1(z)X_2(z)$

$$X(z) = \left(1 - 3z^{-1} + 2z^{-2}\right)\left(1 + 2z^{-1} + z^{-2}\right)$$

$$= 1 - z^{-1} - 3z^{-2} + z^{-3} + 2z^{-4}$$

Step 3: Obtain the inverse z-transform of $X(z)$

$$x(n) = Z^{-1}[1 - z^{-1} - 3z^{-2} + z^{-3} + 2z^{-4}] = \{1, -1, -3, 1, 2\}.$$

3.6.5   Inverse z-Transform via Power Series Expansion Using MATLAB

The M-file impz can be used to find the inverse z-transform using the power series expansion. The coefficients of the numerator and denominator polynomial for the Example 3.25 can be written as

$$\text{num} = \begin{bmatrix} 1 & 2 & 1 \end{bmatrix};$$
$$\text{den} = \begin{bmatrix} 1 & 0.4 & -0.12 \end{bmatrix};$$

The following statement can be run to obtain the coefficients of the inverse z-transform

$$h = \text{impz(num, den)};$$

where h is the vector containing the coefficients of the inverse z-transform. The first 11 coefficients of the inverse z-transform of the Example 3.25 obtained after execution of the above MATLAB statements are:

Columns 1 through 9
1.0000   1.6000   0.4800   0   0.0576 − 0.0230   0.0161 − 0.0092   0.0056
Columns 10 through 11
− 0.0034   0.0020

3.6.6   Solution of Difference Equations Using the z-Transform

Example 3.28 Determine the impulse response of the system described by the difference equation:

$$y(n) - 3y(n - 1) - 4y(n - 2) = x(n) + 2x(n - 1).$$

Assume that the system is relaxed initially.

Solution Let $X(z) = Z[x(n)]$ and $Y(z) = Z[y(n)]$. Taking z-transform on both sides and using the shifting property, we get

$$(1 - 3z^{-1} - 4z^{-2})Y(z) = (1 + 2z^{-1})X(z)$$

Since $X(z) = 1$, we have

$$Y(z) = \frac{1 + 2z^{-1}}{1 - 3z^{-1} + 4z^{-2}}$$

$$\frac{Y(z)}{z} = \frac{z+2}{(z-4)(z+1)} = \frac{(6/5)}{z-4} - \frac{(1/5)}{z+1}$$

$$Y(z) = \frac{(6/5)}{1 - 4z^{-1}} - \frac{(1/5)}{1 + z^{-1}}$$

We now take inverse transform of the above and use Table 3.2 to obtain $y(n)$, which is the impulse response of the system as

$$h(n) = y(n) = (6/5)4^n u(n) - (1/5)(-1)^n u(n).$$

Example 3.29 Determine the response $y(n), n \geq 0$, of the system described by the second-order difference equation

$$y(n) - 3y(n-1) - 4y(n-2) = x(n) + 2x(n-1)$$

for the input $x(n) = 4^n u(n)$.

Solution Applying z-transform to both sides of the eqution, we have

$$Y(z)[1 - 3z^{-1} - 4z^{-2}] = X(z)[1 + 2z^{-1}]$$

Given that $x(n) = 4^n u(n)$ we have

$$X(z) = \frac{1}{1 - 4z^{-1}}$$

Substituting for $X(z)$ in the expression for $Y(z)$ and simplifying, we get

$$\frac{Y(z)}{z} = \frac{(z^2 + 2)}{(z-4)^2(z+1)}$$

or

$$\frac{Y(z)}{z} = \frac{-1}{25(z+1)} + \frac{26}{25(z-4)} + \frac{24}{5(z-4)^2}$$

Hence,

$$Y(z) = \frac{-z}{25(z+1)} + \frac{26z}{25(z-4)} + \frac{24z}{5(z-4)^2}$$

By applying inverse z-transforms, we get

$$y(n) = \frac{-1}{25}(-1)^n u(n) + \frac{6}{5}n(4)^n u(n) + \frac{26}{25}(4)^n u(n).$$

Example 3.30 Find the impulse response of the system

$$y(n) = 3y(n-1) + 2y(n-2) + x(n).$$

Solution Taking z-transforms on both sides of the above equation, and using the fact $Z[\delta(n)] = 1$, we get

$$Y(z) = \frac{1}{1 - 3z^{-1} - 2z^{-2}} = \frac{z^2}{z^2 - 3z - 2}$$

$$Y(z) = \frac{0.86}{1 - 3.56z^{-1}} + \frac{0.135}{1 + 0.56z^{-1}}$$

Hence, the impulse response is given by

$$h(n) = y(n) = 0.86(3.56)^n u(n) + 0.135(-0.561)^n u(n).$$

3.7   Analysis of Discrete-Time LTI Systems in the z-Transform Domain

It was stated in Chap. 2 that an LTI system can be completely characterized by its impulse response $h(n)$. The output signal $y(n)$ of a LTI system and the input signal $x(n)$ are related by convolution as

$$y(n) = h(n) * x(n) \tag{3.96}$$

Taking z-transform on both sides of the above equation and using the convolution property, we get

$$Y(z) = H(z)X(z) \tag{3.97}$$

indicating the z-transform of the output sequence $y(n)$ is the product of the z-transforms of the impulse response $h(n)$ and the input sequence $x(n)$. The quantities $h(n)$ and $H(z)$ are two equivalent descriptions of a system in the time domain and z-domain, respectively. The transform $H(z)$ is called the transfer function or the system function and expressed as

$$H(z) = \frac{Y(z)}{X(z)} \tag{3.98}$$

3.7.1 Rational or IIR Transfer Function

Consider a system described by a linear constant coefficient recursive difference equation of the form

$$y(n) = -\sum_{k=1}^{N} a_k y(n-k) + \sum_{k=0}^{M} b_k x(n-k) \tag{3.99}$$

where the constants a_k and b_k are real. Then, the system function can be obtained directly by computing the z-transform of both sides of the above equation. Thus, by applying the linearity property and the time shifting property, the above equation becomes

$$Y(z) = -\sum_{k=1}^{N} a_k z^{-k} Y(z) + \sum_{k=0}^{M} b_k z^{-k} X(z) \tag{3.100}$$

$$Y(z)\left[1 + \sum_{k=1}^{N} a_k z^{-k}\right] = X(z) \sum_{k=0}^{M} b_k z^{-k} \tag{3.101}$$

$$\frac{y(Z)}{x(Z)} = \frac{\sum_{k=0}^{M} b_k z^{-k}}{\left[1 + \sum_{k=1}^{N} a_k z^{-k}\right]} \tag{3.102}$$

Or equivalently,

$$H(z) = \frac{\sum_{k=0}^{M} b_k z^{-k}}{\left[1 + \sum_{k=1}^{N} a_k z^{-k}\right]} \tag{3.103}$$

The above transfer function is a ratio of polynomials in z^{-1} and hence is a rational transfer function or system function. Since the input–output characteristics of an infinite duration impulse response system are described by linear constant coefficient difference equations of recursive nature, the rational transfer function in Eq. (3.103) is also called as IIR transfer function or system function.

3.7.2   FIR Transfer Function

Consider a linear constant coefficient non-recursive difference equation

$$y(n) = \sum_{k=0}^{M} h(k)x(n - k) \tag{3.104}$$

Taking z-transform on both sides of the above equation, we get

$$Y(z) = \sum_{k=0}^{M} h(k)z^{-k}X(z) \tag{3.105}$$

$$\frac{y(Z)}{x(Z)} = \sum_{k=0}^{M} h(k)z^{-k} \tag{3.106}$$

or

$$H(z) = \sum_{k=0}^{M} h(k)z^{-k} = \frac{1}{z^M} \sum_{k=0}^{M} h(k)z^{M-k} \tag{3.107}$$

The above transfer function has a pole of order M at the origin and M finite zeros. Finite duration impulse response (FIR) systems are characterized by linear constant coefficient difference equations of a non-recursive nature. Hence, the transfer function obtained as in Eq. (3.107) is called an FIR transfer function.

3.7.3   Poles and Zeros of a Rational Transfer Function

As mentioned earlier, the zeros of a system function $H(z)$ are the values of z for which $H(z) = 0$, while the poles are the values of z for which $H(z) = \infty$. Since $H(z)$ is a rational transfer function, the number of finite zeros and the number of finite poles are equal to the degrees of the numerator and denominator polynomials, respectively.

In MATLAB, tf2zp command can be used to find the zeros, poles, and gains of a rational transfer function. z-plane command can be used for plotting pole-zero plot of a rational transfer function.

Example 3.31 Determine the pole-zero plot using MATLAB for the system described by the system function

Fig. 3.6 Pole-zero plot of
Example 3.31

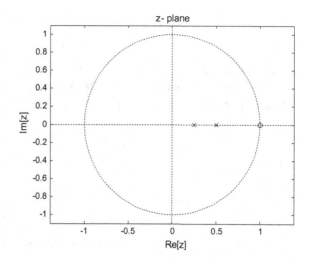

$$H(z) = \frac{Y(z)}{X(z)} = \frac{z-1}{8z^2 - 6z + 1}.$$

Solution The coefficients of the numerator and denominator polynomial can be written as

$$\text{numerator} = \begin{bmatrix} 0 & 1 & -1 \end{bmatrix};$$
$$\text{denominator} = \begin{bmatrix} 8 & -6 & 1 \end{bmatrix};$$

The following MATLAB statement yields the poles and zeros and gain of the system

$$[z, p, \text{gain}] = \text{tf2zp (numerator, denominator)}$$
$$\text{zeros, } z = 1$$
$$\text{poles, } p = \begin{bmatrix} 0.500 & 0.250 \end{bmatrix} \text{ and gain} = 0.1250$$

The MATLAB command z-plane (z, p) plots the poles and zeros as shown in Fig. 3.6.

3.7.4 Frequency Response from Poles and Zeros

By factorizing the numerator and denominator polynomials of Eq. (3.103), the transfer function can be written in pole-zero form as

$$H(z) = b_0 z^{(N-M)} \frac{\prod_{i=1}^{M}(z - z_i)}{\prod_{i=1}^{N}(z - p_i)} \tag{3.108}$$

where z_i and p_i are the zeros and poles of $H(z)$. It should be noted that the zeros are either real or occur in conjugate pairs, The frequency response of the system can be obtained by letting $z = e^{j\omega}$ in the transfer function $H(z)$, that is,

$$H(e^{j\omega}) = H(z)|_{z=e^{j\omega}}$$

Hence,

$$H(e^{j\omega}) = b_0 e^{j\omega(N-M)} \frac{\prod_{i=1}^{M}(e^{j\omega} - z_i)}{\prod_{i=1}^{N}(e^{j\omega} - p_i)} \tag{3.109}$$

The contribution of the zeros and poles to the system frequency response can be visualized from the above expression.

The magnitude of the frequency response can be expressed by

$$\left| H(e^{j\omega}) \right| = |b_0| |e^{j\omega}|^{(N-M)} \frac{\prod_{i=1}^{M}\left| (e^{j\omega} - z_i) \right|}{\prod_{i=1}^{N}\left| (e^{j\omega} - p_i) \right|} \tag{3.110}$$

The zeros contribute to pulling down the magnitude of the frequency response, whereas the poles contribute to pushing up the magnitude of the frequency response. The size of decrease or increase in the magnitude response depends on how far the zero or the pole is from the unit circle. A peak in $\left| H(e^{j\omega}) \right|$ appears at the frequency of a pole very close to the unit circle.

To illustrate this, consider the following example.

Example 3.32 Consider a system with the transfer function

$$H(z) = \frac{0.1(z^2 + 2z + 1)}{1.2z^2 + 1} \tag{3.111}$$

The numerator and denominator polynomials coefficients in descending powers of z can be written as

$$\text{num} = [1 \quad 2 \quad 1];$$
$$\text{den} = [1 \quad 0 \quad 1];$$

Then, as used in Example 3.31, using the MATLAB commands tf2zp and z-plane, the pole-zero plot can be obtained as shown in Fig. 3.7a. The magnitude and phase responses of the above system transfer function are obtained using the above num and den vectors using the MATLAB command freqz. The magnitude and phase responses are shown in Fig. 3.7b and c, respectively.

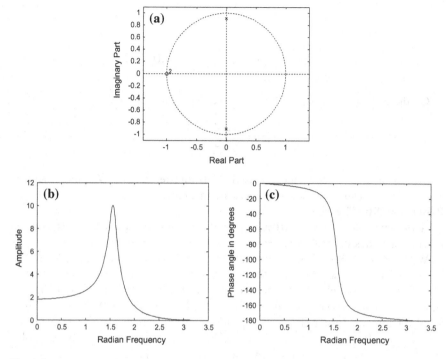

Fig. 3.7 **a** Pole-zero plot. **b** Magnitude response. **c** Phase response

Figure 3.7a indicates that the system has zeros of order 2 at $z = -1$ and two poles on the imaginary axis close to the unit circle. In the magnitude response of Fig. 3.7b, a peak occurs at $\omega = \pi/2$. This can be attributed to the fact that the frequency of the poles is $\pi/2$. The magnitude response is small at high frequencies due to the zeros.

3.7.5 Stability and Causality

The stability of a LTI system can be expressed in terms of the transfer function or the impulse response of the system. It is known from Sect. 2.5.5 that a necessary and sufficient condition for a LTI system to be bounded-input bounded-output (BIBO) stable is that its impulse response be absolutely summable, i.e.,

$$\sum_{n=-\infty}^{\infty} |h(n)| < \infty \tag{3.112}$$

$$H(z) = \sum_{n=-\infty}^{\infty} h(n)z^{-n} \qquad (3.113)$$

$$|H(z)| \le \sum_{n=-\infty}^{\infty} |h(n)z^{-n}| = \sum_{n=-\infty}^{\infty} |h(n)||z^{-n}| \qquad (3.114)$$

On the unit circle (i.e., $|z| = 1$), the above expression becomes

$$|H(z)| \le \sum_{n=-\infty}^{\infty} |h(n)| \qquad (3.115)$$

Therefore, for a stable system, the ROC of its transfer function $H(z)$ must include the unit circle. Thus, we have the following theorem.

BIBO Stability Theorem

A discrete LTI system is BIBO stable if and only if the ROC of its system function includes the unit circle, $|z| = 1$.

We know from Sect. 2.5.5 that for a discrete LTI system to be causal $h(n) = 0$ for $n < 0$. Thus, the sequence should be right-sided. We also know from Sect. 3.2 that the ROC of a right-sided sequence is the exterior of a circle whose radius is equal to the magnitude of the pole that is farthest from the origin. At the same time, we also know that for a right-sided sequence the ROC may or may not include the point $z = \infty$. But we know from Sect. 3.2 that a causal system cannot have a pole at infinity. Thus, in a causal system, the ROC should include the point $z = \infty$. Thus, we may summarize the result for causality by the following theorem:

Causality Theorem

A discrete LTI system is causal if and only if the ROC of its system function is the exterior of a circle including $z = \infty$. An alternate way of stating this result is that a system is causal if and only if its ROC contains no poles, finite, or infinite.

Thus, the conditions for stability and causality are quite different. A causal system could be stable or unstable, just as a non-causal system could be stable or unstable. Also, a stable system could be causal or non-causal just as an unstable system could be causal or non-causal. However, we can conclude from the above two theorems that a causal stable system must have a system function whose ROC is $|z| = r$, where $r < 1$. Hence, we can summarize this result as follows.

Condition for a System to be both Causal and Stable

A causal LTI system is BIBO stable if and only if all its poles are within the unit circle.

As a consequence, for a LTI system with a system function $H(z)$ to be stable and causal, it is necessary that the degree of the numerator polynomial in z not exceed that of the denominator polynomial. As such, an FIR system is always stable, whereas if an IIR system is not designed properly, it may be unstable.

Example 3.33 Given the system function

$$H(z) = \frac{z(4z - 3)}{(z - \frac{1}{3})(z - 4)}$$

Find the various regions of convergence for $H(z)$, and state whether the system is stable and/or causal in each of these regions. Also, find the impulse response $h(n)$ in each case.

Solution The system function can be expressed in partial fraction in the form

$$H(z) = \frac{z}{(z - \frac{1}{3})} + \frac{3z}{(z - 4)} = \frac{1}{(1 - \frac{1}{3}z^{-1})} + 3\frac{1}{1 - 4z^{-1}}$$

The system function has two zeros, viz. $z = 0, \frac{3}{4}$ and two poles at $z = \frac{1}{3}, 4$. Hence, there are three regions of convergence: (i) $|z| < \frac{1}{3}$, (ii) $\frac{1}{3} < |z| < 4$, and (iii) $|z| > 4$. Let us consider each of these regions separately.

(i) $|z| < \frac{1}{3}$

In this region, there are no poles including the origin, but has poles exterior to it. Hence, the system is non-causal. Also, it is an unstable system, since the ROC does not include the unit circle. By using Table 3.2, we get

$$h(n) = -\left[\left(\frac{1}{3}\right)^n + 3(4)^n\right]u(-n - 1)$$

(ii) $\frac{1}{3} < |z| < 4$

This region includes the unit circle and hence the system is stable. However, since the pole $|z| = 4$ is exterior to this region, it is non-causal, and the corresponding sequence is two-sided. Again by using Table 3.2, we have

$$h(n) = \left(\frac{1}{3}\right)^n u(n) - 3(4)^n u(-n - 1)$$

(iii) $|z| > 4$

This region does not include the unit circle and hence the system is unstable. However, in this region there are no poles, finite, or infinite, and hence, the system is causal. The impulse response of the system is obtained from $H(z)$ using Table 3.2 as

$$h(n) = \left(\frac{1}{3}\right)^n u(n) + 3(4)^n u(n).$$

Example 3.34 The rotational motion of a satellite was described by the difference equation

$$y(n) = y(n-1) - 0.5y(n-2) + 0.5x(n) + 0.5x(n-1)$$

Is the system stable? Is the system causal? Justify your answer.

Solution Taking the z-transform on both sides of the given difference equation, we get

$$Y(z) = z^{-1}Y(z) - 0.5z^{-2}Y(z) + 0.5X(z) + 0.5z^{-1}X(z)$$

$$H(z) = \frac{Y(z)}{X(z)} = \frac{0.5(1+z^{-1})}{1 - z^{-1} + 0.5z^{-2}} = \frac{0.5(z+1)z}{(z^2 - z + 0.5)}$$

The poles of the system are at $z = 0.5 \pm 0.5j$ as shown in Fig. 3.8.

All poles of the system are inside the unit circle. Hence, the system is stable. It is causal since the output only depends on the present and past inputs.

Example 3.35 Consider the difference equation

$$y(n) - \frac{7}{3}y(n-1) + \frac{2}{3}y(n-2) = x(n)$$

(a) Determine the possible choices for the impulse response of the system. Each choice should satisfy the difference equation. Specifically indicate which choice corresponds to a stable system and which choice corresponds to a causal system.

(b) Can you find a choice which implies that the system is both stable and causal? If not, justify your answer.

Fig. 3.8 Poles of Example 3.34

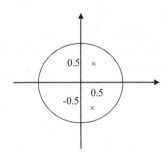

Solution (a) Taking the z-transform on both sides and using the shifting theorem, we get

$$\left(1 - \frac{7}{3}z^{-1} + \frac{2}{3}z^{-2}\right) Y(z) = X(z)$$

$$\frac{Y(z)}{X(z)} = \frac{1}{1 - \frac{7}{3}z^{-1} + \frac{2}{3}z^{-2}}$$

$$H(z) = \frac{z^2}{(z-2)\left(z - \frac{1}{3}\right)}$$

The system function $H(z)$ has a zero of order 2 at $z = 0$ and two poles at $z = 1/3, 2$. Hence, there are three regions of convergence and thus, there are three possible choices for the impulse response of the system. The regions are:

(i) $R_1 : |z| < \frac{1}{3}$, (ii) $R_2 : \frac{1}{3} < |z| < 2$, and (iii) $R_3 : |z| > 2$.

The region R_1 is devoid of any poles including the origin, and hence corresponds to an anti-causal system, which is not stable since it does not include the unit circle. Region R_2 does include the unit circle and hence corresponds to a stable system; however, it is not causal in view of the presence of the pole $z = 2$. Finally, the region R_3 does not have any poles including at infinity and hence corresponds to a causal system; however, since R_3 does not include the unit circle, the system is not stable.

(b) There is no ROC that would imply that the system is both stable and causal. Therefore, there is no choice for $h(n)$ which make the system both stable and causal.

Example 3.36 A system is described by the difference equation

$$y(n) + y(n-1) = x(n), y(n) = 0, \text{ for } n < 0.$$

(i) Determine the transfer function and discuss the stability of the system.
(ii) Determine the impulse response $h(n)$ and show that it behaves according to the conclusion drawn from (i)
(iii) Determine the response when $x(n) = 10$ for $n \geq 0$. Assume that the system is initially relaxed.

Solution (i) Taking the z-transforms on both sides of the given equation, we get

$$Y(z) + Y(z)z^{-1} = X(z)$$

Hence,

$$H(z) = \frac{Y(z)}{X(z)} = \frac{z}{z+1}$$

The pole is at $z = -1$, that is, on the unit circle. So the system is marginally stable or oscillatory

(ii) Since $h(n) = 0$ for $n < 0$,

$$h(n) = Z^{-1}\left[\frac{z}{z+1}\right] = (-1)^n u(n)$$

This impulse response confirms that the impulse response is oscillatory

(iii) Since

$$x(n) = 10 \quad \text{for } n \geq 0,$$

$$X(z) = \frac{10z}{z-1}$$

Thus,

$$Y(z) = H(z)X(z) = \frac{z}{z+1}\frac{10}{z-1}$$

or

$$\frac{Y(z)}{z} = \frac{5}{z+1} + \frac{5}{z-1}$$

Therefore,

$$y(n) = Z^{-1}[Y(z)] = [5(-1)^n + 5]u(n).$$

3.7.6 Minimum-Phase, Maximum-Phase, and Mixed-Phase Systems

A causal stable transfer function with all its poles and zeros inside the unit circle is called a minimum-phase transfer function. A causal stable transfer function with all its poles inside the unit circle and all the zeros outside the unit circle is called a maximum-phase transfer function. A causal stable transfer function with all its poles inside the unit circle and with zeros inside and outside the unit circle is called a mixed-phase transfer function. For example, consider the systems with the following transfer functions

$$H_1(z) = \frac{Y(z)}{X(z)} = \frac{z + 0.4}{z + 0.3} \tag{3.116}$$

$$H_2(z) = \frac{Y(z)}{X(z)} = \frac{0.4z + 1}{z + 0.5} \tag{3.117}$$

$$H_3(z) = \frac{Y(z)}{X(z)} = \frac{(0.4z + 1)(z + 0.4)}{(z + 0.5)(z + 0.3)} \tag{3.118}$$

The pole-zero plot of the above transfer functions are shown in Fig. 3.9a, b, and c, respectively. The transfer function $H_1(z)$ has a zero at $z = -0.4$ and a pole at $z = -0.3$ and they are both inside the unit circle. Hence, $H_1(z)$ is a minimum-phase function. The transfer function $H_2(z)$ has a pole inside the unit circle, at $z = -0.5$ and a zero at $z = -2.5$, outside the unit circle. Thus, $H_2(z)$ is a maximum-phase function. The transfer function $H_3(z)$ has two poles one at $z = -0.3$ and the other at $z = -0.5$, and two zeros one at $z = -0.4$, inside the unit circle and the other at $z = -2.5$, outside the unit circle. Hence, $H_3(z)$ is a mixed-phase function.

3.7.7 Inverse System

Let $H(z)$ be the system function of a linear time-invariant system. Then, its inverse system function $H_I(z)$ is defined, if and only if the overall system function is unity when $H(z)$ and $H_I(z)$ are connected in cascade, that is $H(z) H_I(z) = 1$, implying

$$H_I(z) = \frac{1}{H(z)} \tag{3.119}$$

In the time domain, this is equivalently expressed as

$$h_I(n) * h(n) = \delta(n) \tag{3.120}$$

If $H(z)$ is a rational transfer function represented by

$$H(z) = \frac{N(z)}{D(z)} \tag{3.121}$$

then the inverse transfer function

$$H_I(z) = \frac{D(z)}{N(z)} \tag{3.122}$$

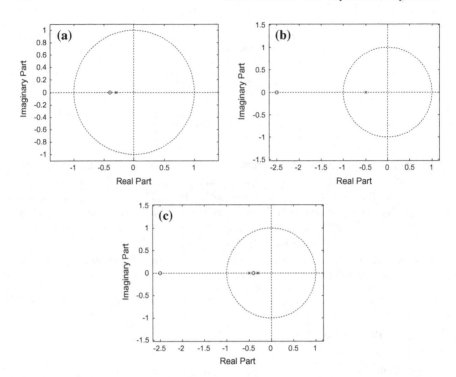

Fig. 3.9 Pole-zero plot of **a** a minimum-phase function, **b** a maximum-phase function, and **c** a mixed-phase function

For $H_I(z)$ to be stable and causal, all its poles must be inside the unit circle, i.e., all the zeros of $H(z)$ must lie inside the unit circle.

If $H(z)$ is an FIR system with all its zeros inside the unit circle, then $H_I(z)$ becomes an all-pole system with all its poles lying inside the unit circle. If $H(z)$ is an all-pole system, then $H_I(z)$ becomes a FIR system. Hence, $H(z)$ must be a minimum-phase system for the existence of its inverse system $H_I(z)$.

Example 3.37 A system is described by the following difference equation

$$y(n) = x(n) - e^{-8\alpha}x(n-8)$$

where the constant $\alpha > 0$. Find the corresponding inverse system function to recover $x(n)$ from $y(n)$. Check for the stability and causality of the resulting recovery system, justifying your answer.

Solution

$$Y(z) = X(z) - e^{-8\alpha}z^{-8}X(z); \quad \frac{Y(z)}{X(z)} = \left(1 - e^{-8\alpha}z^{-8}\right)$$

The corresponding inverse system

$$H_1(z) = \frac{1}{\left(1 - e^{-8\alpha}z^{-8}\right)} = \frac{X(z)}{Y(z)}$$

The recovery system is both stable and causal, since all the poles of the system $H_I(z)$ are inside the unit circle.

3.7.8   Allpass System

Consider a causal stable Nth-order transfer function of the form

$$H(z) = \pm \frac{a_N + a_{N-1}z^{-1} + \cdots + z^{-N}}{1 + a_1 z^{-1} + \cdots + a_N z^{-N}} = \pm \frac{M(z)}{D(z)} \tag{3.123}$$

Now,

$$\begin{aligned} D(z^{-1}) &= 1 + a_1 z + a_2 z^2 + \cdots + a_N z^N \\ &= z^N \left[a_N + a_{N-1}z^{-1} + \cdots + z^{-N}\right] \\ &= z^N M(z) \end{aligned} \tag{3.124}$$

or

$$M(z) = z^{-N}D(z^{-1}) \tag{3.125}$$

Hence,

$$H(z) = \pm z^{-N}\left[\frac{D(z^{-1})}{D(z)}\right] \tag{3.126}$$

and

$$H(z^{-1}) = \pm z^N \left[\frac{D(z)}{D(z^{-1})}\right] \tag{3.127}$$

Therefore,

$$H(z)H(z^{-1}) = 1 \qquad (3.128)$$

Thus,

$$|H(\omega)|^2 = H(e^{j\omega})H(e^{-j\omega}) = 1 \quad \text{for all values of } \omega. \qquad (3.129)$$

In other words, $H(z)$ given by (3.123) passes all the frequencies contained in the input signal to the system, and hence such a transfer function is an all pass transfer function, and the corresponding system is an allpass system. It is also seen from (3.123) that if $z = p_i$ is a zero of $D(z)$, then $z = (1/p_i)$ is a zero of $M(z)$. That is, the poles and zeros of an allpass function are reciprocals of one another. Since all the poles of $H(z)$ are located within the unit circle, all the zeros are located outside the unit circle.

If $x(n)$ is the input sequence and $y(n)$ the output sequence for an allpass system, then

$$Y(z) = H(z)X(z). \qquad (3.130)$$

Thus,

$$Y(e^{j\omega}) = H(e^{j\omega})X(e^{j\omega}). \qquad (3.131)$$

Since $|H(e^{j\omega})| = 1$, we get

$$|Y(e^{j\omega})| = |X(e^{j\omega})| \qquad (3.132)$$

We know from Parseval's relation that the output energy of a LTI system is given by

$$\sum_{n=-\infty}^{\infty} |y(n)|^2 = \frac{1}{2\pi} \int_{-\pi}^{\pi} |Y(e^{j\omega})|^2 d\omega \qquad (3.133)$$

$$= \frac{1}{2\pi} \int_{-\pi}^{\pi} |X(e^{j\omega})|^2 d\omega \qquad (3.134)$$

Hence,

$$\sum_{n=-\infty}^{\infty} |y(n)|^2 = \sum_{n=-\infty}^{\infty} |x(n)|^2 \qquad (3.135)$$

Thus, the output energy is equal to the input energy for an allpass system. Hence, an allpass system is a lossless system.

Fig. 3.10 Pole-zero plot of a second-order allpass system

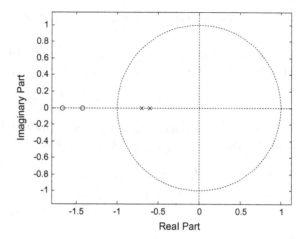

For example, consider a second-order allpass system

$$H_{ap}(z) = \frac{0.42 + 1.3z^{-1} + z^{-2}}{1 + 1.3z^{-1} + 0.42z^{-2}} \tag{3.136}$$

The pole-zero plot of the second-order allpass system is shown in Fig. 3.10. It can be seen from this figure that the poles and zeros occur in reciprocal pairs.

Example 3.38 A causal LTI system is described by the following difference equation

$$y(n) - \frac{1}{4}y(n-2) = x(n-2) - \frac{1}{4}x(n)$$

Determine whether the system is an allpass system.

Solution Taking the z-transform of both sides

$$Y(z) = \frac{1}{4}z^{-2}Y(z) + z^{-2}X(z) - \frac{1}{4}X(z)$$

$$H(z) = \frac{Y(z)}{X(z)} = \frac{z^{-2} - (1/4)}{1 - (1/4)z^{-2}}$$

Since the poles and zeros occur in conjugate reciprocal pairs, the system is all pass. Hence, $|H(e^{j\omega})| = 1$, that is, the magnitude of the frequency response of the system is unity.

3.7.9   *Allpass and Minimum-Phase Decomposition*

Consider an Nth-order mixed-phase system function $H(z)$ with m zeros outside the unit circle and $(n - m)$ zeros inside the unit circle. Then, $H(z)$ can be expressed as

$$H(z) = H_1(z)(z^{-1} - a_1^*)(z^{-1} - a_2^*) \ldots (z^{-1} - a_m^*) \qquad (3.137)$$

where $H_1(z)$ is a minimum-phase function as its N poles and $(n - m)$ zeros are inside the unit circle. Equation (3.137) can be equivalently expressed as

$$
\begin{aligned}
H(z) &= H_1(z)(1 - z^{-1}a_1)(1 - z^{-1}a_2) \ldots \\
&\quad (1 - z^{-1}a_m) \frac{(z^{-1} - a_1^*)(z^{-1} - a_2^*) \ldots (z^{-1} - a_m^*)}{(1 - z^{-1}a_1)(1 - z^{-1}a_2) \ldots (1 - z^{-1}a_m)}
\end{aligned}
\qquad (3.138)
$$

In the above equation, the factor $H_1(z)(1 - z^{-1}a_1)(1 - z^{-1}a_2) \ldots (1 - z^{-1}a_m)$ is also a minimum-phase function, since $|a_1|, |a_2|, \ldots, |a_m|$ are less than 1 and the zeros are inside the unit circle. Hence, the factor $\frac{(z^{-1} - a_1^*)(z^{-1} - a_2^*) \ldots (z^{-1} - a_m^*)}{(1 - z^{-1}a_1)(1 - z^{-1}a_2) \ldots (1 - z^{-1}a_m)}$ is allpass. Thus, any transfer function $H(z)$ can be written as

$$H(z) = H_{\min}(z)H_{\text{ap}}(z) \qquad (3.139)$$

$H_{\min}(z)$ has all the poles and zeros of $H(z)$ that are inside unit circle in addition to the zeros that are conjugate reciprocals of the zeros of $H(z)$ that are outside the unit circle, while $H_{\text{ap}}(z)$ is an allpass function that has all the zeros of $H(z)$ that lie outside the unit circle along with poles to cancel the conjugate reciprocals of the zeros of $H(z)$ that lie outside the unit circle, which are now contained as zeros in $H_{\min}(z)$.

Example 3.39 In the system shown in Fig. 3.11, if S_1 is a causal LTI system with system function

$$H(z) = \left(1 - \frac{1}{2}z^{-1}\right)\left(1 - \frac{3}{4}z^{-1}\right)(1 - 3z^{-1})$$

determine the system function for a system S_2 so that the overall system is an allpass system.

Solution $H(z) = \left(1 - \frac{1}{2}z^{-1}\right)\left(1 - \frac{3}{4}z^{-1}\right)(1 - 3z^{-1})$.

Decompose $H(z)$ as

$$H(z) = H_{\min}(z)H_{\text{ap}}(z)$$

$$H_{\text{ap}}(z) = \frac{z^{-1} - \left(\frac{1}{3}\right)}{1 - \left(\frac{1}{3}\right)z^{-1}}$$

Fig. 3.11 Cascade connection of two systems S_1 and S_2

Thus,

$$H(z) = -3\left(1 - \left(\frac{1}{3}\right)z^{-1}\right)\left(1 - \frac{1}{2}z^{-1}\right)\left(1 - \frac{3}{4}z^{-1}\right)\frac{z^{-1} - \left(\frac{1}{3}\right)}{1 - \left(\frac{1}{3}\right)z^{-1}}$$

Let the system function of S_2 be $H_C(z)$. For the overall system functions to be allpass, we should have

$$H(z)H_C(z) = H_{ap}(z)$$

$$H(z)H_C(z) = H_{\min}(z)H_{ap}(z)\frac{1}{H_{\min}(z)} = H_{ap}(z)$$

Hence,

$$H_C(z) = \frac{1}{H_{\min}(z)} = \frac{1}{(-3)\left(1 - \left(\frac{1}{3}\right)z^{-1}\right)\left(1 - \frac{1}{2}z^{-1}\right)\left(1 - \frac{3}{4}z^{-1}\right)}.$$

Example 3.40 A signal $x(n)$ is transmitted across a distorting digital channel characterized by the following system function

$$H_d(z) = \frac{\left(1 - 0.5z^{-1}\right)\left(1 - 1.25e^{j0.8\pi}z^{-1}\right)\left(1 - 1.25e^{-j0.8\pi}z^{-1}\right)}{\left(1 - 0.81z^{-2}\right)}$$

consider the compensating system shown in Fig. 3.12. Find $H_{1C}(z)$ such that the overall system function $G_1(z)$ is an allpass system.

Solution $H_d(z) = H_{d\min 1}(z)H_{ap}(z)$

$$H_{d\min 1}(z) = \frac{\left(1 - 0.5z^{-1}\right)}{\left(1 - 0.8z^{-2}\right)}(1.25)^2\left(1 - 0.8e^{j0.8\pi}z^{-1}\right)\left(1 - 0.8e^{-j0.8\pi}z^{-1}\right)$$

$$H_{ap}(z) = \frac{\left(z^{-1} - 0.8e^{-j0.8\pi}\right)\left(z^{-1} - 0.8e^{j0.8\pi}\right)}{\left(1 - 0.8e^{-j0.8\pi}z^{-1}\right)\left(1 - 0.8e^{j0.8\pi}z^{-1}\right)}$$

$$H_{1C}(z) = \frac{1}{H_{d\min 1}(z)} = \frac{\left(1 - 0.81z^{-2}\right)}{(1.25)^2\left(1 - 0.5z^{-1}\right)\left(1 - 0.8e^{-j0.8\pi}z^{-1}\right)\left(1 - 0.8e^{j0.8\pi}z^{-1}\right)}$$

Fig. 3.12 Compensating
system

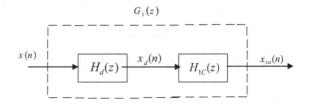

Then

$$G_1(z) = H_d(z)H_{1C}(z) = H_{\text{ap}}(z)$$

is an allpass system.

3.8   One-Sided z-Transform

The unilateral or one-sided z-transform, which is appropriate for problems
involving causal signals and systems, is evaluated using the portion of a signal
associated with nonnegative values of time index $(n \geq 0)$. It gives considerable
meaning to assume causality in many applications of the z-transforms.

Definition The one-sided z-transform of a signal $x[n]$ is defined as

$$Z_+[x(n)] = X^+(z) = \sum_{n=0}^{\infty} x(n)z^{-n} \tag{3.140}$$

which depends only on $x(n)$ for $(n \geq 0)$. It should be mentioned that the two-sided
z-transform is not useful in the evaluation of the output of a non-relaxed system.
The one-sided transform can be used to solve for systems with nonzero initial
conditions or for solving difference equations with nonzero initial conditions. The
following special properties of $X^+(z)$ should be noted.

1. The one-sided transform $X^+(z)$ of $x(n)$ is identical to the two-sided transform
   $X(z)$ of the sequence $x(n)u(n)$. Also, since $x(n)u(n)$ is always causal, its ROC
   and hence that of $X^+(z)$ is always the exterior of a circle. Hence, it is not
   necessary to indicate the ROC of a one-sided z-transform.
2. $X^+(z)$ is unique for a causal signal, since such a signal is zero for $n < 0$.
3. Almost all the properties of the two-sided transform are applicable to the
   one-sided transform, one major exception being the shifting property.

Shifting Theorem for $X^+(z)$ When the Sequence is Delayed by k
If

$$Z_+[x(n)] = X^+(z),$$

then

$$Z_+[x(n-k)] = z^{-k}\left[X^+(z) + \sum_{n=1}^{k} x(-n)z^n\right], \quad k > 0 \qquad (3.141)$$

However, if $x(n)$ is a causal sequence, then the result is the same as in the case of the two-sided transform and

$$Z_+[x(n-k)] = z^{-k}X^+(z) \qquad (3.142)$$

Proof By definition

$$Z_+[x(n-k)] = \sum_{n=0}^{\infty} x(n-k)z^{-n}$$

Letting $(n-k) = m$, the above equation may be written as

$$Z_+[x(n-k)] = z^{-k}\left[\sum_{m=0}^{\infty} x(m)z^{-m} + \sum_{m=-k}^{-1} x(m)z^{-m}\right]$$

$$= z^{-k}\left[X^+(z) + \sum_{n=1}^{k} x(-n)z^n\right]$$

which proves (3.141). If the sequence $x(n)$ is causal, then the second term on the right side of the above equation is zero, and hence we get the result (3.142).
Shifting Theorem for $X^+(z)$ When the Sequence is Advanced by k
If

$$Z_+[x(n)] = X^+(z),$$

then

$$Z_+[x(n+k)] = z^k\left[X^+(z) - \sum_{n=0}^{k-1} x(n)z^{-n}\right], \quad k > 0 \qquad (3.143)$$

Proof By definition

$$Z_+[x(n+k)] = \sum_{n=0}^{\infty} x(n+k)z^{-n}$$

Letting $(n + k) = m$, the above equation may be written as

$$Z_+ [x(n+k)] = z^k \left[\sum_{m=k}^{\infty} x(m)z^{-m} \right]$$

$$= z^k \left[\sum_{m=0}^{\infty} x(m)z^{-m} - \sum_{m=0}^{k-1} x(m)z^{-m} \right]$$

$$= z^k \left[X^+ (z) - \sum_{n=0}^{k-1} x(n)z^{-n} \right]$$

thus establishing the result (3.143).

Final Value Theorem

If a sequence $x(n)$ is causal, i.e., $x(n) = 0$ for $n < 0$, then

$$\lim_{n \to \infty} x(n) = \lim_{z \to 1} (z - 1)X(z) \qquad (3.144)$$

The above limit exists only if the ROC of $(z - 1)X(z)$ exists.

Proof Since the sequence $x(n)$ is causal, we can write its z-transform as follows:

$$Z[x(n)] = \sum_{n=0}^{\infty} x(n)z^{-n} = x(0) + x(1)z^{-1} + x(2)z^{-2} + \cdots. \qquad (3.145)$$

Also,

$$Z[x(n+1)] = \sum_{n=0}^{\infty} x(n+1)z^{-n} = x(1) + x(2)z^{-1} + x(3)z^{-2}.\ldots \qquad (3.146)$$

Hence, we see that

$$Z[x(n+1)] = z[Z[x(n)] - x(0)] \qquad (3.147)$$

Thus,

$$Z[x(n+1)] - Z[x(n)] = (z - 1)Z[x(n)] - zx(0)$$

Substituting (3.144) and (3.145) for the LHS, we have

$$[x(1) - x(0)] + [x(2) - x(1)]z + [x(3) - x(2)]z^2 + \cdots = (z - 1)X(z) - zx(0)$$

Taking the limit as $z \to 1$, we get

$$[x(1) - x(0)] + [x(2) - x(1)] + [x(3) - x(2)] + \cdots = \lim_{z \to 1}(z - 1)X(z) - x(0)$$

Thus

$$-x(0) + x(\infty) = \lim_{z \to 1}(z - 1)X(z) - x(0)$$

or

$$x(\infty) = \lim_{z \to 1}(z - 1)X(z)$$

Hence,

$$\lim_{n \to \infty} x(n) = \lim_{z \to 1}(z - 1)X(z)$$

It should be noted that the limit exists only if the function $(z - 1)X(z)$ has an ROC that includes the unit circle; otherwise, system would not be stable and the $\lim_{n \to \infty} x(n)$ would not be finite.

Example 3.41 Find the final value of $x(n)$ if its z-transform $X(z)$ is given by

$$X(z) = \frac{0.5z^2}{(z - 1)(z^2 - 0.85z + 0.35)}.$$

Solution The final value or steady value of $x(n)$ is given by

$$x(n) = \lim_{z \to 1}(z - 1)X(z) = \frac{0.5}{(1 - 0.85 + 0.35)} = 1$$

The result can be directly verified by taking the inverse transform of the given $X(z)$.

3.8.1 Solution of Difference Equations with Initial Conditions

The one-sided z-transform is very useful in obtaining solutions for difference equations which have initial conditions. The procedure is illustrated with an example.

Example 3.42 Find the step response of the system

$$y(n) - \left(\frac{1}{2}\right)y(n-1) = x(n)$$

with the initial condition $y(-1) = 1$.

Solution Taking one-sided z-transforms on both sides of the given equation and using (3.141), we have

$$Y^+(z) - \left(\frac{1}{2}\right)\left[z^{-1}Y^+(z) + y(-1)\right] = X^+(z)$$

Substituting for $X^+(z)$ and $y(-1)$, we have

$$\left[1 - \left(\frac{1}{2}\right)z^{-1}\right]Y^+(z) = \frac{1}{2} + \frac{1}{1-z^{-1}}$$

Hence,

$$Y^+(z) = \frac{1}{2}\frac{1}{\left(1-\frac{1}{2}z^{-1}\right)} + \frac{1}{\left(1-\frac{1}{2}z^{-1}\right)\left(1-z^{-1}\right)}$$

$$= \frac{2}{1-z^{-1}} - \frac{1}{2}\frac{1}{\left(1-\frac{1}{2}z^{-1}\right)}$$

Taking the inverse transform, we get

$$Z_+^{-1}[Y(z)] = y(n) = \left[2 - \left(\frac{1}{2}\right)^{n+1}\right]u(n).$$

3.9　Problems

1. Find the z-transform and the ROC of the causal sequence $x(n) = \{2, 0, 1, -3, 2\}$
2. Find the z-transform and the ROC of the anti-causal sequence $x(n) = \{-2, -1, 0, 1, 2, 3\}$
3. Find the z-transform of the signal $x(n) = [3(3)^n - 4(2)^n]$
4. Find the z-transform of the sequence $x(n) = (1/3)^{n-1}-u(n-1)$.
5. Find the z-transform of the sequence

$$x(n) = \begin{cases} 1, & 0 \le n \le N-1 \\ 0, & \text{otherwise} \end{cases}$$

6. Find the z-transform of the following discrete-time signals, and find the ROC for each.

   (i) $x(n) = \left(-\frac{1}{2}\right)^n u(n) + 3\left(\frac{1}{4}\right)^{-n} u(-n-1)$

   (ii) $x(n) = \left(\frac{1}{4}\right)\delta(n) + \delta(n-2) - \left(\frac{1}{3}\right)\delta(n-3)$

   (iii) $x(n) = (n+0.5)\left(\frac{1}{2}\right)^n u(n-1) - \left(\frac{1}{3}\right)\delta(n-3)$

7. Find the z-transform of the sequence $x(n) = na^{n-1}u(n-1)$

8. Find the z-transform of the sequence $x(n) = (1/4)^{n+1}u(n)$.

9. Find the z-transform of the signal $x(n) = \left[(4)^{n+1} - 3(2)^{n-1}\right]$

10. Determine the z-transform and the ROC for the following time signals. Sketch the ROC, poles and zeros in the z-plane.

    (i) $x(n) = \sin\left(\frac{3\pi}{4}n - \frac{\pi}{8}\right)u[n-1]$

    (ii) $x(n) = (n+1)\sin\left(\frac{3\pi}{2}n + \frac{\pi}{4}\right)u[n+2]$.

11. Find the inverse the z-transform of the following, using partial fraction expansions:

    (i) $X(z) = \frac{z+0.5}{(z+0.2)(z-2)}$, $|z| > 2$

    (ii) $X(z) = \frac{1+z^{-1}}{1+3z^{-1}+2z^{-2}}$, $|z| > 2$

    (iii) $X(z) = \frac{z^2+z}{(z-3)(z-2)}$, $|z| > 3$

    (iv) $X(z) = \frac{z(z+1)}{\left(z-\frac{1}{2}\right)\left(z-\frac{1}{3}\right)}$, $|z| > \frac{1}{2}$

12. Find the inverse z-transform of the following using the partial fraction expansion.

    (i) $X(z) = \frac{z}{(z-1)(z-4)}$, $|z| < 1$

    (ii) $X(z) = \frac{z^2+2z-3}{(z-1)(z-3)(z-4)}$,   for (a) $|z| > 4$ and (b) $|z| < 1$

    (iii) $X(z) = \frac{z}{3z^2-4z+1}$, $|z| < \frac{1}{3}$

13. Determine all the possible signals that can have the following z-transform

$$X(z) = \frac{z^2}{z^2 - 0.8z + 0.15}$$

14. Find the stability of the system with the following transfer function

$$H(z) = \frac{z}{z^3 - 1.4z^2 + 0.65z - 0.1}$$

Fig. P3.1 Compensating
system

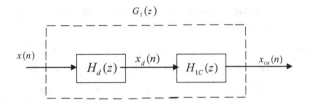

15. The transfer function of a system is given as

$$H(z) = \frac{z + 0.5}{(z + 0.4)(z - 2)}$$

Specify the ROC of $H(z)$ and determine $h(n)$ for the following conditions:

  (i)   The system is causal
  (ii) The system is stable
  (iii) Can the given system be both causal and stable?

16. A signal $x(n)$ is transmitted across a distorting digital channel characterized by
the following system function

$$H_d(z) = \frac{(z - 3)(z + 4)}{\left(z + \frac{1}{2}\right)\left(z - \frac{1}{3}\right)}$$

consider the compensating system shown in Fig. P3.1. Find $H_{1C}(z)$ such that
the overall system function $G_1(z)$ is an all pass system.

17. The transfer function of a system is given by

$$H(z) = \frac{1}{z^2 + 5z + 6}$$

Determine the response when $x(n) = u(n)$. Assume that the system is initially
relaxed.

18. Using the one-sided z-transform, solve the following difference equation

$$y(n) - \left(\frac{1}{9}\right)y(n - 2) = u(n), \ y(-1) = 0, \ y(-2) = 2$$

3.10   MATLAB Exercises

1. Write a MATLAB program using the command *residuez* to find the inverse of
the following by partial fraction expansion

$$X(z) = \frac{16 - 4z^{-1} + z^{-2}}{8 + 2z^{-1} - 2z^{-2}}$$

2. Write a MATLAB program using the command *impz* to find the inverse of the following by power series expansion

$$X(z) = \frac{15z^3}{15z^3 + 5z^2 - 3z - 1}$$

3. Write a MATLAB program using the command *z*-plane to obtain a pole-zero plot for the following system

$$H(z) = \frac{1 + \frac{1}{3}z^{-1} + \frac{5}{7}z^{-2} - \frac{3}{2}z^{-3}}{1 + \frac{5}{2}z^{-1} - \frac{1}{3}z^{-2} - \frac{3}{5}z^{-3}}$$

4. Write a MATLAB program using the command freqz to obtain magnitude and phase responses of the following system

$$H(z) = \frac{1 - 3.0538z^{-1} + 3.8281z^{-2} - 2.2921z^{-3} + 0.5507z^{-4}}{1 - 4z^{-1} + 6z^{-2} - 4z^{-3} + z^{-4}}$$

Reference

1. R.V. Churchill, J.W. Brown, in *Introduction to Complex Variables and Applications*, 5th edn. (McGraw-Hill, New York, NY, 1990)

Chapter 4
The Discrete Fourier Transform

The DTFT of a discrete-time signal is a continuous function of the frequency (ω), and hence, the relation between $X(e^{j\omega})$ and $x(n)$ is not a computationally convenient representation. However, it is possible to develop an alternative frequency representation called the discrete Fourier transform (DFT) for finite duration sequences. The DFT is a discrete-time sequence with equal spacing in frequency. We first obtain the discrete-time Fourier series (DTFS) expansion of a periodic sequence. Next, we define the DFT of a finite length sequence and consider its properties in detail. We also show that the DTFS represents the DFT of a finite length sequence. Further, evaluation of linear convolution using the DFT is discussed. Finally, some fast Fourier transform (FFT) algorithms for efficient computation of DFT are described.

4.1 The Discrete-Time Fourier Series

If a sequence $x(n)$ is periodic with period N, then $x(n) = x(n+N)$ for all n. In analogy with the Fourier series representation of a continuous periodic signal, we can look for a representation of $x(n)$ in terms of the harmonics corresponding to the fundamental frequency of $(2\pi/N)$. Hence, we may write $x(n)$ in the form

$$x(n) = \sum_k b_k e^{j2\pi kn/N} \qquad (4.1a)$$

It can easily be verified from Eq. (4.1a) that $x(n) = x(n + N)$. Also, we know that there are only N distinct values for $e^{j2\pi kn/N}$ corresponding to $k = 0, 1, \ldots, N - 1$, these being $1, e^{j2\pi n/N}, \ldots, e^{j2\pi k(N-1)/N}$. Hence, we may rewrite Eq. (4.1a) as

© Springer Nature Singapore Pte Ltd. 2018
K. D. Rao and M. N. S. Swamy, *Digital Signal Processing*,
https://doi.org/10.1007/978-981-10-8081-4_4

$$x(n) = \sum_{k=0}^{N-1} a_k e^{j2\pi kn/N} \tag{4.1b}$$

It should be noted that the summation could be taken over any N consecutive values of k. Equation (4.1b) is called the discrete-time Fourier series (DTFS) of the periodic sequence $x(n)$ and a_k as the Fourier coefficients. We will now obtain the expression for the Fourier coefficients a_k. It can easily be shown that $\{e^{j2\pi kn/N}\}$ is an orthogonal sequence satisfying the relation

$$\sum_{n=0}^{N-1} e^{j2\pi kn/N} e^{-j2\pi ln/N} = \begin{cases} 0 & k \neq l \\ N & k = l \end{cases} \quad (0 \leq k,\, l \leq (N-1) \tag{4.2}$$

Now, multiplying both sides of Eq. (4.1b) by $e^{-j2\pi ln/N}$ and summing over n between 0 and $(N-1)$, we get

$$\sum_{n=0}^{N-1} x(n) e^{-j2\pi ln/N} = \sum_{n=0}^{N-1} \sum_{k=0}^{N-1} a_k e^{j2\pi kn/N} e^{-j2\pi ln/N}$$

$$= \sum_{k=0}^{N-1} a_k \sum_{n=0}^{N-1} e^{j2\pi kn/N} e^{-j2\pi ln/N}$$

$$= a_l N, \text{ using (4.2)}.$$

Hence,

$$a_k = \frac{1}{N} \sum_{n=0}^{N-1} x(n) e^{-j2\pi kn/N}, \quad k = 0, 1, 2, \ldots, N-1 \tag{4.3}$$

It is common to associate the factor $(1/N)$ with $x(n)$ rather than a_k. This can be done by denoting $N a_k$ by $X(k)$; in such a case, we have

$$x(n) = \frac{1}{N} \sum_{k=0}^{N-1} X(k) e^{j2\pi kn/N} \tag{4.4}$$

where the Fourier coefficients $X(k)$ are given by

$$X(k) = \sum_{n=0}^{N-1} x(n) e^{-j2\pi kn/N}, \quad k = 0, 1, 2, \ldots, N-1 \tag{4.5}$$

It is easily seen that $X(k+N) = X(k)$ that is, the Fourier coefficient sequence $X(k)$, is also periodic of period N. Hence, the spectrum of a signal $x(n)$ that is periodic with period N is also a periodic sequence with the same period. It is also noted that since the Fourier series of a discrete periodic signal is a finite sequence, the series always converges and the Fourier series gives an exact alternate representation of the discrete sequence $x(n)$.

4.1.1   Periodic Convolution

In the case of two periodic sequences $x_1(n)$ and $x_2(n)$ having the same period N, linear convolution as defined by Eq. (2.38) does not converge. Hence, we define a different form of convolution for periodic signals by the relation

$$y(n) = \sum_{m=0}^{N-1} x_1(m)x_2(n-m) = \sum_{m=0}^{N-1} x_1(n-m)x_2(m) \qquad (4.6)$$

The above convolution is called *periodic convolution*. It may be observed that $y(n) = y(n+N)$, that is, the periodic convolution is itself periodic of period N.

Some important properties of the DTFS are given in Table 4.1. In this table, it is assumed that $x_1(n)$ and $x_2(n)$ are periodic sequences having the same period N. The proofs are omitted here, since they are similar to the ones that will be given in Sect. 4.3 for the corresponding properties of the DFT.

Example 4.1 Obtain the DTFS representation of the periodic sequence shown in Fig. 4.1.

Table 4.1 Some important properties of DTFS

Property	Periodic sequence	DTFS coefficients
Linearity	$ax_1(n) + bx_2(n)$ a and b are constants	$aX_1(k) + bX_2(k)$
Time shifting	$x(n-m)$	$e^{-j\left(\frac{2\pi}{N}\right)km}X(k)$
Frequency shifting	$e^{j\left(\frac{2\pi}{N}\right)ln}x(n)$	$X(k-l)$
Periodic convolution	$\sum_{m=0}^{N-1} x_1(m)x_2(n-m)$	$X_1(k)X_2(k)$
Multiplication	$x_1(n)x_2(n)$	$\frac{1}{N}\sum_{l=0}^{N-1} X_1(l)X_2(k-l)$
Symmetry properties	$x^*(n)$	$X^*(-k)$
	$x^*(-n)$	$X^*(k)$
	Re $\{x(n)\}$ jIm $\{x(n)\}$	$X_e(k) = \frac{1}{2}(X(k) + X^*(-k))$ $X_o(k) = \frac{1}{2j}(X(k) - X^*(k))$
	$x_e(n) = \frac{1}{2}[x(n) + x^*(-n)]$ $x_o(n) = \frac{1}{2}[x(n) - x^*(-n)]$	Re$\{X(k)\}$ jIm$\{X(k)\}$
	If $x(n)$ is real $x_e(n) = \frac{1}{2}[x(n) + x(-n)]$ $x_o(n) = \frac{1}{2}[x(n) - x(-n)]$	Re$\{X(k)\}$ jIm$\{X(k)\}$

Fig. 4.1 Periodic sequence with period $N = 5$

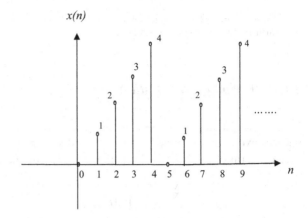

Solution The sequence is periodic with period $N = 5$. Using Eq. (4.5), the DTFS coefficients are computed as

$$X(0) = \sum_{n=0}^{N-1} x(n)e^0 = 0 + 1 + 2 + 3 + 4 = 10$$

$$X(1) = \sum_{n=0}^{4} x(n)e^{-j2\pi n/5} = 0 + e^{-j2\pi/5} + 2e^{-j4\pi/5} + 3e^{-j6\pi/5} + 4e^{-j8\pi/5}$$
$$= -2.5000 + j3.4410$$

$$X(2) = \sum_{n=0}^{4} x(n)e^{-j4\pi n/5} = 0 + e^{-j4\pi/5} + 2e^{-j8\pi/5} + 3e^{-j12\pi/5} + 4e^{-j16\pi/5}$$
$$= -2.5000 + j0.8123$$

$$X(3) = \sum_{n=0}^{4} x(n)e^{-j6\pi n/5} = 0 + e^{-j6\pi/5} + 2e^{-j12\pi/5} + 3e^{-j18\pi/5} + 4e^{-j24\pi/5}$$
$$= -2.5000 - j0.8123$$

$$X(4) = \sum_{n=0}^{4} x(n)e^{-j8\pi n/5} = 0 + e^{-j8\pi/5} + 2e^{-j16\pi/5} + 3e^{-j24\pi/5} + 4e^{-j32\pi/5}$$
$$= -2.5000 - j3.4410$$

Hence, from Eq. (4.4), the DTFS for $x(n)$ is given by

$$x(n) = 2 + ((-2.5000 + j3.4410)/5)e^{j-2\pi n/5} + ((-2.5000 + j0.8123))/5e^{j-4\pi n/5}$$
$$+ ((-2.5000 - j0.8123))/5e^{j-6\pi n/5} + ((-2.5000 - j3.4410))/5e^{j-8\pi n/5}$$

Example 4.2 Find the Fourier coefficients in DTFS representation of the sequence
$x(n) = \sin\left(\frac{5\pi}{4}\right)n$.

Solution It is clear that the sequence is periodic with period $N = 8$. We may rewrite
$x(n)$ in exponential form as

$$x(n) = \frac{1}{2j}e^{\frac{j2\pi 5n}{8}} - \frac{1}{2j}e^{-\frac{j2\pi 5n}{8}} = \frac{1}{2j}e^{\frac{j2\pi 5n}{8}} - \frac{1}{2j}e^{\frac{j2\pi 3n}{8}}$$

Hence, the Fourier coefficients are

$$X(0) = X(1) = X(2) = 0, \ X(3) = -\frac{1}{2j}, \ X(4) = \frac{1}{2j}, \ X(5) = X(6) = X(7) = 0$$

4.2   The Discrete Fourier Transform

Consider a finite discrete sequence $x(n), 0 \le n \le N - 1$. It is known from Eq. (2.69)
that the DTFT of the sequence $x(n)$ is given by

$$X(\omega) = \sum_{n=0}^{N-1} x(n)e^{-j\omega n}$$

where $X(\omega)$ is a continuous function of ω in the range $-\pi$ to π or 0–2π. When
$X(\omega)$ is computed at a finite number of values ω_k that are uniformly spaced, we have

$$X(\omega_k) = \sum_{n=0}^{N-1} x(n)e^{-j\omega_k n}, \quad k = 0, 1, 2, \ldots, M - 1$$

where $\omega_k = (2\pi k/M)$. The number of frequency samples may take any value;
however, it is chosen as equal N, the length of the discrete sequence $x(n)$. Rewriting
$X(\omega_k)$ as $X(k)$, the above equation can be written as

$$X(k) = \sum_{n=0}^{N-1} x(n)e^{-j2\pi nk/N}, \quad k = 0, 1, 2, \ldots, N - 1 \tag{4.7}$$

Equation (4.7) is called the discrete Fourier transform of the N-point sequence
$x(n)$. One of the main reasons as to why DFT is used to such a great extent is in
view of the existence of fast and efficient algorithms for its computation. These
algorithms are called fast Fourier transforms (FFTs). Later, in this chapter we
consider two of the FFTs.

Given $X(k)$, we now find an expression for $x(n)$ in terms of $X(k)$. For this
purpose, we multiply both sides of Eq. (4.7) by $e^{j2\pi lk/N}$ to get

$$X(k)e^{j2\pi lk/N} = \sum_{n=0}^{N-1} x(n)e^{j2\pi lk/N}e^{-j2\pi nk/N}$$

Hence,

$$\sum_{k=0}^{N-1} X(k)e^{j2\pi lk/N} = \sum_{n=0}^{N-1}\sum_{k=0}^{N-1} x(n)e^{j2\pi lk/N}e^{-j2\pi nk/N} \qquad (4.8)$$

Using Eq. (4.2), we have

$$\sum_{k=0}^{N-1} x(n)e^{j2\pi lk/N}e^{-j2\pi nk/N} = \begin{cases} 0 & n \neq l \\ N & n = l \end{cases}$$

Substituting the above in Eq. (4.8), we get

$$\sum_{k=0}^{N-1} X(k)e^{j2\pi lk/N} = Nx(l)$$

or

$$x(n) = \frac{1}{N}\sum_{k=0}^{N-1} X(k)e^{j2\pi nk/N}, \quad n = 0, 1, 2, \ldots, N-1 \qquad (4.9)$$

The above equation is called the inverse discrete Fourier transform (IDFT). It is seen that $X(k)$ as defined by Eq. (4.7) is periodic with a period N, since $X(k) = X(k+N)$; that is, the IDFT operation results in a periodic sequence of which only the first N values corresponding to one period are evaluated. Also, from Eq. (4.9), we see that $x(n) = x(n+N)$. In other words, we are replacing in effect the finite sequence $x(n)$ by its periodic extension in all the operations that involve DFT and IDFT. In fact, if we now compare Eqs. (4.4) and (4.5) with Eqs. (4.9) and (4.7), we see that the DFT $X(k)$ of a finite sequence of length N can be interpreted as the Fourier coefficient in the DFS expansion of its periodic extension $\tilde{x}(n)$.

If we now define

$$W_N = e^{-j2\pi/N} \qquad (4.10)$$

then the DFT and IDFT defined in Eqs. (4.7) and (4.9) can be rewritten as

$$X(k) = \sum_{n=0}^{N-1} x(n) W_N^{nk}, \quad k = 0, 1, 2, \ldots, N - 1 \tag{4.11}$$

and

$$x(n) = \frac{1}{N} \sum_{k=0}^{N-1} X(k) W_N^{-nk}, \quad n = 0, 1, 2, \ldots, N - 1 \tag{4.12}$$

For notational convenience, the above DFT and IDFT equations are denoted as

$$X(k) = \text{DFT}\{x(n)\}$$
$$x(n) = \text{IDFT}\{X(k)\}$$

In the DFT expression, W_N^{nk} for $0 \leq n$, $k \leq N - 1$, are called the twiddle factors of the DFT. The twiddle factors are periodic and define points on the unit circle in the complex plane. Also, they possess some interesting symmetry properties. Some basic properties of W_N are given below.

1. $W_N^k = W_N^{(k+N)}$
2. $W_N^{N/4} = j$
3. $W_N^{N/2} = -1$
4. $W_N^{3N/4} = j$
5. $W_N^{N/N} = 1$
6. $W_N^{kN} = 1$
7. $W_N^{kN+r} = W_N^r$
8. $W_N^{k+N/2} = -W_N^k$
9. $W_N^{2k} = W_{N/2}^k$
10. $W_N^* = W_N^{-1}$

Example 4.3 Find the twiddle factors for an eight-point DFT.

Solution For $N = 8$, $W_8^k = e^{-j2\pi k/8}$. Hence, the twiddle factors are:

$$W_8^0 = 1, W_8^1 = 0.707 - j0.707, W_8^2 = j, W_8^3 = -0.707 - j0.707$$
$$W_8^4 = -1, W_8^5 = -W_8^1, W_8^6 = -W_8^2, W_8^7 = -W_8^3, \text{ and}$$
$$W_8^{k+N} = W_8^k.$$

Example 4.4 Find the DFT of the sequence $x(n) = \{1, 0, 1, 0\}$.

Solution

$$X(k) = \sum_{n=0}^{N-1} x(n)W_N^{nk} \quad k = 0, 1, \ldots, N-1$$

$$= \sum_{n=0}^{3} x(n)W_4^{kn} \quad k = 0, 1, \ldots, 3;$$

$$X(0) = \sum_{n=0}^{3} x(n) = \{1 + 0 + 1 + 0\} = 2;$$

$$X(1) = \sum_{n=0}^{3} x(n)W_4^n = \{1 + 0 - 1 + 0\} = 0;$$

$$X(2) = \sum_{n=0}^{3} x(n)W_4^{2n} = \{1 + 0 + 1 + 0\} = 2;$$

$$X(3) = \sum_{n=0}^{3} x(n)W_4^{3n} = \{1 + 0 - 1 + 0\} = 0;$$

Example 4.5 Determine the eight-point DFT of the sequence $x(n) = \{1, 1, 1, 1, 0, 0, 1, 1\}$.

Solution

$$X(k) = \sum_{n=0}^{N-1} x(n)W_N^{nk} \quad k = 0, 1, \ldots, N-1.$$

$$= \sum_{n=0}^{8} x(n)W_8^{kn} \quad k = 0, 1, \ldots, 7.$$

$$X(0) = \sum_{n=0}^{7} x(n) = \{1 + 1 + 1 + 1 + 0 + 0 + 1 + 1\} = 6;$$

$$X(1) = \sum_{n=0}^{7} x(n) W_8^n = \{1 + 0.707 - j0.707 - j - 0.707 - j0.707$$

$$+ 0 + 0 + j + 0.707 + j0.707\} = 1.707 - j0.707;$$

$$X(2) = \sum_{n=0}^{7} x(n) W_8^{2n} = \{1 - j - 1 + j + 0 + 0 - 1 + j\} = -1 + j;$$

$$X(3) = \sum_{n=0}^{7} x(n) W_8^{3n} = \{1 - 0.707 - j0.707 + j + 0.707 - j0.707$$

$$+ 0 + 0 - j - 0.707 + j0.707\} = 0.293 - j0.707;$$

$$X(4) = \sum_{n=0}^{7} x(n) W_8^{4n} = \{1 - 1 + 1 - 1 + 0 + 0 + 1 - 1\} = 0;$$

$$X(5) = \sum_{n=0}^{7} x(n) W_8^{5n} = \{1 - 0.707 + j0.707 - j + 0.707 + j0.707 + 0$$

$$+ 0 + j - 0.707 - j0.707\} = 0.293 + j0.707;$$

$$X(6) = \sum_{n=0}^{7} x(n) W_8^{6n} = \{1 + j - 1 - j + 0 + 0 - 1 - j\} = -1 - j;$$

$$X(7) = \sum_{n=0}^{7} x(n) W_8^{7n} = \{1 + 0.707 + j0.707 + j - 0.707 + j0.707 + 0$$

$$+ 0 - j + 0.707 - j0.707\} = 1.707 + j0.707;$$

Example 4.6 Find the N-point DFT of the signal $x(n) = b^n$.

Solution

$$X(k) = \sum_{n=0}^{N-1} b^n e^{-j2\pi nk/N}$$

$$= \sum_{n=0}^{N-1} \left(be^{-j2\pi k/N} \right)^n$$

Hence,

$$X(k) = \frac{1 - b^N e^{-j2\pi k}}{1 - be^{-j2\pi k/N}}.$$

Example 4.7 A finite duration sequence of length N is given as

$$x(n) = \begin{cases} 1 & 0 \le n \le M - 1 \\ 0 & \text{otherwise} \end{cases}$$

Determine the N-point DFT of this sequence.

Solution

$$X(k) = \sum_{n=0}^{M-1} e^{-j2\pi kn/N}$$

$$= \frac{1 - e^{-j2\pi kM/N}}{1 - e^{-j2\pi k/N}} = \frac{\sin(\pi kM/N)}{\sin(\pi k/N)} e^{-j2\pi k(M-1)/N}, \quad k = 0, 1, \ldots, N - 1$$

Example 4.8 A finite duration sequence $x(n)$ of length eight has the DFT $X(k)$ as shown in Fig. 4.2. A new sequence $y(n)$ of length 16 is defined by

$$y(n) = x\left(\frac{n}{2}\right) \quad \text{for } n \text{ even}$$

$$= 0 \quad \text{for } n \text{ odd}.$$

Sketch the DFT $Y(k)$ as a function of k.

Solution The 16-point DFT of $y(n)$ is

$$Y(k) = \sum_{n=0}^{15} x(n) W_{16}^{nk}, \quad 0 \le k \le 15$$

$$= \sum_{n=0}^{7} x(n) W_{16}^{2nk}$$

Fig. 4.2 DFT $X(k)$ of $x(n)$ of Example 4.8

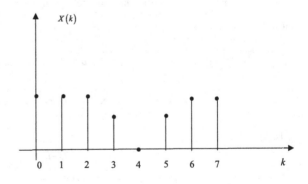

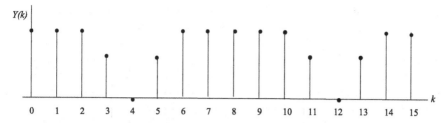

Fig. 4.3 DFT $Y(k)$ of $y(n)$ of Example 4.8

Since $W_N^{2k} = W_{N/2}^k$, the above reduces to

$$Y(k) = \sum_{n=0}^{7} x(n) W_8^{nk}, \quad 0 \le k \le 15$$

Thus, the 16-point DFT $Y(k)$ contains two copies of the eight-point DFT of $x(n)$, and $Y(k)$ has a period of 8. The DFT $Y(k)$ as a function of k is shown in Fig. 4.3.

4.2.1 Circular Operations on a Finite Length Sequence

Circular Shift

Consider a sequence $x(n)$ of length N, $0 \le n \le N - 1$. For such a sequence $x(n) = 0$ for $n < 0$ and $n > N - 1$. In such a case, if we shift the sequence by an arbitrary integer m, then the shifted sequence is no longer be defined in the range $0 \le n \le N - 1$. In order to make sure that the shifted sequence always stays in the range $0 \le n \le N - 1$, we define what is known as the *circular shift*, by the relation

$$x_c(n) = x(n - m)_N \tag{4.13a}$$

where

$$(n - m)_N = (n - m) \text{ modulo } N \tag{4.13b}$$

This way, any integer n is related to the modulo N as

$$n = (n)_N + \gamma N \tag{4.14}$$

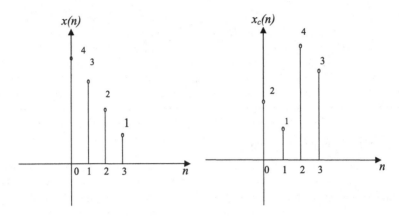

Fig. 4.4 Illustration of circular shift

where γ is an integer and $(n)_N$ is always such that $0 \le n \le N - 1$. Consequently,

$$x(n - m)_N = \begin{cases} x(n - m) & \text{if } 0 \le (n - m) \le N - 1 \\ x(\pm N + n - m) & \text{otherwise} \end{cases} \qquad (4.15)$$

where $+N$ is used if $m > 0$, and $-N$ is used if $m < 0$.

The circular shift for $m = 2$ is illustrated in Fig. 4.4.

The sequence $x_c(n)$ is related to $x(n)$ by a circular shift of two samples. The samples of $x_c(n)$ can be evaluated using $x_c(n) = x(n - m)_4$. Hence,

$$x_c(0) = x(-2)_4 = x(2); \ x_c(1) = x(-1)_4 = x(3);$$
$$x_c(2) = x(0)_4 = x(0); \ x_c(3) = x(1)_4 = x(1);$$

Circular Time Reversal

For a length-N sequence $x(n)$, $0 \le n \le N - 1$, the circular time-reversal sequence is also of length-N sequence given by

$$x(-n)_N = x(N - n)_N \qquad (4.16)$$

Circular Convolution

Consider two sequences $x(n)$ and $h(n)$, each of length N. Then, the circular convolution of $x(n)$ and $h(n)$ is defined as the length-N sequence $y_c(n)$ given by

$$y_c(n) = \sum_{m=0}^{N-1} x(m)h(n-m)_N \tag{4.17}$$

It is often called as the N-point circular convolution and is denoted by

$$x(n)\ \textcircled{N}\ h(n) \tag{4.18}$$

The circular convolution is also commutative like the linear convolution; that is,

$$x(n)\ \textcircled{N}\ h(n) = \quad h(n)\ \textcircled{N}\ x(n) \tag{4.19}$$

Example 4.9 Find the circular convolution of the three-point sequences $x(n) = \{1, 3, -4\}$ and $h(n) = \{-2, 1, 2\}$.

Solution From Eq. (4.17), $y_c(n) = \sum_{m=0}^{2} x(m)h(n-m)_3$.
Hence,

$$\begin{aligned}
y_c(0) &= x(0)h(0) + x(1)h(-1)_3 + x(2)h(-2)_3 \\
&= -2 + 3h(2) - 4h(1) = -2 + 6 - 4 = 0 \\
y_c(1) &= x(0)h(1) + x(1)h(0) + x(2)h(-1)_3 \\
&= 1 + 3h(0) - 4h(2) = 1 - 6 - 8 = -13 \\
y_c(2) &= x(0)h(2) + x(1)h(1) + x(2)h(0) \\
&= 2 + 3 - 8 = -3
\end{aligned}$$

Thus, $y_c(n) = (0, -13, -3)$.

It can also be verified that $\sum_{m=0}^{2} h(m)x(n-m)_3$ leads to the same result, showing that the circular convolution operation is commutative.

Circular Correlation:

Consider two complex-valued sequences $x_1(n)$ and $x_2(n)$, each of length N. Then, the circular correlation of $x_1(n)$ and $x_2(n)$ is defined as the N-point sequence

$$r_{x_1 x_2}(m) = \sum_{n=0}^{N-1} x_1(n)x_2^*(n-m)_N \tag{4.20}$$

where $x_2^*(n)$ is the complex conjugate of $x_2(n)$.

4.3   Basic Properties of the Discrete Fourier Transform

In this section, we state and prove some properties of the DFT, which play an important role in digital signal processing applications. We will denote an N-point DFT pair $x(n)$ and $X(k)$ by the following notation

$$x(n) \overset{\text{DFT}}{\underset{N}{\leftrightarrow}} X(k)$$

Linearity:

Consider a sequence $a_1x_1(n) + a_2x_2(n)$ that is a linear combination of $x_1(n)$ and $x_2(n)$, each sequence being of length N, where a_1 and a_2 are arbitrary constants. If the sequences are not of the same length, then the sequence with the lower length is augmented by zeros so that its length is now equal to that of the other sequence. In such a case,

$$a_1x_1(n) + a_2x_2(n) \overset{\text{DFT}}{\underset{N}{\leftrightarrow}} a_1X_1(k) + a_2X_2(k) \tag{4.21}$$

Proof By the definition of the DFT,

$$
\begin{aligned}
\mathbf{DFT}(a_1x_1(n) + a_2x_2(n)) &= \sum_{n=0}^{N-1} [a_1x_1(n) + a_2x_2(n)] W_N^{kn} \\
&= \sum_{n=0}^{N-1} [a_1x_1(n)] W_N^{kn} + \sum_{n=0}^{N-1} [a_2x_2(n)] W_N^{kn} \\
&= a_1 \sum_{n=0}^{N-1} x_1(n) W_N^{kn} + a_2 \sum_{n=0}^{N-1} x_2(n) W_N^{kn} \\
&= a_1X_1(k) + a_2X_2(k)
\end{aligned}
$$

Hence, we can write

$$a_1x_1(n) + a_2x_2(n) \overset{\text{DFT}}{\underset{N}{\leftrightarrow}} a_1X_1(k) + a_2X_2(k)$$

Time Reversal of a Sequence:

If $x(n)$ and $X(k)$ are an N-point DFT pair, then

$$x(N - n) \overset{\text{DFT}}{\underset{N}{\leftrightarrow}} X(N - k) \tag{4.22}$$

Proof

$$\mathrm{DFT}\{x(N-n)\} = \sum_{n=0}^{N-1} x(N-n)e^{-j2\pi kn/N}$$

Changing the index from n to $m = N - n$ in the RHS of the above equation, we can rewrite it as

$$\mathrm{DFT}\{x(N-n)\} = \sum_{m=0}^{N-1} x(m)e^{-j2\pi k(N-m)/N}$$

$$= \sum_{m=0}^{N-1} x(m)e^{j2\pi km/N} = \sum_{m=0}^{N-1} x(m)e^{-j2\pi m(N-k)/N} = X(N-k)$$

Circular Time Shifting:

The DFT of a circularly time-shifted sequence $x(n-m)_N$ is given by $W_N^{km}X(k)$, that is,

$$x\big[(n-m)_N\big] \overset{\mathrm{DFT}}{\underset{N}{\longleftrightarrow}} W_N^{km} X(k) \qquad (4.23)$$

Proof By the definition of DFT,

$$\mathrm{DFT}\{x(n-m)_N\} = \sum_{n=0}^{N-1} x(n-m)_N W_N^{km}$$

$$= \sum_{n=0}^{m-1} x(n-m)_N W_N^{km} + \sum_{n=m}^{N-1} x(n-m) W_N^{km}$$

Since $x(n-m)_N = x(N-m+n)$, we can write the above equation as

$$\mathrm{DFT}\{x(n-m)_N\} = \sum_{n=0}^{m-1} x(N-m+n)e^{-j2\pi kn/N} + \sum_{l=0}^{N-1-m} x(l)e^{-j2\pi k(l+m)/N}$$

$$= \sum_{l=N-m}^{N-1} x(l)e^{-j2\pi k(l+m+N)/N} + \sum_{l=0}^{N-1-m} x(l)e^{-j2\pi k(l+m)/N}$$

$$= \sum_{l=N-m}^{N-1} x(l)e^{-j2\pi k(l+m)/N} + \sum_{l=0}^{N-1-m} x(l)e^{-j2\pi k(l+m)/N}$$

$$= \sum_{l=0}^{N-1} x(l)e^{-j2\pi k(l+m)/N} = e^{-j2\pi km/N} \sum_{l=0}^{N-1} x(l)e^{-j2\pi kl/N}$$

$$= W_N^{km} X(k)$$

Circular Frequency Shifting:

If $x(n)$ and $X(k)$ are an N-point DFT pair, then

$$W_N^{-mn}x(n) \overset{\mathrm{DFT}}{\underset{N}{\leftrightarrow}} X\big[(k-m)_N\big] \qquad (4.24)$$

where $X\big[(k-m)_N\big]$ is a circularly frequency-shifted version of $X(k)$.

Proof

$$\mathrm{DFT}\{W_N^{-mn}x(n)\} = \sum_{n=0}^{N-1} W_N^{-mn}x(n)W_N^{kn}$$

$$= \sum_{n=0}^{N-1} x(n)W_N^{n(k-m)} = \sum_{n=0}^{N-1} x(n)W_N^{n(N+k-m)}$$

Circular Convolution:

The DFT of the circular convolution of two length-N sequences is the product of their N-point DFTs, i.e.,

$$x_1(n) \;\textcircled{\scriptsize N}\; x_2(n) \overset{\mathrm{DFT}}{\underset{N}{\leftrightarrow}} X_1(k)X_2(k) \qquad (4.25)$$

Proof Let $y_c(n)$ represent the circular convolution of the sequences $x_1(n)$ and $x_2(n)$, i.e.,

$$y_c(n) = \sum_{l=0}^{N-1} x_1(l)x_2(n-l)_N$$

Then, the DFT of $y_c(n)$ is

$$Y_c(k) = \sum_{n=0}^{N-1} y_c(n)W_N^{kn} = \sum_{n=0}^{N-1}\left[\sum_{l=0}^{N-1} x_1(l)x_2(n-l)_N\right]W_N^{kn}$$

By interchanging the order of the summation, we obtain

$$Y_c(k) = \sum_{l=0}^{N-1} x_1(l)\left[\sum_{n=0}^{N-1} x_2(n-l)_N\right]W_N^{kn}$$

Substituting $(n - l) = m$, where m is integer with $0 \leq m \leq N - 1$, we get

$$Y_c(k) = \sum_{l=0}^{N-1} x_1(l) \left[\sum_{m=0}^{N-1} x_2(m) \right] W_N^{k(l+m)} = \sum_{l=0}^{N-1} x_1(l) \left[\sum_{m=0}^{N-1} x_2(m) W_N^{km} \right] W_N^{kl}$$

$$= \sum_{l=0}^{N-1} x_1(l)[X_2(k)] W_N^{kl} = \left[\sum_{l=0}^{N-1} x_1(l) W_N^{kl} \right] [X_2(k)]$$

$$= X_1(k)X_2(k)$$

Circular Correlation:

The DFT of the circular correlation of two complex-valued N-point sequences $x_1(n)$ and $x_2(n)$ is given by $X_1(k)X_2^*(k)$, i.e.,

$$r_{x_1 x_2}(m) = \sum_{n=0}^{N-1} x_1(n)x_2^*[(n-m)]_N \overset{\text{DFT}}{\underset{N}{\longleftrightarrow}} X_1(k)X_2^*(k) \tag{4.26}$$

Proof From Eq. (4.20), we know that

$$r_{x_1 x_2}(m) = \sum_{n=0}^{N-1} x_1(n)x_2^*(n-m)_N = \sum_{n=0}^{N-1} x_1(n)x_2^*(-(m-n)_N) \tag{4.27a}$$

Also, the circular convolution of two sequences $x_1(m)$ and $x_2(m)$ is given by

$$y_c(m) = \sum_{l=0}^{N-1} x_1(l)x_2(m-l)_N = \sum_{n=0}^{N-1} x_1(n)x_2(m-n)_N \tag{4.27b}$$

Comparing Eqs. (4.27a) and (4.27b), we see that $r_{x_1 x_2}(m)$ can be considered as the circular convolution of $x_1(m)$ and $x_2^*(-m)_N$. Hence, DFT$[r_{x_1 x_2}(m)] = [\text{DFT}\{x_1(m)\}][\text{DFT}\{x_2^*(-m)_N\}]$. It can be shown that (see Eq. (4.41)) DFT$\{x_2^*(-m)_N\} = X_2^*(k)$. Thus,

$$\text{DFT}[r_{x_1 x_2}(m)] = R_{x_1 x_2}(k) = X_1(k)X_2^*(k) \tag{4.28}$$

If $x_1(n) = x_2(n) = x(n)$, then

$$R_{x_1 x_2}(k) = |X(k)|^2 \tag{4.29}$$

Parseval's Theorem:

If $x_1(n)$ and $x_2(n)$ are two complex-valued N-point sequences with DFTs $X_1(k)$ and $X_2(k)$, then

$$\sum_{n=0}^{N-1} x_1(n)x_2^*(n) = \frac{1}{N}\sum_{k=0}^{N-1} X_1(k)X_2^*(k) \qquad (4.30)$$

Proof From Eq. (4.28), we have $R_{x_1x_2}(k) = X_1(k)X_2^*(k)$. Hence,

$$r_{x_1x_2}(m) = \frac{1}{N}\sum_{k=0}^{N-1} X_1(k)X_2^*(k)W_N^{-km}$$

Evaluating the above at $m = 0$ gives

$$r_{x_1x_2}(0) = \frac{1}{N}\sum_{k=0}^{N-1} X_1(k)X_2^*(k)$$

Hence,

$$\sum_{n=0}^{N-1} x_1(n)x_2^*(n) = \frac{1}{N}\sum_{k=0}^{N-1} X_1(k)X_2^*(k)$$

If $x_1(n) = x_2(n) = x(n)$, then we have

$$\sum_{n=0}^{N-1} |x(n)|^2 = \frac{1}{N}\sum_{k=0}^{N-1} |X(k)|^2 \qquad (4.31)$$

The above expression gives a relationship between the energy in a finite duration sequence to the power in the frequency components.

Multiplication of two Sequences:

The DFT of the product of two sequences $x_1(n)$ and $x_2(n)$, each of length N, is given by the circular convolution of their DFTs $X_1(k)$ and $X_2(k)$ divided by N, i.e.,

$$x_1(n)x_2(n) \underset{N}{\overset{\text{DFT}}{\longleftrightarrow}} \frac{1}{N}X_1(k) \left(\text{N}\right) X_2(k) \qquad (4.32)$$

This property is dual of the circular convolution property and is left as an exercise for the student.

The above properties are summarized in Table 4.2.

Table 4.2 Basic properties of the discrete Fourier transform

Property	Sequence	DFT
Linearity	$a_1x_1(n)+a_2x_2(n)$	$a_1X_1(k)+a_2X_2(k)$
Periodicity	$x(n+N)=x(n)$	$X(k+N)=X(k).$
Time reversal	$x(N-n)$	$X(N-k)$
Circular time shifting	$x\left[(n-m)_N\right]$	$W_N^{km}X(k)$
Circular frequency shifting	$W_N^{-mn}x(n)$	$X\left[(k-m)_N\right]$
N-point circular convolution	$x_1(n) \ \text{Ⓝ} \ x_2(n)$	$X_1(k)X_2(k)$
Circular correlation	$x_1(n) \ \text{Ⓝ} \ x_2^*(-n)$	$X_1(k)X_2^*(k)$
Multiplication of two sequences	$x_1(n)x_2(n)$	$\frac{1}{N}X_1(k) \ \text{Ⓝ} \ X_2(k)$

Parseval's theorem $\sum_{n=0}^{N-1}|x(n)|^2 = \frac{1}{N}\sum_{k=0}^{N-1}|X(k)|^2$

4.4 Symmetry Relations of DFT

4.4.1 Symmetry Relations of DFT of Complex-Valued Sequences

Consider a complex-valued sequence $x(n)$, which is expressed as

$$x(n) = x_R(n)+jx_I(n), \qquad 0 \le n \le N-1 \tag{4.33}$$

The DFT of $x(n)$ is given by

$$X(k) = \sum_{n=0}^{N-1}x(n)W_N^{kn} = \sum_{n=0}^{N-1}[x_R(n)+jx_I(n)]\left[\cos\frac{2\pi kn}{N}-j\sin\frac{2\pi kn}{N}\right]$$

$$= \sum_{n=0}^{N-1}\left[x_R(n)\cos\frac{2\pi kn}{N}+x_I(n)\sin\frac{2\pi kn}{N}\right] -j\sum_{n=0}^{N-1}\left[x_R(n)\sin\frac{2\pi kn}{N}+x_I(n)\cos\frac{2\pi kn}{N}\right]$$

$$\tag{4.34}$$

If

$$X(k) = X_R(k)+jX_I(k) \tag{4.35}$$

then

$$X_R(k) = \sum_{n=0}^{N-1}\left[x_R(n)\cos\frac{2\pi kn}{N}+x_I(n)\sin\frac{2\pi kn}{N}\right] \tag{4.36a}$$

and

$$X_I(k) = -\sum_{n=0}^{N-1}\left[x_R(n)\sin\frac{2\pi kn}{N} + x_I(n)\cos\frac{2\pi kn}{N}\right] \tag{4.36b}$$

Similarly, we can show that

$$x_R(n) = \frac{1}{N}\sum_{k=0}^{N-1}\left[X_R(k)\cos\frac{2\pi kn}{N} - X_I(nk)\sin\frac{2\pi kn}{N}\right] \tag{4.37a}$$

and

$$x_I(n) = \frac{1}{N}\sum_{k=0}^{N-1}\left[X_R(k)\sin\frac{2\pi kn}{N} + X_I(k)\cos\frac{2\pi kn}{N}\right] \tag{4.37b}$$

Let us now consider a length-N complex conjugate sequence $x^*(n)$. Taking the complex conjugate on both sides of Eq. (4.11), we get

$$X^*(k) = \left[\sum_{n=0}^{N-1}x(n)e^{-j2\pi nk/N}\right]^*$$

which can be rewritten as

$$X^*(k) = \sum_{n=0}^{N-1}x^*(n)e^{j2\pi nk/N} \tag{4.38}$$

Hence,

$$X^*\left((-k)_N\right) = X^*(N-k) = \sum_{n=0}^{N-1}x^*(n)e^{j2\pi n(N-k)/N}$$

$$= \sum_{n=0}^{N-1}x^*(n)e^{-j2\pi nk/N} = \text{DFT}\{x^*(n)\}$$

Therefore,

$$\text{DFT}\{x^*(n)\} = X^*\left((-k)_N\right) \tag{4.39}$$

Now, we find the DFT of $x^*((-n)_N)$ as follows:

$$
\begin{aligned}
\mathrm{DFT}\{x^*((-n)_N)\} &= \sum_{n=0}^{N-1} x^*((-n)_N)e^{-j2\pi nk/N} \\
&= \sum_{n=0}^{N-1} x^*(N-n)e^{-j2\pi nk/N}
\end{aligned}
\tag{4.40a}
$$

Replacing n by $(N-n)$ in Eq. (4.38), we have

$$
X^*(k) = \sum_{n=0}^{N-1} x^*(N-n)e^{j2\pi(N-n)k/N} = \sum_{n=0}^{N-1} x^*(N-n)e^{-j2\pi nk/N}
\tag{4.40b}
$$

It is seen from Eq. (4.40a) and Eq. (4.40b) that

$$
\mathrm{DFT}\{x^*((-n)_N)\} = X^*(k)
\tag{4.41}
$$

Since a complex sequence $x(n)$ can be decomposed into a sum of its real and imaginary parts as

$$
x(n) = x_R(n) + jx_I(n)
\tag{4.42}
$$

where

$$
x_R(n) = \frac{1}{2}[x(n) + x^*(n)]
\tag{4.43a}
$$

and

$$
jx_I(n) = \frac{1}{2}[x(n) - x^*(n)]
\tag{4.43b}
$$

it can be easily shown that the DFTs of the real and imaginary parts of complex sequence are given by

$$
\mathrm{DFT}\{x_R(n)\} = \frac{1}{2}[X(k) + X^*((-k)_N)] = \frac{1}{2}[X(k) + X^*(N-k)]
\tag{4.44a}
$$

and

$$
\mathrm{DFT}\{jx_I(n)\} = \frac{1}{2}[X(k) - X^*((-k)_N)] = \frac{1}{2}[X(k) - X^*(N-k)]
\tag{4.44b}
$$

A complex sequence $x(n)$ can be represented as the sum of a *circular conjugate symmetric sequence* $x_e(n)$ and a *circular conjugate antisymmetric sequence* $x_o(n)$:

Sequence	DFT
$x^*(n)$	$X^*((-k)_N) = X^*(N-k)$
$x^*((-n)_N)$	$X^*(k)$
$x_R(n)$	$\frac{1}{2}[X(k) + X^*(N-k)]$
$jx_I(n)$	$\frac{1}{2}[X(k) - X^*(N-k)]$
$x_e(n)$	$X_R(k)$
$x_o(n)$	$jX_I(k)$

Table 4.3 Symmetry properties of DFT of a complex sequence

$$x(n) = x_e(n) + x_o(n) \tag{4.45}$$

where

$$x_e(n) = \frac{1}{2}\left[x(n) + x^*(-n)_N\right] \tag{4.46a}$$

and

$$x_o(n) = \frac{1}{2}\left[x(n) - x^*(-n)_N\right] \tag{4.46b}$$

Then, the DFTs of $x_e(n)$ and $x_o(n)$ can be easily obtained, using Eq. (4.39), as

$$\text{DFT}\{x_e(n)\} = \frac{1}{2}[X(k) + X^*(k)] = X_R(k) \tag{4.47a}$$

and

$$\text{DFT}\{x_o(n)\} = \frac{1}{2}[X(k) - X^*(k)] = jX_I(k) \tag{4.47b}$$

The symmetry properties of the DFT of a complex sequence are summarized in Table 4.3.

4.4.2   Symmetry Relations of DFT of Real-Valued Sequences

For a real-valued sequence $x(n)$, $x_I(n) = 0$. Hence, from Eq. (4.34), we get

$$X(k) = \sum_{n=0}^{N-1}\left[x(n)\cos\frac{2\pi kn}{N} - jx(n)\sin\frac{2\pi kn}{N}\right]$$

From symmetry,

$$X\big((-k)_N\big) = X(n-k) = \sum_{n=0}^{N-1}\left[x(n)\cos\frac{2\pi(n-k)n}{N} - jx(n)\sin\frac{2\pi(n-k)n}{N}\right]$$

$$= \sum_{n=0}^{N-1}\left[x(n)\cos\frac{2\pi kn}{N} + jx(n)\sin\frac{2\pi kn}{N}\right] = X^*(k)$$

Hence, we have the symmetry relation

$$X(n-k) = X\big((-k)_N\big) = X^*(k) \tag{4.48}$$

Also, from Eqs. (4.36a), we have

$$X_R(k) = \sum_{n=0}^{N-1}\left[x(n)\cos\frac{2\pi kn}{N}\right]$$

Hence,

$$X_R\big((-k)_N\big) = X_R(N-k) = \sum_{n=0}^{N-1}\left[x(n)\cos\frac{2\pi(N-k)n}{N}\right] = X_R(k)$$

Thus,

$$X_R(k) = X_R\big((-k)_N\big) = X_R(N-k) \tag{4.49a}$$

Similarly, starting with Eqs. (4.36b), we can show that

$$X_I(k) = -X_I\big((-k)_N\big) = -X_I(N-k) \tag{4.49b}$$

From the above relations, we see that the magnitude of $X(k)$ and $X\big((-k)_N\big)$ is equal and that the phase angle of $X(k)$ is negative of that of the phase angle of $X\big((-k)_N\big)$, i.e.,

$$|X(k)| = \big|X((-k)_N)\big| \tag{4.50a}$$

and

$$\angle X(k) = -\angle X((-k)_N) \tag{4.50b}$$

If $x(n)$ is real and even, that is,

$$x(n) = x(N - n) \qquad 0 \leq n \leq N - 1 \tag{4.51}$$

then, from Eq. (4.36a) and Eq. (4.36b), we see that $X_I(k) = 0$ and that the N-point DFT reduces to

$$X(k) = \sum_{n=0}^{N-1} \left[x(n) \cos \frac{2\pi kn}{N} \right] = X_R(k) \qquad 0 \leq k \leq N - 1 \tag{4.52a}$$

Hence, the DFT of a real finite even sequence is itself real and even. Furthermore, the IDFT reduces to

$$x(n) = \frac{1}{N} \sum_{k=0}^{N-1} \left[X(k) \cos \frac{2\pi kn}{N} \right] \qquad 0 \leq n \leq N - 1 \tag{4.52b}$$

If $x(n)$ is real and odd, that is,

$$x(n) = -x(N - n) \qquad 0 \leq n \leq N - 1 \tag{4.53}$$

then, from Eq. (4.35a) and (4.35b), we see that $X_R(k) = 0$ and that the N-point DFT reduces to

$$X(k) = -j \sum_{n=0}^{N-1} \left[x(n) \sin \frac{2\pi kn}{N} \right] = jX_j(k) \qquad 0 \leq k \leq N - 1 \tag{4.54a}$$

Hence, the DFT of a real finite odd sequence is purely imaginary and odd. Furthermore, the IDFT reduces to

$$x(n) = j\frac{1}{N} \sum_{k=0}^{N-1} \left[X(k) \sin \frac{2\pi kn}{N} \right] \qquad 0 \leq n \leq N - 1 \tag{4.54b}$$

The symmetry relations of DFT of a real-valued sequence are summarized in Table 4.4.

Table 4.4 Symmetry relations of DFT of a real-valued sequence

Sequence	DFT
Real $x(n)$	$X(n - k) = X((-k)_N) = X^*(k)$
Real $x(n)$	$X_R(k) = X_R((-k)_N) = X_R(N - k)$
Real $x(n)$	$X_I(k) = -X_I((-k)_N) = -X_I(N - k)$
$x(n)$ real and even	$X_R(k)$
$x(n)$ real and odd	$jX_I(k)$
Real $x(n)$	$|X(k)| = |X((-k)_N)|, \angle X(k) = -\angle X((-k)_N)$

4.4.3 DFTs of Two Real Sequences from a Single N-Point DFT

Equations (4.44a) and (4.44b) can be used to advantage in finding the DFTs of two real sequences of length N. Suppose $x_1(n)$ and $x_2(n)$ are two real N-point sequences with DFTs $X_1(k)$ and $X_2(k)$. Let us define a complex sequence $x(n)$ by

$$x(n) = x_1(n) + jx_2(n) \tag{4.55}$$

Using Eqs. (4.44a) and (4.44b), we may write the DFTs of the two real sequences as

$$X_1(k) = \frac{1}{2}\left[X(k) + X^*\left((-k)_N\right)\right] = \frac{1}{2}[X(k) + X^*(N-k)] \tag{4.56a}$$

$$X_2(k) = \frac{1}{2j}\left[X(k) - X^*\left((-k)_N\right)\right] = \frac{1}{2j}[X(k) - X^*(N-k)] \tag{4.56b}$$

Example 4.10 Find the DFTs of the sequences $x_1(n) = (1,2,0,1)$ and $x_2(n) = (1,0,1,0)$ using a single four-point DFT.

Solution

$$x(n) = x_1(n) + jx_2(n) = (1+j, 2, j, 1)$$

Hence,

$$X(k) = x(0) + x(1)W_4^k + x(2)W_4^{2k} + x(3)W_4^{3k}, \quad k = 0, 1, 2, 3$$

Thus,

$$X(k) = (4+2j, 1-j, -2+2j, 1+j)$$

Hence,

$$X^*(N-k) = (4-2j, 1-j, -2-2j, 1+j)$$

Substituting the values of $X(k)$ and $X^*(N-k)$ in Eqs. (4.56a) and (4.56b), we get

$$X_1(k) = (4, 1-j, -2, 1+j) \quad \text{and} \quad X_2(k) = (2, 0, 2, 0)$$

4.5   Computation of Circular Convolution

4.5.1   Circulant Matrix Method

The circular convolution defined by Eq. (4.17) can be written in a matrix form as

$$
\begin{bmatrix} y_c(0) \\ y_c(1) \\ y_c(2) \\ \vdots \\ y_c(N-1) \end{bmatrix} = \begin{bmatrix} x(0) & x(N-1) & x(N-2) & \cdots & x(1) \\ x(1) & x(0) & x(N-1) & \cdots & x(2) \\ x(2) & x(1) & x(0) & \cdots & x(3) \\ \vdots & \vdots & \vdots & \vdots & \vdots \\ x(N-1) & x(N-2) & x(N-3) & \cdots & x(0) \end{bmatrix} \begin{bmatrix} h(0) \\ h(1) \\ h(2) \\ \vdots \\ h(N-1) \end{bmatrix}
$$
$$(4.57)$$

The $(N \times N)$ matrix on the RHS of Eq. (4.57) is called the circular convolution matrix or circulant matrix and denoted by C_x. It may be observed that the first column corresponds to the elements of the sequence $x(n)$, and the rest of the columns are derived from the previous ones in a very simple way.

Example 4.11 Find the circular convolution of the sequences considered in Example 4.9, namely $x(n) = (1, 3, -4,)$ and $h(n) = (-2, 1, 2)$.

Solution The circular convolution matrix C_x is given by

$$
\begin{bmatrix} 1 & -4 & 3 \\ 3 & 1 & -4 \\ -4 & 3 & 1 \end{bmatrix}
$$

Then, the circular convolution of $x(n)$ and $h(n)$ is given by

$$
\begin{bmatrix} y_c(0) \\ y_c(1) \\ y_c(2) \end{bmatrix} = \begin{bmatrix} 1 & -4 & 3 \\ 3 & 1 & -4 \\ -4 & 3 & 1 \end{bmatrix} \begin{bmatrix} -2 \\ 1 \\ 2 \end{bmatrix} = \begin{bmatrix} 0 \\ -13 \\ 13 \end{bmatrix}
$$

Hence, $y_c(n) = (0, -13, 13)$.

4.5.2   Graphical Method

Evaluation of the circular convolution sum at any sample n consists of the following operations:

(i)   The sequences $x(n)$ and $h(n)$ are marked on two concentric circles with one sequence on the inner circle in the clockwise direction and the other on the outer circle in a counter clockwise direction as various points, with equal spacing. For $n = 0$, $y_c(0)$ is obtained by multiplying the two sequences point by point and summing the products.

(ii) Keeping the outer circle stationary, rotate the inner in counterclockwise direction by one sample, multiply the two sequences point by point, and sum the products. This gives $y_c(1)$.

(iii) The procedure is continued to find $y_c(n)$ for other values of n.

The following example illustrates the above procedure:

Example 4.12 Find the circular convolution of the three-point sequences of Example 4.11 with $x(n) = (1, 3, -4)$ and $h(n) = (-2, 1, 2)$.

Solution

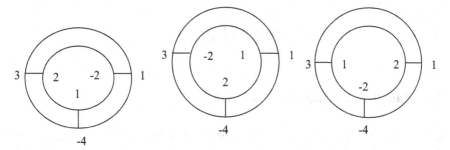

$$y_c(0) = -2.1 + 2.3 + 1(-4) = 0$$
$$y_c(1) = 1.1 + (-2)3 + 2(-4) = -13$$
$$y_c(2) = 2.1 + 1.3 + (-2)(-4) = 13$$

Hence,

$$y_c(n) = x(n) \;\;\text{(N)}\;\; h(n) = (0, -13, 13)$$

4.5.3 DFT Approach

We may obtain the circular convolution $y_c(n)$ of two N-point sequences using the relation given by Eq. (4.25). We first compute the DFTs $X_1(k)$ and $X_2(k)$ of the two sequences and then multiply them to get $Y_c(k) = X_1(k)X_2(k)$, the DFT of the circular convolution. We then perform the IDFT on $Y_c(k)$ to obtain the circular convolution $y_c(n)$. In the next section, we will see how this approach can be used to evaluate linear convolution of two sequences.

Example 4.13 Obtain the circular convolution of the sequences $x_1(n) = (1, 2, 0, 1)$ and $x_2(n) = (1, 0, 1, 0)$ using the DFT approach.

Solution We have already found the DFTs for these two sequences in Example 4.10. These are given by

$$X_1(k) = (4, 1 - j, -2, 1 + j) \quad \text{and} \quad X_2(k) = (2, 0, 2, 0).$$

Hence,

$$Y_c(k) = (8, 0, -4, 0).$$

Using Eq. (4.12), we now compute the IDFT of the above to obtain the circular convolution $y_c(n)$.

$$
\begin{aligned}
y_c(n) &= \frac{1}{N} \left[Y_c(0) + Y_c(2)W_4^{-2n} + Y_c(3)W_4^{-3n} \right] \\
&= \frac{1}{N} \left[8 - 4W_4^{-2n} \right]
\end{aligned}
$$

which gives

$$y_c(n) = (1, 3, 1, 3)$$

4.6 Linear Convolution Using DFT

Linear convolution is an important operation in signal processing applications since it can be used to obtain the response of a linear filter for arbitrary input, once the impulse response of the filter is known. There are efficient algorithms called fast Fourier transforms, two of which will be discussed in the next section, for practical implementation of an N-point DFT. Hence, it is of importance to find methods to implement the linear convolution using the DFT.

4.6.1 Linear Convolution of Two Finite Length Sequences

Consider two sequences $x(n)$ and $h(n)$ of lengths L_1 and L_2, respectively. The linear convolution of these two sequences is a sequence of length $L_1 + L_2 - 1$. Circular convolution cannot be directly used on these two sequences to achieve linear convolution. Now, to obtain linear convolution using circular convolution, we generate two new sequences $x'(n)$ and $h'(n)$, each of length $L_1 + L_2 - 1 = L$ by padding $x(n)$ with $(L_2 - 1)$ zeros and $h(n)$ with $(L_1 - 1)$ zeros. Thus,

$$x'(n) = [x(0), x(1), \ldots, x(L_1 - 1), \underbrace{0, \ldots, 0}_{L_2 - 1}] \tag{4.58}$$

$$h'(n) = [h(0), h(1), \ldots, h(L_2 - 1), \underbrace{0, \ldots, 0}_{L_1 - 1}] \tag{4.59}$$

The linear convolution of $x'(n)$ and $h'(n)$ is given by

$$x'(n) * h'(n) = \sum_{m=}^{L} x'(m)h'(n-m), \quad 0 \le n \le L-1 \tag{4.60}$$

The above expression can be thought of as a circular convolution of the two padded sequences $x'(n)$ and $h'(n)$; hence, we can use any of the methods described in Sect. 4.5 to evaluate it.

Example 4.14 Find the linear convolution of the sequences $x(n) = (1, 2, 3, 1)$ and $x(n) = (1, 1, 1)$.

Solution The two sequences $x(n)$ and $h(n)$ are of lengths 4 and 3, respectively. By appropriately padding the two sequences by zeros, we obtain the padded sequences $x'(n) = (1, 2, 3, 1, 0, 0)$ and $h'(n) = (1, 1, 1, 0, 0, 0)$, each of length $L = 6$. We may now calculate the circular convolution $y_c(n)$ of $x'(n)$ and $h'(n)$ using the circulant matrix Eq. (4.57)

$$\begin{bmatrix} y_c(0) \\ y_c(1) \\ y_c(2) \\ y_c(3) \\ y_c(4) \\ y_c(5) \end{bmatrix} = \begin{bmatrix} 1 & 0 & 0 & 1 & 3 & 2 \\ 2 & 1 & 0 & 0 & 1 & 3 \\ 3 & 2 & 1 & 0 & 0 & 1 \\ 1 & 3 & 2 & 1 & 0 & 0 \\ 0 & 1 & 3 & 2 & 1 & 0 \\ 0 & 0 & 1 & 3 & 2 & 1 \end{bmatrix} \begin{bmatrix} 1 \\ 1 \\ 1 \\ 0 \\ 0 \\ 0 \end{bmatrix} = \begin{bmatrix} 1 \\ 3 \\ 6 \\ 6 \\ 4 \\ 1 \end{bmatrix}$$

Thus, $y_c(n) = (1, 3, 6, 6, 4, 1)$, and therefore, the linear convolution

$$y_l(n) = x(n) * h(n) = (1, 3, 6, 6, 4, 1).$$

Instead of using the circulant matrix, we could have used the DFT approach to find the circular convolution. In this case, we would first find the $L = (L_1 + L_2 - 1)$-point DFTs $X'(k)$ and $H'(k)$ of $x'(n)$ and $h'(n)$. Then, the L-point IDFT of the product $X'(k)H'(k)$ would yield the linear convolution of $x(n)$ and $h(n)$.

The following MATLAB fragments illustrate as to how to obtain the linear convolution using the DFT:

For the above example,

```
x=[1 2 3 1 0 0]; % sequence x(n)
h=[1 1 1 0 0 0];% sequence h(n)
L=length(x)+length(h)-1;%length of convolution sequence
XE=fft(x,L); % DFT of sequence x(n) with zero padding
HE=fft(h,L); % DFT of sequence h(n) with zero padding
yl=ifft(XE.*HE); % linear convolution of sequences x(n) and h(n)
```

After execution of the above MATLAB commands, the linear convolution of $x(n)$ and $h(n)$ is given by

$$y_l(n) = x(n) * h(n) = \{1, 3, 6, 6, 4, 1\}.$$

4.6.2   Linear Convolution of a Finite Length Sequence with a Long Duration Sequence

There are two methods for the evaluation of the linear convolution using the DFT, called the overlap-add and the overlap-save, when one sequence is of finite length and the other is of infinite length or much greater than the length of the finite length sequence.

(a) Overlap-Add Method

Let $x(n)$ be a sequence of long duration and $h(n)$ of finite length L_2. Let the sequence $x(n)$ be divided into a set of subsequences, each having a finite length L, and let each subsequence be padded with L_2-1 zeros to make its length equal to $L + L_2 - 1$. Then, we have

$$x_1(n) = [x(0), x(1), \ldots, x(L-1), \underbrace{0, \ldots, 0}_{L_2-1}]$$

$$x_2(n) = [x(L), x(L+1), \ldots, x(2L-1), \underbrace{0, \ldots, 0}_{L_2-1}]$$

$$x_3(n) = [x(2L), x(2L+1), \ldots, x(3L-1), \underbrace{0, \ldots, 0}_{L_2-1}] \tag{4.61}$$

$$.$$

$$.$$

$$x_m(n) = [(x((m-1)L), x((m-1)L+1), \ldots, x(mL-1), \underbrace{0, \ldots, 0}_{L_2-1}]$$

Also, the sequence $h(n)$ is padded with $L-1$ zeros to form the sequence $h'(n)$. Each of the subsequences is now convolved with $h'(n)$ of length $L + L_2-1$. Since each subsequence is terminated with $L_2 - 1$ zeros, the last $L_2 - 1$ points from each subsequence convolution output are to be overlapped and added to the first $L_2 - 1$ points of the succeeding subsequence convolution output. Hence, this procedure is called the overlap-add method. The following example illustrates this method.

Example 4.15 If the impulse response of a filter is $h(n) = \{1, 0, 1\}$, find its output $y(n) = x(n) * h(n)$ for the input sequence $x(n) = \{3, -1, 0, 1, 2, 1, 0, 1, 2\}$, by using overlap-add method.

Solution Let each subblock of the data be of length 3. Since $L_2 = 3$, two zeros are added to bring the length of each subblock to 5. Two zeros are added to $h(n)$ so that $h'(n)$ is also of length 5. Hence, the sub sequences are

$$x_1(n) = \{3, -1, 0, 0, 0\}; \quad x_2(n) = \{1, 2, 1, 0, 0\}; \quad x_3(n) = \{0, 1, 2, 0, 0\}.$$
and

$$h'(n) = \{1, 0, 1, 0, 0\}$$

Then, the circular convolutions of the subsequences with $h'(n)$ are given by

$$y_1(n) = x_1(n) \,\text{\large N}\, h'(n) \quad = \{3, -1, 3, -1, 0\}$$

$$y_2(n) = x_2(n) \,\text{\large N}\, h'(n) \quad = \{1, 2, 2, 2, 1\}$$

$$y_3(n) = x_3(n) \,\text{\large N}\, h'(n) \quad = \{0, 1, 2, 1, 2\}$$

Hence, the linear convolution of $x(n)$ and $h(n)$ is given by

$$y_l(n) = x(n) * h(n) = (3, -1, 3, 0, 2, 2, 2, 2, 2, 1, 2)$$

The above process is illustrated in Fig. 4.5.

The above procedure can be implemented by using the MATLAB command fftfilt.

h = [1 0 1 0 0];

x = [3 − 1 0 1 2 1 0 1 2 0 0];

y = fftfilt(h, x);
Thus after the execution of the above MATLAB statements, we get $y_l(n)$ as

$$y_l(n) = \{3, -1, 3, 0, 2, 2, 2, 2, 2, 1, 2\}.$$

(b) **Overlap-Save Method**

In this method, the sequence $x(n)$ is divided into a set of overlapping subsequences, each having a finite length $L + L_2 - 1$. Each subsequence contains the last $L_2 - 1$ samples of the previous subsequence, followed by the next L samples of $x(n)$. The first $L_2 - 1$ samples of the first subsequence are set to zero. Hence, the subsequences are

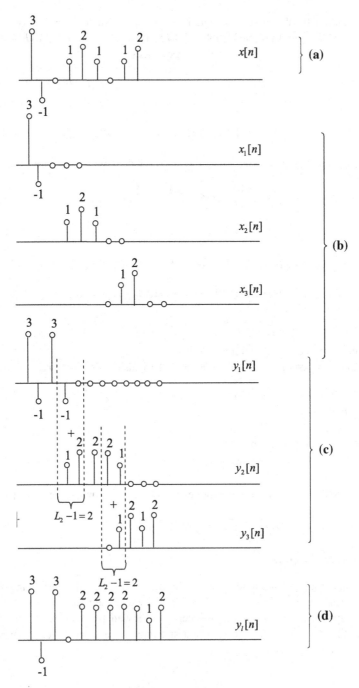

Fig. 4.5 **a** Original signal $x(n)$, **b** subblocks of $x(n)$, **c** circular convolution of the subblocks of $x(n)$ and $h'(n)$, and **d** linear convolution of $x(n)$ and $h(n)$

$$x_1(n) = [\underbrace{0,\ldots,0}_{L_2-1},x(0),x(1),\ldots,x(L-1)]$$

$$x_2(n) = [\underbrace{x(L+1-L_2),\ldots,x(L-1)}_{L_2-1\,\text{samples from}\,x_1(n)},\underbrace{x(L+1),\ldots,x(2L-1)}_{L\,\text{new samples}}]$$

$$x_3(n) = [\underbrace{x(2L+1-L_2),\ldots,x(2L-1)}_{L_2-1\,\text{samples from}\,x_2(n)},\underbrace{x(2L),\ldots,x(3L-1)}_{L\,\text{new samples}}]$$

and so on. Now, the length of the sequence $h(n)$ is increased to $L + L_2 - 1$ by padding it with $L - 1$ zeros to form the sequence $h'(n)$. Then, each of the subsequences is convolved with $h'(n)$. The first $L_2 - 1$ points of the circular convolution of each of the subsequences with $h'(n)$ do not agree with the linear convolution output of each subsequence with $h'(n)$ due to aliasing, and the remaining L points are in agreement with the linear convolution output. Hence, the first $L_2 - 1$ points of the circular convolution of each subsequence with $h'(n)$ output are to be discarded and the remaining L points from each subsequence convolution output are to be abutted to obtain the linear convolution output of $x(n)$ and $h(n)$. The following example illustrates this method:

Example 4.16 Find the filter output $y(n) = x(n) * h(n)$ for the input $x(n)$ and the impulse response $h(n)$ of Example 4.15.

Solution The subsequences of $x(n)$ are

$$x_1(n) = \{0,0,3,-1,0\}, \quad x_2(n) = \{-1,0,1,2,1\},$$
$$x_3(n) = \{2,1,0,1,2\}, \quad x_4(n) = \{1,2,0,0,0\}$$

and

$$h'(n) = \{1,0,1,0,0\}$$

Then, the circular convolution of the subsequences with $h(n)$ is given by

$$y_1(n) = x_1(n) \, \text{\textcircled{N}} \, h'(n) \quad = \{0,\,0,\,3,-1,\,3\}$$

$$y_2(n) = x_2(n) \, \text{\textcircled{N}} \, h'(n) \quad = \{-1,\,0,\,0,\,2,\,2\}$$

$$y_3(n) = x_3(n) \, \text{\textcircled{N}} \, h'(n) \quad = \{2,\,1,\,2,\,2,\,2\}$$

$$y_4(n) = x_4(n) \, \text{\textcircled{N}} \, h'(n) \quad = \{1,\,2,\,1,\,2,\,0\}$$

Hence, the linear convolution of $x(n)$ and $h(n)$ is given by

$$y_l(n) = \{3, -1, 3, 0, 2, 2, 2, 2, 2, 1, 2\}$$

This process is illustrated in Fig. 4.6a, b.

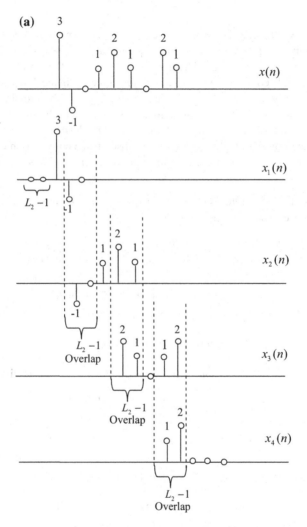

Fig. 4.6 **a** Original input $x(n)$ and subsections of $x(n)$ and **b** circular convolution of subsections of $x(n)$ and $h'(n)$, and the linear convolution of $x(n)$ and $h(n)$

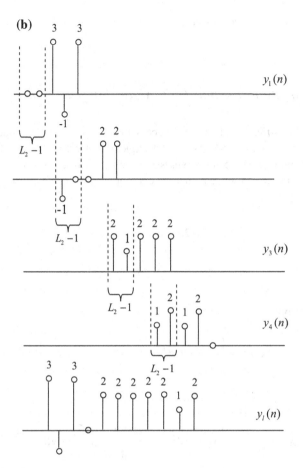

Fig. 4.6 (continued)

4.7 Fast Fourier Transform

It is evident from Eqs. (4.11) that a direct evaluation of each value of $X(k)$ requires N complex multiplications and $(N - 1)$ complex additions. As such, N^2 complex multiplications and $N(N - 1)$ complex additions are necessary for the computation of an N-point DFT. Consequently, for large N, the computational complexity in terms of the arithmetic operations is high in direct evaluation of the DFT. Therefore, a number of efficient algorithms have been developed for the computation of the DFT. These efficient algorithms collectively have become known *fast Fourier transforms*. The FFT algorithms decompose successively the computation of the discrete Fourier transform of a sequence of length N into smaller and smaller discrete Fourier transforms. The two most basic FFT algorithms are the

decimation-in-time and decimation-in frequency [1, 2], and these are considered in
the following sections.

4.7.1   Decimation-in-Time FFT Algorithm with Radix-2

The *decimation-in-time* (DIT) is the process that decomposes the input sequence
successively into smaller and smaller subsequences. Here, the radix-2 means the
number of output points N can be expressed as a power of 2; that is, $N = 2^v$, where
v is an integer. Let the input sequence be decomposed into an even sequence $g_1(n)$
and an odd sequence $g_2(n)$ as

$$g_1(n) = x(2n), \ n = 0, 1, \ldots, \frac{N}{2} - 1 \tag{4.62}$$

$$g_2(n) = x(2n), \ n = 0, 1, \ldots, \frac{N}{2} - 1 \tag{4.63}$$

We know from Eq. (4.11) that

$$X(k) = \sum_{n=0}^{N-1} x(n) W_N^{nk}, \quad k = 0, 1, \ldots, N - 1 \tag{4.64}$$

Substituting Eqs. (4.62) and (4.63) in (4.64), we get

$$X(k) = \sum_{n=0}^{(N/2)-1} x(2n) W_N^{2nk} + \sum_{n=0}^{(N/2)-1} x(2n+1) W_N^{(2n+1)k} \tag{4.65}$$

Using $W_N^2 = W_{N/2}$ in Eq. (4.65) yields

$$X(k) = \sum_{n=0}^{(N/2)-1} x(2n) W_{N/2}^{nk} + W_N^k \sum_{n=0}^{(N/2)-1} x(2n+1) W_{N/2}^{nk} \tag{4.66}$$

The RHS may be identified as the sum of two ($N/2$)-point DFTs, $G_1(k)$ and
$G_2(k)$ of the even and odd sequences $g_1(n)$ and $g_2(n)$:

$$G_1(k) = \sum_{n=0}^{(N/2)-1} g_1(n) W_{N/2}^{nk} \tag{4.67}$$

$$G_2(k) = \sum_{n=0}^{(N/2)-1} g_2(n) W_{N/2}^{nk} \tag{4.68}$$

Hence, $X(k)$ in Eq. (4.66) can be written as

$$X(k) = G_1(k) + W_N^k G_2(k) \quad k = 0, 1, \ldots, N - 1 \tag{4.69}$$

Also, since $G_1(k)$ and $G_2(k)$ are periodic with a period of $(N/2)$, $G_1(k + N/2) = G_1(k)$ and $G_2(k + N/2) = G_2(k)$, and the twiddle constant $W_N^{k+N/2} = -W_N^k$. Hence, Eq. (4.69) can be written as

$$X(k) = G_1(k) + W_N^k G_2(k) \qquad k = 0, 1, \ldots, (N/2) - 1 \tag{4.70a}$$

$$X(k + N/2) = G_1(k) - W_N^k G_2(k) \quad k = 0, 1, \ldots, (N/2) - 1 \tag{4.70b}$$

Repeating the process for each of the sequences $g_1(n)$ and $g_2(n)$, $g_1(n)$ yields two $(N/4)$-point sequences

$$\begin{aligned} g_{11}(n) &= g_1(2n) & n = 0, 1, \ldots, (N/4) - 1 \\ g_{12}(n) &= g_1(2n + 1) & n = 0, 1, \ldots, (N/4) - 1 \end{aligned} \tag{4.71a}$$

and $g_2(n)$ yields

$$\begin{aligned} g_{21}(n) &= g_2(2n) & n = 0, 1, \ldots, (N/4) - 1 \\ g_{22}(n) &= g_2(2n + 1) & n = 0, 1, \ldots, (N/4) - 1 \end{aligned} \tag{4.71b}$$

and their DFTs satisfy

$$\begin{aligned} G_1(k) &= G_{11}(k) + W_{N/2}^k G_{12}(k) & k = 0, 1, \ldots, (N/4) - 1 \\ G_1\left(k + \frac{N}{4}\right) &= G_{11}(k) - W_{N/2}^k G_{12}(k) & k = 0, 1, \ldots, (N/4) - 1 \end{aligned} \tag{4.72a}$$

$$\begin{aligned} G_2(k) &= G_{21}(k) + W_{N/2}^k G_{22}(k) & k = 0, 1, \ldots, (N/4) - 1 \\ G_2\left(k + \frac{N}{4}\right) &= G_{21}(k) - W_{N/2}^k G_{22}(k) & k = 0, 1, \ldots, (N/4) - 1 \end{aligned} \tag{4.72b}$$

This process can be continued until we are left with only two-point transforms. For example, for $N = 4$, Eqs. (4.70a) and (4.70b) become

$$\begin{aligned} X(k) &= G_1(k) + W_N^k G_2(k) & k = 0, 1 \\ X(k + 2) &= G_1(k) - W_N^k G_2(k) & k = 0, 1 \end{aligned} \tag{4.73}$$

Equation (4.73) can be represented by the flow graph as shown in Fig. 4.7. This is usually referred to as the butterfly diagram for four-point DFT. In the first stage, two 2-point DFTs and, in the second stage, one 4-point DFT are computed.

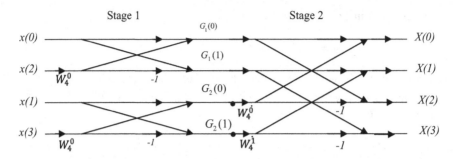

Fig. 4.7 Decomposition of a four-point DFT using DIT

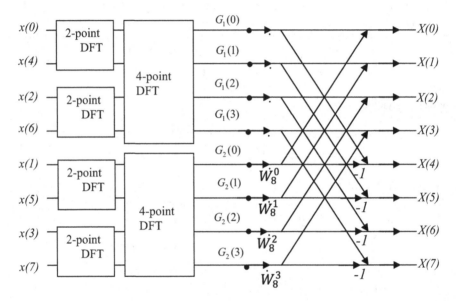

Fig. 4.8 Decomposition of an eight-point DFT using DIT

For $N = 8$, Eqs. (4.70a) and (4.70b) become

$$\begin{aligned}
X(k) &= G_1(k) + W_N^k\, G_2(k) \qquad k = 0, 1, 2, 3 \\
X(k+2) &= G_1(k) - W_N^k G_2(k) \qquad k = 0, 1, 2, 3
\end{aligned} \tag{4.74}$$

The computation of an eight-point DFT is performed in three stages as shown in Fig. 4.8.

It is observed from the flow graph that in the first stage, four 2-point DFTs, in the second stage, two 4-point DFTs, and finally, in the third stage, one 8-point DFT are computed. Also, the number of complex multiplications carried out at each stage is equal to $4 = N/2$, and the number of additions performed is N. Hence, the total

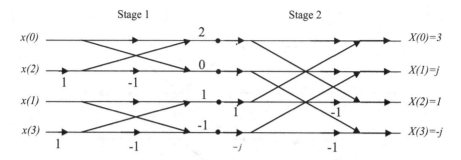

Fig. 4.9 Decomposition of the four-point DFT of Example 4.17 using the DIT algorithm

number of complex multiplications and additions in computing all the 8 samples is 12 and 24, respectively. Following the same argument, it can be observed that in the general case of $N = 2^v$, the number of stages of computation will be $v = \log_2 N$; hence, the total number of complex multiplications and additions needed in computing all the N DFT samples is $(N/2) \log_2 N$, and the number of complex additions is $N \log_2 N$.

Example 4.17 Find the four-point FFT of $x(n) = \{1, 0, 1, 1\}$ using the decimation-in-time algorithm.

Solution With $N = 4$, the two twiddle factors are

$$W_4^0 = 1 \quad \text{and} \quad W_4^1 = e^{-j2\pi/4} = \cos(\pi/2) - j \sin(\pi/2) = -j.$$

Since it is a four-point DFT, the DIT flow graph consists of two stages as shown in Fig. 4.9. The outputs of the first and second stages are computed as follows:

Stage 1

$$x_1(0) = x(0) + W_4^0 x(2) = 1 + 1 = 2;$$
$$x_1(2) = x(0) - W_4^0 x(2) = 1 - 1 = 0;$$
$$x_1(1) = x(1) + W_4^0 x(3) = 0 + 1 = 1;$$
$$x_1(3) = x(1) - W_4^0 x(3) = 0 - 1 = -1;$$

where the sequence $x_1(n)$ represents the intermediate output after the first stage and becomes the input to the second (final) stage.

Stage 2

$$X(0) = x_1(0) + W_4^0 x_1(1) = 2 + 1 = 3;$$
$$X(2) = x_1(0) - W_4^0 x_1(1) = 2 - 1 = 1;$$

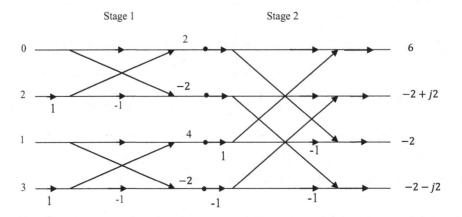

Fig. 4.10 Decomposition of the four-point DFT of Example 4.18 using the DIT algorithm

$$X(1) = x_1(2) + W_4^1 x_1(3) = 1 + (-j)(-1) = j;$$
$$X(3) = x_1(2) - W_4^1 x_1(3) = 0 - (-j)(-1) = -j;$$

Example 4.18 Consider an input data string of $x(n) = (0, 1, 2, 3)$. Draw the butterfly diagram of the FFT showing the input, intermediate outputs, and the final output to compute the DFT of $x(n)$.

Solution By computing the outputs of the first and second stages as was done in the previous example, the required butterfly diagram is shown in Fig. 4.10.

Example 4.19 Find the eight-point FFT of $x(n) = \{1, 0, 1, 1, 1, 1, 1, 0\}$ using the DIT algorithm.

Solution With $N = 8$, the four twiddle factors are

$$W_8^0 = 1;$$
$$W_8^1 = e^{-j2\pi/8} = \cos(\pi/4) - j\sin(\pi/4) = 0.707 - j0.707;$$
$$W_8^2 = e^{-j4\pi/8} = -j;$$
$$W_8^3 = e^{-j6\pi/8} = -0.707 - j0.707;$$

Since it is an eight-point DFT with radix-2, the DIT flow graph consists of three stages as shown in Fig. 4.11. The outputs of the three stages are computed as follows:

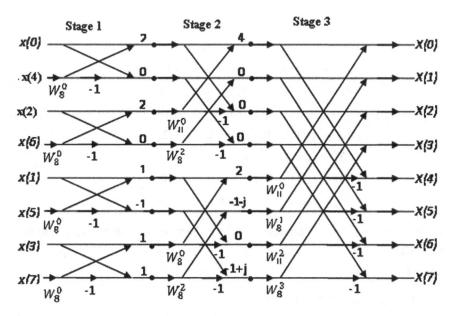

Fig. 4.11 Decomposition of the eight-point DFT of Example 4.19 using the DIT algorithm

Stage 1

$$x_1(0) = x(0) + W_8^0 x(4) = 1 + 1 = 2;$$
$$x_1(4) = x(0) - W_8^0 x(4) = 1 - 1 = 0;$$
$$x_1(2) = x(2) + W_8^0 x(6) = 1 + 1 = 2;$$
$$x_1(6) = x(2) - W_8^0 x(6) = 1 - 1 = 0;$$
$$x_1(1) = x(1) + W_8^0 x(5) = 0 + 1 = 1;$$
$$x_1(5) = x(1) - W_8^0 x(5) = 0 - 1 = -1;$$
$$x_1(3) = x(3) + W_8^0 x(7) = 1 + 0 = 1;$$
$$x_1(7) = x(3) - W_8^0 x(7) = 1 - 0 = 1;$$

where the sequence $x_1(n)$ represents the intermediate output after the first stage and becomes the input to the second stage.

Stage 2

$$x_2(0) = x_1(0) + W_8^0 x_1(2) = 2 + 2 = 4;$$
$$x_2(4) = x_1(4) + W_8^2 x_1(6) = 0 + (-j)0 = 0;$$
$$x_2(2) = x_1(0) - W_8^0 x_1(2) = 2 - 2 = 0;$$
$$x_2(6) = x_1(4) - W_8^2 x_1(6) = 0 + (-j)0 = 0;$$
$$x_2(1) = x_1(1) + W_8^0 x_1(3) = 1 + 1 = 2;$$
$$x_2(5) = x_1(5) + W_8^2 x_1(7) = -1 + (-j) = -1 - j;$$
$$x_2(3) = x_1(1) - W_8^0 x_1(3) = 1 - 1 = 0;$$
$$x_2(7) = x_1(5) - W_8^2 x_1(7) = -1 - (-j) = -1 + j;$$

where the second-stage output sequence $x_2(n)$ becomes the input sequence to the final stage.

Stage 3

$$X(0) = x_2(0) + W_8^0 x_2(1) = 4 + 2 = 6;$$
$$X(1) = x_2(4) + W_8^1 x_2(5) = 0 + (0.707 - j0.707)(-1 - j) = -1.414;$$
$$X(2) = x_2(2) + W_8^2 x_2(3) = 0 + (-j)0 = 0;$$
$$X(3) = x_2(6) + W_8^3 x_2(7) = 0 + (-0.707 - j0.707)(-1 + j) = 1.414;$$
$$X(4) = x_2(0) - W_8^0 x_2(1) = 4 - 2 = 2;$$
$$X(5) = x_2(4) - W_8^1 x_2(5) = 0 - (0.707 - j0.707)(-1 - j) = 1.414;$$
$$X(6) = x_2(2) - W_8^2 x_2(3) = 0 - (-j)(0) = 0;$$
$$X(7) = x_2(6) - W_8^3 x_2(7) = 0 - (-0.707 - j0.707)(-1 + j) = -1.414;$$

Example   4.20 Find   the   16-point   FFT   of   the   sequence $x(n) =$ $\{1, 0, 1, 1, 0, 1, 1, 0, 1, 0, 0, 1, 1, 1, 1, 0\}$ using the DIT algorithm.

Solution With $N = 16$, eight twiddle factors need to be calculated; these are

$$W_{16}^0 = 1; W_{16}^1 = e^{-j2\pi/16} = 0.9238 - j0.3826;$$

$$W_{16}^2 = e^{-j4\pi/16} = 0.707 - j0.707;$$

$$W_{16}^3 = e^{-j6\pi/16} = 0.3826 - j0.9238;$$

$$W_{16}^4 = e^{-j8\pi/16} = 0 - j;$$

$$W_{16}^5 = e^{-j10\pi/16} = -0.3826 - j0.9238;$$

$$W_{16}^6 = e^{-j12\pi/16} = -0.707 - j0.707;$$

$$W_{16}^7 = e^{-j14\pi/16} = -0.9238 - j0.3826.$$

Since it is a 16-point DFT with radix-2, the DIT flow graph consists of four stages as shown in Fig. 4.12. The outputs of the four stages are computed as follows:

Stage 1

$$x_1(0) = x(0) + W_{16}^0 x(8) = 1 + 1 = 2;$$

$$x_1(8) = x(0) - W_{16}^0 x(8) = 1 - 1 = 0;$$

$$x_1(4) = x(4) + W_{16}^0 x(12) = 0 + 1 = 1;$$

$$x_1(12) = x(4) - W_{16}^0 x(12) = 0 - 1 = -1;$$

$$x_1(2) = x(2) + W_{16}^0 x(10) = 1 + 0 = 1;$$

$$x_1(10) = x(2) - W_{16}^0 x(10) = 1 - 0 = 1;$$

$$x_1(6) = x(6) + W_{16}^0 x(14) = 1 + 1 = 2;$$

$$x_1(14) = x(6) - W_{16}^0 x(14) = 1 - 1 = 0;$$

$$x_1(1) = x(1) + W_{16}^0 x(9) = 0 + 0 = 0;$$

$$x_1(9) = x(1) - W_{16}^0 x(9) = 0 - 0 = 0;$$

$$x_1(5) = x(5) + W_{16}^0 x(13) = 1 + 1 = 2;$$

$$x_1(13) = x(5) - W_{16}^0 x(13) = 1 - 1 = 0;$$

$$x_1(3) = x(3) + W_{16}^0 x(11) = 1 + 1 = 2;$$

$$x_1(11) = x(3) - W_{16}^0 x(11) = 1 - 1 = 0;$$

$$x_1(7) = x(7) + W_{16}^0 x(15) = 0 + 0 = 0;$$

$$x_1(15) = x(7) - W_{16}^0 x(15) = 0 - 0 = 0;$$

where the sequence $x_1(n)$ represents the intermediate output after the first iteration and becomes the input to the second stage.

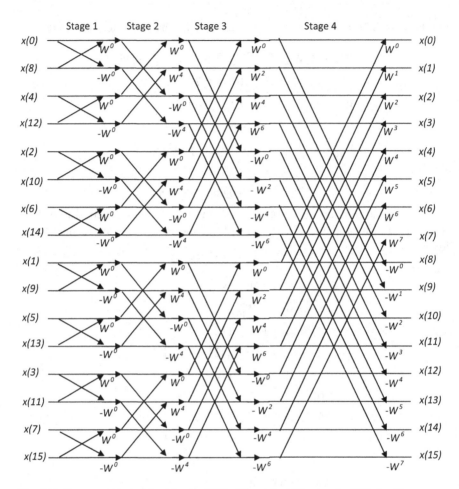

Fig. 4.12 Decomposition of the 16-point DFT of Example 4.20 using the DIT algorithm

Stage 2

$$x_2(0) = x_1(0) + W_{16}^0 x_1(4) = 2 + 1 = 3;$$
$$x_2(8) = x_1(8) + W_{16}^4 x_1(12) = 0 + (-j)(-1) = j;$$
$$x_2(4) = x_1(0) - W_{16}^0 x_1(4) = 2 - 1 = 1;$$
$$x_2(12) = x_1(8) - W_{16}^4 x_1(12) = 0 - (-j)(-1) = -j;$$
$$x_2(2) = x_1(2) + W_{16}^0 x_1(6) = 1 + 2 = 3;$$
$$x_2(10) = x_1(10) + W_{16}^4 x_1(14) = 1 - (-j)0 = 1;$$
$$x_2(6) = x_1(2) - W_{16}^0 x_1(6) = 1 - 2 = -1;$$
$$x_2(14) = x_1(10) - W_{16}^4 x_1(14) = 1 - (-j)0 = 1;$$
$$x_2(1) = x_1(1) + W_{16}^0 x_1(5) = 0 + 2 = 2;$$
$$x_2(9) = x_1(9) + W_{16}^4 x_1(13) = 0 + (-j)0 = 0;$$
$$x_2(5) = x_1(1) - W_{16}^0 x_1(5) = 0 - 2 = -2;$$
$$x_2(13) = x_1(9) - W_{16}^4 x_1(13) = 0 - (-j)0 = 0;$$
$$x_2(3) = x_1(3) + W_{16}^0 x_1(7) = 2 + 0 = 2;$$
$$x_2(11) = x_1(11) + W_{16}^4 x_1(15) = 0 + (-j)0 = 0;$$
$$x_2(7) = x_1(3) - W_{16}^0 x_1(7) = 2 - 0 = 2;$$
$$x_2(15) = x_1(11) - W_{16}^4 x_1(15) = 0 - (-j)0 = 0;$$

where the intermediate second-stage output sequence $x_2(n)$ becomes the input sequence to the next one.

Stage 3

$$x_3(0) = x_2(0) + W_{16}^0 x_2(2) = 3 + 3 = 6;$$

$$x_3(8) = x_2(8) + W_{16}^2 x_2(10) = j + (0.707 - j0.707)(1) = 0.707 + j0.2929;$$

$$x_3(4) = x_2(4) + W_{16}^4 x_2(6) = 1 + (-j)(-1) = 1 + j;$$

$$x_3(12) = x_2(12) + W_{16}^6 x_2(14) = (-j) + (-0.707 - j0.707)(1) = -0.707 - j1.707;$$

$$x_3(2) = x_2(0) - W_{16}^0 x_2(2) = 3 - 3 = 0;$$

$$x_3(10) = x_2(8) - W_{16}^2 x_2(10) = j - (0.707 - j0.707)(1) = -0.707 + j1.707;$$

$$x_3(6) = x_2(4) - W_{16}^4 x_2(6) = 1 - (-j)(-1) = 1 - j;$$

$$x_3(14) = x_2(12) - W_{16}^6 x_2(14) = (-j) - (-0.707 - j0.707)(1) = 0.707 - j0.2929;$$

$$x_3(1) = x_2(1) + W_{16}^0 x_2(3) = 2 + 2 = 4;$$

$$x_3(9) = x_2(9) + W_{16}^2 x_2(11) = 0 + (0.707 - j0.707)0 = 0;$$

$$x_3(5) = x_2(5) + W_{16}^4 x_2(7) = -2 + (-j)2 = -2 - 2j;$$

$$x_3(13) = x_2(13) + W_{16}^6 x_2(15) = 0 - (-0.707 - j0.707)0 = 0;$$

$$x_3(3) = x_2(1) - W_{16}^0 x_2(3) = 2 - 2 = 0;$$

$$x_3(11) = x_2(9) - W_{16}^2 x_2(11) = 0 - (0.707 - j0.707)0 = 0;$$

$$x_3(7) = x_2(5) - W_{16}^4 x_2(7) = -2 - 0 = -2;$$

$$x_3(15) = x_2(13) - W_{16}^6 x_2(15) = 0 - (-j)0 = 0;$$

where the intermediate third-stage output sequence $x_3(n)$ becomes the input sequence to the final stage.

Stage 4

$$X(0) = x_3(0) + W_{16}^0 x_3(1) = 6 + 4 = 10;$$
$$X(1) = x_3(8) + W_{16}^1 x_3(9) = 0.707 + j0.2929;$$
$$X(2) = x_3(4) + W_{16}^2 x_3(5) = -1.8284 + j;$$
$$X(3) = x_3(12) + W_{16}^3 x_3(13) = -0.707 - j1.707;$$
$$X(4) = x_3(2) + W_{16}^4 x_3(3) = 0;$$
$$X(5) = x_3(10) + W_{16}^5 x_3(11) = -0.707 + j1.707;$$
$$X(6) = x_3(6) + W_{16}^6 x_3(7) = 3.8284 - j;$$
$$X(7) = x_3(14) + W_{16}^7 x_3(15) = 0.707 - j0.2929;$$
$$X(8) = x_3(0) - W_{16}^0 x_3(1) = 2;$$
$$X(9) = x_3(8) - W_{16}^1 x_3(9) = 0.7071 + j0.2929;$$
$$X(10) = x_3(4) - W_{16}^2 x_3(5) = 3.8284 + j;$$
$$X(11) = x_3(12) - W_{16}^3 x_3(13) = -0.707 - j1.707;$$
$$X(12) = x_3(2) - W_{16}^4 x_3(3) = 0;$$
$$X(13) = x_3(10) - W_{16}^5 x_3(11) = -0.707 + j1.707;$$
$$X(14) = x_3(6) - W_{16}^6 x_3(7) = -1.8284 - j;$$
$$X(15) = x_3(14) - W_{16}^7 x_3(15) = 0.707 - j0.2929;$$

4.7.2 In-Place Computation

In the implementation of the DIT FFT algorithm, only one complex array of N storage registers is physically necessary, since the complex numbers resulting from the mth stage can be stored in the same registers that had stored the complex numbers resulting from the $(m - 1)$th stage, once the output variables of the mth stage have been determined from the output numbers of the $(m - 1)$th stage. This type of computation is referred to as *in-place computation*. Thus, for in-place computation in the DIT algorithm in which the DFT samples appear in the natural order (i.e., $X(k)$, $k = 0, 1, ..., N - 1$), the input sequence samples are to be stored in index *bit-reversed* order. If $x(b_2 b_1 b_0)$ represents the sample $x(n)$ in the index bit-reversed binary form, then the sample $x(b_2 b_1 b_0)$ would appear in the location of the sample $x(b_0 b_1 b_2)$ of the input sequence to the DIT algorithm. For an eight-point DFT, the bit-reversal process is shown in Table 4.5.

Table 4.5 Bit-reversal process for $N = 8$

Input sequence samples	Input sequence samples with index binary representation	Input sequence samples with bit-reversed binary index	Index bit-reversed samples
$x(0)$	$x(000)$	$x(000)$	$x(0)$
$x(1)$	$x(001)$	$x(100)$	$x(4)$
$x(2)$	$x(010)$	$x(010)$	$x(2)$
$x(3)$	$x(011)$	$x(110)$	$x(6)$
$x(4)$	$x(100)$	$x(001)$	$x(1)$
$x(5)$	$x(101)$	$x(101)$	$x(5)$
$x(6)$	$x(110)$	$x(011)$	$x(3)$
$x(7)$	$x(111)$	$x(111)$	$x(7)$

4.7.3 Decimation-in-Frequency FFT Algorithm with Radix-2

The basic idea in the decimation-in-time (DIT) algorithm was to decompose the input sequence successively into smaller and smaller subsequences. In the case of *decimation-in-frequency* (DIF) algorithm, we decompose the N-point DFT sequence $X(k)$ successively into smaller and smaller subsequences. Consider an input sequence $x(n)$, and divide it into two halves. Then, the DFT of $x(n)$ can be written as

$$X(k) = \sum_{n=0}^{(N/2)-1} x(n)W_N^{nk} + \sum_{n=N/2}^{(N/2)-1} x(n)W_N^{nk} \qquad (4.75a)$$

The above equation can be rewritten as

$$X(k) = \sum_{n=0}^{(N/2)-1} x(n)W_N^{nk} + W_N^{kN/2}\sum_{n=N/2}^{(N/2)-1} x\left(n + \frac{N}{2}\right)W_N^{nk} \qquad (4.75b)$$

Since $W_N^{Nk/2} = (-1)^k$, Eq. (4.75b) becomes

$$X(k) = \sum_{n=0}^{(N/2)-1}\left[x(n) + (-1)^k x\left(n + \frac{N}{2}\right)\right]W_N^{nk} \qquad (4.75c)$$

Now, splitting $X(k)$ into even-indexed and odd-indexed samples, Eq. (4.75c) can be written as consisting of two $(N/2)$-point DFTs for $k = 0,1, \ldots, (N/2)-1$.

$$X(2k) = \sum_{n=0}^{(N/2)-1} \left[x(n) + x\left(n + \frac{N}{2} \right) \right] W_{N/2}^{nk} \qquad (4.76\text{a})$$

$$X(2k+1) = \sum_{n=0}^{(N/2)-1} \left[x(n) - x\left(n + \frac{N}{2} \right) \right] W_N^n W_{N/2}^{nk} \qquad (4.76\text{b})$$

Let

$$x_1(n) = x(n) + x\left(n + \frac{N}{2} \right) \quad n = 0, 1, 2, \ldots, \left(\frac{N}{2} \right) - 1 \qquad (4.77\text{a})$$

$$x_2(n) = x(n) - x\left(n + \frac{N}{2} \right) W_N^n \quad n = 0, 1, 2, \ldots, \left(\frac{N}{2} \right) - 1 \qquad (4.77\text{b})$$

Then, the even- and odd-indexed $X(k)$'s are found from the $(N/2)$-point transforms of $x_1(n)$ and $x_2(n)$ as

$$X(2k) = \sum_{n=0}^{(N/2)-1} x_1(n) W_{N/2}^{nk} \qquad (4.78\text{a})$$

and

$$X(2k+1) = \sum_{n=0}^{(N/2)-1} x_2(n) W_{N/2}^{nk} \qquad (4.78\text{b})$$

Repeating the process for each of the sequences $x_1(n)$ and $x_2(n)$ yields the two $(N/4)$-point sequences

$$\begin{aligned} x_{11}(n) &= x_1(n) + x_1\left(n + \frac{N}{4} \right) & n = 0, 1, \ldots, (N/4) - 1 \\ x_{12}(n) &= \left(x_1(n) - x_1\left(n + \frac{N}{4} \right) \right) W_N^{2n} & n = 0, 1, \ldots, (N/4) - 1 \end{aligned} \qquad (4.79\text{a})$$

and $x_2(n)$ yields

$$\begin{aligned} x_{21}(n) &= x_2(n) + x_2\left(n + \frac{N}{4} \right) & n = 0, 1, \ldots, (N/4) - 1 \\ x_{22}(n) &= \left(x_2(n) - x_2\left(n + \frac{N}{4} \right) \right) W_N^{2n} & n = 0, 1, \ldots, (N/4) - 1 \end{aligned} \qquad (4.79\text{b})$$

Then, the even- and odd-indexed $X(k)$'s are found from the $(N/4)$-point transforms of $x_{11}(n), x_{12}(n), x_{21}(n)$ and $x_{22}(n)$ as

$$X(4k) = \sum_{n=0}^{(N/4)-1} x_{11}(n) W_{N/4}^{nk}$$

$$X(4k+2) = \sum_{n=0}^{(N/4)-1} x_{12}(n) W_{N/4}^{nk}$$

(4.80)

$$X(4k+1) = \sum_{n=0}^{(N/4)-1} x_{21}(n) W_{N/4}^{nk}$$

$$X(4k+3) = \sum_{n=0}^{(N/4)-1} x_{22}(n) W_{N/4}^{nk}$$

The process is to be continued until they reduce to two-point transforms.

For example, for $N = 4$, the two twiddle factors needed are $W_4^0 = 1$ and $W_4^1 = -j$. The DIF flow graph for a four-point DFT contains two stages as shown in Fig. 4.13. The outputs of the two stages are computed as follows:

Stage 1

$$x_1(0) = x(0) + x(2)$$
$$x_1(1) = x(1) + x(3)$$
$$x_1(2) = [x(0) - x(2)]w_4^0$$
$$x_1(3) = [x(1) - x(3)]w_4^1$$

where $x_1(0)$, $x_1(1)$, $x_1(2)$ and $x_1(3)$ represent the intermediate output sequence after the first stage, which become the input to the second stage.

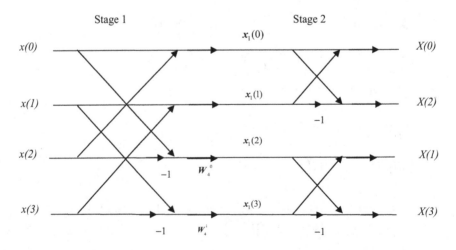

Fig. 4.13 Decomposition of a four-point DFT using the DIF algorithm

Stage 2

$$X(0) = x_1(0) + x_1(1)$$
$$X(1) = x_1(2) + x_1(3)$$
$$X(2) = x_1(0) - x_1(1)$$
$$X(3) = x_1(2) - x_1(3)$$

For $N = 8$, the decomposition of an 8-point DFT into two 4-point DFTS with DIF algorithm is shown in Fig. 4.14.

Example 4.21 Find the DFT of the sequence $x(n) = (1, 2, 3, 4)$ using the DIF algorithm.

Solution The two twiddle factors needed are $W_4^0 = 1$ and $W_4^1 = -j$.

The DIF flow graph for four-point DFT consists of two stages as shown in Fig. 4.15. The outputs of the two stages are computed as follows:

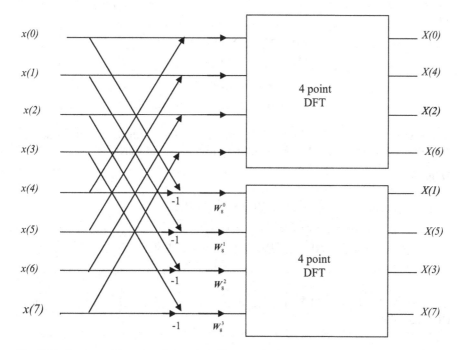

Fig. 4.14 Decomposition of an eight-point DFT using the DIF algorithm decimation-in-frequency

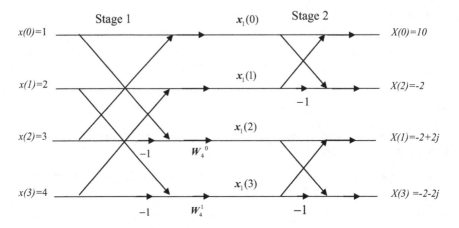

Fig. 4.15 Flow graph for the four-point FFT of Example 4.21 using the DIF algorithm

Stage 1

$$x_1(0) = x(0) + x(2) = 4$$
$$x_1(1) = x(1) + x(3) = 6$$
$$x_1(2) = [x(0) - x(2)]W_4^0 = -2$$
$$x_1(3) = [x(1) - x(3)]W_4^1 = 2j$$

where $x_1(0)$, $x_1(1)$, $x_1(2)$ and $x_1(3)$ represent the intermediate output sequence after the first stage, which become the input to the second stage.

Stage 2

$$X(0) = x_1(0) + x_1(1) = 10$$
$$X(2) = x_1(0) - x_1(1) = -2$$
$$X(1) = x_1(2) + x_1(3) = -2 + 2j$$
$$X(3) = x_1(2) - x_1(3) = -2 - 2j$$

Example 4.22 Find the DFT of a sequence $x(n) = (1,1,1,1,1,1,0,0)$ using the DIF algorithm.

Solution With $N = 8$, the four twiddle factors needed are

$$W_8^0 = 1;$$
$$W_8^1 = e^{-j2\pi/8} = e^{-j\pi/4} = 0.707 - j0.707;$$
$$W_8^2 = e^{-j4\pi/8} = e^{-j\pi/2} = -j;$$
$$W_8^3 = e^{-j6\pi/8} = e^{-j3\pi/4} = -0.707 - j0.707;$$

Stage 1

$$x_1(0) = x(0) + x(4) = 2;$$
$$x_1(1) = x(1) + x(5) = 2;$$
$$x_1(2) = x(2) + x(6) = 1;$$
$$x_1(3) = x(3) + x(7) = 1;$$
$$x_1(4) = [x(0) - x(4)]W_8^0 = 0;$$
$$x_1(5) = [x(1) - x(5)]W_8^1 = 0;$$
$$x_1(6) = [x(2) - x(6)]W_8^2 = -j;$$
$$x_1(7) = [x(3) - x(7)]W_8^3 = -0.707 - j0.707;$$

where $x_1(0), x_1(1), \ldots, x_1(7)$ represent the intermediate output sequence after the first stage, which become the input to the second stage.

Stage 2

$$x_2(0) = x_1(0) + x_1(2) = 3;$$
$$x_2(1) = x_1(1) + x_1(3) = 3;$$
$$x_2(2) = [x_1(0) - x_1(2)]W_8^0 = 1;$$
$$x_2(3) = [x_1(1) - x_1(3)]W_8^2 = -j;$$
$$x_2(4) = x_1(4) + x_1(6) = -j;$$
$$x_2(5) = x_1(5) + x_1(7) = -0.707 - j0.707;$$
$$x_2(6) = [x_1(4) - x_1(6)]W_8^0 = j;$$
$$x_2(7) = [x_1(5) - x_1(7)]W_8^2 = 0.707 - j0.707;$$

where $x_2(0), x_2(1), \ldots, x_2(7)$ represent the intermediate output sequence after the second stage, which become the input to the final stage.

Stage 3

We now use the notation of X's to represent the final output sequence. The values $X(0), X(1), \ldots, X(7)$ form the output sequence.

$$X(0) = x_2(0) + x_2(1) = 6;$$
$$X(4) = x_2(0) - x_2(1) = 0;$$
$$X(2) = x_2(2) + x_2(3) = 1 - j1;$$
$$X(6) = x_2(2) - x_2(3) = 1 + j1;$$
$$X(1) = x_2(4) + x_2(5) = -0.707 - j1.707;$$
$$X(5) = x_2(4) - x_2(5) = 0.707 - j0.2929;$$
$$X(3) = x_2(6) + x_2(7) = 0.707 + j0.2929;$$
$$X(7) = x_2(6) - x_2(7) = -0.707 + j1.707;$$

The DIF flow graph for eight-point DFT consists of three stages as shown in Fig. 4.16. The outputs of the three stages are computed in Fig. 4.16.

It should be noted that flow graph representing the DIF FFT may be considered as an in-place computation, just as in the case of the DIT FFT. Further, it should be noted that the input sequence $x(n)$ is in order, while the output sequence $X(k)$ is in bit-reversed order. The number of multiplications and additions for computing an N-point by DIF FFT is the same as in the case of the DIT FFT, namely $(N/2)\log_2 N$ and $N\log_2 N$, respectively.

It is worth pointing out that the flow graphs of DIT FFT and DIF FFT algorithms are transposes of one another.

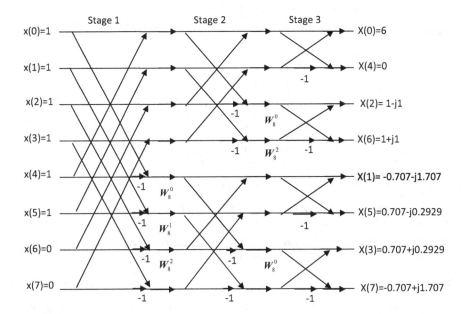

Fig. 4.16 Flow graph of the eight-point FFT for the Example 4.22 using DIF algorithm

4.7.4 Radix-4 DIF FFT Algorithm

If $N = 2^{2\nu}$, then we can use radix-4 algorithms rather than radix-2 algorithms, and this gives us a reduction in the number of multiplications to be performed. Here, we will consider the radix-4 DIF algorithm. Radix-4 DIT algorithm can be developed in a way similar to that of the radix-2 DIT algorithm.

Consider a sequence $x(n)$, and divide it into four parts so that the DFT of $x(n)$ can be written as

$$X(k) = \sum_{n=0}^{(N/4)-1} x(n)W_N^{nk} + \sum_{n=N/4}^{(N/2)-1} x(n)W_N^{nk} + \sum_{n=N/2}^{(3N/4)-1} x(n)W_N^{nk} + \sum_{n=3N/4}^{N-1} x(n)W_N^{nk}$$

$$(4.81)$$

The above equation can be rewritten as

$$X(k) = \sum_{n=0}^{(N/4)-1} x(n)W_N^{nk} + W^{kN/4}\sum_{n=0}^{(N/4)-1} x(n+N/4)W_N^{nk}$$

$$+ W^{kN/2}\sum_{n=0}^{(N/4)-1} x(n+N/2)W_N^{nk} + W^{k3N/4}\sum_{n=0}^{(N/4)-1} x(n+3N/4)W_N^{nk}$$

$$(4.82)$$

Substituting

$$W_N^{kN/4} = e^{-jk\pi/2} = (-j)^k; \; W_N^{kN/2} = e^{-jk\pi} = (-1)^k; \; W_N^{3kN/4} = (j)^k$$

in the above equation, we get

$$X(k) = \sum_{n=0}^{\frac{N}{4}-1}\left[x(n) + (-j)^k x\left(n+\frac{N}{4}\right) + (-1)^k x\left(n+\frac{N}{2}\right) + (j)^k x\left(n+\frac{3N}{4}\right)\right]W_N^{nk}$$

$$(4.83)$$

Since the twiddle factor depends on N, the above relation is not $N/4$-point DFT. To represent it as an $N/4$-point DFT, the DFT sequence is divided into four $N/4$-point subsequences, $X(4k)$, $X(4k+1)$, $X(4k+2)$ and $X(4k+3)$ for $k = 0, 1, \ldots \left(\frac{N}{4}-1\right)$. Thus, the DIF FFT with radix-4 can be represented as

$$X(4k) = \sum_{n=0}^{(N/4)-1} [x(n) + x(n+N/4) + x(n+N/2) + x(n+3N/4)]W_{N/4}^{nk} \quad (4.84)$$

$$X(4k+1) = \sum_{n=0}^{(N/4)-1} [x(n) - jx(n+N/4) - x(n+N/2) + jx(n+3N/4)] W_N^n W_{N/4}^{nk}$$

$$(4.85)$$

$$X(4k+2) = \sum_{n=0}^{(N/4)-1} [x(n) - x(n+N/4) + x(n+N/2) - x(n+3N/4)] W_N^{2n} W_{N/4}^{nk}$$

$$(4.86)$$

$$X(4k+3) = \sum_{n=0}^{(N/4)-1} [x(n) + jx(n+N/4) - x(n+N/2) - jx(n+3N/4)] W_N^{3n} W_{N/4}^{nk}$$

$$(4.87)$$

The following example illustrates a 16-point radix-4 FFT using the DIF procedure.

Example 4.23 Find the DFT of a sequence $x(n) = \{1, 1, 0, 1, 1, 0, 1, 1, 0, 1, 1, 1, 1, 1, 1, 1\}$ using the radix-4 DIF algorithm.

Solution The twiddle factors for 16-point radix-4 FFT are

$$W_{16}^0 = 1; W_{16}^1 = 0.9238 - j0.3826; W_{16}^2 = 0.707 - j0.707;$$
$$W_{16}^3 = 0.3826 - j0.9238; W_{16}^4 = 0 - j; W_{16}^5 = -0.3826 - j0.9238;$$
$$W_{16}^6 = -0.707 - j0.707; W_{16}^7 = -0.9238 - j0.3826.$$
$$W_4^0 = 1; W_4^1 = -j; W_4^2 = -1; W_4^3 = +j; W_4^4 = 1; W_4^5 = -j;$$
$$W_4^6 = -1; W_4^7 = +j;$$

The outputs of the two stages are computed as follows:

Stage 1

$$x_1(0) = [x(0) + x(4) + x(8) + x(12)]W_{16}^0 = 1 + 1 + 0 + 1 = 3;$$
$$x_1(1) = [x(1) + x(5) + x(9) + x(13)]W_{16}^0 = 1 + 0 + 1 + 1 = 3;$$
$$x_1(2) = [x(2) + x(6) + x(10) + x(14)]W_{16}^0 = 0 + 1 + 1 + 1 = 3;$$
$$x_1(3) = [x(3) + x(7) + x(11) + x(15)]W_{16}^0 = 1 + 1 + 1 + 1 = 4;$$
$$x_1(4) = [x(0) - jx(4) - x(8) + jx(12)]W_{16}^0 = 1 - j - 0 + j = 1;$$
$$x_1(5) = [x(1) - jx(5) - x(9) + jx(13)]W_{16}^1 = (1 - 0 - 1 + j)W_{16}^1 = 0.3826 + j0.9238;$$
$$x_1(6) = [x(2) - jx(6) - x(10) + jx(14)]W_{16}^2 = (0 - j - 1 + j)W_{16}^2 = -0.707 + j0.707;$$
$$x_1(7) = [x(3) - jx(7) - x(11) + jx(15)]W_{16}^3 = (1 - j - 1 + j)W_{16}^3 = 0;$$
$$x_1(8) = [x(0) - x(4) + x(8) - x(12)]W_{16}^0 = 1 - 1 + 0 - 1 = -1;$$
$$x_1(9) = [x(1) - x(5) + x(9) - x(13)]W_{16}^2 = (1 - 0 + 1 - 1)W_{16}^2 = 0.707 - j0.707;$$
$$x_1(10) = [x(2) - x(6) + x(10) - x(14)]W_{16}^4 = (0 - 1 + 1 - 1)W_{16}^4 = j;$$
$$x_1(11) = [x(3) - x(7) + x(11) - x(15)]W_{16}^6 = (1 - 1 + 1 - 1)W_{16}^6 = 0;$$
$$x_1(12) = [x(0) + jx(4) - x(8) - jx(12)]W_{16}^0 = (1 + j - 0 - j)W_{16}^0 = 1;$$
$$x_1(13) = [x(1) + jx(5) - x(9) - jx(13)]W_{16}^3 = (1 - 0 - 1 - j)W_{16}^3 = -0.9238 - j0.3826;$$
$$x_1(14) = [x(2) + jx(6) - x(10) - jx(14)]W_{16}^6 = (0 + j - 1 - j)W_{16}^6 = 0.707 + j0.707;$$
$$x_1(15) = [x(3) + jx(7) - x(11) - jx(15)]W_{16}^9 = (1 + j - 1 - j)W_{16}^9 = (1 + j - 1 - j)W_{16}^{-1} = 0;$$

Stage 2

$$X(0) = [x_1(0) + x_1(1) + x_1(2) + x_1(3)]W_{16}^0 = 3 + 3 + 3 + 4 = 13;$$
$$X(1) = [x_1(4) + x_1(5) + x_1(6) + x_1(7)]W_{16}^0 = 0.6756 + j1.6310;$$
$$X(2) = [x_1(8) + x_1(9) + x_1(10) + x_1(11)]W_{16}^0 = -0.2929 + j0.2929;$$
$$X(3) = [x_1(12) + x_1(13) + x_1(14) + x_1(15)]W_{16}^0 = 0.7832 + j0.3244;$$
$$X(4) = x_1(0) + jx_1(1) - x_1(2) + jx_1(3) = j;$$
$$X(6) = x_1(8) + jx_1(9) - x_1(10) + jx_1(11) = -1.7071 - j1.7071;$$
$$X(7) = x_1(12) + jx_1(13) - x_1(14) + jx_1(15) = -0.0898 + j0.2168;$$
$$X(8) = x_1(0) + x_1(1) - x_1(2) - x_1(3) = -1;$$
$$X(9) = x_1(4) + x_1(5) - x_1(6) - x_1(7) = -0.0898 - j0.2168;$$
$$X(10) = x_1(8) + x_1(9) - x_1(10) - x_1(11) = -1.7071 + j1.7071;$$
$$X(11) = x_1(12) + x_1(13) - x_1(14) - x_1(15) = 2.6310 + j1.0898;$$
$$X(12) = x_1(0) - jx_1(1) + x_1(2) - jx_1(3) = -j;$$
$$X(13) = x_1(4) - jx_1(5) + x_1(6) - jx_1(7) = 0.7832 - j0.3244;$$
$$X(14) = x_1(8) - jx_1(9) + x_1(10) - jx_1(11) = -0.2929 - j0.2929;$$
$$X(15) = x_1(12) - jx_1(13) + x_1(14) - jx_1(15) = 0.6756 - j1.6310;$$

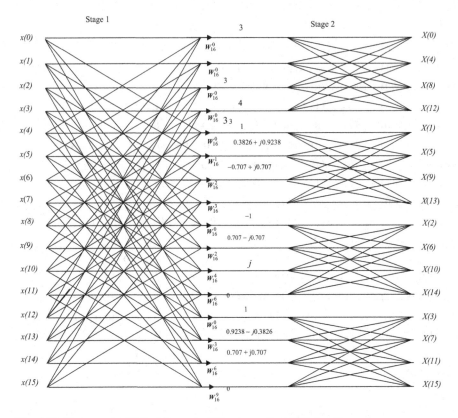

Fig. 4.17 Sixteen-point DFT of Example 4.23 using radix-4 DIF algorithm

The flow graph for the 16-point radix-4 DIF FFT is shown in Fig. 4.17. The ($\pm$) j and -1 are not shown in stage 2 for the four-point butterfly of the flow graph.

4.8   Comparison of Computational Complexity

As mentioned earlier, the number of complex multiplications required in the radix-2 FFT of an N-point sequence is $(N/2) \log_2 N$ while the number of complex additions needed is $N \log_2 N$.

In the radix-4 FFT of an N-point sequence, there are $\log_4 N = (1/2) \log_2 N$ stages and $(N/4)$ butterflies per stage. Each radix-4 butterfly requires three complex multiplications and eight complex additions. Thus, it requires $(3N/4)(1/2)$ $\log_2 N = (3N/8) \log_2 N$ complex multiplications and $(8N/4)(1/2) \log_2 N = N \log_2 N$ complex additions.

A comparison of the computational complexity in terms of the number of complex multiplications needed to compute the DFT of an N-point sequence

Table 4.6 Comparison of the computational complexity for direct DFT and FFT

Number of points N	Number of complex multiplications			FFT speed improvement factor	
	Direct DFT N^2	Radix-2 FFT $(\frac{N}{2})\log_2 N$	Radix-4 FFT $(\frac{3N}{8})\log_2 N$	Radix-2	Radix-4
16	256	32	24	8	10.6667
64	4096	192	144	21.3333	28.4444
256	65536	1024	768	64	85.3333
1024	1,048,576	5120	3840	204.8	273.0667

directly is compared to that required using radix-2 and radix-4 FFTs as given in Table 4.6.

4.9   DFT Computation Using the Goertzel Algorithm and the Chirp Transform

While the fast Fourier transform's various incarnations have gained considerable popularity, careful selection of an appropriate algorithm for computing the DFT in practice need not be limited to choosing between these so-called fast implementations. In this section, it is focused on two other techniques, namely the Goertzel algorithm and the chirp transform for computing the DFT.

4.9.1   The Goertzel Algorithm

The Goertzel algorithm [3] uses the periodicity of the sequence W_N^{nk} to reduce the computational complexity. From the definition of DFT, it is known that

$$X(k) = \sum_{n=0}^{N-1} x(n) W_N^{nk}, \quad W_N = e^{\frac{-j2\pi}{N}} \tag{4.88a}$$

Equation (4.88a) can be rewritten as

$$X(k) = \sum_{n=0}^{N-1} x(n) W_N^{-k(N-n)}, \quad W_N^{-kN} = 1$$

If a sequence $y_k(n)$ is defined as

$$y_k(n) = \sum_{r=0}^{N-1} x(r) W_N^{-k(n-r)} \tag{4.88b}$$

implying that passing a signal $x(n)$ through an LTI filter with impulse response $h(n) = W_N^{-nk} u(n)$ and evaluating the result, $y_k(n)$ at $n = N$ will give the corresponding N-point DFT coefficient $X(k) = y_k(n)$.

Representing the filter by its z-transform, we obtain

$$\begin{aligned} H_k(z) &= \sum_{n=0}^{\infty} W_N^{-nk} z^{-n} \\ &= \frac{1}{1 - W_N^{-k} z^{-1}} \end{aligned} \tag{4.89}$$

having a pole on the unit circle at the frequency $\omega_k = \frac{2\pi k}{N}$. Hence, the DFT can be computed by passing the block of input data into a parallel bank of N filters each filter having a pole at the frequency of the corresponding DFT. The DFT can be computed by using the following difference equation corresponding to the filter expressed by Eq. (4.89)

$$y_k(n) = W_N^{-k} y_k(n-1) + x(n) y_k(-1) = 0. \tag{4.90a}$$

The inherent complex multiplications and addition in Eq. (4.90a) can be avoided by using the following two-pole filter having complex conjugate pole pairs equivalent to the filtering operation represented by Eq. (4.89).

$$\begin{aligned} H_k(z) &= \frac{1 - W_N^k z^{-1}}{1 - W_N^k z^{-1}} \frac{1}{1 - W_N^{-k} z^{-1}} \\ &= \frac{1 - W_N^k z^{-1}}{1 - \left(2\cos\frac{2\pi k}{N}\right) z^{-1} + z^{-2}} \\ &= \frac{Y_k(z)}{X(z)} \end{aligned} \tag{4.90b}$$

where $= H_{1k}(z) H_{2k}(z)$

$$H_{2k}(z) = \frac{Y_k(z)}{v_k(z)} = 1 - W_N^k z^{-1}$$

$$H_{1k}(z) = \frac{v_k(z)}{X(z)} = \frac{1}{1 - \left(2\cos\frac{2\pi k}{N}\right) z^{-1} + z^{-2}}$$

From $H_{1k}(z)$ and $H_{2k}(z)$, we obtain the following difference equations

$$v_k(n) = 2\cos\frac{2\pi k}{N}v_k(n-1) - v_k(n-2) + x(n) \tag{4.91a}$$

$$y_k(n) = v_k(n) - W_N^k v_k(n-1) \tag{4.91b}$$

with initial conditions $v_k(-1) = v_k(-2) = 0$.

The Goertzel algorithm evaluates $X(k)$ at any M values of k instead of evaluating at all N values of k. Hence, it is more efficient than FFT [4] for computing DFT, when $M \leq \log_2(N)$.

Example 4.24 Considering the sequence $x(n) = \{1, 2, 1, 1\}$, compute DFT coefficient $X(1)$ and the corresponding spectral amplitude at the frequency bin $k = 1$ using the Goertzel algorithm.

Solution We have $k = 1$, $N = 4$, $x(0) = 1$, $x(1) = 2$, $x(2) = 1$, $x(3) = 1$.

$$2\cos\frac{2\pi}{4} = 0,\ W_4^1 = e^{-\frac{j2\pi}{4}} = \cos\frac{\pi}{2} - j\sin\frac{\pi}{2} = -j$$

For $n = 0, 1, \ldots, 4$

$$v_1(n) = -v_1(n-2) + x(n)$$
$$y_1(n) = v_1(n) + jv_1(n-1)$$

Then, $X(1) = y_1(4)$ $|X(1)|^2 = v_1^2(4) + v_1^2(3)$

$$X(1) = y_1(4) = v_1(4) + jv_1(3)$$
$$v_1(0) = -v_1(-2) + x(0) = 1$$
$$y_1(0) = v_1(0) + jv_1(-1) = 1$$
$$v_1(1) = -v_1(-1) + x(1) = 2$$
$$y_1(1) = v_1(1) + jv_1(0) = 2 + j1$$
$$v_1(2) = -v_1(0) + x(2) = 0$$
$$y_1(2) = v_1(2) + jv_1(1) = j2$$
$$v_1(3) = -v_1(1) + x(3) = -1$$
$$y_1(3) = v_1(3) + jv_1(2) = -1$$
$$v_1(4) = -v_1(2) + x(4) = 0$$
$$y_1(4) = v_1(4) + jv_1(3) = -j$$
$$X(1) = y_1(4) = -j$$
$$|X(1)|^2 = v_1^2(4) + v_1^2(3) = 1$$

4.9.2   The Chirp Transform Algorithm

The chirp transform algorithm [5] is also based on expressing DFT as a convolution. As it can be used to compute the Fourier transform of any set of equally spaced samples on the unit circle, it is more flexible than the FFT.

If it is desired to compute the values of the z-transform of $x(n)$ at a set of points $\{z_k\}$, then,

$$X(z_k) = \sum_{n=0}^{N-1} x(n) z_k^{-n} \qquad k = 0, 1, \ldots, M-1 \qquad (4.92\text{a})$$

Equation (4.92a) can be rewritten as

$$X\left(e^{j\omega_k}\right) = \sum_{n=0}^{N-1} x(n) e^{-j\omega_k n} \qquad k = 0, 1, \ldots, M-1 \qquad (4.92\text{b})$$

where

$$\omega_k = \omega_0 + k\Delta\omega \qquad k = 0, 1, \ldots, M-1 \qquad (4.92\text{c})$$

Equation (4.92b) can be rewritten as,

$$X\left(e^{j\omega_k}\right) = \sum_{n=0}^{N-1} x(n) e^{-j(\omega_0 + k\Delta\omega)n} \qquad k = 0, 1, \ldots, M-1 \qquad (4.92\text{d})$$

For the DFT computation, $\omega_0 = 0$, $\Delta\omega = \frac{2\pi}{N}$ and $M = N$.
Hence, Eq. (4.92d) becomes

$$X\left(e^{j\omega_k}\right) = \sum_{n=0}^{N-1} x(n) e^{-j\frac{2\pi}{N}nk} \qquad k = 0, 1, \ldots, M-1 \qquad (4.93\text{a})$$

$$X\left(e^{j\omega_k}\right) = \sum_{n=0}^{N-1} x(n) W_N^{nk} \qquad k = 0, 1, \ldots, M-1 \qquad (4.93\text{b})$$

Using the identity

$$nk = \frac{1}{2}\left(n^2 + k^2 - (k-n)^2\right)$$

Equation (4.93b) can be written as

$$X(z_k) = W_N^{\frac{k^2}{2}} \sum_{n=0}^{N-1} g(n) W_N^{\frac{-(k-n)^2}{2}} \qquad k = 0, 1, \ldots, M - 1 \qquad (4.93c)$$

where

$$g(n) = x(n) W_N^{\frac{n^2}{2}}$$

For notation convenience, replacing n by k and k by n in Eq. (4.93c), it can be rewritten as

$$X(z_n) = W_N^{\frac{n^2}{2}} \sum_{n=0}^{N-1} g(k) W_N^{\frac{-(n-k)^2}{2}} \qquad n = 0, 1, \ldots, M - 1 \qquad (4.94a)$$

Equation (4.94a) can also be expressed as

$$X\left(e^{j\omega_n}\right) = W_N^{\frac{n^2}{2}} \sum_{n=0}^{N-1} g(k) W_N^{\frac{-(n-k)^2}{2}} \qquad n = 0, 1, \ldots, M - 1 \qquad (4.94b)$$

implying that $X(e^{j\omega_n})$ is the convolution of the sequence $g(n)$ with the sequence $W_N^{\frac{-n^2}{2}}$, premultiplied by the sequence $W_N^{\frac{n^2}{2}}$, and the chirp filter impulse response is

$$h(n) = W_N^{\frac{-n^2}{2}} = \cos\frac{\pi n^2}{N} + j \sin\frac{\pi n^2}{N} \qquad (4.95)$$

Thus, the block diagram of chirp transform system for DFT computation is shown in Fig. 4.18.

4.10 Decimation-in-Time FFT Algorithm for a Composite Number

In the previous sections, we discussed FFT algorithms for radix-2 and radix-4 cases. However, it may not be possible in all cases to choose N to be a power of 2 or 4. We now consider the case where N is a composite number composed of a product

Fig. 4.18 Block diagram of chirp transform system for DFT computation

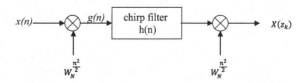

of two factors n_1 and n_2, i.e., $N = n_1 n_2$, so that we can divide the sequence $x(n)$ into n_1 subsequences of length n_2. Then, $X(K)$ can be written as

$$X(k) = \sum_{n=0}^{N-1} x(n) W_N^{kn} \tag{4.88}$$

$$= \sum_{i=0}^{n_2-1} x(n_1 i) W_N^{n_1 ik} + \sum_{i=0}^{n_2-1} x(n_1 i + 1) W_N^{k} W_N^{n_1 ik} + \cdots \\ + \sum_{i=0}^{n_2-1} x(n_1 i + n_1 - 1) W_N^{(n_1-1)k} W_N^{n_1 ik} \tag{4.89}$$

The above equation can be rewritten as

$$X(k) = \sum_{j=0}^{n_1-1} W_N^{jk} \sum_{i=0}^{n_2-1} x(n_1 i + j) W_N^{n_1 ik} \tag{4.90}$$

Define

$$F_j(k) = \sum_{i=0}^{n_2-1} x(n_1 i + j) W_N^{n_1 ik} \tag{4.91}$$

Then, $X(k)$ can be expressed in terms of n_1 DFTs of sequences of length n_2 samples as

$$X(k) = \sum_{j=0}^{n_1-1} F_j(k) W_N^{jk} \tag{4.92}$$

For illustration, consider computation of a 12-point DIT FFT ($N = 12 = 3.4$). The original sequence is divided into three sequences, each of length 4.
First sequence: $x(0)x(3)x(6)x(9)$; second sequence: $x(1)x(4)x(7)x(10)$;
Third sequence: $x(2)x(5)x(8)x(11)$. Then, $X(k)$ can be expressed as

$$X(k) = \sum_{j=0}^{2} W_{12}^{jk} \sum_{i=0}^{3} x(3i + j) W_{12}^{3ik} \\ = F_0(k) + W_{12}^{k} F_1(k) + W_{12}^{2k} F_2(k) \tag{4.93}$$

The flow graph of the 12-point DFT is shown in Fig. 4.19.

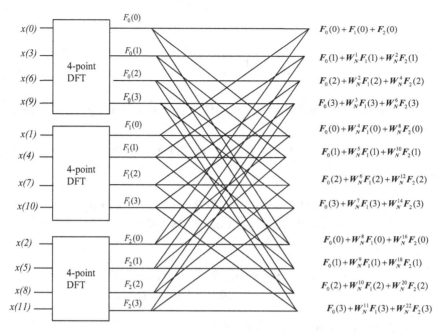

Fig. 4.19 Flow graph of a 12-point DIT FFT

4.11 The Inverse Discrete Fourier Transform

An FFT algorithm for computing the DFT can be effectively used to compute the inverse DFT. The inverse of an N-point DFT $X(k)$ is given by

$$x(n) = \frac{1}{N}\sum_{k=0}^{N-1} X(k)W_N^{-nk} \qquad (4.94)$$

where $W = e^{-j2\pi/N}$. Multiplying both sides of the above expression by N and taking complex conjugates, we obtain

$$Nx^*(n) = \sum_{k=0}^{N-1} X^*(k)\, W_N^{nk} \qquad (4.95)$$

The RHS of Eq. (4.94) is the DFT of the sequence $X^*(k)$ and can be rewritten as

$$Nx^*(n) = \text{DFT}\{X^*(k)\} \qquad (4.96)$$

Taking the complex conjugate on both sides of Eq. (4.96) and using the FFT for the computation of DFT yield

$$Nx(n) = \{FFT\{X^*(k)\}\}^*$$

Hence,

$$x(n) = \frac{1}{N}\{FFT\{X^*(k)\}\}^* \tag{4.97}$$

The following example illustrates the IDFT computation using the DIT FFT algorithm:

Example 4.25 Find the eight-point IDFT using DIT algorithm.

Solution Let the input be

$$X(k) = \{20, -5.828 - j2.279, 0, -0.172 - j0.279, 0,$$
$$-0.172 + j0.279, 0, -5.828 + j2.279\}$$

Hence,

$$X^*(k) = \{20, -5.828 + j2.279, 0, -0.172 + j0.279, 0,$$
$$-0.172 - j0.279, 0, -5.828 - j2.279\}$$

With $N = 8$, the four twiddle factors are

$$W_8^0 = 1; W_8^1 = e^{-j2\pi/8} = \cos(\pi/4) - j\sin(\pi/4) = 0.707 - j0.707;$$
$$W_8^2 = e^{-j4\pi/8} = -j; W_8^3 = e^{-j6\pi/8} = -0.707 - j0.707;$$

The flow diagram for the eight-point inverse DFT using the DIT algorithm is shown in Fig. 4.20.

Stage 1

$$x_1(0) = X^*(0) + W_8^0 X^*(4) = 20 + 0 = 20$$
$$x_1(4) = X^*(0) - W_8^0 X^*(4) = 20 - 0 = 20$$
$$x_1(2) = X^*(2) + W_8^0 X^*(6) = 0 + 0 = 0$$
$$x_1(6) = X^*(2) - W_8^0 X^*(4) = 0 - 0 = 0$$
$$x_1(1) = X^*(1) + W_8^0 X^*(5) = -5.828 + j2.279 - 0.172 - j0.279 = -6 + j2$$
$$x_1(5) = X^*(1) - W_8^0 X^*(5) = -5.828 + j2.279 + 0.172 + j0.279 = -5.656 + j2.558$$
$$x_1(3) = X^*(3) + W_8^0 X^*(7) = -0.172 + j0.279 - 5.828 - j2.279 = -6 - j2$$
$$x_1(7) = X^*(3) - W_8^0 X^*(7) = -0.172 + j0.279 + 5.828 + j2.279 = 5.656 + j2.558$$

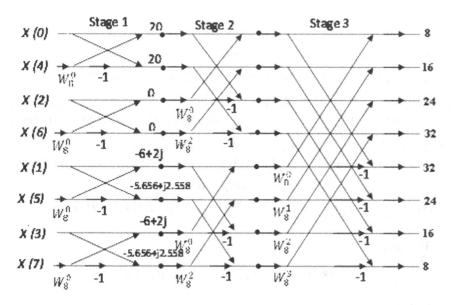

Fig. 4.20 Eight-point inverse DFT of Example 4.24 using the DIT algorithm

Stage 2

$x_2(0) = x_1(0) + W_8^0 x_1(2) = 20 + 0 = 20;$

$x_2(4) = x_1(4) + W_8^2 x_1(6) = 20 + 0 = 20;$

$x_2(2) = x_1(0) - W_8^0 x_1(2) = 20;$

$x_2(6) = x_1(4) - W_8^2 x_1(6) = 20;$

$x_2(1) = x_1(1) + W_8^0 x_1(3) = -6 + 2j - 6 - 2j = -12;$

$x_2(5) = x_1(5) + W_8^2 x_1(7) = -5.656 + j2.558 + (-j)(5.656 + j2.558) = -3.098 - j3.098;$

$x_2(3) = x_1(1) - W_8^0 x_1(3) = -6 + 2j + 6 + 2j = 4j;$

$x_2(7) = x_1(5) - W_8^2 x_1(7) = -5.656 + j2.558 + j5.656 - 2.558 = -8.214 + j8.224;$

Stage 3

$x_3(0) = x_2(0) + W_8^0 x_2(1) = 20 - 12 = 8;$

$x_3(1) = x_2(4) + W_8^1 x_2(5) = 20 + (-3.098 - j3.098)(0.707 - j0.707) = 16.0006;$

$x_3(2) = x_2(2) + W_8^2 x_2(3) = 20 + (-j)(4j) = 24;$

$x_3(3) = x_2(6) + W_8^3 x_2(7) = 20 + (-0.707 - j0.707)(-8.214 + j8.214) = 31.9982;$

$x_3(4) = x_2(0) - W_8^0 x_2(1) = 20 + 12 = 32;$

$x_3(5) = x_2(4) - W_8^1 x_2(5) = 20 - (-3.098 - j3.098)(0.707 - j0.707) = 23.9994;$

$x_3(6) = x_2(2) - W_8^2 x_2(3) = 20 - (-j)(4j) = 16;$

$x_3(7) = x_2(6) - W_8^3 x_2(7) = 20 - (-0.707 - j0.707)(-8.214 + j8.214) = 8.0018;$

Therefore,

$$8x^*(n) = \{8, 16, 24, 32, 32, 24, 16, 8\}$$

Hence,

$$x(n) = \{1, 2, 3, 4, 4, 3, 2, 1\}$$

4.12 Computation of DFT and IDFT Using MATLAB

The built-in MATLAB functions **fft(x)** and **ifft(x)** can be used for the computation of the DFT and the IDFT, respectively. The functions use computationally efficient FFT algorithms.

Example 4.26 Consider the input sequence $x(n) = \{1, 1, 1, 1, 0, 0, 1, 1\}$ of Example 4.5. Compute the DFT using MATLAB.

Solution Execution of fft(x) yields the DFT of $x(n)$ as

6.000 $1.7071 - 0.7071i$ $-1.0000 + 1.0000i$ $0.2929 - 0.7071i$ 0

$0.2929 + 0.7071i$ $-1.0000 - 1.0000i$ $1.7071 + 0.7071i$

which is equivalent to the DFT computed using the definition of DFT as in Example 4.3.

Example 4.27 Consider the input

$$X(k) = \{20, -5.828 - j2.279, 0, -0.172 - j0.279, 0,$$
$$-0.172 + j0.279, 0, -5.828 + j2.279\}$$

of Example 4.24. Compute IDFT using MATLAB.

Solution Execution of ifft(X) yields the IDFT of X as

1.0   2.0   3.0   4.0   4.0   3.0   2.0   1.0

which is the same as the result obtained in Example 4.24.

4.13 Application Examples

4.13.1 Detection of Signals Buried in Noise

One of the applications of the DFT-based spectral analysis is to detect the signals buried in noise. For example, consider a noisy signal with K sinusoidal components with unknown frequencies $f_1, f_2, \ldots, f_K$ given by

$$x(n) = \sum_{i=1}^{K} \frac{2\pi n f_i}{F_T} + \eta(n) \quad 0 \leq n \leq N \tag{4.98}$$

where $\eta(n)$ is additive white noise. The unknown frequencies $f_1, f_2, \ldots, f_K$ can be detected by using DFT. For simulation, a signal with two $(K = 2)$ sinusoidal components $N = 1024$ and the sampling frequency $F_T = 1000$ Hz are assumed. The following MATLAB program is used to generate the noisy signal and to detect the unknown frequencies by applying the DFT on the generated noisy signal.

Program 4.1 Detection of signals buried in noise

```
clear;clc;
N = 1024;
K=2;
x =randn(1,N);% random noise generation
FT =1000; % sampling frequency
T = 1/FT; % sampling time period
k=1:N;
f=(FT/2)*rand(1,K); %random generation of unknown frequencies
for i=1:K
    x=x+sin(2*pi*f(i)*k*T); % noisy signal with sinusoidal components
```

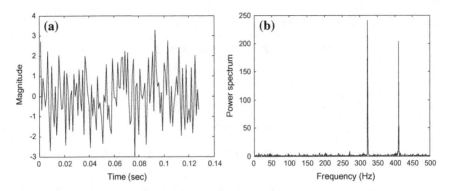

Fig. 4.21 **a** Noisy signal and **b** power spectrum density of the noisy signal

```
end
t = k*T;
figure(1),plot (t(1:N/8),x(1:N/8))
xlabel('Time(sec)');ylabel('Magnitude');
% Compute and plot power density spectrum
figure(2),
X= abs(fft(x));
S = X.^2/N;
f = linspace (0,(N-1)*Fs/N,N);
plot (f(1:N/2),S(1:N/2))
set(gca,'Xlim',[0,Fs/2])
xlabel('Frequency (Hz)');
ylabel('Power spectrum')
% Finding frequencies
s = f_prompt ('Enter threshold for locating peaks',0,max(S),.7*max(S));
for i = 1: N/2
   if (S(i)>s)
     fprintf ('f = %.0f Hz\n',f(i))
   end
end
```

For a random run of the above program, the noisy signal and its power spectral
density are shown in Fig. 4.21a, b, respectively, and the two unknown frequencies
are identified as $f_1 = 322$ Hz and $f_2 = 411$ Hz.

4.13.2 Denoising of a Speech Signal

The DFT can be applied to Fourier domain filtering which is equivalent to circular
convolution of a sequence of finite length with an ideal impulse response of finite
length. This approach is useful in denoising a signal for suppressing high-frequency

noise from a low-frequency signal corrupted with noise. For purpose of illustration, we considered the speech signal 'To take good care of yourself' from sound file 'goodcare.wav'. The following MATLAB program is used to read the speech signal from the wav file and to add noise to the speech signal and to reconstruct the original speech signal by performing circular convolution of the noisy speech signal with finite length impulse response.

%**Program 4.2** Denoising using circular convolution

```
clear;clc;
[x,fs]=wavread('goodcare.wav');
wavplay(x,fs)% listen to original speech signal
no=0.075*randn(1,length(x));% noise generation
xn=x+no';%add noise to original speech signal
wavplay(xn,fs)%listen to noisy speech signal
figure(1),plot(x);xlabel('Number of samples');ylabel('Amplitude');
figure(2),plot(xn);xlabel('Number of samples');ylabel('Amplitude');
h=ones(1,64)/64;y=fftfilt(h,xn);%perform denoising
wavplay(12*y,fs);% listen to recovered speech signal
figure(3),plot(12*y);xlabel('Number of samples');ylabel('Amplitude');
```

The speech signal, the noisy speech signal, and the recovered speech signal after denoising, obtained from the above MATLAB program, are shown in Figs. 4.22a–c, respectively. From these figures, it can be observed that the recovered speech signal after denoising is nearly same as the original signal.

4.13.3 DTMF Tone Detection Using Goertzel Algorithm

Dual-tone multifrequency (DTMF) signaling is widely used worldwide for voice communications in modern telephony to dial numbers and configure switch boards. It is also used in voice mail, electronic mail, and telephone banking.

DTMF signaling uses two tones to represent each key on the touch pad. There are 12 distinct tones. When any key is pressed, the tone of the column and the tone of the row are generated. As an example, pressing the '5' button generates the tones 770 Hz and 1336 Hz. In this example, use the number 10 to represent the '*' key and 11 to represent the '#' key.

The frequencies were chosen to avoid harmonics: No frequency is a multiple of another, the difference between any two frequencies does not equal any of the frequencies, and the sum of any two frequencies does not equal any of the frequencies.

The industry standard frequency specifications for all the keys are listed in Fig. 4.23.

The DTMF signals for each button on telephone pad are shown in Fig. 4.24. The MATLAB program to generate the DTMF signals is listed in Program 4.3.

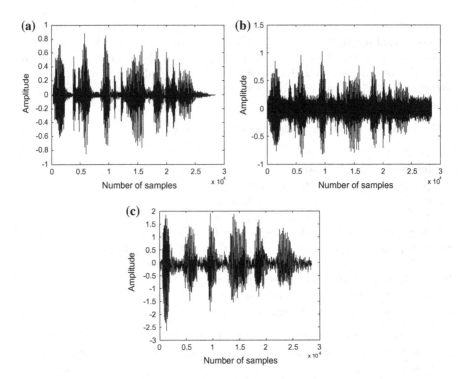

Fig. 4.22 **a** Speech signal, **b** noisy speech signal, and **c** recovered speech signal after denoising

Fig. 4.23 DTMF tone specifications

	1209 Hz	1336 Hz	1477 Hz
697 Hz	1	ABC 2	DEF 3
770 Hz	GHI 4	JKL 5	MNO 6
852 Hz	PRS 7	TUV 8	WXY 9
941 Hz	*	0	#

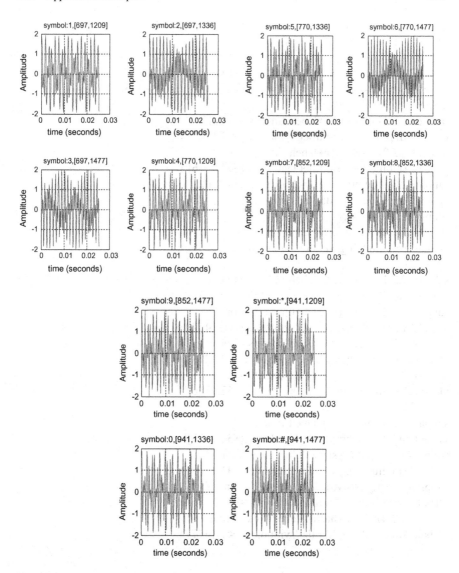

Fig. 4.24 Time responses of each tone of the telephone pad

Program 4.3

```
%MATLAB program DTMF tones generation
clear all;clc;
Fs = 8000;N = 205;t=[0:1:204]/Fs;
lf=[697;770;852;941];hf=[1209;1336;1477];
ylf1=sin(2*pi*lf(1)*(0:N-1)/Fs);ylf2=sin(2*pi*lf(2)*(0:N-1)/Fs);
ylf3=sin(2*pi*lf(3)*(0:N-1)/Fs);ylf4=sin(2*pi*lf(4)*(0:N-1)/Fs);
```

```
yhf1=sin(2*pi*hf(1)*(0:N-1)/Fs);yhf2=sin(2*pi*hf(2)*(0:N-1)/Fs);
yhf3=sin(2*pi*hf(3)*(0:N-1)/Fs);
y1=ylf1+yhf1;y2=ylf1+yhf2;y3=ylf1+yhf3;y4=ylf2+yhf1;
y5=ylf2+yhf2;y6=ylf2+yhf3;y7=ylf3+yhf1;y8=ylf3+yhf2;
y9=ylf3+yhf3;ystar=ylf4+yhf1;y0=ylf4+yhf2;yhash=ylf4+yhf3;
figure(1)
subplot(2,2,1);plot(t,y1);xlabel('time (seconds)')
ylabel('Amplitude');grid;title('symbol:1,[697,1209]');
subplot(2,2,2);plot(t,y2);xlabel('time (seconds)')
ylabel('Amplitude');grid;title('symbol:2,[697,1336]');
subplot(2,2,3);plot(t,y3);xlabel('time (seconds)')
ylabel('Amplitude');grid;title('symbol:3,[697,1477]');
subplot(2,2,4);plot(t,y4);
xlabel('time (seconds)')
ylabel('Amplitude');grid;title('symbol:4,[770,1209]');
figure(2)
subplot(2,2,1);plot(t,y5);xlabel('time (seconds)')
ylabel('Amplitude');grid;title('symbol:5,[770,1336]');
subplot(2,2,2);plot(t,y6);xlabel('time (seconds)')
ylabel('Amplitude');grid;title('symbol:6,[770,1477]');
subplot(2,2,3);plot(t,y7);xlabel('time (seconds)')
ylabel('Amplitude');grid;title('symbol:7,[852,1209]');
subplot(2,2,4);plot(t,y8);xlabel('time (seconds)')
ylabel('Amplitude');grid;title('symbol:8,[852,1336]');
figure(3)
subplot(2,2,1);plot(t,y9);xlabel('time (seconds)')
ylabel('Amplitude');grid;title('symbol:9,[852,1477]');
subplot(2,2,2);plot(t,ystar);xlabel('time (seconds)')
ylabel('Amplitude');grid;title('symbol:*,[941,1209]');
subplot(2,2,3);plot(t,y0);xlabel('time (seconds)')
ylabel('Amplitude');grid;title('symbol:0,[941,1336]');
subplot(2,2,4);plot(t,yhash);xlabel('time (seconds)')
ylabel('Amplitude');grid;title('symbol:#,[941,1477]');
```

DTMF tone detection

The DTMF detection relies on the Goertzel algorithm (Goertzel filter). The main purpose of using the Goertzel filters is to calculate the spectral value at the specified frequency index using the filtering method. Its advantage includes the reduction of the required computations and avoidance of complex algebra. The detection of frequencies using Goertzel algorithm contained in each tone of the telephone pad is shown in Fig. 4.25. The MATLAB program for the tones detection using the Goertzel algorithm is listed in Program 4.4.

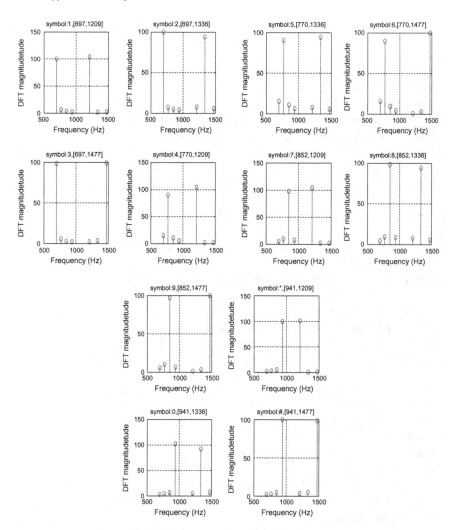

Fig. 4.25 DTMF tone detection using Goertzel algorithm

Program 4.4

```
clear all;clc;
Fs = 8000;N = 205;load DTMFdata
f = [697 770 852 941 1209 1336 1477];
freq_indices = round(f/Fs*N) + 1;
for tonechoice=1:12
tonedata=DTMFs(tonechoice,:);
dft_data(tonechoice,:) = goertzel(tonedata,freq_indices);
end
```

```
figure(1)
subplot(2,2,1);stem(f,abs(dft_data(1,:)));ax = gca;ax.XTick = f;
xlabel('Frequency(Hz')
ylabel('DFT magnitudetude');grid;title('symbol:1,[697,1209]');
subplot(2,2,2);stem(f,abs(dft_data(2,:)));ax = gca;ax.XTick = f;
xlabel('Frequency(Hz')
ylabel('DFT magnitudetude');grid;title('symbol:2,[697,1336]');
subplot(2,2,3);stem(f,abs(dft_data(3,:)));ax = gca;ax.XTick = f;
xlabel('Frequency(Hz')
ylabel('DFT magnitudetude');grid;title('symbol:3,[697,1477]');
subplot(2,2,4);stem(f,abs(dft_data(4,:)));ax = gca;ax.XTick = f;
xlabel('Frequency(Hz')
ylabel('DFT magnitudetude');grid;title('symbol:4,[770,1209]');
figure(2)
subplot(2,2,1);stem(f,abs(dft_data(5,:)));ax = gca;ax.XTick = f;
xlabel('Frequency(Hz')
ylabel('DFT magnitudetude');grid;title('symbol:5,[770,1336]');
subplot(2,2,2);stem(f,abs(dft_data(6,:)));ax = gca;ax.XTick = f;
xlabel('Frequency(Hz')
ylabel('DFT magnitudetude');grid;title('symbol:6,[770,1477]');
subplot(2,2,3);stem(f,abs(dft_data(7,:)));ax = gca;ax.XTick = f;
xlabel('Frequency(Hz')
ylabel('DFT magnitudetude');grid;title('symbol:7,[852,1209]');
subplot(2,2,4);stem(f,abs(dft_data(8,:)));ax = gca;ax.XTick = f;
xlabel('Frequency(Hz')
ylabel('DFT magnitudetude');grid;title('symbol:8,[852,1336]');
figure(3)
subplot(2,2,1);stem(f,abs(dft_data(9,:)));ax = gca;ax.XTick = f;
xlabel('Frequency(Hz')
ylabel('DFT magnitudetude');grid;title('symbol:9,[852,1477]');
subplot (2,2,2);stem(f,abs(dft_data(10,:)));ax = gca;ax.XTick = f;
xlabel('Frequency(Hz')
ylabel('DFT magnitudetude');grid;title('symbol:*,[941,1209]');
subplot (2,2,3);stem(f,abs(dft_data(11,:)));ax = gca;ax.XTick = f;
xlabel('Frequency(Hz')
ylabel('DFT magnitudetude');grid;title('symbol:0,[941,1336]');
subplot (2,2,4);stem(f,abs(dft_data(12,:)));ax = gca;ax.XTick = f;
xlabel('Frequency(Hz')
ylabel('DFT magnitudetude');grid;title('symbol:#,[941,1477]');
```

4.14 Problems

1. Determine the Fourier series representation for the following discrete-time signals:

   (a) $x(n) = 3\sin\left(\frac{\pi n}{4}\right)\sin\left(\frac{2\pi n}{5}\right)$
   (b) $x(n)$ is periodic of period 8, and $x(n) = n$ for $0 \le n \le 3$, and $x(n) = n$ for $4 \le n \le 7$

2. Compute the eight-point DFT of $(-1)^n$
3. Find the four-point DFT of the following sequences

   (i) $x(n) = \{1, 2, 1, 1\}$
   (ii) $x(n) = \sin(n+1)\pi/4$
   (iii) $x(n) = \{2, -1, 1, -2\}$.

4. Find eight-point DFT of the following sequences

   (i) $x(n) = \{1, 0, 1, 0, 0, 1, 1, 0\}$
   (ii) $x(n) = \cos(n+1)\pi/2$
   (iii) $x(n) = \{1, 1, 0, 0, 1, 0, 1, 1\}$

5. Compute the eight-point DFT of the square-wave sequence:

$$x(n) = \begin{cases} 2 & 0 \le n \le (N/2) \\ -2 & (N/2) \le n < N-1 \end{cases}$$

6. Find 16-point DFT of the following sequence:

$$x(n) = \begin{cases} 1 & 0 \le n \le 7 \\ 0 & 7 < n < 15 \end{cases}$$

7. Compute the eight-point circular convolution of

$$x_1(n) = \{1, 1, 0, 1, 0, 1, 1, 0\} \text{ and } x_2(n) = \sin(3\pi/4), 0 \le n \le 7.$$

8. Find the output $y(n)$ of a filter whose impulse response is $h(n) = \{0, 1, 1\}$ and the input signal is $x(n) = \{1, -2, 0, 1, 0, 2, 1, 2, 2, 1\}$ using the overlap-add method.
9. Using linear convolution, find $y(n) = x(n) * h(n)$ for the sequences $x(n) = (2, -3, 1, 2, 1, 1, -1, -3, 1, 2, 1, -1)$ and $h(n) = (2, 1)$. Compare the result by solving the problem using overlap-save method.
10. Compute the eight-point DFT of the following sequence using the radix-2 DIT algorithm for the following sequences:

(i) $x(n) = \{1, 1, -1, 0, 1, 0, 1, -1\}$
(ii) $x(n) = \{1, 2, 1, -1, 2, 1, -1, 1\}$
(iii) $x(n) = \{0.5, 0, 1, 0.5, 1, 0, 0.5, 0.5\}$

11. Compute the eight-point DFT of the sequence $x(n) = \{1, 1, -1, 0, 1, 0, 1, -1\}$ using the DIF algorithm

12. Find the 16-point DFT of the following sequence using radix-4 DIF algorithm.

$$x(n) = \{1, 1, 0, 0, 1, 1, 1, 1, 0, 0, 0, 0, 1, 1, 1, 1\}$$

13. Compute DFT of the sequence $x(n) = \{1, 2, 3, 4\}$ using the Goertzel algorithm
14. Develop the FFT algorithm for the composite number 18, and show the flow graph.
15. Find the IDFT of $Y(k) = \{1, 0, 0, 1\}$.
16. Compute the IDFT of the sequence $X(k) = \{3, j, 1 + 2j, 1 - j, 1 + 2j, 1, 0, -j\}$ using (a) DIT algorithm and (b) DIF algorithm.

4.15 MATLAB Exercises

1. Verify the results of Problem 10 of Sect. 4.13 using MATLAB.
2. Verify the results of Problem 14 of Sect. 4.13 using MATLAB.
3. Write a MATLAB program using the command circshift to compute circular convolution of two sequences and verify the result of Problem 7 of Sect. 4.13.
4. Verify the results of Problem 8 of Sect. 4.13 using MATLAB.

References

1. J.W. Cooley, P.A.W. Lewis, P.D. Welch, Historical notes on the fast Fourier transform. IEEE Trans. Audio Electroacoust. **55**(10), 1675–1677 (1967)
2. J.W. Cooley, P.A.W. Lewis, P.D. Welch, Historical notes on the fast Fourier transform. Proc. IEEE **55**(10), 1675–1677 (1967)
3. G. Goertzel, An algorithm for the evaluation of finite trignometric series. Am. Math Monthly **65**, 34–35 (1958)
4. A.V. Oppenheim, R.W. Schafer, J.R. Buck, *Discrete-Time Signal Processing*, 2nd edn. (Prentice Hall, 1999)
5. L. Rabiner, R. Schafer, C. Rader, IEEE Trans. Audio Electroacoust. **17**(2) (1969)

Chapter 5
IIR Digital Filter Design

Filtering is an important aspect of signal processing. It allows desired frequency components of a signal to pass through the system without distortion and suppresses the undesired frequency components. One of the most important steps in the design of a digital filter is to obtain a realizable transfer function $H(z)$, satisfying the given frequency response specifications. In the case of the design of an IIR filter, it is required to confirm that $H(z)$ is stable. The most common technique used in designing IIR digital filters involves first designing an analog prototype lowpass filter and then transforming the prototype to a digital filter. In this chapter, the design of analog lowpass filters is first described. Second, frequency transformations for transforming analog lowpass filter into bandpass, bandstop, or highpass analog filters are considered. Next, the design of IIR filters is discussed and illustrated with numerical examples. Further, the design of IIR filters using MATLAB is demonstrated with a number of examples Also, the design of IIR filters using graphical user interface MATLAB filter design SPTOOL is discussed and illustrated with examples. Finally, some application examples of IIR filters for audio processing are included.

5.1 Analog Lowpass Filter Design

A number of approximation techniques for the design of analog lowpass filters are well established [1–4]. The design of analog lowpass filter using Butterworth, Chebyshev I, Chebyshev II (inverse Chebyshev), and elliptic approximations is discussed in this section.

© Springer Nature Singapore Pte Ltd. 2018
K. D. Rao and M. N. S. Swamy, *Digital Signal Processing*,
https://doi.org/10.1007/978-981-10-8081-4_5

5.1.1   Filter Specifications

The specifications for an analog lowpass filter with tolerances are depicted in Fig. 5.1, where

Ω_p—passband edge frequency
Ω_s—stopband edge frequency
δ_p—peak ripple value in the passband
δ_s—peak ripple value in the stopband
Peak passband ripple in dB $= \alpha_p = -20\log_{10}\left(1 - \delta_p\right)$ dB
Minimum stopband ripple in dB $= \alpha_s = -20\log_{10}(\delta_s)$ dB
Peak ripple value in passband $\delta_p = 1 - 10^{-\alpha_p/20}$
Peak ripple value in stopband $\delta_s = 10^{-\alpha_s/20}$

5.1.2   Butterworth Analog Lowpass Filter

The magnitude-square response of an Nth-order analog lowpass Butterworth filter is given by

$$|H_a(j\Omega)|^2 = \frac{1}{1 + (\Omega/\Omega_c)^{2N}} \tag{5.1}$$

Two parameters completely characterizing a Butterworth lowpass filter are Ω_c and N. These are determined from the specified band edges Ω_p and Ω_c, and peak

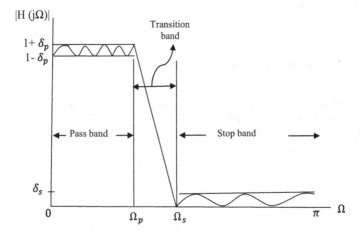

Fig. 5.1 Specifications of a lowpass analog filter

passband ripple α_p, and minimum stopband attenuation α_s. The first $(2N - 1)$ derivatives of $|H_a(j\Omega)|^2$ at $\Omega = 0$ are equal to zero. Thus, the Butterworth lowpass filter is said to have a maximally flat magnitude at $\Omega = 0$. The gain in dB is given by $10 \log_{10} |H_a(j\Omega)|^2$. At $\Omega = \Omega_c$, the gain is $10 \log_{10}(0.5) = -3$ dB; therefore, Ω_c is called the 3 dB cutoff frequency. The loss in dB in a Butterworth filter is given by

$$\alpha = 10 \log(1 + (\Omega/\Omega_c)^{2N}) \tag{5.2}$$

For $\Omega = \Omega_p$, the passband attenuation is given by

$$\alpha_p = 10 \log(1 + (\Omega_p/\Omega_c)^{2N}) \tag{5.3}$$

For $\Omega = \Omega_s$, the stopband attenuation is

$$\alpha_s = 10 \log(1 + (\Omega_s/\Omega_c)^{2N}) \tag{5.4}$$

Equations (5.3) and (5.4) can be rewritten as

$$(\Omega_p/\Omega_c)^{2N} = 10^{0.1\alpha_p} - 1 \tag{5.5}$$

$$(\Omega_s/\Omega_c)^{2N} = 10^{0.1\alpha_s} - 1 \tag{5.6}$$

From Eqs. (5.5) to (5.6), we obtain

$$(\Omega_s/\Omega_p) = \left(\frac{10^{0.1\alpha_s} - 1}{10^{0.1\alpha_p} - 1}\right)^{1/2N} \tag{5.7}$$

Equation (5.7) can be rewritten as

$$\log(\Omega_s/\Omega_p) = \frac{1}{2N} \log\left(\frac{10^{0.1\alpha_s} - 1}{10^{0.1\alpha_p} - 1}\right) \tag{5.8}$$

From Eq. (5.8), solving for N we get

$$N \geq \frac{\log\left(\frac{10^{0.1\alpha_s} - 1}{10^{0.1\alpha_p} - 1}\right)}{2 \log(\Omega_s/\Omega_p)} \tag{5.9}$$

Since the order N must be an integer, the value obtained is rounded up to the next higher integer. This value of N is used in either Eq. (5.5) or Eq. (5.6) to determine the 3-dB cutoff frequency Ω_c. In practice, Ω_c is determined by Eq. (5.6) that exactly satisfies stopband specification at Ω_c, while the passband specification is exceeded with a safe margin at Ω_p [2]. We know that $|H(j\Omega)|^2$ may be evaluated by letting $s = j\Omega$ in $H(s)H(-s)$, which may be expressed as

$$H(s)H(-s) = \frac{1}{1 + \left(-s^2/\Omega_c^2\right)^N} \qquad (5.10)$$

If $\Omega_c = 1$, the magnitude response $|H_N(j\Omega)|$ is called the normalized magnitude response. Now, we have

$$1 + \left(-s^2\right)^N = \prod_{k=1}^{2N} (s - s_k) \qquad (5.11)$$

where

$$s_k = \begin{cases} e^{j(2k-1)\pi/(2N)} & \text{for } n \text{ even} \\ e^{j(k-1)\pi/N} & \text{for } n \text{ odd} \end{cases} \qquad (5.12)$$

Since $|s_k| = 1$, we can conclude that there are $2N$ poles placed on the unit circle in the s-plane. The normalized transfer function can be formed as

$$H_N(s) = \frac{1}{\prod_{l=1}^{N} (s - p_l)} \qquad (5.13)$$

where p_l for $l = 1, 2, \ldots, N$ are the left half s-plane poles. The complex poles occur in conjugate pairs.

For example, in the case of $N = 2$, from Eq. (5.12), we have

$$s_k = \cos\left(\frac{(2k-1)\pi}{4}\right) + j\sin\left(\frac{(2k-1)\pi}{4}\right), \quad k = 1, \ldots, 2$$

The poles in the left half of the s-plane are

$$s_2 = -\frac{1}{\sqrt{2}} + \frac{j}{\sqrt{2}}; \quad s_3 = -\frac{1}{\sqrt{2}} - \frac{j}{\sqrt{2}}$$

Hence,

$$p_1 = -\frac{1}{\sqrt{2}} + \frac{j}{\sqrt{2}}; \quad p_2 = -\frac{1}{\sqrt{2}} - \frac{j}{\sqrt{2}}$$

and

$$H_N(s) = \frac{1}{s^2 + \sqrt{2}s + 1}$$

In the case of $N = 3$,

$$s_k = \cos\left(\frac{(k-1)\pi}{3}\right) + j\sin\left(\frac{(k-1)\pi}{3}\right), \quad k = 1,\ldots,2$$

The left half of s-plane poles are

$$s_3 = -\frac{1}{2} + \frac{j\sqrt{3}}{2}; \quad s_4 = -1; \quad s_5 = -\frac{1}{2} - \frac{j\sqrt{3}}{2}.$$

Hence

$$p_1 = -\frac{1}{2} + \frac{j\sqrt{3}}{2}; \quad p_2 = -1; \quad p_3 = -\frac{1}{2} - \frac{j\sqrt{3}}{2}.$$

and

$$H_N(s) = \frac{1}{(s+1)(s^2+s+1)}$$

The following MATLAB Program 5.1 can be used to obtain the Butterworth normalized transfer function for various values of N.

Program 5.1 Analog Butterworth lowpass filter normalized transfer function

```
N = input('enter order of the filter');
[z,p,k] = buttap(N)% determines poles and zeros
disp('Poles are at');disp(p);
[num,den] = zp2tf(z,p,k);
%Print coefficients in powers of s
disp('Numerator polynomial');disp(num);
disp('Denominator polynomial');disp(den);
sos = zp2sos(z,p,k);%determines coefficients of second order sections
```

The normalized Butterworth polynomials generated from the above program for typical values of N are tabulated in Table 5.1.

Table 5.1 List of normalized Butterworth polynomials

N	Denominator of $H_N(s)$
1	$s+1$
2	$s^2 + \sqrt{2}s + 1$
3	$(s+1)(s^2+s+1)$
4	$(s^2 + 0.76537s + 1)(s^2 + 1.8477s + 1)$
5	$(s+1)(s^2 + 0.61803s + 1)(s^2 + 1.61803s + 1)$
6	$(s^2 + 1.931855s + 1)(s^2 + \sqrt{2}s + 1)(s^2 + 0.51764s + 1)$
7	$(s+1)(s^2 + 1.80194s + 1)(s^2 + 1.247s + 1)(s^2 + 0.445s + 1)$

Fig. 5.2 Magnitude response
of typical Butterworth
lowpass filter

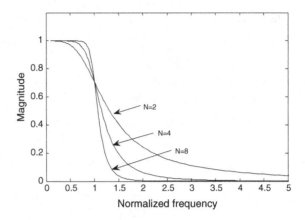

The magnitude response of the normalized Butterworth lowpass filter for some
typical values of N is shown in Fig. 5.2. From this figure, it can be seen that the
response monotonically decreases in both the passband and the stopband as Ω
increases. As the filter order N increases, the magnitude responses in both
the passband and the stopband are improved with a corresponding decrease in the
transition width. Since the normalized transfer function corresponds to $\Omega_c = 1$,
the transfer function of the lowpass filter corresponding to the actual Ω_c can be
obtained by replacing s by (s/Ω_c) in the normalized transfer function.

Example 5.1 Design a Butterworth analog lowpass filter with 1 dB passband ripple, passband edge frequency $\Omega_p = 2000\pi$ rad/s, stopband edge frequency
$\Omega_s = 10{,}000\pi$ rad/s, and a minimum stopband ripple of 40 dB.

Solution Since $\alpha_s = 40$ dB, $\alpha_p = 1$ dB, $\Omega_p = 2000\pi$, and $\Omega_s = 10{,}000\pi$

$$\log\left(\frac{10^{0.1\alpha_s} - 1}{10^{0.1\alpha_p} - 1}\right) = \log\left(\frac{10^4 - 1}{10^{0.1} - 1}\right) = 4.5868.$$

Hence from Eq. (5.9),

$$N \geq \frac{\log\left(\frac{10^4 - 1}{10^{0.1} - 1}\right)}{2\log(5/1)} = \frac{4.5868}{1.3979} = 3.2811$$

Since the order must be an integer, we choose $N = 4$.

The normalized lowpass Butterworth filter for $N = 4$ can be formulated as

$$H_N(s) = \frac{1}{(s^2 + 0.76537s + 1)(s^2 + 1.8477s + 1)}$$

From Eq. (5.6), we have

$$\Omega_c = \frac{\Omega_s}{(10^4 - 1)^{1/2N}} = \frac{10{,}000\pi}{(10^4 - 1)^{1/8}} = 9935$$

The transfer function for $\Omega_c = 9935$ can be obtained by replacing s by $\left(\frac{s}{\Omega_c}\right) = \left(\frac{s}{9935}\right)$ in $H_N(s)$

$$H_a(s) = \frac{1}{\left(\frac{s}{9935}\right)^2 + 0.76537\left(\frac{s}{9935}\right) + 1} \times \frac{1}{\left(\frac{s}{9935}\right)^2 + 1.8477\left(\frac{s}{9935}\right) + 1}$$

$$= \frac{9.7425 \times 10^{15}}{(s^2 + 7.604 \times 10^3 s + 9.8704225 \times 10^7)(s^2 + 1.8357 \times 10^4 s + 9.8704225 \times 10^7)}$$

5.1.3 Chebyshev Analog Lowpass Filter

Type 1 Chebyshev lowpass filter

The magnitude-square response of an Nth-order analog lowpass Type 1 Chebyshev filter is given by

$$|H(\Omega)|^2 = \frac{1}{1 + \varepsilon^2 T_N^2(\Omega/\Omega_p)} \tag{5.14}$$

where $T_N(\Omega)$ is the Chebyshev polynomial of order N

$$T_N(\Omega) = \begin{cases} \cos(N\cos^{-1}\Omega), & |\Omega| \leq 1 \\ \cosh\left(N\cosh^{-1}\Omega\right), & |\Omega| > 1 \end{cases} \tag{5.15}$$

The loss in dB in a Type 1 Chebyshev filter is given by

$$\alpha = 10\log\left(1 + \varepsilon^2 T_N^2(\Omega/\Omega_p)\right) \tag{5.16}$$

For $\Omega = \Omega_p$, $T_N(\Omega) = 1$, and the passband attenuation is given by

$$\alpha_p = 10\log(1 + \varepsilon^2) \tag{5.17}$$

From Eq. (5.17), ε can be obtained as

$$\varepsilon = \sqrt{10^{0.1\alpha_p} - 1} \tag{5.18}$$

For $\Omega = \Omega_s$, the stopband attenuation is

$$\alpha_s = 10\log\left(1 + \epsilon^2 T_n^2\left(\frac{\Omega_s}{\Omega_p}\right)\right) \tag{5.19}$$

Since $(\Omega_s/\Omega_p) > 1$, the above equation can be written as

$$\alpha_s = 10\log\left[1 + \epsilon^2\cosh^2\left(N\cosh^{-1}\left(\Omega_s/\Omega_p\right)\right)\right] \tag{5.20}$$

Substituting Eq. (5.18) for ε in the above equation and solving for N, we get

$$N \geq \frac{\cosh^{-1}\sqrt{\frac{10^{0.1\alpha_s}-1}{10^{0.1\alpha_p}-1}}}{\cosh^{-1}\left(\Omega_s/\Omega_p\right)} \tag{5.21}$$

We choose N to be the lowest integer satisfying (5.21). In determining N using the above equation, it is convenient to evaluate $\cosh^{-1}(x)$ by applying the identity $\cosh^{-1}(x) = \ln\left(x + \sqrt{x^2-1}\right)$.

The poles of the normalized Type 1 Chebyshev filter transfer function lie on an ellipse in the s-plane and are given by [5]

$$x_k = -\sinh\left\{\frac{1}{N}\sinh^{-1}\left(\frac{1}{\epsilon}\right)\right\} \cdot \sin\left\{\frac{(2k-1)\pi}{2N}\right\}, \quad \text{for } k = 1, 2, \ldots, N \tag{5.22}$$

$$y_k = \cosh\left\{\frac{1}{N}\sinh^{-1}\left(\frac{1}{\epsilon}\right)\right\} \cdot \cos\left\{\frac{(2k-1)\pi}{2N}\right\}, \quad \text{for } k = 1, 2, \ldots, N \tag{5.23}$$

Also, the normalized transfer function is given by

$$H_N(s) = \frac{H_0}{\prod_k (s - p_k)} \tag{5.24}$$

where

$$\begin{aligned}
p_k = &-\sinh\left\{\frac{1}{N}\sinh^{-1}\left(\frac{1}{\epsilon}\right)\right\} \cdot \sin\left\{\frac{(2k-1)\pi}{2N}\right\} \\
&+j\cosh\left\{\frac{1}{N}\sinh^{-1}\left(\frac{1}{\epsilon}\right)\right\} \cdot \cos\left\{\frac{(2k-1)\pi}{2N}\right\}
\end{aligned} \tag{5.25a}$$

and

$$H_0 = \frac{1}{2^{N-1}}\frac{1}{\varepsilon} \tag{5.25b}$$

As an illustration, consider the case of $N = 2$ with a passband ripple of 1 dB. From Eq. (5.18), we have

$$\frac{1}{\varepsilon} = \frac{1}{\sqrt{10^{0.1\alpha_p} - 1}} = 1.965227$$

Hence

$$\sinh^{-1}\left(\frac{1}{\varepsilon}\right) = \sinh^{-1}(1.965227) = 1.428$$

Therefore, from (5.25a), the poles of the normalized Chebyshev transfer function are given by

$$p_k = -\sinh(0.714)\sin\left\{\frac{(2k-1)\pi}{4}\right\}$$
$$+ j\cosh(0.714)\cos\left\{\frac{(2k-1)\pi}{4}\right\}, \quad k = 1, 2$$

Hence

$$p_1 = -0.54887 + j0.89513, \quad p_2 = -0.54887 - j0.89513$$

Also, from (5.25b), we have

$$H_0 = \frac{1}{2}(1.965227) = 0.98261$$

Thus for $N = 2$, with a passband ripple of 1 dB, the normalized Chebyshev transfer function is

$$H_N(s) = \frac{0.98261}{(s - p_1)(s - p_2)} = \frac{0.98261}{(s^2 + 1.098s + 1.103)}$$

Similarly for $N = 3$, for a passband ripple of 1 dB, we have

$$p_k = -\sinh(1.428/3)\sin\frac{(2k-1)\pi}{6}$$
$$+ j\cosh(1.428/3)\cos\frac{(2k-1)\pi}{6}, \quad k = 1, 2, 3$$

Thus,

$$p_1 = -0.24709 + j0.96600; \quad p_2 = -0.49417; \quad p_3 = -0.24709 - j0.966.$$

Also, from (5.25b),

$$H_0 = \frac{1}{4}(1.965227) = 0.49131$$

Hence, the normalized transfer function of Type 1 Chebyshev lowpass filter for $N = 3$ is given by

$$H_N(s) = \frac{0.49131}{(s - p_1)(s - p_2)(s - p_3)}$$
$$= \frac{0.49131}{(s^3 + 0.988s^2 + 1.238s + 0.49131)}$$

The following MATLAB Program 5.2 can be used to form the Type 1 Chebyshev normalized transfer function for a given order and passband ripple.

Program 5.2 Analog Type 1 Chebyshev lowpass filter normalized transfer function

```
N = input('enter order of the filter');
Rp = input('enter passband ripple in dB');
[z,p,k] = cheb1ap(N,Rp)% determines poles and zeros
disp('Poles are at');disp(p);
[num,den] = zp2tf(z,p,k);
%Print coefficients in powers of s
disp('Numerator polynomial');disp(num);
disp('Denominator polynomial');disp(den);
```

The normalized Type 1 Chebyshev polynomials generated from the above program for typical values of N and passband ripple of 1 dB are tabulated in Table 5.2.

The typical magnitude responses of a Type 1 Chebyshev lowpass filter for $N = 3, 5$, and 8 with 1 dB passband ripple are shown in Fig. 5.3. From this figure, it

Table 5.2 List of normalized Type 1 Chebyshev transfer functions for passband ripple = 1 dB

N	Denominator of $H_N(s)$	H_0
1	$s + 1.9652$	1.9652
2	$s^2 + 1.0977s + 1.1025$	0.98261
3	$s^3 + 0.98834s^2 + 1.2384s + 0.49131$	0.49131
4	$s^4 + 0.95281s^3 + 1.4539s^2 + 0.74262s + 0.27563$	0.24565
5	$s^5 + 0.93682s^4 + 1.6888s^3 + 0.9744s^2 + 0.58053s + 0.12283$	0.12283

Fig. 5.3 Magnitude response of typical Type 1 Chebyshev lowpass filter with 1 dB passband ripple

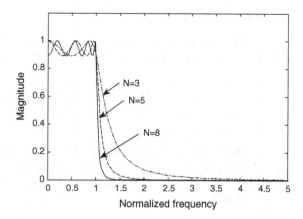

is seen that Type 1 Chebyshev lowpass filter exhibits equiripple in the passband with a monotonic decrease in the stopband.

Example 5.2 Design a Type 1 Chebyshev analog lowpass filter for the specifications given in Example 5.1.

Solution Since $\alpha_s = 40$ dB, $\alpha_p = 1$ dB, $\Omega_p = 2000\pi$, and $\Omega_s = 10000\pi$

$$\cosh^{-1} \sqrt{\frac{10^{0.1\alpha_s} - 1}{10^{0.1\alpha_p} - 1}} = \cosh^{-1} \sqrt{\frac{10^4 - 1}{10^{0.1} - 1}} = \cosh^{-1}(196.52)$$

$$\cosh^{-1}\left(\Omega_s/\Omega_p\right) = \cosh^{-1}(5) = 2.2924$$

$$N \geq \frac{\cosh^{-1}\sqrt{\frac{10^4-1}{10^{0.1}-1}}}{\cosh^{-1}(5)} = 2.6059$$

Since the order of the filter must be an integer, we choose the next higher integer value 3 for N. The normalized Type 1 Chebyshev lowpass filter for $N = 3$ with a passband ripple of 1 dB is given from Table 5.2 as

$$H_N(s) = \frac{0.49131}{s^3 + 0.988s^2 + 1.238s + 0.49131}$$

The transfer function for $\Omega_p = 2000\pi$ is obtained by substituting $s = \left(s/\Omega_p\right) = (s/2000\pi)$ in $H_N(s)$

$$\begin{aligned} H_a(s) &= \frac{0.49131}{\left(\frac{s}{2000\pi}\right)^3 + 0.988\left(\frac{s}{2000\pi}\right)^2 + 1.238\left(\frac{s}{2000\pi}\right) + 0.49131} \\ &= \frac{1.2187 \times 10^{11}}{s^3 + 6.2099 \times 10^3 s^2 + 4.889 \times 10^7 s + 1.2187 \times 10^{11}} \end{aligned}$$

Type 2 Chebyshev Filter

The squared-magnitude response of Type 2 Chebyshev lowpass filter, which is also known as the inverse Chebyshev filter, is given by

$$|H(\Omega)|^2 = \frac{1}{1 + \varepsilon^2 \left(\dfrac{T_N^2(\Omega_s/\Omega_p)}{T_N^2(\Omega_s/\Omega)} \right)} \tag{5.26}$$

The order N can be determined using Eq. (5.21). The Type 2 Chebyshev filter has both poles and zeros, and the zeros are on the $j\Omega$ axis. The normalized Type 2 Chebyshev lowpass filter, or the normalized inverse Chebyshev filter (normalized to $\Omega_s = 1$), may be formed as [4]

$$H_N(s) = \frac{H_0 \sum_k (s - z_k)}{\sum_k (s - p_k)}, \quad k = 1, 2, \ldots, N \tag{5.27}$$

where

$$z_k = j \frac{1}{\cos \frac{(2k-1)\pi}{N}} \quad \text{for } k = 1, 2, \ldots, N \tag{5.28a}$$

$$p_k = \frac{\sigma_k}{\sigma_k^2 + \Omega_k^2} + j \frac{\Omega_k}{\sigma_k^2 + \Omega_k^2} \quad \text{for } k = 1, 2, \ldots, N \tag{5.28b}$$

$$\sigma_k = -\sinh\left\{ \frac{1}{N} \sinh^{-1}\left(\frac{1}{\delta_s} \right) \right\} \sin \frac{(2k-1)\pi}{2N} \quad \text{for } k = 1, 2, \ldots, N \tag{5.28c}$$

$$\Omega_k = \cosh\left\{ \frac{1}{N} \sinh^{-1}\left(\frac{1}{\delta_s} \right) \right\} \cos \frac{(2k-1)\pi}{2N} \quad \text{for } k = 1, 2, \ldots, N \tag{5.28d}$$

$$\delta_s = \frac{1}{\sqrt{10^{0.1\alpha_s} - 1}} \tag{5.28e}$$

$$H_0 = \frac{\prod_k (-z_k)}{\prod_k (-p_k)} \tag{5.28f}$$

For example, if we consider $N = 3$ with a stopband ripple of 40 dB, then from (5.28e),

$$\frac{1}{\delta_s} = \sqrt{10^{0.1\alpha_s} - 1} = \sqrt{10^4 - 1} = 99.995$$

Hence,

$$\sinh^{-1}\left(\frac{1}{\delta_s}\right) = 5.28829$$

Using (5.28c) and (5.28d), we have

$$\sigma_k = -\sinh(5.28829/3)\sin\frac{(2k-1)\pi}{6} \quad \text{for } k = 1, 2, 3$$

$$\Omega_k = \cosh(5.28829/3)\cos\frac{(2k-1)\pi}{6} \quad \text{for } k = 1, 2, 3$$

Hence,

$$\sigma_1 = -1.41927, \sigma_2 = -2.83854, \sigma_3 = -1.41927$$

$$\Omega_1 = -2.60387, \Omega_2 = -2.83854, \Omega_3 = 2.60387$$

Thus, from (5.28b), the poles are

$$p_1 = -0.16115 + j0.29593, p_2 = -0.3523, p_3 = -0.16115 + j0.29593$$

Also, using (5.28a), the zeros are given by

$$z_1 = -j(2/\sqrt{3}), z_2 = j(2/\sqrt{3})$$

Finally, from (5.28f),

$$H_0 = 0.03$$

Therefore, the normalized Type 2 Chebyshev lowpass filter for $N = 3$ with a stopband ripple of 40 dB is given by

$$H_N(s) = \frac{0.03(s - z_1)(s - z_2)}{(s - p_1)(s - p_2)(s - p_3)}$$

$$= \frac{0.03(s^2 + 1.3333)}{(s^3 + 0.6746s^2 + 0.22709s + 0.04)}$$

The following MATLAB Program 5.3 can be used to form the Type 2 Chebyshev normalized transfer function for a given order and stopband ripple.

Program 5.3 Analog Type 2 Chebyshev lowpass filter normalized transfer function

```
N = input('enter order of the filter');
Rs = input('enter stopband attenuation in dB');
[z,p,k] = cheb2ap(N,Rs);% determines poles and zeros
disp('Poles are at');disp(p);
[num,den] = zp2tf(z,p,k);
%Print coefficients in powers of s
disp('Numerator polynomial');disp(num);
disp('Denominator polynomial');disp(den);
```

The normalized Type 2 Chebyshev transfer functions generated from the above program for typical values of N with a stopband ripple of 40 dB are tabulated in Table 5.3.

The typical magnitude response of a Type 2 Chebyshev lowpass filter for $N = 2$ and 7 with 20 dB stopband ripple is shown in Fig. 5.4. From this figure, it is seen that Type 2 Chebyshev lowpass filter exhibits monotonicity in the passband and equiripple in the stopband.

Example 5.3 Design a Type 2 Chebyshev lowpass filter for the specifications given in Example 5.1.

Table 5.3 List of normalized Type 2 Chebyshev transfer functions for stopband ripple = 40 dB

Order N	$H_N(s)$
1	$\dfrac{0.01}{s+0.01}$
2	$\dfrac{0.01s^2+0.02}{s^2+0.199s+0.02}$
3	$\dfrac{0.03s^2+0.04}{s^3+0.6746s^2+0.2271s+0.04}$
4	$\dfrac{0.01s^4+0.08s^2+0.08}{s^4+1.35s^3+0.9139s^2+0.3653s+0.08}$
5	$\dfrac{0.05s^4+0.2s^2+0.16}{s^5+2.1492s^4+2.3083s^3+1.5501s^2+0.6573s+0.16}$
6	$\dfrac{0.01s^6+0.18s^4+0.48s^2+0.32}{s^6+3.0166s^5+4.5519s^4+4.3819s^3+2.8798s^2+1.2393s+0.32}$

Fig. 5.4 Magnitude response of typical Type 2 Chebyshev lowpass filter with 20 dB stopband ripple

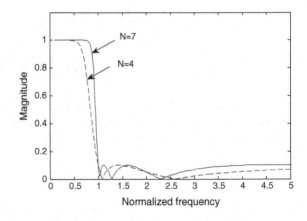

Solution The order N is chosen as 3, as in Example 5.2, since the equation for order finding is the same for both Type 1 and Type 2 Chebyshev filters. The normalized Type 2 Chebyshev lowpass filter for $N = 3$ with a stopband ripple of 40 dB has already been found earlier and is given by

$$H_N(s) = \frac{0.03(s^2 + 1.3333)}{(s^3 + 0.6746s^2 + 0.2271s + 0.04)}$$

For $\Omega_s = 10{,}000\pi$, the corresponding transfer function can be obtained by substituting $s = (s/\Omega_s) = (s/10{,}000\pi)$ in the above expression for $H_N(s)$. Thus, the required filter transfer function is

$$H_a(s) = \frac{0.03\left(\frac{s}{10{,}000\pi}\right)^2 + 0.04}{\left(\frac{s}{10{,}000\pi}\right)^3 + 0.6746\left(\frac{s}{10{,}000\pi}\right)^2 + 0.22709\left(\frac{s}{10{,}000\pi}\right) + 0.04}$$

$$= \frac{9.4252 \times 10^2 s^2 + 1.2403 \times 10^{12}}{s^3 + 2.1193 \times 10^4 s^2 + 2.2413 \times 10^8 s + 1.2403 \times 10^{12}}$$

5.1.4   Elliptic Analog Lowpass Filter

The square-magnitude response of an elliptic lowpass filter is given by

$$|H_a(j\Omega)|^2 = \frac{1}{1 + \varepsilon^2 U_N(\Omega/\Omega_p)} \tag{5.29}$$

where $U_N(x)$ is the Jacobian elliptic function of order N and ε is a parameter related to the passband ripple. In an elliptic filter, a constant k, called the selectivity factor, representing the sharpness of the transition region is defined as

$$k = \frac{\Omega_p}{\Omega_s} \tag{5.30}$$

A large value of k represents a wide transition band, while a small value indicates a narrow transition band.

For a given set of Ω_p, Ω_s, α_p and α_s, the filter order can be estimated using the formula [5]

$$N \cong \frac{\log\left(16 \times \frac{10^{0.1\alpha_s} - 1}{10^{0.1\alpha_p} - 1}\right)}{\log_{10}(1/\rho)} \tag{5.31}$$

where ρ can be computed using

$$\rho_0 = \frac{1 - \sqrt{k'}}{2\left(1 + \sqrt{k'}\right)} \tag{5.32}$$

$$k' = \sqrt{1 - k^2} \tag{5.33}$$

$$\rho = \rho_0 + 2(\rho_0)^5 + 15(\rho_0)^9 + 150(\rho_0)^{13} \tag{5.34}$$

The following MATLAB Program 5.4 can be used to form the elliptic normalized transfer function for given filter order, and passband ripple and stopband attenuation. The normalized passband edge frequency is set to 1.

Program 5.4 Analog elliptic lowpass filter normalized transfer function

```
N = input('enter order of the filter');
Rp = input('enter passband ripple in dB');
Rs = input('enter stopband attenuation in dB');
[z,p,k] = ellipap(N,Rp,Rs)% determines poles and zeros
disp('Poles are at');disp(p);
[num,den] = zp2tf(z,p,k);
%Print coefficients in powers of s
disp('Numerator polynomial');disp(num);
disp('Denominator polynomial');disp(den);
```

The normalized elliptic transfer functions generated from the above program for typical values of N and stopband ripple of 40 dB are tabulated in Table 5.4.

The magnitude response of a typical elliptic lowpass filter is shown in Fig. 5.5, from which it can be seen that it exhibits equiripple in both the pass and the stopbands.

For more details on elliptic filters, readers may refer to [2, 4, 6].

Example 5.4 Design an elliptic analog lowpass filter for the specifications given in Example 5.1.

Solution

$$k = \frac{\Omega_p}{\Omega_s} = \frac{2000\pi}{10,000\pi} = 0.2$$

Order N	$H_N(s)$
1	$\frac{1.9652}{s + 1.9652}$
2	$\frac{0.01s^2 + 0.9876}{s^2 + 1.0915s + 1.1081}$
3	$\frac{0.0692s^2 + 0.5265}{s^3 + 0.9782s^2 + 1.2434s + 0.5265}$
4	$\frac{0.01s^4 + 0.1502s^2 + 0.3220}{s^4 + 0.9391s^3 + 1.5137s^2 + 0.8037s + 0.3612}$
5	$\frac{0.0470s^4 + 0.2201s^2 + 0.2299}{s^5 + 0.9234s^4 + 1.84715s^3 + 1.1292s^2 + 0.7881s + 0.2299}$
6	$\frac{0.01s^6 + 0.1172s^4 + 0.28s^2 + 0.186}{s^6 + 0.9154s^5 + 2.2378s^4 + 1.4799s^3 + 1.4316s^2 + 0.5652s + 0.2087}$

Table 5.4 List of normalized elliptic transfer functions for passband ripple = 1 dB and stopband ripple = 40 dB

Fig. 5.5 Magnitude response
of typical elliptic lowpass
filter with 1 dB passband
ripple and 30 dB stopband
ripple

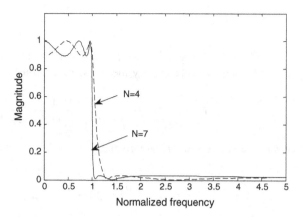

and

$$k' = \sqrt{1 - k^2} = \sqrt{1 - 0.04} = 0.979796.$$

Substituting these values in Eqs. (5.32) and (5.34), we get

$$\rho_0 = 0.00255135,$$

$$\rho = 0.0025513525$$

and hence

$$N = \frac{\log\left(16 \times \frac{10^4 - 1}{10^{0.1} - 1}\right)}{\log_{10}\left(\frac{1}{0.0025513525}\right)} = 2.2331.$$

Choose $N = 3$. Then, for $N = 3$, a passband ripple of 1 dB, and a stopband ripple
of 40 dB, the normalized elliptic transfer function is as given in Table 5.4. For
$\Omega_p = 2000\pi$, the corresponding transfer function can be obtained by substituting
$s = (s/\Omega_p) = (s/2000\pi)$ in the expression for $H_N(s)$. Thus, the required filter
transfer function is

$$H_a(s) = \frac{0.0692\left(\frac{s}{2000\pi}\right)^2 + 0.5265}{\left(\frac{s}{2000\pi}\right)^3 + 0.97825\left(\frac{s}{2000\pi}\right)^2 + 1.2434\left(\frac{s}{2000\pi}\right) + 0.5265}$$

$$= \frac{4.348 \times 10^2 s^2 + 1.306 \times 10^{11}}{s^3 + 6.1465 \times 10^3 s^2 + 4.9087 \times 10^7 s + 1.306 \times 10^{11}}$$

5.1.5   Bessel Filter

Bessel filter is a class of all-pole filters that provide linear phase response in the passband and characterized by the transfer function [5]

$$H_a(s) = \frac{1}{a_0 + a_1 s + a_2 s^2 + \cdots + a_{N-1} s^{N-1} + a_N s^N} \tag{5.35}$$

where the coefficients a_N are given by

$$a_n = \frac{(2N - n)!}{2^{N-n} n! (N - n)!} \tag{5.36}$$

The magnitude responses of a third-order Bessel filter and Butterworth filter are shown in Fig. 5.6, and the phase responses of the same filters with the same order are shown in Fig. 5.7. From these figures, it is seen that the magnitude response of the Bessel filter is poorer than that of the Butterworth filter, whereas the phase response of the Bessel filter is more linear in the passband than that of the Butterworth filter.

5.1.6   Comparison of Various Types of Analog Filters

The magnitude response and phase response of the normalized Butterworth, Chebyshev Type 1, Chebyshev Type 2, and elliptic filters of the same order are compared with the following specifications:
filter order = 8, maximum passband ripple = 1 dB, and minimum stopband ripple = 35 dB.

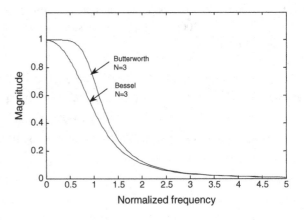

Fig. 5.6 Magnitude responses of Bessel and Butterworth filters of order $N = 3$

Fig. 5.7 Phase responses of Bessel and Butterworth filters of order $N = 3$

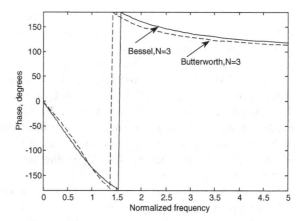

The following MATLAB program is used to generate the magnitude and phase responses for these specifications.

Program 5.5 Magnitude and phase responses of analog filters of order 8 with a passband ripple of 1 dB and a stopband ripple of 35 dB.

```
clear all;clc;
[z,p,k] = buttap(8);
[num1,den1] = zp2tf(z,p,k);[z,p,k] = cheb1ap(8,1);
[num2,den2] = zp2tf(z,p,k);[z,p,k] = cheb2ap(8,35);
[num3,den3] = zp2tf(z,p,k);[z,p,k] = ellipap(8,1,35);
[num4,den4] = zp2tf(z,p,k);
omega = [0:0.01:5];
h1 = freqs(num1,den1,omega);h2 = freqs(num2,den2,omega);
h3 = freqs(num3,den3,omega);h4 = freqs(num4,den4,omega);
ph1 = angle(h1);ph1 = unwrap(ph1);
ph2 = angle(h2);ph2 = unwrap(ph2);
ph3 = angle(h3);ph3 = unwrap(ph3);
ph4 = angle(h4);ph4 = unwrap(ph4);
Figure (1),plot(omega,20*log10(abs(h1)),'-');hold on
plot(omega,20*log10(abs(h2)),'-');hold on
plot(omega,20*log10(abs(h3)),':');hold on
plot(omega,20*log10(abs(h4)),'-.');
xlabel('Normalized frequency');ylabel('Gain,dB');axis([0 5-80 5]);
legend('Butterworth','Chebyshev Type 1','Chebyshev Type 2','Elliptic');hold off
Figure(2),plot(omega,ph1,'-');hold on
plot(omega,ph2,'-');hold on
plot(omega,ph3,':');hold on
plot(omega,ph4,'-.')
xlabel('Normalized frequency');ylabel('Phase,radians');axis([0 5 -8 0]);
legend('Butterworth','Chebyshev Type 1','Chebyshev Type 2','Elliptic');
```

The magnitude and phase responses for the above specifications are shown in Fig. 5.8. The magnitude response of Butterworth filter decreases monotonically in both the passband and the stopband with wider transition band. The magnitude response of the Chebyshev Type 1 exhibits ripples in the passband, whereas the Chebyshev Type 2 has approximately the same magnitude response to that of the Butterworth filter. The transition band of both the Type 1 and Type 2 Chebyshev filters is the same, but less than that of the Butterworth filter. The elliptic filter exhibits an equiripple magnitude response both in the passband and in the stopband with a transition width smaller than that of the Chebyshev Type 1 and Type 2 filters. But the phase response of the elliptic filter is more nonlinear in the passband than that of the phase response of the Butterworth and Chebyshev filters. If linear phase in the passband is the stringent requirement, then the Bessel filter is preferred, but with a poor magnitude response.

Another way of comparing the various filters is in terms of the order of the filter required to satisfy the same specifications. Consider a lowpass filter that meets the passband edge frequency of 450 Hz, stopband edge frequency of 550 Hz, passband ripple of 1 dB, and stopband ripple of 35 dB. The orders of the Butterworth, Chebyshev Type 1, Chebyshev Type 2, and elliptic filters are computed for the above specifications and listed in Table 5.5. From this table, we can see that elliptic filter can meet the specifications with the lowest filter order.

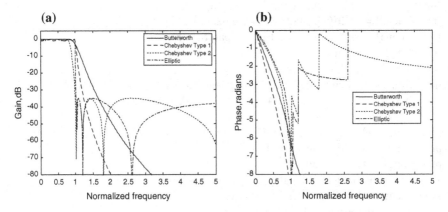

Fig. 5.8 Comparison of various types of analog lowpass filters **a** magnitude response and **b** phase response

Table 5.5 Comparison of orders of various types of filters

Filter	Order
Butterworth	24
Chebyshev Type 1	9
Chebyshev Type 2	9
Elliptic	5

5.1.7 Design of Analog Highpass, Bandpass, and Bandstop Filters

The analog highpass, bandpass, and bandstop filters can be designed using analog frequency transformations. In this design process, first, the analog prototype low-pass filter specifications are derived from the desired specifications of the analog filter using suitable analog-to-analog transformation. Next, by using the specifications so obtained, a prototype lowpass filter is designed. Finally, the transfer function of the desired analog filter is determined from the transfer function of the prototype analog lowpass transfer function using the appropriate analog-to-analog frequency transformation. The lowpass-to-lowpass, lowpass-to-highpass, lowpass-to-bandpass, and lowpass-to-bandstop analog transformations are considered next.

Lowpass to Lowpass:

Let $\Omega_p = 1$ and $\hat{\Omega}_p$ be the passband edge frequencies of the normalized prototype low pass filter and the desired lowpass filter, as shown in Fig. 5.9. The transformation from the prototype lowpass to the required lowpass must convert $\hat{\Omega} = 0$ to $\Omega = 0$ and $\hat{\Omega} = \pm\infty$ to $\Omega = \pm\infty$. The transformation such as $s = k\hat{s}$ or $\Omega = k\hat{\Omega}$ achieves the above transformation for any positive value of k. If k is chosen to be $(1/\hat{\Omega}_p)$, then $\hat{\Omega}_p$ gets transformed to $\Omega_p = 1$, and $\hat{\Omega}_s$ to $\Omega_s = \hat{\Omega}_s / \hat{\Omega}_p$. Since $\Omega_p = 1$ is the passband edge frequency for the normalized Type I Chebyshev and elliptic lowpass filters, we have the design equations for these filters as

$$\Omega_p = 1, \Omega_s = \hat{\Omega}_s/\hat{\Omega}_p. \tag{5.37a}$$

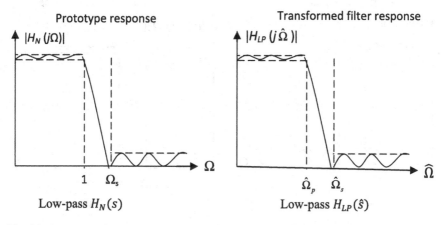

Fig. 5.9 Lowpass-to-lowpass frequency transformation

Also, the transfer function $H_{\text{LP}}(\hat{s})$ for these filters is related to the corresponding normalized transfer function $H_N(s)$ by

$$H_{\text{LP}}(\hat{s}) = H_N(s)_{s=\frac{\hat{s}}{\hat{\Omega}_p}} \qquad (5.37\text{b})$$

However, in the case of a Butterworth filter, since $\Omega = 1$ corresponds to the cutoff frequency of the filter, the transfer function $H_{\text{LP}}(\hat{s})$ for the Butterworth filter is related to the normalized lowpass Butterworth transfer function $H_N(s)$ by

$$H_{\text{LP}}(\hat{s}) = H_N(s)_{s=\hat{s}/\hat{\Omega}_c} \qquad (5.37\text{c})$$

where $\hat{\Omega}_c$ is the cutoff frequency of the desired Butterworth filter and is given by Eq. (5.5). For similar reasons, the transfer function $H_{\text{LP}}(\hat{s})$ for the Type 2 Chebyshev filter is related to the normalized transfer function $H_N(s)$ by

$$H_{LP}(\hat{s}) = H_N(s)_{s=\hat{s}/\hat{\Omega}_s} \qquad (5.37\text{d})$$

Lowpass to Highpass:

Let the passband edge frequencies of the prototype lowpass and the desired highpass filters be $\Omega_p = 1$ and $\hat{\Omega}_p$, as shown in Fig. 5.10. The transformation from prototype lowpass to the desired highpass must transform $\hat{\Omega} = 0$ to $\Omega = \infty$ and $\hat{\Omega} = \infty$ to $\Omega = 0$. The transformation such as $s = k/\hat{s}$ or $\hat{\Omega} = \infty$ achieves the above transformation for any positive value of k. By transforming $\hat{\Omega}_p$ to $\Omega_p = 1$, the constant k can be determined as $k = \hat{\Omega}_p$. Thus, design equations are

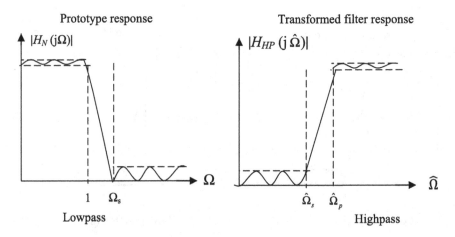

Fig. 5.10 Lowpass-to-highpass frequency transformation

$$\Omega_p = 1, \Omega_s = \hat{\Omega}_p / \hat{\Omega}_s, \tag{5.38a}$$

and the desired transfer function $H_{HP}(\hat{s})$ is related to the normalized lowpass transfer function $H_N(s)$ by

$$H_{HP}(\hat{s}) = H_N(s)\big|_{s = \hat{\Omega}_p / \hat{s}} \tag{5.38b}$$

Equations (5.38a) and (5.38b) hold for all filters except for Butterworth and Type 2 Chebyshev filter. For Butterworth

$$H_{LP}(\hat{s}) = H_N(s)_{s = \hat{s}/\hat{\Omega}_c} \tag{5.38c}$$

$$H_{HP}(\hat{s}) = H_N(s)\big|_{s = \hat{\Omega}_p / \hat{s}} \tag{5.38d}$$

For Type 2 Chebyshev filter, the design equations are

$$\Omega_p = \hat{\Omega}_s / \hat{\Omega}_p, \Omega_s = 1 \tag{5.39a}$$

and

$$H_{HP}(\hat{s}) = H_N(s)_{s = \hat{\Omega}_s / \hat{s}} \tag{5.39b}$$

Lowpass to Bandpass

The prototype lowpass and the desired bandpass filters are shown in Fig. 5.11. In this figure, $\hat{\Omega}_{p1}$ is the lower passband edge frequency, $\hat{\Omega}_{p2}$ the upper passband edge frequency, $\hat{\Omega}_{s1}$ the lower stopband edge frequency, and $\hat{\Omega}_{s2}$ the upper stopband

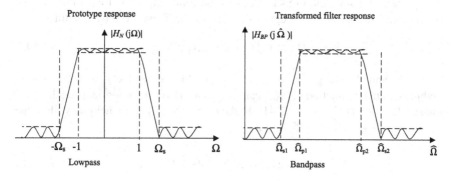

Fig. 5.11 Lowpass-to-bandpass frequency transformation

edge frequency of the desired bandpass filter. Let us denote by B_p, the bandwidth of the passband, and by $\hat{\Omega}_{mp}$ the geometric mean between the passband edge frequencies of the bandpass filter, i.e.,

$$B_p = \hat{\Omega}_{p2} - \hat{\Omega}_{p1} \tag{5.40a}$$

$$\hat{\Omega}_{mp} = \sqrt{\hat{\Omega}_{p1}\hat{\Omega}_{p2}} \tag{5.40b}$$

Now, consider the transformation

$$s = \frac{\left(\hat{s}^2 + \hat{\Omega}_{mp}^2\right)}{B_p\hat{s}} \tag{5.41}$$

As a consequence of this transformation, it is seen that $\hat{\Omega} = 0$, $\hat{\Omega}_{p1}$, $\hat{\Omega}_{mp}$, $\hat{\Omega}_{p2}$, and ∞ transform to the frequencies $\Omega = -\infty$, -1, 0, $+1$, and ∞, respectively, for the normalized lowpass filter. Also, the transformation (5.41) transforms the frequencies $\hat{\Omega}_{s1}$ and $\hat{\Omega}_{s2}$ to Ω'_s and Ω''_s, respectively, where

$$\Omega'_s = \frac{\hat{\Omega}_{s1}^2 - \hat{\Omega}_{p1}\hat{\Omega}_{p2}}{\left(\hat{\Omega}_{p2} - \hat{\Omega}_{p1}\right)\hat{\Omega}_{s1}} = A_1(\text{say}) \tag{5.42}$$

and

$$\Omega''_s = \frac{\hat{\Omega}_{s2}^2 - \hat{\Omega}_{p1}\hat{\Omega}_{p2}}{\left(\hat{\Omega}_{p2} - \hat{\Omega}_{p1}\right)\hat{\Omega}_{s2}} = A_2(\text{say}) \tag{5.43}$$

In order to satisfy the stopband requirements and to have symmetry of the stopband edges in the lowpass filter, we choose Ω_s to be the min $\{|A_1|, |A_2|\}$. Thus, the spectral transformation (5.41) leads to the following design equations for the normalized lowpass filter (except in the case of the Type 2 Chebyshev filter)

$$\Omega_p = 1, \ \Omega_s = \min\{|A_1|, |A_2|\} \tag{5.44a}$$

where A_1 and A_2 are given by (5.42) and (5.43), respectively, and the desired highpass transfer function $H_{BP}(\hat{s})$ can be obtained from the normalized lowpass transfer function $H_N(s)$ using (5.41). In the case of the Type 2 Chebyshev filter, the equation corresponding to (5.44a) is

$$\Omega_p = \max\{1/|A_1|, 1/|A_2|\}, \Omega_s = 1 \tag{5.44b}$$

Lowpass to Bandstop

The prototype lowpass and the desired bandstop filters are shown in Fig. 5.12. In this figure, $\hat{\Omega}_{p1}$ is the lower passband edge frequency, $\hat{\Omega}_{p2}$ the upper passband edge frequency, $\hat{\Omega}_{s1}$ the lower stopband edge frequency, and $\hat{\Omega}_{s2}$ the upper stopband edge frequency of the desired bandstop filter. Let us now consider the transformation

$$s = \frac{k\hat{s}}{\left(\hat{s}^2 + \hat{\Omega}_{ms}^2\right)} \tag{5.45}$$

where $\hat{\Omega}_{ms}$ is the geometric mean between the stopband edge frequencies of the bandstop filter, i.e.,

$$\hat{\Omega}_{ms} = \sqrt{\hat{\Omega}_{s1}\hat{\Omega}_{s2}} \tag{5.46}$$

As a consequence of this transformation, it is seen that $\hat{\Omega} = 0$ and ∞ transform to the frequency $\hat{\Omega} = 0$ for the normalized lowpass filter. Now, we transform the lower stopband edge frequency $\hat{\Omega}_{s1}$ to the stopband edge frequency Ω_s of the normalized lowpass filter; hence,

$$\Omega_S = \frac{k}{\hat{\Omega}_{s2} - \hat{\Omega}_{s1}} = \frac{k}{B_S} \tag{5.47a}$$

where $B_S = \left(\hat{\Omega}_{s2} - \hat{\Omega}_{s1}\right)$ is the bandwidth of the stopband. Also, the upper stopband edge frequency $\hat{\Omega}_{s2}$ is transformed to

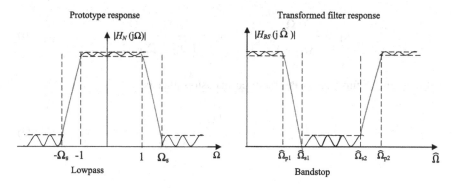

Fig. 5.12 Lowpass-to-bandstop frequency transformation

$$-\frac{k}{\hat{\Omega}_{s2} - \hat{\Omega}_{s1}} = -\frac{k}{B_S} = -\Omega_S \qquad (5.47b)$$

Hence, the constant k is given by

$$k = B_S \Omega_S = \left(\hat{\Omega}_{s2} - \hat{\Omega}_{s1}\right)\Omega_S \qquad (5.47c)$$

As a consequence, the passband edge frequencies $\hat{\Omega}_{p1}$ and $\hat{\Omega}_{p2}$ are transformed to

$$\Omega'_p = \frac{\left(\hat{\Omega}_{s2} - \hat{\Omega}_{s1}\right)\hat{\Omega}_{p1}}{\hat{\Omega}_{s1}\hat{\Omega}_{s2} - \hat{\Omega}_{p1}^2}\Omega_S = \frac{1}{A_1}\Omega_S \qquad (5.48a)$$

and

$$\Omega''_p = \frac{\left(\hat{\Omega}_{s2} - \hat{\Omega}_{s1}\right)\hat{\Omega}_{p2}}{\hat{\Omega}_{s1}\hat{\Omega}_{s2} - \hat{\Omega}_{p2}^2}\Omega_S = \frac{1}{A_2}\Omega_S \qquad (5.48b)$$

In order to satisfy the passband requirement as well as to satisfy the symmetry requirement of the passband edge of the normalized lowpass filter, we have to choose the higher of $\left|\Omega'_p\right|$ and $\left|\Omega''_p\right|$ as Ω_p. Since for the normalized filter (except for the case of Type 2 Chebyshev filter), $\Omega_p = 1$, we have to choose Ω_s to be the lower of $\{|A_1|, |A_2|\}$. Hence, the design equations for the normalized lowpass filter (except for the Type 2 Chebyshev) are

$$\Omega_p = 1, \ \Omega_s = \min\{|A_1|, |A_2|\} \qquad (5.49a)$$

where

$$A_1 = \frac{\hat{\Omega}_{s1}\hat{\Omega}_{s2} - \hat{\Omega}_{p1}^2}{\left(\hat{\Omega}_{s2} - \hat{\Omega}_{s1}\right)\hat{\Omega}_{p1}}, A_2 = \frac{\hat{\Omega}_{s1}\hat{\Omega}_{s2} - \hat{\Omega}_{p2}^2}{\left(\hat{\Omega}_{s2} - \hat{\Omega}_{s1}\right)\hat{\Omega}_{p2}} \qquad (5.49b)$$

and the transfer function of the required bandstop filter is

$$H_{BS}(\hat{s}) = H_N(s)_s = \frac{\left(\hat{\Omega}_{s2} - \hat{\Omega}_{s1}\right)\Omega_S \hat{s}}{\hat{s}^2 + \hat{\Omega}_{s1}\hat{\Omega}_{s2}} \qquad (5.49c)$$

For the Type 2 Chebyshev filter, Eq. (5.49a) would be replaced by

$$\Omega_p = \max\{1/|A_1|, 1/|A_2|\}, \Omega_S = 1 \qquad (5.50)$$

For further details on analog frequency transformations, readers may refer to [7].

5.2 Design of Digital Filters from Analog Filters

5.2.1 Digital Filter Specifications

The digital filter frequency response specifications are often in the form of a tolerance scheme. The specifications for a low pass filter are depicted in Fig. 5.13.
The following parameters are usually used as the specifications.

ω_p—passband edge frequency
ω_s—stopband edge frequency
δ_p—peak ripple value in the passband
δ_s—peak ripple value in the stopband

Generally, the passband edge frequency (f_p), the stopband edge frequency (f_s), and the sampling frequency (F_T) are represented in Hz. But, the digital filter design methods require normalized angular edge frequencies in radians. The normalized angular edge frequencies ω_p and ω_s can be obtained using the following relations

$$\omega_p = \frac{2\pi f_p}{F_T} = 2\pi f_p T \qquad (5.51a)$$

$$\omega_s = \frac{2\pi f_s}{F_T} = 2\pi f_s T \qquad (5.51b)$$

where T is the sampling period.

Fig. 5.13 Specifications of a digital lowpass filter

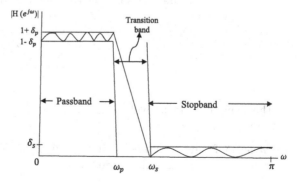

5.2.2 Design of Digital Filters Using Impulse-Invariant Method

In this method, the impulse response of an analog filter is uniformly sampled to obtain the impulse response of the digital filter, and hence, this method is called the impulse-invariant method. The process of designing an IIR filter using this method is as follows:

Step 1 Design an analog filter to meet the given frequency specifications. Let $H_a(s)$ be the transfer function of the designed analog filter. We assume for simplicity that $H_a(s)$ has only simple poles. In such a case, the transfer function of the analog filter can be expressed in partial fraction form as

$$H_a(s) = \sum_{k=1}^{N} \frac{A_k}{s - p_k} \tag{5.52}$$

where A_k is the residue of $H(s)$ at the pole p_k.

Step 2 Calculate the impulse response $h(t)$ of this analog filter by applying the inverse Laplace transformation on $H(s)$. Hence,

$$h_a(t) = \sum_{k=1}^{N} A_k e^{p_k t} u_a(t) \tag{5.53}$$

Step 3 Sample the impulse response of the analog filter with a sampling period T. Then, the sampled impulse response $h(n)$ can be expressed as

$$h(n) = h_a(t)|_{t=nT}$$
$$= \sum_{k=1}^{N} (A_k e^{p_k t})^n u(n) \tag{5.54}$$

Step 4 Apply the z-transform on the sampled impulse response obtained in Step 3, to form the transfer function of the digital filter, i.e., $H(z) = Z[h(n)]$. Thus, the transfer function $H(z)$ for the impulse-invariant method is given by

$$H(z) = \sum_{k=1}^{N} \frac{A_k}{1 - e^{p_k T} z^{-1}} \tag{5.55}$$

This impulse-invariant method can be extended for the case when the poles are not simple [8].

Example 5.5 Design a third-order Butterworth digital filter using impulse-invariant technique. Assume a sampling period of $T = 1$ s.

Solution For $N = 3$, the transfer function of a normalized Butterworth filter is given by

$$H(s) = \frac{1}{(s+1)(s^2+s+1)}$$

$$H(s) = \frac{1}{s+1} + \frac{-0.5+0.288j}{s+0.5+j0.866} + \frac{-0.5-0.288j}{s+0.5-j0.866}$$

Hence, from (5.55), we have

$$H(z) = \frac{1}{1-e^{-1}z^{-1}} + \frac{-0.5+j0.288}{1-e^{-0.5}e^{-j0.866}z^{-1}} + \frac{-0.5-j0.288}{1-e^{-0.5}e^{j0.866}z^{-1}}$$

$$= \frac{1}{1-0.368z^{-1}} + \frac{-1+0.66z^{-1}}{1-0.786z^{-1}+0.368z^{-2}}$$

Example 5.6 Design a Butterworth filter using the impulse-invariant method for the following specifications:

$$0.8 \le \left|H\left(e^{j\omega}\right)\right| \le 1 \qquad 0 \le \omega \le 0.2\pi$$

$$\left|H\left(e^{j\omega}\right)\right| \le 0.2 \qquad 0.6\pi \le \omega \le \pi$$

Solution From (5.1), the magnitude-squared function of the Butterworth filter is

$$\left|H_a(j\Omega)\right|^2 = \frac{1}{1+(\Omega/\Omega_c)^{2N}}$$

Substituting the requirements in the above magnitude function, we get

$$1 + \left(\frac{0.2\pi}{\Omega_c}\right)^{2N} = \left(\frac{1}{0.8}\right)^2$$

$$1 + \left(\frac{0.6\pi}{\Omega_c}\right)^{2N} = \left(\frac{1}{0.2}\right)^2$$

The solution of the above two equations leads to

$$N = \frac{\log\frac{24}{0.5625}}{2\log 3} = \frac{1.6301}{0.9542} = 1.71$$

Approximating to the nearest higher value, we have $N = 2$. Substituting $N = 2$ in

$$1 + \left(\frac{0.2\pi}{\Omega_C}\right)^{2N} = \left(\frac{1}{0.8}\right)^2$$

we get $\Omega_c = 0.231\pi$. Also, for $N = 2$ the transfer function of the normalized Butterworth filter is

$$H_N(s) = \frac{1}{s^2 + \sqrt{2}s + 1}$$

Hence, from (5.37c),

$$
\begin{aligned}
H_a(s) &= H_N(s)_{s=s/\Omega_c} \\
&= \frac{0.5266}{s^2 + 1.03s + 0.5266} \\
&= \frac{0.516j}{s + 0.51 + j0.51} - \frac{0.516j}{s + 0.51 - j0.51}
\end{aligned}
$$

$$H(z) = \frac{0.516j}{1 - e^{-0.51T}e^{-j0.51T}z^{-1}} - \frac{0.516j}{1 - e^{-0.51T}e^{j0.51T}z^{-1}}$$

Since $T = 1$, we have

$$H(z) = \frac{0.3019z^{-1}}{1 - 1.048z^{-1} + 0.36z^{-2}}$$

Disadvantage of Impulse-Invariant Method

The frequency responses of the digital and analog filters are related by

$$H(e^{j\omega}) = \frac{1}{T}\sum_{k=-\infty}^{\infty} H_a\left(j\frac{\omega + 2\pi k}{T}\right) \qquad (5.56)$$

From Eq. (5.56), it is evident that the frequency response of the digital filter is not identical to that of the analog filter due to aliasing in the sampling process. If the analog filter is band-limited with

$$H_a\left(j\frac{\omega}{T}\right) = 0 \qquad \left|\frac{\omega}{T}\right| = |\Omega| \geq \pi/T \qquad (5.57)$$

then the digital filter frequency response is of the form

$$H(e^{j\omega}) = \frac{1}{T}H_a\left(j\frac{\omega}{T}\right) \quad |\omega| \le \pi \tag{5.58}$$

In the above expression, if T is small, the gain of the filter becomes very large. This can be avoided by introducing a multiplication factor T in the impulse-invariant transformation. In such a case, the transformation would be

$$h(n) = T\,h_a(nT) \tag{5.59}$$

and $H(z)$ would be

$$H(z) = T\sum_{k=1}^{N}\frac{A_k}{1 - e^{p_k T}z^{-1}} \tag{5.60}$$

Also, the frequency response is

$$H(e^{j\omega}) = \frac{1}{T}H_a\left(j\frac{\omega}{T}\right) \quad |\omega| \le \pi \tag{5.61}$$

Hence, the impulse-invariant method is appropriate only for band-limited filters, i.e., lowpass and bandpass filters, but not suitable for highpass or bandstop filters where additional band limiting is required to avoid aliasing. Thus, there is a need for another mapping method such as bilinear transformation technique which avoids aliasing.

5.2.3   Design of Digital Filters Using Bilinear Transformation

In order to avoid the aliasing problem mentioned in the case of the impulse-invariant method, we use the bilinear transformation, which is a one-to-one mapping from the s-plane to the z-plane; that is, it maps a point in the s-plane to a unique point in the z-plane and vice versa. This is the method that is mostly in designing an IIR digital filter from an analog filter. This approach is based on the trapezoidal rule, and for details, one could refer to [8]. Consider the bilinear transformation given by

$$s = \frac{(z-1)}{(z+1)} \tag{5.62}$$

Then, a transfer function $H_a(s)$ in the analog domain is transformed in the digital domain as

$$H(z) = H_a(s)\big|_{s=\frac{(z-1)}{(z+1)}} \tag{5.63}$$

Also, from (5.62) we have

$$z = \frac{(1+s)}{(1-s)} \tag{5.64}$$

We now study the mapping properties of the bilinear transformation. Consider a point $s = -\sigma + j\Omega$ in the left half of the s-plane. Then, from (5.64), we get

$$|z| = \left|\frac{(1-\sigma+j\Omega)}{(1+\sigma-j\Omega)}\right| > 1 \tag{5.65}$$

Hence, the left half of the s-plane maps into the interior of the unit circle in the z-plane (see Fig. 5.14). Similarly, it can be shown that the right half of the s-plane maps into the exterior of the unit circle in the z-plane. For a point z on the unit circle, $z = e^{j\omega}$, we have from (5.62)

$$s = \frac{(e^{j\omega}-1)}{(e^{j\omega}+1)} = j\tan\frac{\omega}{2} \tag{5.66}$$

Thus

$$\Omega = \tan\frac{\omega}{2} \tag{5.67}$$

or

$$\omega = 2\tan^{-1}\Omega \tag{5.68}$$

showing that the positive and negative imaginary axes of the s-plane are mapped respectively into the upper and lower halves of the unit circle in the z-plane. We

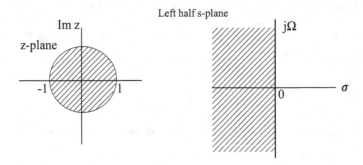

Fig. 5.14 Mapping of the s-plane into the z-plane by the bilinear transformation

thus see that the bilinear transformation avoids the problem of aliasing encountered in the impulse-invariant method, since it maps the entire imaginary axis in the *s*-plane onto the unit circle in the in the *z*-plane. Further, in view of the mapping, this transformation converts a stable analog filter into a stable digital filter.

Warping effect

The price paid, however, is in the introduction of a distortion in the frequency axis due to the nonlinear relation between Ω and ω, exhibited particularly at higher frequencies, as shown in Fig. 5.15. This behavior is called the warping effect. This can be corrected by 'prewarping' the analog filter specifications. The procedure to be followed is as follows:

Step 1   From the digital filter specifications, prewarp the critical frequencies, such as the cutoff frequency, passband edge, stopband edge using Eq. (5.67).

Step 2   From these new critical frequencies, obtain the transfer function $H_a(s)$ of the analog filter using the methods already described.

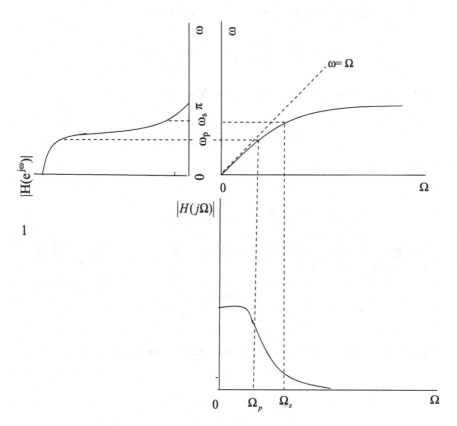

Fig. 5.15 Warping effect due to bilinear transformation

Step 3 Use the bilinear transformation given by Eq. (5.62) to obtain the cor-
        responding digital transfer function $H(z)$.

Example 5.7 Design a low pass Butterworth IIR digital filter using bilinear transformation for the following specifications:

3 dB cutoff frequency $\omega_c = 0.2\pi$ and $|H(e^{j\omega})| \leq 0.317$ for $0.4\,\pi \leq \omega \leq \pi$.

Solution From Eq. (5.1), the magnitude-squared function of the Butterworth filter
is

$$|H_a(j\Omega)|^2 = \frac{1}{1 + (\Omega/\Omega_c)^{2N}}$$

As bilinear transformation is used and $\omega_c = 0.2\pi$, prewarping of the cutoff
frequency yields

$$\Omega_C = \tan\left(\frac{0.2\pi}{2}\right) = \tan(0.1\pi) = 0.325$$

From the magnitude response specification, we obtain

$$1 + \left(\frac{\tan(0.4\pi/2)}{\tan(0.2\pi/2)}\right)^{2N} = \left(\frac{1}{0.317}\right)^2$$

Solving the above equation gives $N = 2$. Hence,

$$H_a(s) = \frac{\Omega_c^2}{s^2 + \sqrt{2}\Omega_c s + \Omega_c^2}$$

Thus,

$$H_a(s) = \frac{0.10563}{s^2 + 0.46s + 0.10563}$$

The digital transfer function $H(z)$ is now obtained by using Eq. (5.62) in the
above transfer function $H_a(s)$.

$$H(z) = \frac{0.068(z+1)^2}{z^2 - 1.142z + 0.413}$$

Example 5.8 Consider the following analog transfer function

$$H_a(s) = \frac{s^2 - 3s + 3}{s^2 + 3s + 3}$$

(i) Is it possible to obtain the corresponding digital transfer function using the impulse-invariant method?

(ii) Is it possible to obtain the corresponding digital transfer function using bilinear transformation?

Solution

(i) $H_a(s) = \frac{s^2 - 3s + 3}{s^2 + 3s + 3}$ represents an allpass system.

According to the impulse-invariant design, using Eq. (5.56),

$$H(e^{j\omega}) = \frac{1}{T} \sum_{k=-\infty}^{\infty} H_a\left(j\frac{\omega + 2\pi k}{T}\right)$$

The aliasing terms will destroy the allpass nature of the continuous time filter. Therefore, one cannot design a corresponding digital system using the impulse-invariant method.

(ii) The bilinear transformation only warps the frequency axis. The magnitude response is not affected; therefore, an allpass filter will map to an allpass filter. Thus, one can design a corresponding digital system using the bilinear transformation.

$$H(z) = H_a(s)|_{s=(z-1)/(z+1)}$$

Example 5.9 Design a low pass IIR digital filter using the bilinear transformation for the following specifications:

$$0.9 \le |H(e^{j\omega})| \le 1, \qquad 0 \le \omega \le 0.2\pi$$

$$|H(e^{j\omega})| \le 0.25, \qquad 0.3\pi \le \omega \le \pi$$

Solution Prewarping the critical frequencies, we have the passband and stopband edge frequencies of the analog filter to be

$$\Omega_p = \tan\left(\frac{0.2\,\pi}{2}\right) = 0.325$$

$$\Omega_s = \tan\left(\frac{0.3\pi}{2}\right) = 0.51$$

Since

$$|H_a(j\Omega)|^2 = \frac{1}{1 + (\Omega/\Omega_c)^{2N}}$$

we have

$$1 + \left(\frac{\Omega_p}{\Omega_c}\right)^{2N} = 1 + \left(\frac{0.325}{\Omega_c}\right)^{2N} = \left(\frac{1}{0.9}\right)^2$$

and

$$1 + \left(\frac{\Omega_s}{\Omega_c}\right)^{2N} = 1 + \left(\frac{0.51}{\Omega_c}\right)^{2N} = \left(\frac{1}{0.25}\right)^2$$

Solving the above two equations, we get $N = 4.6$; hence, we choose $N = 5$. Using this value of N, we can calculate Ω_c to be $\Omega_c = 0.398$. Hence, we have

$$H_a(s) = \frac{\Omega_c^5}{(s + \Omega_c)(s^2 + 0.61803\Omega_c s + \Omega_c^2)(s^2 + 1.61803\Omega_c s + \Omega_c^2)}$$

$$= \frac{0.01}{(s + 0.398)(s^2 + 0.246s + 0.1584)(s^2 + 0.644s + 0.1584)}$$

Now substituting for s using Eq. (5.62), we get the required digital transfer function to be

$$H(z) = \frac{0.01(z + 1)}{(1.398z - 0.602)(1.404z^2 - 1.683z + 0.9124)(1.802z^2 - 1.683z + 0.5144)}$$

Example 5.10 Design a lowpass digital filter with 3 dB cutoff frequency at 50 Hz and attenuation of at least 10 dB for frequency larger than 100 Hz. Assume a suitable sampling frequency.

Solution Assume the sampling frequency as 500 Hz. Then,

$$\omega_c = \frac{2\pi f_c}{F_T} = \frac{2\pi \times 50}{500} = 0.2\pi$$

$$\omega_s = \frac{2\pi f_s}{F_T} = \frac{2\pi \times 100}{500} = 0.4\pi$$

Prewarping of the above-normalized frequencies yields

$$\Omega_C = \tan\left(\frac{0.2\pi}{2}\right) = \tan(0.1\pi) = 0.325$$

$$\Omega_s = \tan\left(\frac{0.4\pi}{2}\right) = \tan(0.2\pi) = 0.727$$

Substituting these values in $(\Omega_s/\Omega_c)^{2N} = 10^{0.1\alpha_s} - 1$ and solving for N, we get

$$N = \frac{\log(10^1 - 1)}{2\log(0.727/0.325)} = \frac{0.9542}{0.6993} = 1.3643.$$

Hence, the order of the Butterworth filter is 2. The normalized lowpass Butterworth filter for $N = 2$ is given by

$$H_N(s) = \frac{1}{s^2 + \sqrt{2}s + 1}$$

The transfer function $H_c(s)$ corresponding to $\Omega_c = 0.325$ is obtained by substituting

$$s = \frac{s}{\Omega_c} = \frac{s}{0.325}$$

in the expression for $H_N(s)$; hence,

$$H_a(s) = \frac{0.1056}{s^2 + 0.4595s + 0.1056}$$

The digital transfer function $H(z)$ of the desired filter is now obtained by using the bilinear transformation (5.52) in the above expression:

$$H(z) = H_c(s)\big|_{s=\frac{(z-1)}{(z+1)}}$$

$$H(z) = \frac{0.1056z^2 + 0.2112z + 0.1056}{1.5651z^2 - 1.7888z + 0.646}$$

Example 5.11 Design a lowpass Butterworth IIR filter for the following specifications:

Passband edge frequency: 1000 Hz
Stopband edge frequency: 3000 Hz
Passband ripple: 2 dB
Stopband ripple: 20 dB

Assume a suitable sampling frequency and use the bilinear transformation.

Solution Assuming the sampling frequency as 8 kHz, the normalized angular band edge frequencies are given by

$$\omega_p = \frac{2\pi f_p}{F_T} = \frac{2\pi \times 1000}{8000} = 0.25\pi$$

$$\omega_s = \frac{2\pi f_s}{F_T} = \frac{2\pi \times 3000}{8000} = 0.75\pi$$

By prewarping these frequencies, we get

$$\hat{\Omega}_p = \tan(\omega_p/2) = 0.4142; \hat{\Omega}_s = \tan(\omega_s/2) = 2.4142.$$

For the prototype analog lowpass filter, we get

$$\Omega_p = 1, \Omega_s = \hat{\Omega}_s/\hat{\Omega}_p = 2.4142/0.41422 = 5.8286, \alpha_p = 2\,\text{dB}, \alpha_s = 20\,\text{dB}$$

Using these values, the order of the filter is computed using Eq. (5.9) as

$$N \geq \frac{\log\left(\frac{10^2 - 1}{10^{0.2} - 1}\right)}{2\log\left(\frac{5.8286}{1}\right)} = 1.4556$$

Hence, we choose $N = 2$. The normalized lowpass Butterworth filter for $N = 2$ is given by

$$H_N(s) = \frac{1}{s^2 + \sqrt{2}s + 1}$$

Substituting the values of $\hat{\Omega}_s$ and N in Eq. (5.6), we obtain

$$\left(2.4142/\hat{\Omega}_c\right)^4 = 10^2 - 1$$

Solving for $\hat{\Omega}_c$, we get $\hat{\Omega}_c = 0.7654$. The transfer function corresponding to $\hat{\Omega}_c = 0.7654$ is obtained by substituting $s = \left(s/\hat{\Omega}_c\right) = (s/0.7654)$ in $H_N(s)$; hence,

$$H_a(s) = \frac{0.5858}{s^2 + 1.0824s + 0.5858}$$

The digital transfer function $H(z)$ of the desired filter is now obtained as

$$H(z) = H_a(s)\big|_{s = \frac{(z-1)}{(z+1)}}$$

$$H(z) = \frac{0.2195z^2 + 0.439z + 0.2195}{z^2 - 0.31047z + 0.1887}$$

Example 5.12 Design a lowpass Chebyshev Type 1 IIR filter for the following specifications:

Passband edge frequency: 1 kHz
Stopband edge frequency: 3 kHz
Sampling frequency: 10 kHz
Passband ripple: 1 dB
Stopband ripple: 40 dB

Solution The normalized angular band edge frequencies are given by

$$\omega_p = \frac{2\pi f_p}{F_T} = \frac{2\pi \times 1000}{10000} = 0.2\pi$$

$$\omega_s = \frac{2\pi f_s}{F_T} = \frac{2\pi \times 3000}{10000} = 0.6\pi$$

By prewarping these frequencies, we get

$$\hat{\Omega}_p = \tan(\omega_p/2) = 0.32492; \hat{\Omega}_s = \tan(\omega_s/2) = 1.3764.$$

For the prototype analog lowpass filter

$$\Omega_p = 1, \Omega_s = \hat{\Omega}_s/\hat{\Omega}_p = 1.3764/0.32492 = 4.236, \alpha_p = 1\,\text{dB}, \alpha_s = 40\,\text{dB}$$

Hence from (5.21), we have

$$N \geq \frac{\cosh^{-1}\sqrt{\frac{10^4-1}{10^{0.1}-1}}}{\cosh^{-1}(4.236)} = 2.45$$

Hence, we choose $N = 3$. For $N = 3$, from Table 5.2, the normalized transfer function is given by

$$H_N(s) = \frac{0.49131}{s^3 + 0.988s^2 + 1.238s + 0.49131}$$

The transfer function corresponding to $\hat{\Omega}_p = 0.32492$ is obtained by substituting $s = (s/\hat{\Omega}_p) = (s/0.32492)$ in $H_N(s)$; hence,

$$H_a(s) = \frac{0.016849}{s^3 + 0.32099s^2 + 0.13068s + 0.016849}$$

The digital transfer function $H_{LP}(z)$ of the desired lowpass filter is now obtained as

$$H_{LP}(z) = H_a(s)\big|_{s=\frac{(z-1)}{(z+1)}}$$

$$H_{\text{LP}}(z) = \frac{0.011474z^3 + 0.034421z^2 + 0.034421z + 0.011474}{z^3 - 2.178z^2 + 1.7698z + 0.53976}$$

Example 5.13 Design a lowpass Chebyshev Type 2 IIR digital filter for the specifications given in Example 5.12.

Solution The order of the filter required is the same as in Example 5.12, i.e., $N = 3$. For, $N = 3$, from Table 5.3, the normalized transfer function is given by

$$H_N(s) = \frac{0.03(s^2 + 1.3333)}{(s^3 + 0.6746s^2 + 0.22709s + 0.04)}$$

for which the stopband edge is at $\Omega_s = 1$. The transfer function corresponding to the stopband edge $\hat{\Omega}_S = 1.3764$ is obtained by substituting $s = \left(s/\hat{\Omega}_S\right) = (s/1.3764)$ in the expression for $H_N(s)$; hence,

$$H_a(s) = \frac{0.041292s^2 + 0.10430}{s^3 + 0.92852s^2 + 0.43022s + 0.10430}$$

The digital transfer function $H_{\text{LP}}(z)$ of the desired lowpass filter is now obtained as

$$H(z) = H_a(s)\big|_{s=\frac{(z-1)}{(z+1)}}$$

$$H_{\text{LP}}(z) = \frac{0.059111z^3 + 0.11028z^2 + 0.11028z + 0.059111}{z^3 - 1.2933z^2 + 0.7934z - 0.16134}$$

Example 5.14 Design an elliptic lowpass IIR filter for the following specifications:

Passband edge frequency: 800 Hz
Stopband edge frequency: 1600 Hz
Sampling frequency: 4 kHz
Passband ripple: 1 dB
Stopband ripple: 40 dB

Solution The normalized angular bandedge frequencies are given by

$$\omega_p = \frac{2\pi \times 800}{4000} = 0.4\pi$$

$$\omega_s = \frac{2\pi \times 1600}{4000} = 0.8\pi$$

Prewarping these frequencies, we get

$$\hat{\Omega}_p = \tan(\omega_p/2) = 0.72654; \hat{\Omega}_s = \tan(\omega_s/2) = 3.0777.$$

For the prototype analog lowpass filter

$$\Omega_p = 1, \Omega_s = \hat{\Omega}_s/\hat{\Omega}_p = 4.236, \quad \alpha_p = 1\,\text{dB}, \quad \alpha_s = 40\,\text{dB}$$

$$k = \frac{\Omega_p}{\Omega_s} = \frac{0.72654}{3.0777} = 0.23607$$

$$k' = \sqrt{1 - k^2} = 0.97174$$

$$\rho_0 = \frac{1 - \sqrt{k'}}{2(1 + \sqrt{k'})} = 0.0035837$$

$$\rho = \rho_0 + 2(\rho_0)^5 + 15(\rho_0)^9 + 150(\rho_0)^{13} = 0.0035837$$

$$N \cong \frac{\log\left(16 \times \frac{10^{0.1\alpha_s}-1}{10^{0.1\alpha_p}-1}\right)}{\log_{10}(1/\rho)} = 2.3678$$

Hence, we choose $N = 3$. For $N = 3$, from Table 5.4, the normalized transfer function is

$$H_N(s) = \frac{0.0692s^2 + 0.5265}{s^3 + 0.9782s^2 + 1.2434s + 0.5265}$$

The transfer function corresponding to the passband edge $\hat{\Omega}_p = 0.72654$ is obtained by substituting $s = \left(s/\hat{\Omega}_p\right) = (s/0.72654)$ in the expression for $H_N(s)$; hence,

$$H_a(s) = \frac{0.050277s^2 + 0.10430}{s^3 + 0.7107s^2 + 0.65634s + 0.20192}$$

The digital transfer function $H_{LP}(z)$ of the desired lowpass filter is now obtained as

$$H_{LP}(z) = H_a(s)\big|_{s=\frac{(z-1)}{(z+1)}}$$

$$H_{LP}(z) = \frac{0.09817z^3 + 0.21622z^2 + 0.21622z + 0.09817}{z^3 - 0.95313z^2 + 0.87143z - 0.2895}$$

Example 5.15 Design a Butterworth IIR digital highpass filter for the following specifications:

Passband edge frequency: 40 Hz
Stopband edge frequency: 25 Hz
Sampling frequency: 100 Hz
Passband ripple: 1 dB
Stopband ripple: 20 dB

Solution The normalized angular bandedge frequencies are

$$\omega_s = \frac{2\pi \times 25}{100} = 0.5\pi$$

$$\omega_p = \frac{2\pi \times 40}{100} = 0.8\pi$$

Prewarping these frequencies, we get

$$\hat{\Omega}_p = \tan(\omega_p/2) = 3.0777$$
$$\hat{\Omega}_s = \tan(\omega_s/2) = 1.0$$

For the prototype analog lowpass filter, we have

$$\Omega_p = 1, \Omega_s = \hat{\Omega}_p/\hat{\Omega}_s = 3.0777, \alpha_p = 1\,dB, \alpha_s = 20\,dB$$

Substituting these values in Eq. (5.9), the order of the filter is given by

$$N \geq \frac{\log\left(\frac{10^2-1}{10^{0.1}-1}\right)}{2\log\left(\frac{3.077}{1}\right)} = 2.6447$$

Hence, we choose $N = 3$. From Table 5.1, the third-order normalized Butterworth lowpass filter transfer function is given by

$$H_N(s) = \frac{1}{(s+1)(s^2+s+1)}$$

Substituting the values of Ω_s and N in Eq. (5.6), we obtain

$$\left(\frac{3.0777}{\Omega_c}\right)^6 = 10^2 - 1$$

Solving for Ω_c, we get $\Omega_c = 1.4309$.

The analog transfer function of the lowpass filter is obtained from the above transfer function by substituting $s = \frac{s}{\Omega_c} = \frac{s}{1.4309}$; hence,

$$H_{LP}(s) = \frac{2.93}{s^3 + 2.8619s^2 + 4.0952s + 2.93}$$

From the above transfer function, the analog transfer function of the highpass filter can be obtained by substituting $s = \frac{\hat{\Omega}_p}{s} = \frac{3.0777}{s}$

$$H_{HP}(s) = \frac{s^3}{s^3 + 4.3017s^2 + 9.2521s + 9.9499}$$

The digital transfer function of the required highpass filter is obtained by using the bilinear transformation:

$$H_{HP}(z) = H_{HP}(s)\big|_{s=\frac{z-1}{z+1}}$$

Thus,

$$H_{HP}(z) = \frac{0.0408z^3 - 0.1224z^2 + 0.1224z - 0.0408}{z^3 + 1.2978z^2 + 0.7875z + 0.1632}$$

Example 5.16 Design a Type 1 Chebyshev IIR digital highpass filter for the following specifications:

Passband edge frequency: 700 Hz
Stopband edge frequency: 500 Hz
Sampling frequency: 2 kHz
Passband ripple: 1 dB
Stopband ripple: 40 dB

Solution Normalized angular bandedge frequencies are

$$\omega_s = \frac{2\pi \times 500}{2000} = 0.5\pi; \quad \omega_p = \frac{2\pi \times 700}{2000} = 0.7\pi$$

Prewarping these frequencies, we get

$$\hat{\Omega}_p = \tan(\omega_p/2) = 1.9626105, \hat{\Omega}_s = \tan(\omega_s/2) = 1$$

For the prototype analog lowpass filter

$$\Omega_p = 1, \Omega_s = \hat{\Omega}_p/\hat{\Omega}_s = 1.9626105, \alpha_p = 1\,dB, \alpha_s = 40\,dB$$

Substituting these values in Eq. (5.21), the order of the filter is given by

$$N \geq \frac{\cosh^{-1}\sqrt{\frac{10^4 - 1}{10^{0.1} - 1}}}{\cosh^{-1}(1.9626)} = 4.6127$$

Hence, we choose $N = 5$. From (5.18), we have

$$\frac{1}{\varepsilon} = \frac{1}{\sqrt{10^{0.1\alpha_p} - 1}} = 1.965227$$

$$\sinh^{-1}\left(\frac{1}{\varepsilon}\right) = \sinh^{-1}(1.965227) = 1.428$$

Using Eqs. (5.24) and (5.25a, 5.25b), the normalized transfer function is given by

$$H_N(s) = \frac{H_0}{\prod_k (s - p_k)}$$

where

$$p_k = -\sinh\left\{\frac{1}{N}\sinh^{-1}\left(\frac{1}{\epsilon}\right)\right\} \cdot \sin\left\{\frac{(2k - 1)\pi}{2N}\right\}$$
$$+ j \cosh\left\{\frac{1}{N}\sinh^{-1}\left(\frac{1}{\epsilon}\right)\right\} \cdot \cos\left\{\frac{(2k - 1)\pi}{2N}\right\}$$

and

$$H_0 = \frac{1}{2^{N-1}}\frac{1}{\varepsilon}$$

Substituting $N = 5$ and $k = 1, 2, 3, 4, 5$ in the above equations, we get

$$p_{1,5} = -0.08946 \pm j0.99014, \quad p_{2,4} = -0.23421 \pm j0.61194, \quad p_3 = -0.2895$$

and

$$H_0 = 0.12283$$

Hence,

$$H_N(s) = \frac{0.12283}{s^5 + 0.93682s^4 + 1.6888s^3 + 0.9744s^2 + 0.58053s + 0.12283}$$

The analog transfer function of the highpass filter can be obtained from the above transfer function by substituting $s = (\hat{\Omega}_p/s) = (1.9626105/s)$;

$$H_{HP}(s) = \frac{0.12283s^5}{s^5 + 9.2762s^4 + 30.557s^3 + 103.94s^2 + 113.16s + 237.07}$$

The digital transfer function of the highpass filter can be obtained by using bilinear transformation:

$$H_{HP}(z) = H_{HP}(s)|_{s=\frac{z-1}{z+1}}$$

Thus,

$$H_{HP}(z) = \frac{0.0020202z^5 - 0.010101z^4 + 0.020202z^3 - 0.020202z^2 + 0.010101z - 0.0020202}{z^5 + 3.1624z^4 + 4.7607z^3 + 4.0528z^2 + 1.9344z + 0.41529}$$

Example 5.17 Using bilinear transformation, design a digital bandpass Butterworth filter with the following specifications:

Lower passband edge frequency: 200 Hz
Upper passband edge frequency: 400 Hz
Lower stopband edge frequency: 100 Hz
Upper stopband edge frequency: 500 Hz
Passband ripple: 2 dB
Stopband ripple: 20 dB

Assume a suitable sampling frequency.

Solution Assuming the sampling frequency to be 2000 Hz, the normalized angular bandedge frequencies are given by

$$\omega_{p1} = 0.2\pi, \omega_{p2} = 0.4\pi, \omega_{s1} = 0.1\pi, \omega_{s2} = 0.5\pi$$

The prewarped analog frequencies are given by

$$\hat{\Omega}_{p1} = \tan(\omega_{p1}/2) = \tan(\pi/10) = 0.325$$
$$\hat{\Omega}_{p2} = \tan(\omega_{p2}/2) = \tan(\pi/5) = 0.7265$$
$$\hat{\Omega}_{s1} = \tan(\omega_{s1}/2) = \tan(\pi/20) = 0.1584$$
$$\hat{\Omega}_{s2} = \tan(\omega_{s2}/2) = \tan(\pi/4) = 1$$

We now obtain the corresponding specifications for the normalized analog lowpass filter using the lowpass-to-bandpass transformation. From Eqs. (5.42) to (5.43), we have

$$A_1 = \frac{\hat{\Omega}_{s1}^2 - \hat{\Omega}_{p1}\hat{\Omega}_{p2}}{\left(\hat{\Omega}_{p2} - \hat{\Omega}_{p1}\right)\hat{\Omega}_{s1}} = 1.90258$$

and

$$A_2 = \frac{\hat{\Omega}_{s2}^2 - \hat{\Omega}_{p1}\hat{\Omega}_{p2}}{\left(\hat{\Omega}_{p2} - \hat{\Omega}_{p1}\right)\hat{\Omega}_{s2}} = -3.318$$

Now using (5.44a), we get the specifications for the normalized analog lowpass filter to be

$$\Omega_p = 1, \Omega_s = \min\{|A_1|, |A_2|\} = 1.90258,$$
$$\alpha_p = 2\,\text{dB}, \alpha_s = 20\,\text{dB}$$

Substituting these values in Eq. (5.9), the order of the filter is given by

$$N \geq \frac{\log\left(\frac{10^2 - 1}{10^{0.2} - 1}\right)}{2\log(1.90258)} = 3.9889$$

We choose $N = 4$. The transfer function of the fourth-order normalized Butterworth lowpass filter is given by

$$H_N(s) = \frac{1}{s^4 + 2.6131s^3 + 3.4142s^2 + 2.6131s + 1}$$

Substituting the values of Ω_s and N in Eq. (5.6), we obtain

$$\left(\frac{1.90258}{\Omega_c}\right)^8 = 10^2 - 1$$

Solving for Ω_c, we get $\Omega_c = 1.0712$. The analog transfer function of the lowpass filter is obtained from the above transfer function by substituting $s = \frac{s}{\Omega_c} = \frac{s}{1.0712}$

$$H_{LP}(s) = \frac{1.3169}{s^4 + 2.7993s^3 + 3.9180s^2 + 3.2124s + 1.3169}$$

To arrive at the analog transfer function of the bandpass filter, we use in the above expression the lowpass-to-bandpass transformation given by (5.41), namely

$$s = \frac{s^2 + \hat{\Omega}_{p1}\hat{\Omega}_{p2}}{\left(\hat{\Omega}_{p2} - \hat{\Omega}_{p1}\right)s} = \frac{s^2 + 0.236}{0.402\,s}$$

to obtain

$H_{BP}(s)$

$$= \frac{0.0344s^4}{s^8 + 1.1253s^7 + 1.5772\,s^6 + 1.0054s^5 + 0.6674s^4 + 0.2373s^3 + 0.0878s^2 + 0.0148s + 0.0031}$$

The digital bandpass filter is now obtained by using the bilinear transformation in the above expression. Thus,

$$H_{BP}(z) = \frac{z^8 - 0.0241z^6 + 0.0361z^4 - 0.0241z^2 + 0.0060}{z^8 - 3.8703z^7 + 7.9661z^6 - 10.6337z^5 + 10.0678z^4 - 6.8080z^3 + 3.3529z^2 - 1.002z + 0.1666}$$

Example 5.18 Using bilinear transformation, design a digital bandpass Chebyshev Type 1 filter with the following specifications:

Lower passband edge frequency: 200 Hz
Upper passband edge frequency: 400 Hz
Lower stopband edge frequency: 100 Hz
Upper stopband edge frequency: 500 Hz
Passband ripple: 1 dB
Stopband ripple: 10 dB

Assume a suitable sampling frequency.

Solution The prewarped analog frequencies, as well as the values of A_1 and A_2, are the same as for the above example. Hence, for the prototype analog lowpass filter, the specifications are

$$\Omega_p = 1, \ \Omega_s = \min\{|A_1|, |A_2|\} = 1.90258, \alpha_p = 1 \, dB, \alpha_s = 10 \, dB$$

Substituting these values in Eq. (5.21), the order of the filter is given as

$$N \geq \frac{\cosh^{-1}\sqrt{\frac{10^1 - 1}{10^{0.1} - 1}}}{\cosh^{-1}(1.90258)} = 1.9544$$

We choose $N = 2$. From (5.18), we have

$$\frac{1}{\varepsilon} = \frac{1}{\sqrt{10^{0.1\alpha_p} - 1}} = 1.965227$$

$$\sinh^{-1}\left(\frac{1}{\varepsilon}\right) = \sinh^{-1}(1.965227) = 1.428$$

Using Eqs. (5.25a) and (5.25b), the poles of the normalized lowpass transfer function are given by

$$p_k = -\sinh(0.714)\sin\left\{\frac{(2k-1)\pi}{4}\right\} + j\cosh(0.714)\cos\left\{\frac{(2k-1)\pi}{4}\right\}, \quad k = 1, 2$$

and

$$H_0 = \frac{1}{2}\frac{1}{\varepsilon} = 0.9826$$

Hence,

$$p_{1,2} = -0.54887 \pm j0.89513$$

Thus for $N = 2$, with a passband ripple of 1 dB, the normalized transfer function is

$$H_N(s) = \frac{0.9826}{s^2 + 1.0977s + 1.1025}$$

To arrive at the analog transfer function of the bandpass filter, we use in the above expression the lowpass-to-bandpass transformation given by Eq. (5.41), namely

$$s = \frac{s^2 + \hat{\Omega}_{p1}\hat{\Omega}_{p2}}{\left(\hat{\Omega}_{p2} - \hat{\Omega}_{p1}\right)s} = \frac{s^2 + 0.236}{0.402s}$$

to obtain

$$H_{BP}(z) = \frac{0.1584s^2}{(s^4 + 0.4407s^3 + 0.6497\,s^2 + 0.1040\,s + 0.0557)}$$

We now use the bilinear transformation in the above to obtain the required digital bandpass filter transfer function as

$$H_{BP}(z) = \frac{0.0704z^4 - 0.1408\,z^2 + 0.0704}{z^4 - 1.9779\,z^3 + 2.2375\,z^2 - 1.3793\,z + 0.5158}$$

Example 5.19 Using bilinear transformation, design a digital bandstop Butterworth filter with the following specifications:

Lower passband edge frequency: 35 Hz
Upper passband edge frequency: 215 Hz
Lower stopband edge frequency: 100 Hz
Upper stopband edge frequency: 150 Hz
Passband ripple: 3 dB
Stopband ripple: 15 dB

Assume a suitable sampling frequency.

Solution Assuming a sampling frequency of 500 Hz, the normalized angular bandedge frequencies are given by

$$\omega_{p1} = 0.14\pi, \omega_{p2} = 0.86\pi, \omega_{s1} = 0.4\pi, \omega_{s2} = 0.6\pi$$

The prewarped analog frequencies are given by

$$\hat{\Omega}_{p1} = \tan(\omega_{p1}/2) = \tan(0.14\pi/2) = 0.2235$$

$$\hat{\Omega}_{p2} = \tan(\omega_{p2}/2) = \tan(0.86\pi/2) = 4.4737$$

$$\hat{\Omega}_{s1} = \tan(\omega_{s1}/2) = \tan(0.4\pi/2) = 0.7265$$

$$\hat{\Omega}_{s2} = \tan\left(\frac{\omega_{s2}}{2}\right) = \tan\left(\frac{0.6\pi}{2}\right) = 1.3764$$

We now obtain the corresponding specifications for the normalized analog lowpass filter using the lowpass-to-bandstop transformation. From Eq. (5.49b), we have

$$A_1 = \frac{\hat{\Omega}_{s1}\hat{\Omega}_{s2} - \hat{\Omega}_{p1}^2}{\left(\hat{\Omega}_{s2} - \hat{\Omega}_{s1}\right)\hat{\Omega}_{p1}} = 6.5403, A_2 = \frac{\hat{\Omega}_{s1}\hat{\Omega}_{s2} - \hat{\Omega}_{p2}^2}{\left(\hat{\Omega}_{s2} - \hat{\Omega}_{s1}\right)\hat{\Omega}_{p2}} = -6.5397$$

Now using (5.49a), we get the specifications for the normalized analog lowpass filter to be

$$\Omega_p = 1, \Omega_s = \min\{|A_1|, |A_2|\}, \alpha_p = 3\,\text{dB}, \alpha_s = 15\,\text{dB}$$

Substituting these values in Eq. (5.9), the order of the filter is given by

$$N \geq \frac{\log\left(\frac{10^{1.5}-1}{10^{0.3}-1}\right)}{2\log(6.5397)} = 0.9125$$

We choose $N = 1$. The transfer function of the first-order normalized Butterworth lowpass filter is

$$H_N(s) = \frac{1}{(s+1)}$$

Substituting the values of Ω_s and N in Eq. (5.6), we obtain

$$\left(\frac{6.5397}{\Omega_c}\right)^2 = 10^{1.5} - 1$$

Solving for Ω_c, we get $\Omega_c = 1.1818$. The analog transfer function of the lowpass filter is obtained from $H_N(s)$ by substituting $s = \frac{s}{\Omega_c} = \frac{s}{1.1818}$

$$H_{LP}(s) = \frac{1.1818}{s + 1.1818}$$

To arrive at the analog transfer function of the bandstop filter, we use in the above expression the lowpass-to-bandstop transformation given by (5.49c), namely

$$s = \frac{\left(\hat{\Omega}_{s2} - \hat{\Omega}_{s1}\right)\Omega_s s}{s^2 + \hat{\Omega}_{s1}\hat{\Omega}_{s2}} = \frac{(0.6499)(6.5397)s}{s^2 + 1} = \frac{4.25\,s}{s^2 + 1}$$

to obtain

$$H_{BS}(s) = \frac{s^2 + 1}{s^2 + 3.5964s + 1}$$

The transfer function of the required digital bandstop filter is now obtained by using the bilinear transformation:

$$\begin{aligned} H_{BS}(z) &= H_{BS}(s)\big|_{s=\frac{z-1}{z+1}} \\ &= \frac{0.3574z^2 + 0.3574}{z^2 - 0.2853} \end{aligned}$$

Example 5.20 Design an elliptic IIR digital highpass filter with the specifications given in Example 5.16.

Solution Normalized angular bandedge frequencies are given as

$$\omega_s = \frac{2\pi \times 500}{2000} = 0.5\pi; \quad \omega_p = \frac{2\pi \times 700}{2000} = 0.7\pi$$

Prewarping these frequencies, we get

$$\hat{\Omega}_p = \tan\left(\frac{\omega_p}{2}\right) = 1.9626105, \hat{\Omega}_{s1} = \tan\left(\frac{\omega_s}{2}\right) = 1$$

For the prototype analog lowpass filter

$$\Omega_p = 1, \Omega_s = \hat{\Omega}_p/\hat{\Omega}_s = 1.9626105, \alpha_p = 1\,dB, \alpha_s = 40\,dB$$

From (5.30) and (5.32) to (5.34), we get

$$k = \frac{\Omega_p}{\Omega_s} = \frac{1}{1.9626105} = 0.5095; \quad k' = \sqrt{1-k^2} = 0.8605;$$

$$\rho_0 = \frac{1-\sqrt{k'}}{2(1+\sqrt{k'})} = 0.0188$$

$$\rho = \rho_0 + 2(\rho_0)^5 + 15(\rho_0)^9 + 150(\rho_0)^{13} = 0.0188$$

Substituting these values in Eq. (5.31), the order of the filter is given by

$$N \cong \frac{\log\left(16 \times \frac{10^{0.1\alpha_s}-1}{10^{0.1\alpha_p}-1}\right)}{\log_{10}(1/\rho)} = \frac{\log_{10}\left(16 \times \frac{10^4-1}{10^{0.1}-1}\right)}{\log_{10}(1/0.0188)} = 3.3554$$

Let us choose $N = 4$. Then from Table 5.4, we have

$$H_N(s) = \frac{0.01s^4 + 0.1502s^2 + 0.3220}{s^4 + 0.9391s^3 + 1.5137s^2 + 0.8037s + 0.3612}$$

To arrive at the analog transfer function of the highpass filter, the variable s in the above-normalized transfer function is to be replaced by $(\hat{\Omega}_p/s) = (1.9626105/s)$

$$H_{HP}(s) = \frac{0.322s^4 + 0.5785s^2 + 0.1484}{0.3612s^4 + 1.5774s^3 + 5.8305s^2 + 7.0993s + 14.8367}$$

Then, the required highpass filter in the digital domain is given by

$$H_{HP}(z) = H_{HP}(s)\big|_{s=\frac{z-1}{z+1}}$$

$$H_{HP}(z) = \frac{0.035z^4 - 0.0234z^3 + 0.0561z^2 - 0.0234z + 0.035}{z^4 + 2.321z^3 + 2.6772z^2 + 1.5774z + 0.4158}$$

5.3 Design of Digital Filters Using Digital-to-Digital Transformations

In the design of analog filters, we start with designing a normalized lowpass filter, and then through an appropriate frequency transformation of the lowpass filter, the filter for the given magnitude response specifications is obtained. We can adopt a similar procedure by first designing a digital lowpass filter and then applying frequency transformation $z \to g(\hat{z})$ in the discrete domain to obtain highpass, bandpass, bandstop, or another lowpass filter. The transformation function $g(z)$ has to satisfy certain conditions in order to produce the desired magnitude specifications.

(i)   The transformation function $g(\hat{z})$ should be a rational function of $\hat{z}$.
(ii) The transformation $z \rightarrow g(\hat{z})$ should map the interior of the unit circle in the z-plane into the interior of the unit circle in the $\hat{z}$-plane, the exterior to the exterior, and the unit circle in the z-plane into the unit circle in the $\hat{z}$-plane. Hence, the transformed filter resulting from a stable filter will remain stable.

Table 5.6 shows a set of transformations that can be used to for this purpose, and interested readers may refer to the work of Constantinides [9] for details. For illustration, we consider the cases of lowpass-to-lowpass and lowpass-to-highpass transformations.

Lowpass-to-lowpass transformation

Consider the transformation function

$$z = \frac{\hat{z} - b}{1 - b\hat{z}} = g(\hat{z}) \tag{5.69}$$

where b is real. Then,

$$e^{j\omega} = \frac{e^{j\hat{\omega}} - b}{1 - be^{j\hat{\omega}}}$$

Table 5.6 Digital-to-digital transformations

Filter type	Spectral transformation	Design parameters
Lowpass	$z = \frac{\hat{z}-b}{1-b\hat{z}}$	$b = \dfrac{\sin\left(\frac{\omega_p - \hat{\omega}_p}{2}\right)}{\sin\left(\frac{\omega_p + \hat{\omega}_p}{2}\right)}$ $\hat{\omega}_p$ = new passband edge frequency
Highpass	$z = \frac{\hat{z}+b}{1+b\hat{z}}$	$b = \dfrac{\cos\left(\frac{\omega_p + \hat{\omega}_p}{2}\right)}{\cos\left(\frac{\omega_p - \hat{\omega}_p}{2}\right)}$ $\hat{\omega}_p$ = new passband edge frequency
Bandpass	$z = \frac{\hat{z}^2 - \frac{2ab}{b+1}\hat{z} + \frac{b-1}{b+1}}{\frac{b-1}{b+1}\hat{z}^2 - \frac{2ab}{b+1}\hat{z} + 1}$	$a = \dfrac{\cos\left(\frac{\hat{\omega}_{p2} + \hat{\omega}_{p1}}{2}\right)}{\cos\left(\frac{\hat{\omega}_{p2} - \hat{\omega}_{p1}}{2}\right)}$ $b = \cot\left(\frac{\hat{\omega}_{p2} - \hat{\omega}_{p1}}{2}\right)\tan\left(\frac{\omega_p}{2}\right)$ $\hat{\omega}_{p2}, \hat{\omega}_{p1}$ = desired upper and lower passband edge frequencies
Bandstop	$z = \frac{\hat{z}^2 - \frac{2a}{1+b}\hat{z} + \frac{1-b}{1+b}}{\frac{1-b}{1+b}\hat{z}^2 - \frac{2a}{1+b}\hat{z} + 1}$	$a = \dfrac{\cos\left(\frac{\hat{\omega}_{s2} + \hat{\omega}_{s1}}{2}\right)}{\cos\left(\frac{\hat{\omega}_{s2} - \hat{\omega}_{s1}}{2}\right)}$ $b = \tan\left(\frac{\hat{\omega}_{s2} - \hat{\omega}_{s1}}{2}\right)\tan\left(\frac{\omega_s}{2}\right)$ $\hat{\omega}_{s2}, \hat{\omega}_{s1}$ = desired upper and lower stopband edge frequencies

Note ω_p is the passband edge frequency of the normalized lowpass filter $H_N(z)$

Let the passband edge of the original lowpass filter be ω_p and that of the desired lowpass filter be $\hat{\omega}_p$. It can easily be seen that $g(\hat{z})$ maps $\hat{\omega} = 0$ into $\omega = 0$, and $\hat{\omega} = \pm\pi$ into $\omega = \pm\pi$. We now choose the value of b so that $g(\hat{z})$ maps the frequency $\hat{\omega}_p$ to ω_p; then, we will have the required mapping function. Hence, we should have

$$e^{j\omega_p} = \frac{e^{j\hat{\omega}_p} - b}{1 - be^{j\hat{\omega}_p}}$$

Hence,

$$b = \frac{e^{-j(\omega_p - \hat{\omega}_p)/2} - e^{j(\omega_p - \hat{\omega}_p)/2}}{e^{-j(\omega_p + \hat{\omega}_p)/2} - e^{j(\omega_p + \hat{\omega}_p)/2}} = \frac{\sin(\omega_p - \hat{\omega}_p)/2}{\sin(\omega_p + \hat{\omega}_p)/2} \tag{5.70}$$

Thus, the transformation (5.69) with the value of b given by (5.70) will transform a digital lowpass filter with passband edge at ω_p into another digital lowpass filter with its passband edge at $\hat{\omega}_p$.

Lowpass-to-highpass transformation

Consider the transformation function

$$z = -\frac{\hat{z} + b}{1 + b\hat{z}} = g(\hat{z}) \tag{5.71}$$

where b is real. Then,

$$e^{j\omega} = -\frac{e^{j\hat{\omega}} + b}{1 + be^{j\hat{\omega}}}$$

Let the passband edge of the original lowpass filter be ω_p and that of the desired highpass filter be $\hat{\omega}_p$. It can easily be seen that $g(\hat{z})$ maps $\hat{\omega} = 0$ into $\omega = \pm\pi$, and $\hat{\omega} = \pm\pi$ into $\omega = 0$. We now choose the value of b so that $g(\hat{z})$ maps the frequency $\hat{\omega}_p$ to $-\omega_p$; then, we will have the required mapping function. Hence, we should have

$$e^{-j\omega_p} = -\frac{e^{j\hat{\omega}_p} + b}{1 + be^{j\hat{\omega}_p}}$$

Hence,

$$b = \frac{e^{-j(\omega_p + \hat{\omega}_p)/2} + e^{j(\omega_p + \hat{\omega}_p)/2}}{e^{-j(\omega_p - \hat{\omega}_p)/2} + e^{j(\omega_p - \hat{\omega}_p)/2}} = \frac{\cos(\omega_p + \hat{\omega}_p)/2}{\cos(\omega_p - \hat{\omega}_p)/2} \tag{5.72}$$

Thus, the transformation (5.71) with the value of b given by (5.72) will transform a digital lowpass filter with passband edge at ω_p into a digital highpass filter with its passband edge at $\hat{\omega}_p$.

Similarly, the other transformations given in Table 5.6 can be established [Con70].

These transformations can easily be applied to obtain highpass, bandpass, band reject, or another lowpass filter as follows:

Step 1 Find the normalized transfer function $H_N(z)$ of a lowpass filter using an approximation technique

Step 2 Obtain the passband edge ω_p in $H_N(z)$

Step 3 Find the function $H(z)$ from $H_N(z)$ using the appropriate transformation from Table 5.6.

An important aspect of the filters designed using the above transformations is that the passband edge of the lowpass or the highpass filter can be varied by varying the single parameter b. Similarly, in the case of bandpass or bandstop filters, both the lower and upper passband edges can be varied by varying two parameters, namely a and b [9, 10].

Example 5.21 Consider the second-order lowpass digital filter of Example 5.7 with -3 dB cutoff frequency of 0.2π. Redesign this lowpass filter by applying the lowpass-to-lowpass digital transformation so that the -3 dB cutoff frequency moves from 0.2π to 0.3π.

Solution Since $\omega_p = 0.2\pi$ and $\widehat{\omega}_p = 0.3\pi$, we obtain

$$b = \frac{\sin[(\omega_p - \widehat{\omega}_p)/2]}{\sin[(\omega_p + \widehat{\omega}_p)/2]} = \frac{\sin[(0.2\pi - 0.3\pi)/2]}{\sin[(0.2\pi + 0.3\pi)/2]} = -0.2212$$

From the solution of Example 5.7, the digital transfer function with -3 dB cutoff frequency at 0.2π is

$$H(z) = \frac{0.068(z+1)^2}{z^2 - 1.142z + 0.413}$$

Hence, the desired low pass transfer function with -3 dB cutoff frequency at 0.3π is given by

$$H(z) = \frac{0.068(z+1)^2}{z^2 - 1.142z + 0.413}\Bigg|_{z = \frac{z + 0.2212}{1 + 0.2212z}}$$

$$= \frac{0.1321(z+1)^2}{z^2 - 0.7467z + 0.2727}$$

Example 5.22 Consider the design of a highpass filter by applying lowpass-to-highpass digital transformation to the second-order lowpass digital filter of Example 5.11. The desired passband edge frequency is 0.5π.

Solution To apply the digital lowpass-to-highpass transformation shown in Table 5.6, b is first computed as

$$b = \frac{\cos[(\omega_p + \hat{\omega}_p)/2]}{\cos[(\omega_p - \hat{\omega}_p)/2]} = \frac{-\cos[(0.25\pi + 0.5\pi)/2]}{\cos[(0.25\pi - 0.5\pi)/2]} = -0.4142$$

From the solution of Example 5.11, the digital transfer function with passband edge frequency at 0.25π is

$$H(z) = \frac{0.2195z^2 + 0.439z + 0.2195}{z^2 - 0.31047z + 0.1887}$$

Hence, the desired highpass transfer function with passband edge frequency at 0.5π is given by

$$H(z) = \frac{0.2195z^2 + 0.439z + 0.2195}{z^2 - 0.31047z + 0.1887}\Bigg|_{z = \frac{0.4142-z}{1-0.4142z}}$$

$$= \frac{0.4857z^2 - 0.9714z + 0.4857}{z^2 - 0.6871z + 0.2564}$$

5.4   Design of IIR Digital Filters Using MATLAB

Various types of M-files are included in the signal processing toolbox of MATLAB software for the design of IIR digital filters. The use of these M-files is illustrated by the following examples:

Example 5.23 An IIR digital lowpass filter is required to meet the following specifications:

Passband ripple	≤ 0.5 dB
Passband edge	1.2 kHz
Stopband attenuation	≥ 40 dB
Stopband edge	2 kHz
Sample rate	8 kHz

Design a (i) digital Butterworth filter, (ii) Type 1 Chebyshev digital filter, (iii) Type 2 Chebyshev digital filter, and (iv) digital elliptic filter.

Solution The following MATLAB program is used to design the required filters.

Program 5.6 Butterworth, Chebyshev, and elliptic IIR lowpass filter design

```
flag = input('enter 1 for BWF, 2 for Type 1 CSF, 3 for Type 2 CSF, 4 for
Ellip = ');
%BWT stands for Butterworth filter, CSF for Chebyshev filter, and Ellip for
%Elliptic filter
Wp = input('Normalized passband edge = ');
Ws = input('Normalized stopband edge = ');
Rp = input('Passband ripple in dB = ');
Rs = input('Minimum stopband attenuation in dB = ');
if flag ==1
[N,Wn] = buttord(Wp,Ws,Rp,Rs)
[b,a] = butter(N,Wn);
end
if flag ==2
[N,Wn] = cheb1ord(Wp,Ws,Rp,Rs)
[b,a] = cheby1(N,Rp,Wn);
end
if flag ==3
[N,Wn] = cheb2ord(Wp,Ws,Rp,Rs)
[b,a] = cheby2(N,Rs,Wn);
end
if flag ==4
[N,Wn] = ellipord(Wp,Ws,Rp,Rs)
[b,a] = ellip(N,Rp,Rs,Wn);
end
[h,omega] = freqz(b,a,256);
plot (omega/pi,20*log10(abs(h)));grid;
xlabel('\omega/\pi'); ylabel('Gain, dB');
```

The magnitude responses of the designed lowpass filters are shown in Fig. 5.16.

Example 5.24 An IIR digital highpass filter is required to meet the following specifications:

Passband ripple	≤ 1 dB
Passband edge	800 Hz
Stopband attenuation	≥ 60 dB
Stopband edge	400 Hz
Sample rate	2000 Hz

Design (i) a digital Butterworth filter, (ii) Type 1 Chebyshev digital filter, (iii) Type 2 Chebyshev digital filter, and (iv) digital elliptic filter

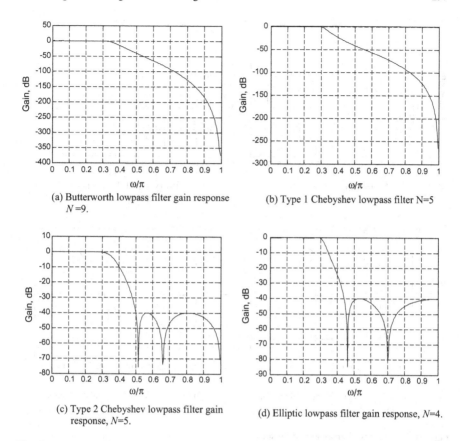

(a) Butterworth lowpass filter gain response
    $N=9$.

(b) Type 1 Chebyshev lowpass filter N=5

(c) Type 2 Chebyshev lowpass filter gain
    response, $N=5$.

(d) Elliptic lowpass filter gain response, $N=4$.

Fig. 5.16 Lowpass filter magnitude responses

Solution Program 5.6 can be used to design highpass filters with the following
MATLAB functions for determining the coefficients b and a.

$[b,a] = butter(N,Wn, 'high');$ $[b,a] = cheby1(N,Rp,Wn, 'high');$
$[b,a] = cheby2(N,Rs,Wn, 'high');$ $[b,a] = ellip(N,Rp,Rs,Wn, 'high');$

The gain responses and filter orders for the Butterworth, Type 1 and Type 2
Chebyshev, and elliptic filters are shown in Fig. 5.17a–d, respectively.

Example 5.25 Design Butterworth, Type 1 Chebyshev bandpass, Type 2
Chebyshev, and elliptic bandpass digital filters satisfying the following
specifications:

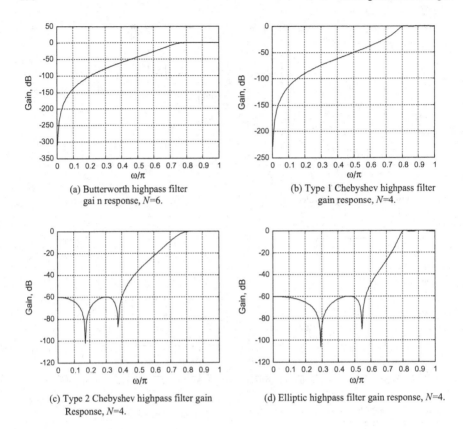

(a) Butterworth highpass filter
gai n response, N=6.

(b) Type 1 Chebyshev highpass filter
gain response, N=4.

(c) Type 2 Chebyshev highpass filter gain
Response, N=4.

(d) Elliptic highpass filter gain response, N=4.

Fig. 5.17 Highpass filter magnitude responses

Lower passband edge $\omega_{p1} = 0.4\pi$ rad
Upper passband edge $\omega_{p2} = 0.6\pi$ rad
Lower stopband edge $\omega_{s1} = 0.25\pi$ rad
Upper stopband edge $\omega_{s2} = 0.7\pi$ rad
Passband ripple = 0.5 dB
Stopband attenuation = 45 dB

Solution The following MATLAB program is used to design the desired filters.

Program 5.7 Butterworth, Chebyshev, and elliptic IIR bandpass Filters Design

```
flag = input('enter 1 for BWF, 2 for Type 1 CSF, 3 for Type 2 CSF, 4 for
Ellip = ');
%BWT stands for Butterworth filter, CSF for Chebyshev filter, and Ellip for
%Elliptic filter
Wp1 = input('Normalized lower passband edge = ');
Wp2 = input('Normalized upper passband edge = ');
Ws1 = input('Normalized lower stopband edge = ');
Ws2 = input('Normalized upper stopband edge = ');
Rp = input('Passband ripple in dB = ');
Rs = input('Minimum stopband attenuation in dB = ');
if flag ==1
[N,Wn] = buttord([Wp1 Wp2],[Ws1 Ws2],Rp,Rs);
[b,a] = butter(N,Wn);
end
if flag ==2
[N,Wn] = cheb1ord([Wp1 Wp2],[Ws1 Ws2],Rp,Rs);
[b,a] = cheby1(N,Rp,Wn);
end
if flag ==3
[N,Wn] = cheb2ord([Wp1 Wp2],[Ws1 Ws2],Rp,Rs);
[b,a] = cheby2(N,Rs,Wn);
end
if flag ==4
[N,Wn] = ellipord([Wp1 Wp2],[Ws1 Ws2],Rp,Rs);
[b,a] = ellip(N,Rp,Rs,Wn);
end
[h,omega] = freqz(b,a,256);
plot (omega/pi,20*log10(abs(h)));
grid;
xlabel('\omega/\pi');
ylabel('Gain, dB');
```

The gain responses and filter orders for the designed filters are shown in Fig. 5.18.

Example 5.26 Design Butterworth, Type 1 Chebyshev bandstop, Type 2 Chebyshev, and elliptic bandstop digital filter satisfying the following specifications:

Lower passband edge $\omega_{p1} = 0.1\pi$ rad
Lower stopband edge $\omega_{s1} = 0.2\pi$ rad
Upper passband edge $\omega_{p2} = 0.5\pi$ rad
Upper stopband edge $\omega_{s2} = 0.4\pi$ rad
Passband ripple 1 dB
Stopband attenuation 40 dB

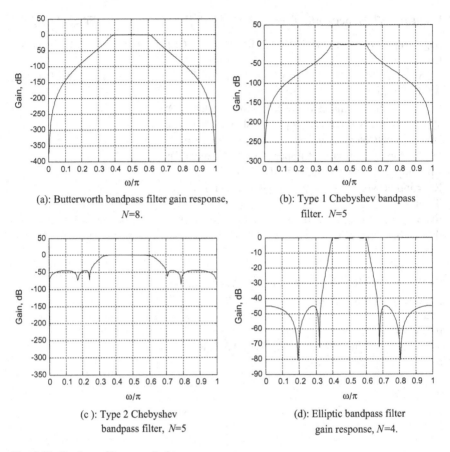

(a): Butterworth bandpass filter gain response, $N=8$.

(b): Type 1 Chebyshev bandpass filter. $N=5$

(c): Type 2 Chebyshev bandpass filter, $N=5$

(d): Elliptic bandpass filter gain response, $N=4$.

Fig. 5.18 Bandpass filter magnitude responses

Solution Program 5.7 can be used to design bandstop filters with the following MATLAB functions for determining the coefficients b and a.

$[b,a] = butter(N,Wn,'stop');$ $[b,a] = cheby1(N,Rp,Wn,'stop');$
$[b,a] = cheby2(N,Rs,Wn,'stop');$ $[b,a] = ellip(N,Rp,Rs,Wn,'stop');$

The gain responses and filter orders for the designed filters are shown in Fig. 5.19.

Example 5.27 Design a filter using digital-to-digital transformation as required in Example 5.21.

Solution To design the desired lowpass filter, the MATLAB command 'iirlp2lp' can be used for digital lowpass-to-lowpass transformation. The following MATLAB program is used to design the desired filter.

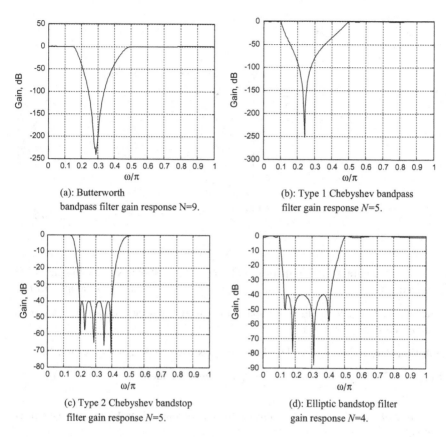

(a): Butterworth
bandpass filter gain response N=9.

(b): Type 1 Chebyshev bandpass
filter gain response *N*=5.

(c) Type 2 Chebyshev bandstop
filter gain response *N*=5.

(d): Elliptic bandstop filter
gain response *N*=4.

Fig. 5.19 Bandstop filter magnitude responses

Program 5.8 Digital lowpass-to-lowpass transformation

```
clear all;
b1 = 0.068*[1 2 1]; % numerator coefficients of original LPF
a1 = [1-1.142 0.413];% denominator coefficients of original LPF
[num,den,anum,aden] = iirlp2lp(b1,a1,0.2,0.3);% coefficients of new LPF
[h1,omega] = freqz(b1,a1,256);
plot (omega/pi,20*log10(abs(h1)));
hold on
[h2,omega] = freqz(num,den,256);
plot (omega/pi,20*log10(abs(h2)),'-');
xlabel('\omega/\pi'); ylabel('Gain, dB');
legend('original lowpass filter','newlowpass filter');grid;
```

The magnitude responses of the original filter and the new transformed filter are
shown in Fig. 5.20. From this figure, it is observed that the requirements of the

Fig. 5.20 Magnitude
responses of the original and
new lowpass filters

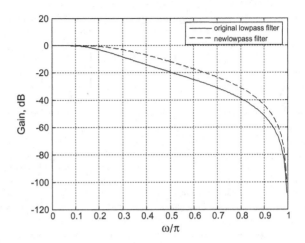

original and the transformed filters are fulfilled. The coefficients of the lowpass filter
obtained using the above program are equal to those obtained in Example 5.21.

Example 5.28 Design a digital highpass filter using digital-to-digital transformation
as required in Example 5.13.

Solution The following MATLAB program is used to design the desired filter.

Program 5.9 Digital lowpass-to-highpass transformation

```
clear all;
b1 = [0.2195 0.439 0.2195]; % numerator coefficients of prototype lowpass filter
HN (z)
a1 = [1 -0.31047 0.1887];% denominator coefficients of prototype lowpass filter
HN (z)
[num,den,anum,aden] = iirlp2hp(b1,a1,0.25,0.5);% coefficients of desired highpass
filter H(z)
[h,omega] = freqz(num,den,256);
plot (omega/pi,20*log10(abs(h)));
xlabel('\omega/\pi'); ylabel('Gain, dB');grid;
```

The magnitude response of the desired filter is shown in Fig. 5.21.

From this figure, it is observed that the requirements of the transformed highpass
filter are fulfilled. The coefficients of the highpass filter obtained using the above
program are equal to those obtained in Example 5.22.

Fig. 5.21 Magnitude
response of the highpass filter

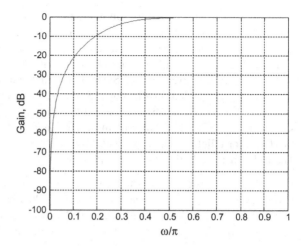

5.5 Design of IIR Filters Using MATLAB GUI Filter Designer SPTOOL

With the aid of MATLAB GUI filter designer SPTOOL, the filter satisfying the specifications can be designed using the following procedure [11]:

Step 1 Access the MATLAB's GUI filter designer SPTOOL for the design of both FIR and IIR filters.
From MATLAB, type the following: sptool

Step 2 From the startup window startup.spt, select a new design and enter the specifications of the filter. Then, the MATLAB's filter designer SPTOOL window with the characteristics of the designed filter is displayed.

Step 3 When finished, access the startup window again. Select → Edit → Name. Change name (enter new variable name).

Step 4 Select File → Export → Export to workspace the new variable name

Step 5 Access MATLAB's workspace and type the following commands:

- new variable name.tf.num;
- round (new variable name.tf.num*2^15).

Example 5.29 Design an IIR lowpass digital filter using the bilinear transformation for the following specifications using (i) Butterworth, (ii) Chebyshev Type 1, and (iii) elliptic approximations:

Passband ripple	≤ 1 dB
Passband edge	4 kHz
Stopband attenuation	≥ 40 dB
Stopband edge	6 kHz
Sample rate	24 kHz

Solution It can be designed by following the above stepwise procedure. After the execution of Steps 1 and 2, the SPTOOL in MATLAB7.0 for Butterworth lowpass filter, the filter characteristics displayed by the window is shown in Fig. 5.22.

From Fig. 5.22, the filter order obtained with MATLAB GUI filter designer SPTOOl is 10. The execution of Steps 3, 4, and 5 will display the designed filter coefficients and coded coefficients. Similarly, execution of the SPTOOL in MATLAB7.0 for Chebyshev Type 1 filter displays the filter characteristics as shown in Fig. 5.23. The order of the filter found to be 6. The execution of the SPTOOL for elliptic filter displays the filter characteristics as shown in Fig. 5.24. The order of the elliptic filter is found to be 4.

Example 5.30 Design a bandstop IIR elliptic digital filter operating at sampling frequency of 2 kHz with the passband edges at 300 and 750 Hz, stopband edges at 450 and 650 Hz, peak passband ripple of 0.5 dB, and minimum stopband attenuation of 30 dB. Use Bilinear transformation method to obtain the transfer function $H(z)$.

Solution Following the stepwise procedure used in the above example and execution of the SPTOOL in MATLAB7.0 for elliptic bandstop filter, the filter characteristics are shown in Fig. 5.25. The order of the designed filter is observed to be 6.

5.6 Design of Specialized Digital Filters by Pole-Zero Placement

There are certain specialized filters often used in digital signal processing applications in addition to the filters designed in the previous sections. These specialized filters can be directly designed based on placement of poles and zeros.

5.6.1 Notch Filter

The notch filter removes a single frequency f_0, called the notch frequency. The magnitude of the notch filter at f_0 can be made zero by placing a zero on the unit circle with $\omega_0 = (2\pi f_0)/F_T$ corresponding to the notch frequency, where F_T is the sampling frequency. The bandwidth B_w of the notch filter can be controlled by placing pole at the same angle with the pole radius $r < 1$. The poles and zeros should occur in complex conjugate pairs. As such, the transfer function of a second-order notch filter can be formed as

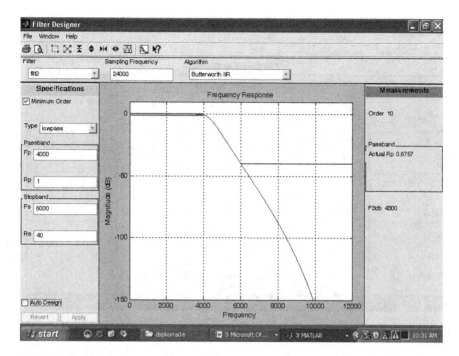

Fig. 5.22 Magnitude response of Butterworth lowpass filter

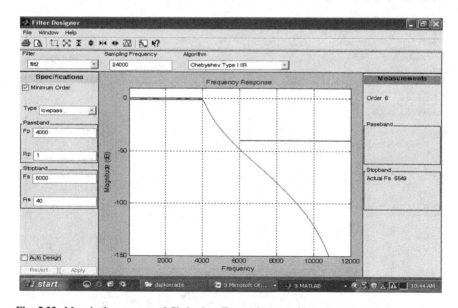

Fig. 5.23 Magnitude response of Chebyshev Type 1 lowpass filter

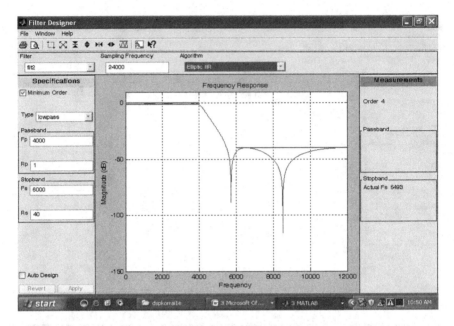

Fig. 5.24 Magnitude response of elliptic lowpass filter

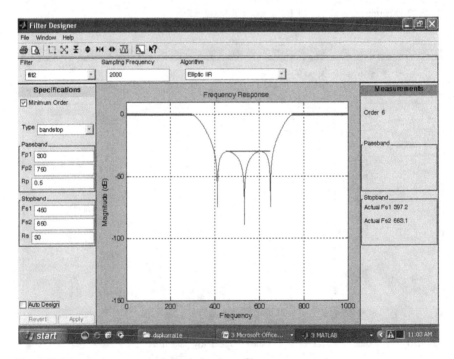

Fig. 5.25 Magnitude response of elliptic bandstop filter

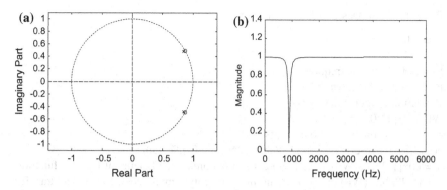

Fig. 5.26 **a** Pole-zero plot and **b** magnitude response of notch filter of Example 5.31

$$H(z) = \frac{b_0(z - e^{j\omega_0})(z - e^{-j\omega_0})}{(z - re^{j\omega_0})(z - re^{-j\omega_0})} \quad (5.73)$$

which can be rewritten as

$$H(z) = \frac{b_0\{z^2 - (2\cos\omega_0)z + 1\}}{z^2 - (2r\cos\omega_0)z + r^2} \quad (5.74)$$

To ensure that the passband gain is unity, the gain factor b_0 is to be chosen so that $|H(1)| = 1$. Hence, b_0 is given by

$$b_0 = \frac{|1 - (2\cos\omega_0) + r^2|}{|2 - 2\cos\omega_0|} \quad (5.75)$$

If $B_W \ll F_T$, the pole radius r can be approximated [12] as

$$r = 1 - (\pi B_w)/F_T \quad (5.76)$$

The following example illustrates the design of a notch filter using MATLAB.

Example 5.31 Design a digital notch filter with notch frequency at 900 Hz, bandwidth of 100 Hz, and the sampling frequency of 11,025 Hz.

Solution The following MATLAB Program 5.10 is used to design the desired notch filter and the pole-zero plot, and the magnitude response of the filter obtained from the program is shown in Fig. 5.26a, b, respectively.

Program 5.10 Design of a notch filter

```
clear;clc;
F_T = 11025;
f0 = 900;% Notch frequency
Bw = 100;% Bandwidth
% Compute filter coefficients
W0 = 2*pi*f0/F_T;
r = 1 - (Bw*pi/ F_T);% pole radius
b0 = abs(1-2*r*cos(W0) + r^2)/abs(2-2*cos(W0));% gain
b = b0*[1-2*cos(W0) 1];% Numerator polynomial coefficients of transfer function
a = [1-2*r*cos(W0) r^2];% denominator polynomial coefficients of transfer
function
% pole-zero plot
[z,p,k] = tf2zp(b,a);
figure(1),zplane(z,p)
% Plot magnitude response
N = 240
[H,f] = freqz (b,a,N, F_T);
A = abs(H);
figure(2),plot (f,A)
xlabel('Frequency (Hz)');ylabel('Magnitude');
```

5.6.2   Comb Filter

Comb filters have a wide range of practical applications such as suppression of interference in LORAN navigation systems [13] and separation of solar and lunar spectral components in ionospheric measurements [14]. Comb filter is a filter with multiple passbands and stopbands with periodic frequency response with periodicity of $(2\pi/N)$ where N is an integer. An Nth-order comb filter can be designed by placing N zeros equally spaced on the unit circle and N poles equally spaced around a circle of radius $r < 1$, but close to the unit circle. Thus, the poles correspond to the N roots of r^N. Hence, the transfer function of an Nth-order comb notch filter is given by

$$H(z) = \frac{b_0(z^N - 1)}{z^N - r^N} \tag{5.77}$$

And the transfer function for comb peaking filter is given by

$$H(z) = \frac{b_0(z^N + 1)}{z^N - r^N} \tag{5.78}$$

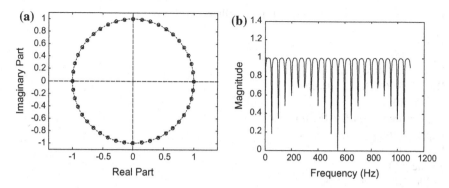

Fig. 5.27 a Pole-zero plot and **b** magnitude response of comb notching filter of Example 5.32

The gain constant b_0 is to be chosen so that the passband gain is unity at $f = \frac{f_0}{2}$, where $F_T = Nf_0$.

Hence, $b_0 = \frac{1+r^N}{2}$ for comb notching filter and $b_0 = \frac{1-r^N}{2}$ for comb peaking filter. The bandwidth B_W is related to the Q-factor of the filter by

$$B_w = \frac{2\pi f_0}{Q} \tag{5.79}$$

The following MATLAB command can be used to design a comb notching filter or a comb peaking filter

$$[b, a] = iircomb(N, Bw, Type)$$

where b and a are the coefficients of the numerator and denominator polynomials of the transfer function of the comb filter, N is the order of the comb filter, B_w is the bandwidth of the comb filter, and *Type* specifies the notch or peak. The following example illustrates the design of comb notching filter using MATLAB.

Example 5.32 Design a comb notch filter to suppress 50 Hz hum of overhead fluorescent lights in biomedical measurements. Choose the sampling frequency of 2200 Hz and the Q-factor of the filter as 35.

Solution For this design, $f_0 = 50, F_T = 2200$. Hence, using Eqs. (5.79) and (5.76), we have

$$B_w = \left(\frac{100\pi}{35}\right) = 8.9760, \quad r = 1 - \frac{8.9760}{2200} = 0.9872$$

The following MATLAB Program 5.11 is used to design the desired notch filter and the pole-zero plot, and the magnitude response of the filter obtained from the program is shown in Fig. 5.27a, b, respectively.

Program 5.11 Design of a comb notch filter

```
clear;clc;
FT = 2200;
f0 = 50;% Notch frequency
Bw = 8.9760;;% Bandwidth
N = FT/f0;
[b,a] = iircomb(N,Bw/Fs,'notch');
[z,p,k] = tf2zp(b,a);
Figure (1),
zplane(z,p)
% Plot magnitude response
N = 240
[H,f] = freqz (b,a,N, FT);
A = abs(H);
figure(2)
plot (f,A)
xlabel('Frequency (Hz)');ylabel('Magnitude');
```

5.7   Some Examples of IIR Filters for Audio Processing Applications

5.7.1   Suppression of Power Supply Hum in Audio Signals

By and large most of the audio processing systems are affected by the interference caused by the power supply hum at 50 Hz or 60 Hz. This can be avoided by using a second-order notch filter whose transfer function is given by

$$H(z) = \frac{b_0 z^2 + b_1 z + b_2}{a_0 z^2 + a_1 z + a_2} \qquad (5.80)$$

For example, consider a speech signal from the sound file 'DT.wav' [15]. The following MATLAB code is used to read the speech signal from the sound file, to add sinusoidal interference at 50 Hz to it, and to plot the power spectra of the interference added signal (x_n) as shown in Fig. 5.28.

```
[x, FT] = wavread('DT.wav');%Reads the wav file to obtain speech signal x and
sampling frequency FT
for i = 1:size(x)
xn(i) = x(i) + 2*sin(2*pi*50*i/ FT); % adds power supply hum at 50 Hz to the
speech signal
end
wavwrite(xn, FT,'DTn50.wav');
```

Fig. 5.28 Power spectrum of the speech signal corrupted with power supply hum

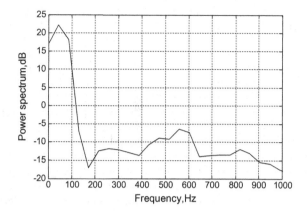

[pxx,f] = psd(xn,256, F$_T$);% power spectrum of the speech signal corrupted with power supply hum plot(f,10*log10(pxx));grid; xlabel('Frequency,Hz');ylabel ('Power spectrum,dB');

In Fig. 5.28, the peak at 50 Hz with large magnitude of the power spectrum is due to the power supply hum at 50 Hz. To suppress the power supply hum, a digital IIR second-order notch filter of the form given in Eq. (5.73) is designed for the notch frequency $f_0 = 50$ Hz and bandwidth $B_w = 100$ Hz. As such, the transfer function given by Eq. (5.80) becomes

$$H(z) = \frac{1.9716z^2 - 3.9415z + 1.9716}{z^2 - 1.9427z + 0.9438} \qquad (5.81)$$

The magnitude response of the notch filter described by the above transfer function is shown in Fig. 5.29.

To recover the original speech signal, the corrupted signal is passed through the designed notch filter, and the power spectrum of the recovered signal is shown in

Fig. 5.29 Magnitude response of IIR digital notch filter

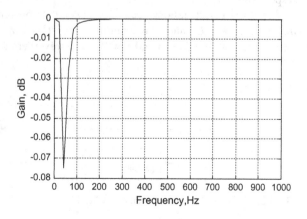

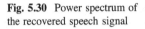

Fig. 5.30 Power spectrum of the recovered speech signal

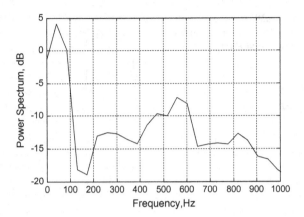

Fig. 5.30. From this figure, it can be observed that the peak at 50 Hz with large magnitude is suppressed and the speech signal is recovered. Also when the recovered signal is connected to a loudspeaker, its audio quality is observed to be the same, as that of the original speech signal.

5.7.2   Generation of Artificial Reverberations

The recorded sounds in a studio are unnatural to the listener, compared to the recorded sounds in a closed room. Digital filtering can be used to generate artificial reverberations, and by adding these reverberations to the studio recorded sounds as shown in Fig. 5.31, one can arrive at a pleasant-sounding reverberation.

An artificial reverberation generator, in general, is an interconnection scheme consisting of parallel connection of IIR filters in cascade with allpass reverberators as shown in Fig. 5.32.

The structures for IIR filters and allpass reverberators are shown in Fig. 5.33a, b, respectively.

Artificial reverberations can be generated by choosing different delays $d_i, i = 1, \ldots, K + L$, and the multiplier constants $a_i, i = 1, \ldots, K + L + 1$, and $b_i, i = 1, \ldots, K$. For $K = 4$ and $L = 2$, the reverberation generator shown in

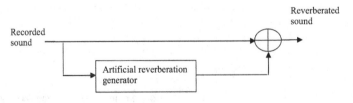

Fig. 5.31 Reverberated sound generation scheme

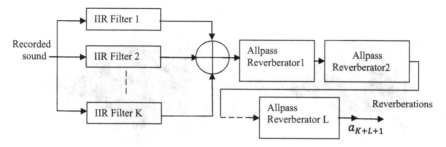

Fig. 5.32 Artificial reverberation generator

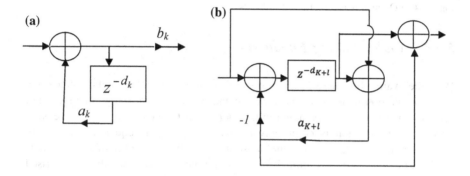

Fig. 5.33 a Structure of IIR filter $(k = 1, 2, \ldots, K)$ and **b** structure of allpass reverberator $(l = 1, 2, \ldots, L)$

Fig. 5.31 corresponds to the Schroeder reverberator with four IIR filters and two allpass reverberators with delays given by

$$d_1 = 800;\ d_2 = 900;\ d_3 = 650;\ d_4 = 700;\ d_5 = 670;\ d_6 = 990;$$

and multiplier constants

$$a_1 = 0.8;\ a_2 = 0.4;\ a_3 = 0.2;\ a_4 = 0.1;\ a_5 = 0.7;\ a_6 = 0.9;\ a_7 = 0.6;$$
$$b_1 = 0.9;\ b_2 = 0.8;\ b_3 = 0.9;\ b_4 = 1.$$

The reverberated sound generation scheme shown in Fig. 5.31 is implemented on music sound from sound file 'utopia.wav'[website4]. The original music sound and the reverberated music sound waveforms are shown in Fig. 5.34a, b, respectively. The reverberated music sound is found to be more pleasant to hear than the original.

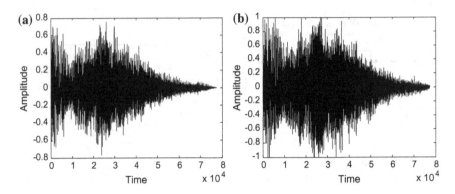

Fig. 5.34 **a** Original music sound and **b** reverberated

5.7.3 Audio Peaking Equalizers

With the availability of low-cost DSPs and recurrent usage of digital sounds, the need for audio equalizers has become crucial. By using a peaking equalizer filter section, a boost or cut is obtained in the vicinity of the center frequency. Peaking equalizer filter section is commonly known as parametric equalizer section, in which the gain is outlying from the boost or cut, so that a number of such sections can be arranged in series. Figure 5.35 shows a typical peaking equalizer comprised of cascaded IIR second-order filters.

The transfer function of an IIR second-order peaking filter is given by Robert [16]

$$H(z) = \frac{b_0 z^2 + b_1 z + b_2}{a_0 z^2 + a_1 z + a_2} \tag{5.82}$$

where

$b_0 = 1 + \alpha\sqrt{K}, b_1 = -2\cos\omega_0, b_2 = 1 - \alpha\sqrt{K}$
$a_0 = 1 + \frac{\alpha}{\sqrt{K}}, a_1 = -2\cos\omega_0, a_2 = 1 - \frac{\alpha}{\sqrt{K}}$
$\alpha = [\sin\omega_0]\sinh\left(\frac{\ln 2}{2} B_w \frac{\omega_0}{\sin\omega_0}\right)$
$\omega_0 = \frac{2\pi f_0}{F_T}, B_w = \frac{2\pi b_w}{F_T}, K = 10^{(G/20)}$
$f_0 = $ Peak frequency in Hz
$b_w = $ Bandwidth in Hz

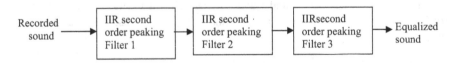

Fig. 5.35 Typical peaking equalizer comprised of cascaded IIR second-order filters

F_T = Sampling frequency
K = Gain at the peak frequency f_0
G = Peak gain in dB
B_w = Bandwidth in octaves given by $\omega_+ = \omega_- 2^{B_w}$, ω_+ and ω_- being the upper and lower edge frequencies, where the gain in dB is $G/2$.

As an example, consider the design of a peaking equalizer (Fig. 5.35) satisfying the following specifications:

Peaking Filter 1: f_0 = 1600 Hz, b_w = 800 Hz, F_T = 44,100 Hz, G = 20 dB
Peaking Filter 2: f_0 = 2400 Hz, b_w = 800 Hz, F_T = 44,100 Hz, G = 20 dB
Peaking Filter 3: f_0 = 3200 Hz, b_w = 800 Hz, F_T = 44,100 Hz, G = 20 dB

The magnitude and phase responses of each of the three peaking equalizers are shown in Fig. 5.36.

The music sound from the sound file 'original.wav' [17] is applied to the peaking equalizer of Fig. 5.35 with the magnitude response and phase response as shown in Fig. 5.36. The original sound signal and the equalized sound signal are shown in Fig. 5.37a, b, respectively.

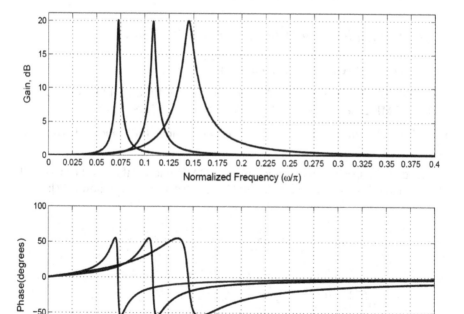

Fig. 5.36 Magnitude and phase responses of the three second-order IIR peaking equalizers

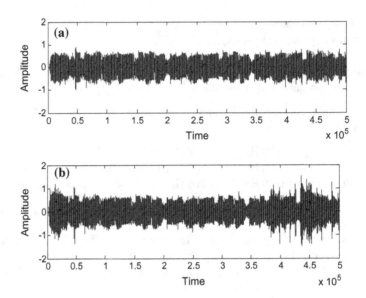

Fig. 5.37 a Original sound and **b** sound obtained after passing through a peaking equalizer comprised of three cascaded IIR second-order filters

5.7.4   Generation and Detection of DTMF Tones

Dual-Tone Multifrequency Tone Generator

The DTMF tone generator can be developed using two IIR digital filters in parallel. The DTMF generator for key '5' is depicted in Fig. 5.38.

Dual-Tone Multifrequency Tone Detection Using the Modified Goertzel Algorithm

Based on the specified frequencies of each DTMF tone and the modified Goertzel algorithm, the stepwise procedure for DTMF tone detection is as follows [18]:

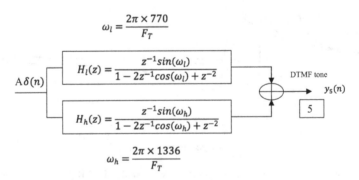

Fig. 5.38 Digital DTMF tone generator for the key '5'

Step 1   For every digitized DTMF tone received, two nonzero frequency components are found from the following seven: 697, 770, 852, 941, 1209, 1336, and 1477 Hz.

Step 2   Apply the modified Goertzel algorithm to compute seven spectral values, which correspond to the seven frequencies mentioned in Step 1. The single-sided amplitude spectrum is computed using the following expression:

$$A_k = \frac{2\sqrt{|X(k)|^2}}{N} \qquad (5.83)$$

Step 3   Determine the Key by using two nonzero spectral components corresponding to the key is pressed.

Step 4   Determine the frequency bin number (frequency index) based on the sampling rate fs and the data size of N using the following relation:

$$k = \frac{f}{F_T} \times N \, (\text{round off to integer}) \qquad (5.84)$$

Since the telephone industry has preset F_T the sampling frequency to 8 kHz and the DTMFs to 697, 770, 852, 941, 1209, 1336, and 1477, the filter length must be large enough to find the desired k value that corresponds to the DTMF frequencies. Therefore, there is a trade-off to be considered between the computation burden and better resolution. For this application report, the filter length, N, was chosen as 105, which is the smallest value that can fulfill DTMF detection. Table 5.7 shows the calculated k values for $N = 105$.

Table 5.7 DTMFs and their frequency bins

DTMF f (Hz)	Frequency bin $k = \frac{f}{F_T} \times N$ (round off to an integer)
697	9
770	10
852	11
941	12
1209	16
1336	18
1477	19

Now, compute the frequency bin k for each DTMF frequency with $f_s = 8000$ Hz and $N = 105$ as tabulated in Table 5.7.

The DTMF detector block diagram is shown in Fig. 5.39.

Step 5. Add all seven spectral values and divide the sum by 4 to obtain the threshold value

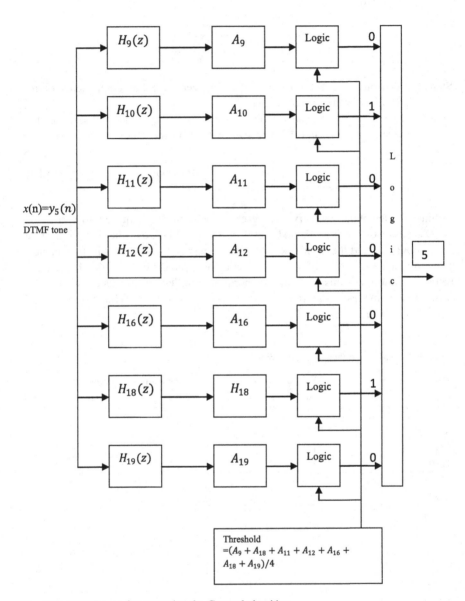

Fig. 5.39 DTMF tone detector using the Goertzel algorithm

Fig. 5.40 Spectral values and threshold for key 5

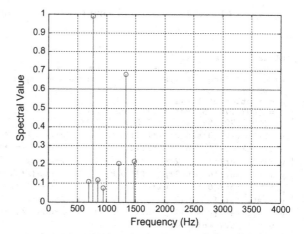

The logic operation outputs logic 1 for the spectrum value greater than the threshold value; otherwise, the logic operation outputs logic 0. The last-stage logic operation decodes the key information based on the 7-bit binary pattern.

The MATLAB simulation for decoding key 5 is shown in Program 5.12. The input is generated as shown in Fig. 5.38. After filtering, the calculated spectral values and the threshold value for decoding key 5 are displayed in Fig. 5.40, where only two spectral values corresponding to the frequencies of 770 and 1336 Hz are above the threshold and are encoded as logic 1. According to the key information in Fig. 5.39, the final logic operation decodes the key as 5.

Program 5.12 DTMF Detection Using Goertzel Algorithm

```
close all;clear all;
N=105;fs=8000; t=[0:1:N-1]/fs; % Sampling rate and time vector
x=zeros(1,length(t));x(1)=1; % Generate the impulse functionn
yDTMF=[];
%Generation of tones
f = [697 770 852 941 1209 1336 1477];
lf=[697;770;852;941];hf=[1209;1336;1477];
ylf1=filter([0 sin(2*pi*lf(1)/F_T)],[1 -2*cos(2*pi*lf(1)/F_T) 1],x);
ylf2=filter([0 sin(2*pi*lf(2)/F_T)],[1 -2*cos(2*pi*lf(2)/F_T) 1],x);
ylf3=filter([0 sin(2*pi*lf(3)/F_T)],[1 -2*cos(2*pi*lf(3)/F_T) 1],x);
ylf4=filter([0 sin(2*pi*lf(4)/F_T)],[1 -2*cos(2*pi*lf(4)/F_T) 1],x);
yhf1=filter([0 sin(2*pi*hf(1)/F_T)],[1 -2*cos(2*pi*hf(1)/F_T) 1],x);
yhf2=filter([0 sin(2*pi*hf(2)/F_T)],[1 -2*cos(2*pi*hf(2)/F_T) 1],x);
yhf3=filter([0 sin(2*pi*hf(3)/F_T)],[1 -2*cos(2*pi*hf(3)/fs) 1],x);
key = input('enter key=') F_T
if key==1 yDTMF=ylf1+yhf1; end if key==2 yDTMF=ylf1+yhf2; end
if key==3 yDTMF=ylf1+yhf3; endif key==4 yDTMF=ylf2+yhf1; end
if key==5 yDTMF=ylf2+yhf2; end if key==6 yDTMF=ylf2+yhf3; end
if key==7 yDTMF=ylf3+yhf1; end if key==8 yDTMF=ylf3+yhf2; end
if key==9 yDTMF=ylf3+yhf3; end
if key==10    % '*'
    yDTMF=ylf4+yhf1; end
if key==11 yDTMF=ylf4+yhf2; end
if key==12    %"#'
    yDTMF=ylf4+yhf3; end
yDTMF=[yDTMF 0]; % DTMF signal appended with a zero
% DTMF detector (use Goertzel algorithm)
a9=[1 -2*cos(2*pi*9/N) 1];a10=[1 -2*cos(2*pi*10/N) 1];
a11=[1 -2*cos(2*pi*11/N) 1];a12=[1 -2*cos(2*pi*12/N) 1];
a16=[1 -2*cos(2*pi*16/N) 1];a18=[1 -2*cos(2*pi*18/N) 1];
a19=[1 -2*cos(2*pi*19/N) 1];y9=filter(1,a9,yDTMF);
y10=filter(1,a10,yDTMF);y11=filter(1,a11,yDTMF);
y12=filter(1,a12,yDTMF);y16=filter(1,a16,yDTMF);
y18=filter(1,a18,yDTMF);y19=filter(1,a19,yDTMF);
% Determine the absolute magnitude of DFT coefficents
m(1)=sqrt(y9(105)^2+y9(104)^2-2*cos(2*pi*9/105)*y9(105)*y9(104));
m(2)=sqrt(y10(105)^2+y10(104)^2-2*cos(2*pi*10/105)*y10(105)*y10(104));
m(3)=sqrt(y11(105)^2+y11(104)^2-2*cos(2*pi*11/105)*y11(105)*y11(104));
m(4)=sqrt(y12(105)^2+y12(104)^2- 2*cos(2*pi*12/105)*y12(105)*y12(104));
m(5)=sqrt(y16(105)^2+y16(104)^2- 2*cos(2*pi*16/105)*y16(105)*y16(104));
m(6)=sqrt(y18(105)^2+y18(104)^2- 2*cos(2*pi*18/105)*y18(105)*y18(104));
m(7)=sqrt(y19(105)^2+y19(104)^2- 2*cos(2*pi*19/105)*y19(105)*y19(104));
m=2*m/105;th=sum(m)/4;%   threshold
f1=[0 F_T/2];th=[ th th];stem(f,m);grid;hold; plot(f1,th);
% xlabel('Frequency (Hz)'); ylabel(' (b) Spectral values');
m=round(m); % Round to the binary pattern
if m==[ 1 0 0 0 1 0 0] disp('Detected Key 1'); end
if m== [ 1 0 0 0 0 1 0] disp('Detected Key 2'); end
if m== [ 1 0 0 0 0 0 1] disp('Detected Key 3'); end
if m== [ 0 1 0 0 1 0 0] disp('Detected Key 4'); end
if m== [ 0 1 0 0 0 1 0] disp('Detected Key 5'); end
if m==[ 0 1 0 0 0 0 1] disp('Detected Key 6'); end
if m== [ 0 0 1 0 1 0 0] disp('Detected Key 7'); end
if m==[ 0 0 1 0 0 1 0] disp('Detected Key 8'); end
if m==[ 0 0 1 0 0 0 1] disp('Detected Key 9'); end
if m== [ 0 0 0 1 1 0 0] disp('Detected Key *'); end
if m== [ 0 0 0 1 0 1 0] disp('Detected Key 0'); end
```

5.8 Problems

1. For the following specifications, design a lowpass IIR digital Butterworth filter using the impulse-invariant method.

$$0.8 \le \left| H(e^{j\omega}) \right| \le 1 \quad \text{for} \ \ 0 \le \omega \le 0.3\pi$$
$$\left| H(e^{j\omega}) \right| \le 0.4 \quad \text{for} \ \ 0.6\pi \le \omega \le \pi$$

2. Using the bilinear transformation, design a lowpass IIR digital Butterworth filter with −3 dB cutoff at 150 Hz and stopband attenuation of 20 dB or greater at 600 Hz. The sampling frequency is 6000 Hz.

3. Design a digital Butterworth highpass filter to meet the following specifications:

   Passband edge frequency: 1000 Hz
   Stopband edge frequency: 400 Hz
   Passband ripple: 3 dB
   Stopband ripple: 10 dB

   Assume a suitable sampling frequency.

4. Design a Butterworth IIR digital bandpass filter for the following specifications:

   Lower passband edge frequency: 500 Hz
   Upper passband edge frequency: 600 Hz
   Lower stopband edge frequency: 100 Hz
   Upper stopband edge frequency: 1000 Hz
   Passband ripple: 2 dB
   Stopband ripple: 10 dB

   Assume 4000 Hz as the sampling frequency. Use bilinear transformation.

5. Design a Chebyshev IIR digital lowpass filter for the following specifications:

   Passband cutoff frequency: 400 Hz
   Stopband cutoff frequency: 600 Hz
   Passband ripple: 1 dB
   Stopband ripple: 10 dB

   Assume a suitable sampling frequency. Use bilinear transformation.

6. Design a Chebyshev IIR digital highpass filter for the following specifications:

   3 dB cutoff frequency: 2000 Hz
   Stopband cutoff frequency: 500 Hz
   Stopband ripple: 10 dB

   Assume a suitable sampling frequency. Use bilinear transformation.

7. Using bilinear transformation, design a digital Chebyshev Type 1 Bandpass filter with the following specifications:

Lower passband edge frequency: 200 Hz
Upper passband edge frequency: 400 Hz
Lower stopband edge frequency: 100 Hz
Upper stopband edge frequency: 500 Hz
Passband ripple: 3 dB
Stopband ripple: 15 dB

Assume a suitable sampling frequency.

8. Using bilinear transformation, design a digital bandstop Chebyshev Type 1 filter with the following specifications:

Lower passband edge frequency: 35 Hz
Upper passband edge frequency: 215 Hz
Lower stopband edge frequency: 100 Hz
Upper stopband edge frequency: 150 Hz
Passband ripple: 2 dB
Stopband ripple: 20 dB

Assume a suitable sampling frequency.

9. Using bilinear transformation, design a digital bandstop elliptic filter with the following specifications:

Lower passband edge frequency: 800 Hz
Upper passband edge frequency: 2000 Hz
Lower stopband edge frequency: 1200 Hz
Upper stopband edge frequency: 1300 Hz
Passband ripple: 1 dB
Stopband ripple: 40 dB

Assume a suitable sampling frequency.

10. A third-order lowpass IIR digital filter with passband edge frequency at 0.25π has a transfer function

$$H(z) = \frac{0.0662272z^3 + 0.1987z^2 + 0.1987z + 0.0662272}{z^3 - 0.9356142z^2 + 0.5671268z - 0.1015911}$$

Design a lowpass filter with passband edge frequency at 0.375π by transforming the above transfer function using lowpass-to-lowpass digital-to digital transformation.

5.9 MATLAB Exercises

1. Write a MATLAB program using the M-file *impinvar* to design a Type 1 Chebyshev IIR digital lowpass filter using the impulse-invariant method for the specifications given in Example 5.9.

2. Write MATLAB code to design a Type 1 Chebyshev bandstop filter using bilinear transformation with the following specifications:

Lower passband edge: 0.3333π
Upper passband edge: 0.75π
Lower stopband edge: 0.45π
Upper stopband edge: 0.75π
Passband ripple: 1 dB
Stopband ripple: 40 dB

3. Write a MATLAB program to design a highpass Butterworth filter using digital-to-digital transformation satisfying the following specifications:

Passband edge frequency: 0.5π
Stopband edge frequency: 0.4π
Passband ripple: 2 dB
Stopband ripple: 20 dB

4. A third-order lowpass IIR digital filter with passband edge frequency at 0.25π has a transfer function

$$H(z) = \frac{0.0662272z^3 + 0.1987z^2 + 0.1987z + 0.0662272}{z^3 - 0.9356142z^2 + 0.5671268z - 0.1015911}$$

Write MATLAB code to design a highpass filter with passband edge frequency at 0.45π by transforming the above transfer function using lowpass-to-highpass digital-to-digital transformation. Show the magnitude responses of the lowpass and highpass filters on the same plot.

5. A first-order lowpass filter with passband edge frequency at $0.1667\ \pi$ has a transfer function as

$$H(z) = \frac{0.5z + 0.5}{z - 0.302}$$

Write MATLAB code to design a bandpass filter with lower passband edge frequency at 0.25π and upper passband edge frequency at 0.75π by transforming the above transfer function using lowpass-to-highpass digital transformation. Show the magnitude responses of the lowpass and bandpass filters.

6. Write a MATLAB program to suppress a sinusoidal interference of 1750 Hz from an audio signal using a second-order IIR digital notch filter. Consider the audio signal 'DT.wav' included in CD, and corrupt it by a sinusoidal signal of 1750 Hz. Implement the notch filter on it and comment on the results.

7. Write a MATLAB program to generate artificial reverberations using the scheme (Fig. 5.31) with six IIR filters and four allpass reverberators as shown in Fig. 5.32, and with structures for IIR filters and allpass reverberators as shown in Fig. 5.32. Implement it with suitable delays and multiplier constants on the music sound 'utopia.wav' included in the CD and comment on the result.

8. Write a MATLAB program for peaking equalizer consisting of three second-order IIR filters in cascade with different center frequencies, bandwidths, and DB gains for the filters. Implement it on the music sound 'original.wav,' and comment on the result.

References

1. G.C. Temes, S.K. Mitra (eds.), *Modern Filter Theory and Design* (Wiley, New York, 1973)
2. G.C. Temes, J.W. LaPatra, *Introduction to Circuit Synthesis and Design* (McGraw Hill, New York, 1977)
3. J. Vlach, *Computerized Approximation and Synthesis of Linear Networks* (Wiley, New York, 1969)
4. A. Antoniou, in *Digital Filters Analysis, Design, and Applications* (McGraw Hill Book Co., 2006)
5. R. Raut, M.N.S. Swamy, *Modern Analog Filter Analysis and Design: A Practical Approach* (Springer, VCH, 2010)
6. T.W. Parks, C.S. Burrus, *Digital Filter Design* (Wiley, New York, NY, 1987)
7. C.T. Chen, in *Digital Signal Processing, Spectral Computation and Filter Design* (Oxford University Press, New York, Oxford, 2001)
8. P.V.A. Mohan, V. Ramachandran, M.N.S. Swamy, *Switched Capacitor Filters: Theory, Analysis and Design* (Prentice Hall, London, 1995)
9. A.G. Constantinides, Proc. IEEE **117**, 1585–1590, 170
10. R.E. Crochier, P. Penfield Jr., On the efficient design of bandpass digital filter structures. IEEE Trans. ASSP **23**, 380–381 (1975)
11. R. Chassaing, in *DSP Applications Using C and the TMS320C6x DSK* (Wiley Interscience Publication, 2002)
12. E.C. Ifeachor, B.W. Jervis, in *Digital Signal Processing—A Practical Approach* (Prentice Hall, 2002)
13. L.B. Jackson, *Digital Filters and Signal Processing* (Kluwer Academic, Norwell, M.A., 1996)
14. P.A. Bernhardt, D.A. Antoniadis, A.V. Da Rosa, Lunar perturbations in columnar electron content and their interpretation in terms of dynamo electrostatic fields. J. Geophys. Res. (1976)
15. www.harmony-central.com
16. B.-J. Robert, in *The Equivalence of Various Methods of Computing Biquad Coefficients for Audio Parametric Equalizers*. Audio Engineering Society Conference, November 1994, paper # 3906
17. Audio Special Effects—Equalization and Artificial Reverberation, http://www.nd.ed
18. L. Tan, J. Jiang, in *Digital Signal Processing Fundamentals and Applications* (Elsevier, Academic Press 2nd, USA, edition, 2013)

Chapter 6
FIR Digital Filter Design

In Chap. 5, the design of IIR filters was considered. In many digital signal processing applications, FIR filters are preferred to IIR filters because of the following advantages of FIR filters.

(i) The FIR filter is always stable since it is described by a non-recursive difference equation and all of its poles are located at the origin of the z-plane.

(ii) Unlike the IIR digital filter design, the FIR filters can be always designed with exact linear phase and constant group delay.

(iii) FIR filters are not sensitive to the finite word length effects like IIR filters.

However, IIR filters are preferred to FIR filters if the linear phase is not a constraint, due to the following disadvantages of the FIR filters.

(i) The order of the FIR filter transfer function is usually much higher than that of an IIR filter transfer function meeting the same frequency response specifications.

(ii) Memory requirement and computation time are high.

In this chapter, the conditions for FIR filters to have linear phase are first described. Second, the design of FIR filters using fixed windows and Kaiser window, and frequency sampling technique is discussed and illustrated with numerical examples. Next, the design of optimal linear phase FIR filters is described. Further, the design of FIR filters using MATLAB is demonstrated with a number of examples. Furthermore, the design of minimum-phase FIR filters is presented. The minimum-phase FIR filter leads to a transfer function with a smaller group delay than that of a linear phase equivalent. Finally, the design of FIR filters using graphical user interface MATLAB filter design SPTOOL is discussed and illustrated with examples for the design of equiripple linear phase FIR filters.

© Springer Nature Singapore Pte Ltd. 2018
K. D. Rao and M. N. S. Swamy, *Digital Signal Processing*,
https://doi.org/10.1007/978-981-10-8081-4_6

6.1   Ideal Impulse Response of FIR Filters

The ideal impulse responses of the lowpass, highpass, bandpass, and bandstop filters are derived below.

Ideal lowpass filter

The frequency response of an ideal lowpass filter is given by

$$H_{\text{LP}}\left(e^{j\omega}\right) = \begin{cases} 1, & |\omega| \leq \omega_c \\ 0, & \omega_c < |\omega| \leq \pi \end{cases} \tag{6.1}$$

The impulse response is given by

$$h_{\text{LP}}(n) = \frac{1}{2\pi} \int_{-\omega_c}^{\omega_c} H_{\text{LP}}\left(e^{j\omega}\right) e^{j\omega n} d\omega$$

From Eq. (6.1), we get

$$h_{\text{LP}}(n) = \frac{1}{2\pi} \int_{-\omega_c}^{\omega_c} e^{j\omega n} d\omega = \frac{\omega_c}{\pi} \quad \text{for } n = 0$$

and

$$\begin{aligned} h_{\text{LP}}(n) &= \frac{1}{2\pi} \frac{e^{j\omega n}}{jn} \bigg|_{-\omega_c}^{\omega_c} \quad \text{for } n \neq 0 \\ &= \frac{1}{j2\pi n} \left[e^{j\omega_c n} - e^{-j\omega_c n} \right] \\ &= \frac{1}{n\pi} \left[\frac{e^{j\omega_c n} - e^{-j\omega_c n}}{2j} \right] \\ &= \frac{\sin \omega_c n}{n\pi} \quad -\infty \leq n \leq \infty \end{aligned}$$

Hence, the impulse response of an ideal lowpass filter is

$$h_{\text{LP}}(n) = \begin{cases} \frac{\omega_c}{\pi}, & n = 0 \\ \frac{\sin \omega_c n}{n\pi}, & n \neq 0 \end{cases} \tag{6.2}$$

Ideal highpass filter

The frequency response of ideal highpass filter is given by

$$H_{HP}(e^{j\omega}) = \begin{cases} 0 & |\omega| \leq \omega_c \\ 1, & \omega_c < |\omega| \leq \pi \end{cases} \tag{6.3}$$

The impulse response is given by

$$h_{HP}(n) = \frac{1}{2\pi} \int\limits_{-\omega_c}^{\omega_c} H_{HP}(e^{j\omega}) e^{j\omega n} d\omega$$

$$h_{HP}(n) = \frac{1}{2\pi} \int\limits_{-\pi}^{-\omega_c} e^{j\omega n} d\omega + \frac{1}{2\pi} \int\limits_{\omega_c}^{\pi} e^{j\omega n} d\omega$$

$$= 1 - \frac{\omega_c}{\pi}, \quad \text{for } n = 0$$

For $n \neq 0$,

$$h_{HP}(n) = \frac{1}{2\pi} \frac{e^{j\omega n}}{jn} \bigg|_{-\pi}^{-\omega_c} + \frac{1}{2\pi} \frac{e^{j\omega n}}{jn} \bigg|_{\omega_c}^{\pi}$$

$$= \frac{1}{j2\pi n} \left[e^{-j\omega_c n} - e^{-j\pi n} \right] + \frac{1}{j2\pi n} \left[e^{j\pi n} - e^{j\omega_c n} \right]$$

$$= -\frac{1}{\pi n} \left[\frac{e^{j\omega_c n}}{2j} - \frac{e^{-j\omega_c n}}{2j} \right] + \frac{1}{\pi n} \left[\frac{e^{j\pi n}}{2j} - \frac{e^{-j\pi n}}{2j} \right]$$

$$= \frac{[\sin \pi n - \sin \omega_c n]}{n\pi} = -\frac{\sin \omega_c n}{\pi n}$$

Hence, the impulse response of an ideal highpass filter is

$$h_{HP}(n) = \begin{cases} 1 - \frac{\omega_c}{\pi}, & n = 0 \\ -\frac{\sin \omega_c n}{\pi n}, & n \neq 0 \end{cases} \tag{6.4}$$

Ideal bandpass filter

The frequency response of an ideal bandpass filter is given by

$$H_{BP}(e^{j\omega}) = \begin{cases} 1, & \omega_{c1} \leq |\omega| \leq \omega_{c2} \\ 0, & -\pi < |\omega| \leq \omega_{c1} \text{ and } \omega_{c2} < |\omega| \leq \pi \end{cases} \tag{6.5}$$

The bandpass filter can be viewed as cascade of a lowpass filter and a highpass filter

$$h_{BP}(n) = \frac{1}{2\pi} \int\limits_{-\omega_{c2}}^{-\omega_{c1}} e^{j\omega n} d\omega + \frac{1}{2\pi} \int\limits_{\omega_{c1}}^{\omega_{c2}} e^{j\omega n} d\omega$$

After mathematical calculations, we get for $n \neq 0$,

$$\begin{aligned} h_{BP}(n) &= \frac{1}{2\pi} \frac{e^{j\omega n}}{jn}\bigg|_{-\omega_{c2}}^{-\omega_{c1}} + \frac{1}{2\pi} \frac{e^{j\omega n}}{jn}\bigg|_{\omega_{c1}}^{\omega_{c2}} \\ &= \frac{1}{j2\pi n}\left[e^{-j\omega_{c1}n} - e^{-j\omega_{c2}n}\right] + \frac{1}{j2\pi n}\left[e^{j\omega_{c2}n} - e^{j\omega_{c1}n}\right] \\ &= -\frac{1}{\pi n}\left[\frac{e^{j\omega_{c1}n}}{2j} - \frac{e^{-j\omega_{c1}n}}{2j}\right] + \frac{1}{\pi n}\left[\frac{e^{j\omega_{c2}n}}{2j} - \frac{e^{-j\omega_{c2}n}}{2j}\right] \\ &= \frac{[\sin \omega_{c2}n - \sin \omega_{c1}n]}{n\pi} \end{aligned}$$

and for $n = 0$

$$h_{BP}(n) = \frac{\omega_{c2}}{\pi} - \frac{\omega_{c1}}{\pi}$$

Thus, the impulse response of an ideal bandpass filter is

$$h_{BP}(n) = \begin{cases} \frac{\sin \omega_{c2}n}{\pi n} - \frac{\sin \omega_{c1}n}{\pi n}, & n \neq 0 \\ \frac{\omega_{c2}}{\pi} - \frac{\omega_{c1}}{\pi}, & n = 0 \end{cases} \tag{6.6}$$

Ideal bandstop filter

The frequency response of ideal bandstop filter is given by

$$H_{BS}(e^{j\omega}) = \begin{cases} 0, & \omega_{c1} \leq |\omega| \leq \omega_{c2} \\ 1, & -\pi < |\omega| \leq \omega_{c1} \text{ and } \omega_{c2} < |\omega| \leq \pi \end{cases} \tag{6.7}$$

The bandstop filter can be viewed as a parallel connection of a lowpass filter and a highpass filter. $H_{BS}(e^{j\omega})$ can be written as

$$H_{BS}(e^{j\omega}) = H_{LP}(e^{j\omega}) + H_{HP}(e^{j\omega}) \tag{6.8}$$

subject to the condition that the cutoff frequency ω_{c2} of the highpass filter is greater than the cutoff frequency ω_{c1} of the lowpass filter. The impulse response of the bandstop filter can be obtained by taking the inverse Fourier transform. Thus, from the properties of Fourier transforms, we have

$$\mathcal{F}^{-1}\left(H_{\text{BS}}\left(e^{j\omega}\right)\right) = \mathcal{F}^{-1}\left(H_{\text{LP}}\left(e^{j\omega}\right) + H_{\text{HP}}\left(e^{j\omega}\right)\right)$$
$$= \mathcal{F}^{-1}\left(H_{\text{LP}}\left(e^{j\omega}\right)\right) + \mathcal{F}^{-1}\left(H_{\text{HP}}\left(e^{j\omega}\right)\right)$$
$$h_{\text{BS}}(n) = h_{\text{LP}}(n) + h_{\text{HP}}(n)$$

Hence, the impulse response $h_{\text{BS}}(n)$ is given by

$$h_{\text{BS}}(n) = \frac{\sin \omega_{c1} n}{\pi n} - \frac{\sin \omega_{c2} n}{\pi n} \quad \text{for } n \neq 0$$

For $n = 0$,

$$h_{\text{BS}}(n) = 1 - \frac{(\omega_{c2} - \omega_{c1})}{\pi}$$

Thus, the impulse response of an ideal bandstop filter is

$$h_{\text{BS}}(n) = \begin{cases} 1 - \dfrac{(\omega_{c2} - \omega_{c1})}{\pi}, & n = 0 \\ \dfrac{\sin \omega_{c1} n}{\pi n} - \dfrac{\sin \omega_{c2} n}{\pi n}, & n \neq 0 \end{cases} \tag{6.9}$$

6.2 Linear Phase FIR Filters

A causal FIR transfer function of length $N + 1$ is given by [see Eq. (3.106)]

$$H(z) = \sum_{n=0}^{N} h(n) z^{-n} \tag{6.10}$$

Substituting $z = e^{j\omega}$ in the above equation, we obtain

$$H\left(e^{j\omega}\right) = \sum_{n=0}^{N} h(n) e^{-j\omega n} \tag{6.11}$$

which is periodic in frequency with a period 2π. Now,

$$H\left(e^{j\omega}\right) = \pm\left|H\left(e^{j\omega}\right)\right| e^{j\theta(\omega)} \tag{6.12}$$

where $\left|H\left(e^{j\omega}\right)\right|$ is the magnitude and $\theta(\omega)$ the phase of $H\left(e^{j\omega}\right)$.

We define the phase delay τ_p and group delay τ_g of a filter as

$$\tau_p = \frac{-\theta(\omega)}{\omega} \tag{6.13a}$$

and

$$\tau_g = \frac{-d\theta(\omega)}{d\omega} \qquad (6.13b)$$

For FIR filters with linear phase, we can define

$$\theta(\omega) = \beta - \alpha\omega \quad 0 \le \omega \le \pi \qquad (6.14)$$

where α and β are real constants.

The tangent of the phase angle of $H(e^{j\omega})$ can be expressed as

$$\frac{-\sum_{n=0}^{N} h(n) \sin \omega n}{\sum_{n=0}^{N} h(n) \cos \omega n} = \frac{\sin(\beta - \alpha\omega)}{\cos(\beta - \alpha\omega)} \qquad (6.15)$$

Cross-multiplying and combining terms lead to the equation

$$\sum_{n=0}^{N} h(n) \sin[(n - \alpha)\omega + \beta] = 0 \quad \text{for all } \omega \qquad (6.16)$$

If $\beta = 0$, Eq. (6.16) becomes

$$\sum_{n=0}^{N} h(n) \sin[(n - \alpha)\omega] = 0 \qquad (6.17)$$

Equation (6.17) is satisfied when

$$h(n) = h(N - n) \qquad (6.18)$$

and

$$\alpha = \frac{N}{2}$$

where N is an integer. Therefore, FIR filters will have constant phase and group delays when the impulse response is symmetrical about $\alpha = \frac{N}{2}$.

If $\beta = \pm\frac{\pi}{2}$, then Eq. (6.16) becomes

$$\sum_{n=0}^{N} h(n) \cos[(n - \alpha)\omega] = 0 \qquad (6.19)$$

The above equation will be satisfied when

$$h(n) = -h(N - n) \tag{6.20}$$

and

$$\alpha = \frac{N}{2}$$

where N is an integer.

Therefore, FIR filters have constant group delay τ_g, but not a constant phase delay, when the impulse response is antisymmetric about $\alpha = \frac{N}{2}$.

Thus, an FIR filter has linear phase, if its impulse response $h(n)$ is either symmetric, i.e.,

$$h(n) = h(N - n), \quad 0 \le n \le N, \tag{6.21}$$

or antisymmetric, i.e.,

$$h(n) = -h(N - n), \quad 0 \le n \le N, \tag{6.22}$$

Since the length of the impulse response can be either even or odd, four types of symmetry can be defined for the impulse response. For an antisymmetric FIR filter of odd length, i.e., N even, $h(N/2) = 0$.

6.2.1 Types of Linear Phase FIR Transfer Functions

Type 1: Symmetric Impulse Response with Odd Length (Even Order)

In this case, the filter order N is even. Assume $N = 6$ for simplicity. The transfer function of the corresponding filter is given by

$$H(z) = h(0) + h(1)z^{-1} + h(2)z^{-2} + h(3)z^{-3} + h(4)z^{-4} + h(5)z^{-5} + h(6)z^{-6} \tag{6.23}$$

For symmetry, $h(0) = h(6)$, $h(1) = h(5)$, and $h(2) = h(4)$. Then, Eq. (6.23) reduces to

$$\begin{aligned} H(z) &= h(0)(1 + z^{-6}) + h(1)(z^{-1} + z^{-5}) + h(2)(z^{-2} + z^{-4}) + h(3)z^{-3} \\ &= z^{-3}\left\{ h(0)\left(z^{3} + z^{-3}\right) + h(1)\left(z^{2} + z^{-2}\right) + h(2)\left(z^{1} + z^{-1}\right) + h(3) \right\} \end{aligned} \tag{6.24}$$

The corresponding frequency response is given by

$$H(e^{j\omega}) = e^{-j3\omega}\{2h(0)\cos(3\omega) + 2h(1)\cos(2\omega) + 2h(2)\cos(\omega) + h(3)\} = e^{-j3\omega}$$

$H_1(\omega)$, since $\left.\frac{(z^m + z^{-m})}{2}\right|_{z=e^{j\omega}} = \cos(m\omega)$. $H_1(\omega)$ is a real function of ω, and we can assume positive or negative values in the range $0 \leq |\omega| \leq \pi$. Sometimes, $H_1(\omega)$ is referred to as the pseudo-magnitude function. Hence, the phase is given by

$$\theta(\omega) = -3\omega + \beta,$$

where β is either 0 or π, and thus, it is a linear function of ω. The group delay is given by

$$\tau_g(\omega) = -\frac{d\theta(\omega)}{d\omega} = 3$$

indicating a constant group delay of three samples.

In the general case for Type 1 FIR filters, the frequency response can be shown to be

$$H(e^{j\omega}) = e^{\frac{-jN\omega}{2}}H_1(\omega) \tag{6.25}$$

where the pseudo-magnitude response $H_1(\omega)$ is given by

$$H_1(\omega) = h\left(\frac{N}{2}\right) + 2\sum_{n=1}^{N/2} h\left(\frac{N}{2} - n\right)\cos(\omega n) \tag{6.26}$$

Type 2: Symmetric Impulse Response with Even Length (Odd Order)

Here, the order N is odd. For illustration, let $N = 7$. By making use of the symmetry of the impulse response coefficients given by Eq. (6.21), the transfer function of the FIR filter can be written as

$$H(z) = h(0)(1 + z^{-7}) + h(1)(z^{-1} + z^{-6}) + h(2)(z^{-2} + z^{-5}) + h(3)(z^{-3} + z^{-4})$$

$$= z^{-7/2}\left\{\begin{array}{l} h(0)\left(z^{\frac{7}{2}} + z^{-\frac{7}{2}}\right) + h(1)\left(z^{\frac{5}{2}} + z^{-\frac{5}{2}}\right) + h(2)\left(z^{\frac{3}{2}} + z^{-\frac{3}{2}}\right) \\ + h(3)\left(z^{1/2} + z^{-1/2}\right) \end{array}\right\}$$

$$\tag{6.27}$$

The frequency response is given by

$$H(e^{j\omega}) = e^{-j7\omega/2} \left\{ \begin{array}{l} 2h(0)\cos\left(\dfrac{7\omega}{2}\right) + 2h(1)\cos\left(\dfrac{5\omega}{2}\right) \\[2mm] + 2h(2)\cos\left(\dfrac{3\omega}{2}\right) + 2h(3)\cos\left(\dfrac{\omega}{2}\right) \end{array} \right\} \tag{6.28}$$

$$= e^{-j7\omega/2}H_1(\omega)$$

where $H_1(\omega)$ is a real function of ω and we can assume positive or negative values in the range $0 \le |\omega| \le \pi$. Hence, the phase is given by

$$\theta(\omega) = -\frac{7}{2}\omega + \beta,$$

where β is either 0 or π, and thus, it is a linear function of ω. The group delay is given by

$$\tau_g(\omega) = -\frac{d\theta(\omega)}{d\omega} = \frac{7}{2}$$

indicating a constant group delay of 7/2 samples.

In general, the expression for the frequency response for Type 2 FIR filter can be shown to be

$$H(e^{j\omega}) = e^{-\frac{jN\omega}{2}}H_1(\omega) \tag{6.29}$$

where the pseudo-magnitude response is given by

$$H_1(\omega) = 2\sum_{n=1}^{(N+1)/2} h\left(\frac{N+1}{2} - n\right)\cos\left(\omega\left(n - \frac{1}{2}\right)\right) \tag{6.30}$$

Type 3: Antisymmetric Impulse Response with Odd Length (Even Order)

Here, the degree N is even. For illustration, we consider $N = 6$. Then, applying the symmetry condition of Eq. (6.22) on the expression for the transfer function, we get

$$H(z) = z^{-3}\{h(0)(z^3 - z^{-3}) + h(1)(z^2 - z^{-2}) + h(2)(z^1 - z^{-1})\} \tag{6.31}$$

The frequency response is given by

$$H(e^{j\omega}) = e^{-j3\omega}e^{j\frac{\pi}{2}}\{2h(0)\sin(3\omega) + 2h(1)\sin(2\omega) + 2h(2)\sin(\omega)\}$$
$$= e^{-j\left(3\omega - \frac{\pi}{2}\right)}H_1(\omega) \tag{6.32}$$

Since $\frac{(z^m - z^{-m})}{2}\Big|_{z=e^{j\omega}} = e^{j\pi/2} \sin(m\omega)$. The linear phase response is given by

$$\theta(\omega) = -3\omega + \beta + \frac{\pi}{2},$$

where β is either 0 or π. The group delay is given by

$$\tau_g(\omega) = -\frac{d\theta(\omega)}{d\omega} = 3$$

indicating a constant group delay of three samples.

In general, the expression for the frequency response for Type 3 FIR filters is given by

$$H(e^{j\omega}) = je^{\frac{-jN\omega}{2}}H_1(\omega) \tag{6.33}$$

where the pseudo-magnitude response is of the form

$$H_1(\omega) = 2\sum_{n=1}^{N/2} h\left(\frac{N}{2} - n\right)\sin(\omega n) \tag{6.34}$$

Type 4: Antisymmetric Impulse Response with Even Length (Odd Order)

Here, the degree N is odd. Let $N = 7$ for illustration purpose. The transfer function can be expressed as:

$$H(z) = z^{-7/2}\left\{ \begin{array}{l} h(0)\left(z^{\frac{7}{2}} - z^{-\frac{7}{2}}\right) + h(1)\left(z^{\frac{5}{2}} - z^{-\frac{5}{2}}\right) + h(2)\left(z^{\frac{3}{2}} - z^{-\frac{3}{2}}\right) \\ + h(3)\left(z^{1/2} - z^{-1/2}\right) \end{array} \right\} \tag{6.35}$$

The frequency response is given by

$$H(e^{j\omega}) = e^{-j7\omega/2}e^{j\pi/2}\left\{ \begin{array}{l} 2h(0)\sin\left(\frac{7\omega}{2}\right) + 2h(1)\sin\left(\frac{5\omega}{2}\right) + 2h(2)\sin\left(\frac{3\omega}{2}\right) \\ + 2h(3)\sin\left(\frac{\omega}{2}\right) \end{array} \right\} \tag{6.36}$$

Thus, the linear phase response is given by

$$\theta(\omega) = -\frac{7}{2}\omega + \beta + \frac{\pi}{2},$$

where β is either 0 or π. The group delay is given by

$$\tau_g(\omega) = -\frac{d\theta(\omega)}{d\omega} = \frac{7}{2}$$

In general, the frequency response for Type 4 FIR filters is

$$H(e^{j\omega}) = je^{\frac{-jN\omega}{2}}H_1(\omega) \qquad (6.37)$$

where the pseudo-amplitude response is given by

$$H_1(\omega) = 2 \sum_{n=1}^{\left(\frac{N+1}{2}\right)} h\left(\frac{N+1}{2} - n\right) \sin\left(\omega\left(n - \frac{1}{2}\right)\right) \qquad (6.38)$$

6.2.2 Zero Locations of Linear Phase FIR Transfer Functions

The constraints on the zeros are important in designing FIR linear phase filters, since they impose limitations on the types of frequency responses that can be achieved. The transfer function $H(z)$ of a linear phase FIR filter is of the form.

$$H(z) = \sum_{n=0}^{N} h(n)z^{-n} \qquad (6.39)$$

For the symmetric impulse response case, Eq. (6.21) is satisfied. Hence, the above equation can be expressed as

$$H(z) = \sum_{n=0}^{N} h(N - n)z^{-n}$$

Letting $m = N - n$, the above equation may be rewritten as

$$H(z) = \sum_{m=N}^{0} h(m)z^{-(N-m)} = z^{-N}\sum_{m=0}^{N} h(m)z^{m} \qquad (6.40)$$
$$= z^{-N}H(z^{-1})$$

Similarly, using Eq. (6.22) for the case of the antisymmetric impulse response, Eq. (6.39) may be rewritten as

$$H(z) = -z^{-N}H(z^{-1}) \qquad (6.41)$$

Hence, whether the impulse response is symmetric or antisymmetric, we see that a zero at $z = z_i$ implies a zero at $z = 1/z_i$. Further, the following observations can be made regarding the zeros of $H(z)$ from Eqs. (6.40) and (6.41), assuming the impulse response to be real.

1. An arbitrary number of zeros can be located at $z_i = \pm 1$, since $z_i^{-1} = \pm 1$.
2. Any number of complex conjugate zeros can be located on the unit circle since

$$(z - z_i)(z - z_i^*) = (z - 1/z_i^*)(z - 1/z_i)$$

3. Real zeros, which are not on the unit circle, must occur in reciprocal pairs, i.e., if $z = a \neq \pm 1$ is a zero, then $z = 1/a$ is also a zero.
4. Complex zeros, not located on the unit circle, must occur in groups of four, i.e., if z_i is a zero, then z_i^*, $1/z_i$, and $1/z_i^*$ are also zeros.

Polynomials with the above properties are called mirror-image polynomials. An example of the zeros of such a polynomial is shown in Fig. 6.1. The presence of zeros at ± 1 leads to some limitations on the use of these linear phase FIR filters in the design of certain types of filters.

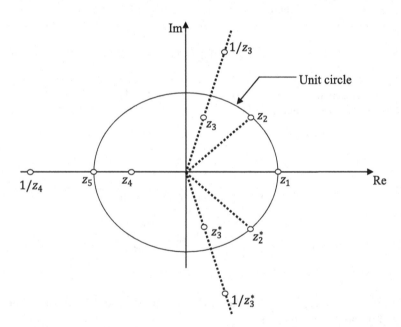

Fig. 6.1 Typical locations of zeros of $H(z)$ for a linear phase filter

Let us first consider Eq. (6.40) corresponding to the symmetric impulse response case.

$$H(-1) = (-1)^{-N}H(-1) \tag{6.42}$$

Thus, for symmetric impulse response with N odd, the system function must be zero at $z = -1$. This implies that the frequency response is constrained to be zero at $\omega = \pi(z = -1)$. Thus, highpass and bandstop filters cannot be designed as Type 2 filter. However, for n even, Eq. (6.40) is always satisfied, and hence, lowpass, highpass, bandpass, and bandstop filters can all be designed using Type 1, since no zeros are necessarily required at $z = \pm 1$.

Similarly, using Eq. (6.41) it is seen that if $z = 1$, we have the constraint

$$H(1) = -H(1) \Rightarrow H(1) = 0 \tag{6.43}$$

Thus, $H(z)$ must have a zero at $z = 1$ for both N even and odd, implying that Type 3 and Type 4 FIR filters have a magnitude response of zero at $\omega = 0$. Thus, Type 3 and Type 4 lowpass and bandstop filters cannot be designed. Also, when $z = -1$ and N is even, Eq. (6.41) reduces to

$$H(-1) = -(1)H(-1) \Rightarrow H(-1) = 0 \tag{6.44}$$

Hence, the response of Type 3 FIR filter is constrained to be zero at $\omega = \pi$, and hence, Type 3 filter cannot be used to design a highpass filter. Table 6.1 summarizes the possibilities for designing the four types of linear phase FIR filters.

The pole-zero locations of a typical linear phase FIR filter are shown in Fig. 6.1.

The MATLAB Program 6.1 is now used to find the zero locations of the four types of transfer functions given by the following expressions.

Type 1:

$$H(z) = 0.14797 + 0.40227z^{-1} + 0.68827z^{-2} + 0.91417z^{-3} + z^{-4}$$
$$+ 0.91417z^{-5} + 0.68827z^{-6} + 0.40227z^{-7} + 0.14797z^{-8}$$

Type 2:

$$H(z) = 0.14797 + 0.44319z^{-1} + 0.76302z^{-2} + 0.9713z^{-3}$$
$$+ 0.9713z^{-4} + 0.76302z^{-5} + 0.44319z^{-6} + 0.14797z^{-7}$$

Table 6.1 Indications of possibilities for realizing LP, HP, BP, and BS filters as Types 1, 2, 3 or 4

Filter	Type			
	Type 1	Type 2	Type 3	Type 4
LP	Yes	Yes	No	No
BP	Yes	Yes	Yes	Yes
HP	Yes	No	No	Yes
BS	Yes	No	No	No

Type 3:

$$H(z) = 0.14797 + 0.40227z^{-1} + 0.68827z^{-2} + 0.91417z^{-3}$$
$$- 0.91417z^{-5} - 0.68827z^{-6} - 0.40227z^{-7} - 0.14797z^{-8}$$

Type 4:

$$H(z) = 0.14797 + 0.44319z^{-1} + 0.76302z^{-2} + 0.9713z^{-3}$$
$$- 0.9713z^{-4} - 0.76302z^{-5} - 0.44319z^{-6} - 0.14797z^{-7}$$

Program 6.1 Determination of the zero locations for the four types of linear phase FIR filters

```
clear;clc;
num=[0.14797 0.40227 0.68827 0.91417 1 0.91417 0.68827 0.40227 0.14797]; %
coefficients of Type1
z=tf2zpk(num)%determine the zeros from the transfer function
figure(1),zplane(z); % plots the zero locations in the z-plane
num=[ 0.14797 0.44319 0.76302 0.9713 0.9713 0.76302 0.44319 0.14797]; %
coefficients of Type2
z=tf2zpk(num);figure (2),zplane(z);
num=[0.14797 0.40227 0.68827 0.91417 0-0.91417 -0.68827 -0.40227 -0.14797];
% coefficients of Type3
z=tf2zpk(num);figure (3),zplane(z);
num=[0.14797 0.44319 0.76302 0.9713 -0.9713 -0.76302 -0.44319 -0.14797];%
coefficients of Type4 z=tf2zpk(num);figure (4), zplane(z);
```

The zero locations of the four types of filters with $N = 8$ and $N = 7$ found from the MATLAB Program 6.1 are shown in Fig. 6.2.

6.3   FIR Filter Design Using Windowing Method

Any periodic function can be expressed as a linear combination of complex exponentials using Fourier series. Since the desired frequency response $H_d(e^{j\omega})$ of a filter is periodic of period 2π, it can be represented by the Fourier series as

$$H_d(e^{j\omega}) = \sum_{n=-\infty}^{\infty} h_d(n)e^{-j\omega n} \qquad (6.45)$$

where the Fourier coefficients $h_d(n)$ are the impulse response coefficients of the desired filter and given by

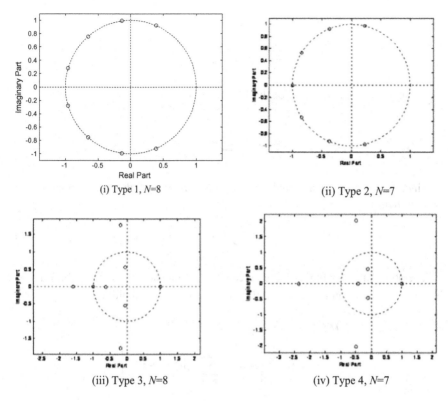

(i) Type 1, N=8   (ii) Type 2, N=7

(iii) Type 3, N=8   (iv) Type 4, N=7

Fig. 6.2 Zero locations of the four types of linear phase FIR filters for $N = 8$ or 7

$$h_d(n) = \frac{1}{2\pi} \int_{-\pi}^{\pi} H_d(e^{j\omega}) e^{j\omega n} d\omega \qquad (6.46)$$

The z-transform of the impulse response sequence is given by

$$H(z) = \sum_{n=-\infty}^{\infty} h_d(n) z^{-n} \qquad (6.47)$$

The transfer function $H(z)$ represents a digital filter of infinite duration. To get an FIR filter transfer function, the impulse response is truncated by multiplying it by a rectangular window defined as

$$w_R(n) = \begin{cases} 1 & \text{for } -\frac{N}{2} \le n \le \frac{N}{2} \\ 0 & \text{otherwise} \end{cases} \qquad (6.48)$$

$$h(n) = h_d(n)w_R(n)$$
$$= \begin{cases} h_d(n) & \text{for } -\frac{N}{2} \leq n \leq \frac{N}{2} \\ 0 & \text{otherwise} \end{cases} \tag{6.49}$$

Then, the transfer function of the FIR filter is

$$H(z) = \sum_{h=-\frac{N}{2}}^{\frac{N}{2}} h(n)z^{-n} \tag{6.50}$$

Since for a symmetrical impulse response, $h(-n) = h(n)$, the above equation can be rewritten as

$$H(z) = h(0) + \sum_{n=1}^{\frac{N}{2}} [h(n)z^{-n} + h(-n)z^{n}] \tag{6.51}$$

The above transfer function is non-causal (i.e., physically not realizable). It can be made causal by introducing a delay of $\frac{N}{2}$ samples, i.e., multiplying it by $z^{-N/2}$.

$$H'(z) = z^{-N/2}H(z)$$
$$= z^{-N/2}\left[h(0) + \sum_{n=1}^{\frac{N}{2}} h(n)[z^{n} + z^{-n}] \right] \tag{6.52}$$

6.3.1   Gibb's Oscillations

From Eq. (6.49), the coefficients of a causal FIR lowpass filter can be obtained by shifting the coefficients of the non-causal FIR lowpass filter to the right by $N/2$. Thus, the coefficients of causal FIR filter are given by

$$h_{\text{LP}}(n) = \frac{\sin \omega_c \left(n - \frac{N}{2}\right)}{\pi \left(n - \frac{N}{2}\right)} \quad \text{for } 0 \leq n \leq N \tag{6.53}$$
$$= 0 \qquad\qquad\qquad \text{otherwise}$$

A lowpass filter with a cutoff frequency $\omega_c = 0.5\pi$ is designed using Eq. (6.53). Its magnitude responses for two different values of filter lengths are shown in Fig. 6.3. Irrespective of the filter length, both of the magnitude responses exhibit an oscillatory behavior with the heights of the largest ripples remaining the same, approximately 11% of the difference between the passband and stopband magnitudes of the ideal filter [1]. These oscillations are more commonly referred to as Gibb's oscillations. Thus, the Gibb's phenomenon can be attributed to the fact that

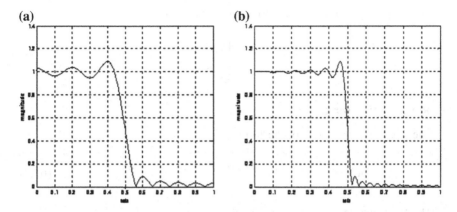

Fig. 6.3 Magnitude responses of lowpass filters designed using truncated impulse response, **a** filter length = 21 and **b** filter length = 51

the rectangular window used for truncation has an abrupt transition to zero outside the range for $-\frac{N}{2} \le n \le \frac{N}{2}$. Thus, the Gibb's phenomenon can be reduced by the use of a tapered window that decays toward zero gradually. The characteristics of a rectangular window and various tapered windows are discussed in the next section.

6.3.2 Fixed Window Functions

The various fixed window functions are given below.

1. **Rectangular window**:

   The rectangular window sequence is given by Eq. (6.48). The frequency response of the rectangular window is given by

$$W_R\left(e^{j\omega}\right) = \sum_{n=-N/2}^{N/2} e^{-j\omega n} = \frac{\sin\frac{(N+1)}{2}\omega}{\sin\frac{\omega}{2}} \qquad (6.54)$$

2. **Triangular or Bartlett window**:

   The N-point triangular window is given by

$$w_T(n) = \begin{cases} 1 - \frac{2|n|}{N} & \text{for } -\frac{N}{2} \le n \le \frac{N}{2} \\ 0 & \text{otherwise} \end{cases} \qquad (6.55)$$

The frequency response of the triangular window is

$$W_T\left(e^{j\omega}\right) = \left[\frac{\sin\left(\frac{N}{4}\right)\omega}{\sin\left(\frac{\omega}{2}\right)}\right]^2 \tag{6.56}$$

3. **Raised cosine window**:

The window sequence is of the form

$$w_\alpha(n) = \begin{cases} \alpha + (1-\alpha)\cos\left(\frac{2\pi n}{N}\right) & \text{for } -\frac{N}{2} \le n \le \frac{N}{2} \\ 0 & \text{otherwise} \end{cases} \tag{6.57}$$

The frequency response of $w_\alpha(n)$ is given by

$$\begin{aligned}
W_\alpha\left(e^{j\omega}\right) &= \sum_{n=-\left(\frac{N}{2}\right)}^{\frac{(N)}{2}} \left[\alpha + (1-\alpha)\cos\left(\frac{2\pi n}{N}\right)\right]e^{-j\omega n} \\
&= \alpha \frac{\sin\left(\frac{\omega(N+1)}{2}\right)}{\sin\left(\frac{\omega}{2}\right)} + \left(\frac{1-\alpha}{2}\right)\frac{\sin\left(\frac{\omega(N+1)}{2} - \frac{\pi(N+1)}{N}\right)}{\sin\left(\frac{\omega}{2} - \frac{\pi}{N}\right)} \\
&\quad + \left(\frac{1-\alpha}{2}\right)\frac{\sin\left(\frac{\omega(N+1)}{2} + \frac{\pi(N+1)}{N}\right)}{\sin\left(\frac{\omega}{2} + \frac{\pi}{N}\right)}
\end{aligned} \tag{6.58}$$

4. **Hanning window**:

The Hanning window sequence can be obtained by substituting $\alpha = 0.5$ in Eq. (6.58)

$$w_{Hn}(n) = \begin{cases} 0.5 + 0.5\cos\left(\frac{2\pi n}{N}\right) & \text{for } -\frac{N}{2} \le n \le \frac{N}{2} \\ 0 & \text{otherwise} \end{cases} \tag{6.59}$$

The frequency response of the Hanning window is

$$\begin{aligned}
W_{Hn}\left(e^{j\omega}\right) &= 0.5\frac{\sin\left(\frac{\omega(N+1)}{2}\right)}{\sin\left(\frac{\omega}{2}\right)} + 0.25\frac{\sin\left(\frac{\omega(N+1)}{2} - \frac{\pi(N+1)}{N}\right)}{\sin\left(\frac{\omega}{2} - \frac{\pi}{N}\right)} \\
&\quad + 0.25\frac{\sin\left(\frac{\omega(N+1)}{2} + \frac{\pi(N+1)}{N}\right)}{\sin\left(\frac{\omega}{2} + \frac{\pi}{N}\right)}
\end{aligned} \tag{6.60}$$

5. **Hamming window**:

The Hamming window sequence can be obtained by substituting $\alpha = 0.54$ in Eq. (6.58)

$$w_{\mathrm{Hm}}(n) = \begin{cases} 0.54 + 0.46\cos\left(\frac{2\pi n}{N}\right) & \text{for } -\frac{N}{2} \le n \le \frac{N}{2} \\ 0 & \text{otherwise} \end{cases} \tag{6.61}$$

The frequency response of the Hamming window is

$$W_{\mathrm{Hm}}\left(e^{j\omega}\right) = 0.54\frac{\sin\left(\frac{\omega(N+1)}{2}\right)}{\sin\left(\frac{\omega}{2}\right)} + 0.23\frac{\sin\left(\frac{\omega(N+1)}{2} - \frac{\pi(N+1)}{N}\right)}{\sin\left(\frac{\omega}{2} - \frac{\pi}{N}\right)}$$

$$+ 0.23\frac{\sin\left(\frac{\omega(N+1)}{2} + \frac{\pi(N+1)}{N}\right)}{\sin\left(\frac{\omega}{2} + \frac{\pi}{N}\right)} \tag{6.62}$$

6. **Blackman window**:

The window sequence is of the form

$$w_B(n) = \begin{cases} 0.42 + 0.5\cos\left(\frac{2\pi n}{N}\right) + 0.08\cos\left(\frac{4\pi n}{N}\right) & \text{for } -\frac{N}{2} \le n \le \frac{N}{2} \\ 0 & \text{otherwise} \end{cases} \tag{6.63}$$

The frequency response of the Blackman window is

$$W_B\left(e^{j\omega}\right) = 0.42\frac{\sin\left(\frac{\omega(N+1)}{2}\right)}{\sin\left(\frac{\omega}{2}\right)} + 0.25\frac{\sin\left(\frac{\omega(N+1)}{2} - \frac{\pi(N+1)}{N}\right)}{\sin\left(\frac{\omega}{2} - \frac{\pi}{N}\right)}$$

$$+ 0.25\frac{\sin\left(\frac{\omega(N+1)}{2} + \frac{\pi(N+1)}{N}\right)}{\sin\left(\frac{\omega}{2} + \frac{\pi}{N}\right)} + 0.04\frac{\sin\left(\frac{\omega(N+1)}{2} - \frac{2\pi(N+1)}{N}\right)}{\sin\left(\frac{\omega}{2} - \frac{2\pi}{N}\right)}$$

$$+ 0.04\frac{\sin\left(\frac{\omega(N+1)}{2} + \frac{2\pi(N+1)}{N}\right)}{\sin\left(\frac{\omega}{2} + \frac{2\pi}{N}\right)} \tag{6.64}$$

6.3.3 *Comparison of the Fixed Windows*

The desirable characteristics of a window are as follows:

1. The main lobe in the frequency response of the window should be narrow and contain most of the energy.
2. The maximum side lobe amplitude in the frequency response of the window should be small so as to have a small ripple ratio. The ripple ratio (RR) of a window is defined as the ratio of the maximum side lobe amplitude to the main lobe amplitude [2].
3. The side lobes of the frequency response should decrease rapidly as ω tends to ∞.

The magnitude responses of the various windows discussed above are shown in Figs. 6.4, 6.5, 6.6, 6.7, and 6.8 for order $N = 50$.

The properties of the different fixed windows are summarized in Table 6.2.

In FIR filter design, the performance of a window can be measured by its main lobe width and ripple ratio or relative side lobe level. The main lobe width is defined as the distance between the first zero crossings on both sides of $\omega = 0$, and

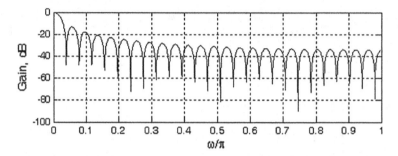

Fig. 6.4 Log magnitude response of the rectangular window

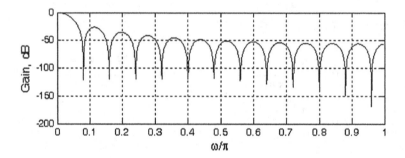

Fig. 6.5 Log magnitude response of the Bartlett window

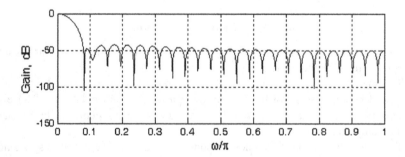

Fig. 6.6 Log magnitude response of the Hamming window

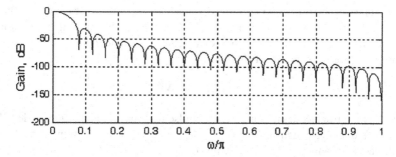

Fig. 6.7 Log magnitude response of the Hanning window

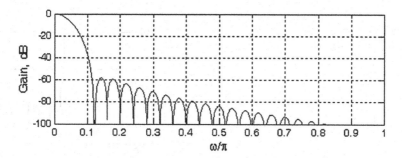

Fig. 6.8 Log magnitude response of the Blackman window

Table 6.2 Comparison of the different fixed windows for $N = 50$

window	Relative side lobe level (dB)	Ripple ratio (RR)	Approximate width of main lobe	Minimum Stopband attenuation in dB
Rectangular	−13	0.22387	$4\pi/(N+1)$	−21
Bartlett	−25	0.056234	$8\pi/(N+1)$	−25
Hamming	−41	0.0089124	$8\pi/(N+1)$	−53
Hanning	−31	0.028184	$8\pi/(N+1)$	−44
Blackman	−57	0.0014126	$12\pi/(N+1)$	−74

the relative side lobe level is the difference in dB between the maximum side lobe amplitude and the main lobe amplitude. For a given filter length, the rectangular window yields the sharpest transitions due to its narrowest main lobe. However, the first side lobe is only about 13 dB below the main peak, resulting in Gibb's oscillations. For the Hamming, Hanning, and Blackman windows, the side lobes are greatly reduced in amplitude and with wider main lobes. As a trade-off between the main lobe width and relative side lobe level, the Hamming window is the best choice.

The following MATLAB Program 6.2 illustrates the effect of each of the above fixed windows on the gain response of an FIR lowpass filter of length 51. In this illustration, the following MATLAB command is used to obtain truncated and windowed impulse response of the filter.

$$b = \mathrm{fir1}(N, Wn, WIN);$$

where b is the truncated and windowed impulse response, N is the filter order, Wn is cutoff frequency, which must be between $0 < Wn < 1.0$, and Win is the N + 1 length vector to window the impulse response.

The gain responses of the designed lowpass filter with N = 50, Wn = 0.5 for the rectangular, Bartlett, Hamming, Hanning, and Blackman windows are shown in Figs. 6.9, 6.10, 6.11, 6.12, and 6.13, respectively.

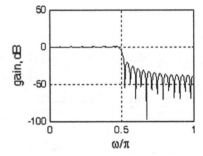

Fig. 6.9 Gain response of the LPF using the rectangular window

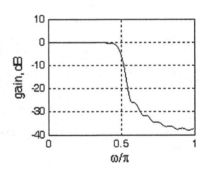

Fig. 6.10 Gain response of the LPF using the Bartlett window

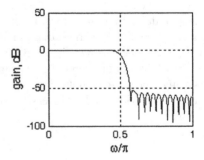

Fig. 6.11 Gain response of the LPF using the Hamming window

Fig. 6.12 Gain response of the LPF using the Hanning window

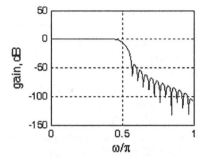

Fig. 6.13 Gain response of the LPF using the Blackman window

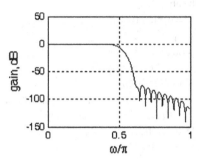

Program 6.2 Gain response of the lowpass filter using various windows

```
clear;clc;
N=50;% filter order
Gain Response of Low pass filter using rectangular window
b =fir1(N,.5,'low',rectwin(N+1))
[Hz,w]=freqz(b,1,512);
h=abs(Hz);
M=20*log10(h);
figure(1)
subplot(2,2,1),plot(w/pi,M,'-');grid;
xlabel('\omega/\pi');
ylabel('gain,dB');
% Gain Response of Low pass filter using Bartlett window
b = fir1(N,.5,'low',bartlett(N+1))
[Hz,w]=freqz(b,1,512);
h=abs(Hz);
M=20*log10(h);
subplot(2, 2, 2),
plot(w/pi,M,'-');grid;
xlabel('\omega/\pi');
ylabel('gain,dB');
```

```
% Gain Response of Low pass filter using Hamming window
b = fir1(N,.5,'low',hamming(N+1))
[Hz,w]=freqz(b,1,512);
h=abs(Hz);
M=20*log10(h);
subplot(2, 2, 3),plot(w/pi,M,'-');grid;
xlabel('\omega/\pi');
ylabel('gain,dB');
% Gain Response of Low pass filter using Hanning window
b = fir1(N,.5,'low',hann(N+1))
[Hz,w]=freqz(b,1,512);
h=abs(Hz);
M=20*log10(h);
subplot(2, 2, 4),
plot(w/pi,M,'-');grid;
xlabel('\omega/\pi');
ylabel('gain,dB');
% Gain Response of Low pass filter using Blackman window
b = fir1(N,.5,'low',blackman(N+1));
[Hz,w]=freqz(b,1,512);
h=abs(Hz);
M=20*log10(h);
figure(2), subplot(2, 2, 1), plot(w/pi,M,'-');grid;
xlabel('\omega/\pi');ylabel('gain,dB');
```

6.3.4 Design of FIR Filters Using Fixed Windows

The various steps involved in the design of FIR filters using fixed windows are as follows:

Step 1: Truncation to obtain impulse response of finite duration
Step 2: Windowing to reduce the effect of Gibb's oscillations
Step 3: Introducing a suitable delay to obtain a realizable transfer function for the filter.

Example 6.1 The desired impulse response of a certain FIR lowpass filter is given by

$$H(f) = 1 \quad \text{for} \ 0 \le f \le 1 \ \text{kHz}$$
$$= 0 \quad \text{for} \ f > 1 \ \text{kHz}$$

Let the sampling rate be $F_T = 10$ kHz. Impulse response is of 1 ms duration. Use Hamming window and compute the impulse response of the FIR filter.

Solution Cutoff frequency $f_c = 1$ kHz. Hence,

$$\omega_c = \frac{2\pi f_c}{F_T} = \frac{2\pi \times 1 \times 10^3}{1 \times 10^4} = 0.2\pi$$

Since the sampling time period is 0.1 ms, the length of the impulse response is 11 (order $N = 10$).

Step 1: The impulse response of an FIR lowpass filter of length 11 is obtained by truncating Eq. (6.2) as

$$h(n) = \frac{\sin(0.2\pi n)}{n\pi} \quad \text{for} \quad -5 \le n \le 5$$

The filter coefficients are

$$h_T(0) = 0.2$$

$$h(1) = h(-1) = \frac{\sin(0.2\pi)}{\pi} = \frac{0.5878}{\pi} = 0.1871$$

$$h(2) = h(-2) = \frac{\sin(0.2\pi \times 2)}{2\pi} = \frac{0.9511}{2\pi} = 0.1514$$

$$h(3) = h(-3) = \frac{\sin(0.2\pi \times 3)}{3\pi} = \frac{0.9511}{3\pi} = 0.1009$$

$$h(4) = h(-4) = \frac{\sin(0.2\pi \times 4)}{4\pi} = \frac{0.5878}{4\pi} = 0.0468$$

$$h(5) = h(-5) = \frac{\sin(0.2\pi \times 5)}{5\pi} = \frac{0}{5\pi} = 0$$

Step 2: The Hamming window sequence for $N = 10$ is given by

$$w_H(n) = 0.54 + 0.46 \cos\left(\frac{\pi n}{5}\right) \quad \text{for} \quad -5 \le n \le 5$$

$$= 0 \quad \text{otherwise}$$

Hence,

$$w_H(0) = 1$$
$$w_H(-1) = w_H(1) = 0.9121$$
$$w_H(-2) = w_H(2) = 0.6821$$
$$w_H(-3) = w_H(3) = 0.3979$$
$$w_H(-4) = w_H(4) = 0.1679$$
$$w_H(-5) = w_H(5) = 0.0800$$

The filter coefficients using Hamming window are

$$h_t(n) = h(n)w_H(n) \quad \text{for } -5 \leq n \leq 5$$
$$= 0 \qquad\qquad \text{otherwise}$$

Thus,

$$h_t(0) = h(0)w_H(0) = 0.20000$$
$$h_t(1) = h_t(-1) = 0.1707$$
$$h_t(2) = h_t(-2) = 0.1033$$
$$h_t(3) = h_t(-3) = 0.0401$$
$$h_t(4) = h_t(-4) = 0.0079$$
$$h_t(5) = h_t(-5) = 0$$

The impulse responses $h(n)$ and $h_t(n)$ are shown in Table 6.3.

Step 3: The transfer function of the filter is

$$H(z) = h_t(0) + \sum_{n=1}^{5} h_t(n)(z^n + z^{-n})$$

Delaying the above non-causal transfer function by $N/2$, the realizable transfer function of the filter is obtained as

$$H(z) = z^{-5}\left[h_t(0) + \sum_{n=1}^{5} h_t(n)(z^n + z^{-n})\right]$$

Example 6.2 Design an FIR bandpass filter of length 9 for the following ideal characteristics

$$H(e^{jw}) = 0 \quad \text{for } 0 \leq |\omega| \leq 0.4\pi$$
$$= 1 \quad \text{for } 0.4\pi \leq |\omega| \leq 0.6\pi$$
$$= 0 \quad \text{for } 0.6\pi \leq |\omega| \leq \pi$$

Use Hamming window.

Table 6.3 Impulse responses $h(n)$ and $h_t(n)$

n	$h(n)$	$w_H(n)$	$h_t(n)$
0	0.2	1.0000	0.2
±1	0.1871	0.9121	0.1707
±2	0.1514	0.6821	0.1033
±3	0.1009	0.3979	0.0401
±4	0.0468	0.1679	0.0079
±5	0	0.0800	0

Solution The lower and upper cutoff frequencies are 0.4π and 0.6π, respectively. The following stepwise procedure is used in the design.

Step 1: The impulse response of an FIR bandpass filter of length 9 is obtained by truncating Eq. (6.6) as

$$h(n) = \frac{\sin(0.6\pi n)}{n\pi} - \frac{\sin(0.4\pi n)}{n\pi} \quad \text{for } -4 \le n \le 4$$

The filter coefficients are

$$h(0) = 0.2$$

$$h(1) = h(-1) = \frac{\sin(0.6\pi)}{\pi} - \frac{\sin(0.4\pi)}{\pi} = 0$$

$$h(2) = h(-2) = \frac{\sin(0.6\pi \times 2)}{2\pi} - \frac{\sin(0.4\pi \times 2)}{2\pi} = -0.1871$$

$$h(3) = h(-3) = \frac{\sin(0.6\pi \times 3)}{3\pi} - \frac{\sin(0.4\pi \times 3)}{3\pi} = 0$$

$$h(4) = h(-4) = \frac{\sin(0.6\pi \times 4)}{4\pi} - \frac{\sin(0.4\pi \times 4)}{4\pi} = 0.1514$$

Step 2: The Hamming window sequence for $N = 8$ is given by

$$w_H(n) = 0.54 + 0.46 \cos\left(\frac{\pi n}{4}\right) \quad \text{for } -4 \le n \le 4$$
$$= 0 \qquad\qquad\qquad\qquad \text{otherwise}$$

Hence,

$$w_H(0) = 1$$
$$w_H(-1) = w_H(1) = 0.8653$$
$$w_H(-2) = w_H(2) = 0.5400$$
$$w_H(-3) = w_H(3) = 0.2147$$
$$w_H(-4) = w_H(4) = 0.0800$$

The filter coefficients using Hamming window are

$$h_t(n) = h(n)w_H(n) \quad \text{for } -4 \le n \le 4$$
$$= 0 \qquad\qquad\qquad \text{otherwise}$$

Thus,

$$h_t(0) = h(0)w_H(0) = 0.2000$$
$$h_t(1) = h_t(-1) = 0$$
$$h_t(2) = h_t(-2) = -0.1010$$
$$h_t(3) = h_t(-3) = 0$$
$$h_t(4) = h_t(-4) = 0.0121$$

The impulse responses $h(n)$ and $h_t(n)$ are shown in Table 6.4.

Step 3: The transfer function of the filter is

$$H(z) = h_t(0) + \sum_{n=1}^{4} h_t(n)(z^n + z^{-n})$$

Delaying the above transfer function by N/2, the realizable transfer function of the filter is

$$H(z) = z^{-4}\left[h_t(0) + \sum_{n=1}^{4} h_t(n)(z^n + z^{-n}) \right]$$

Example 6.3 Design a linear phase FIR lowpass filter of length 11 to meet the following characteristics cutoff frequency that is 100 Hz. Use Hamming window. Assume a suitable sampling frequency.

Solution Assume the sampling frequency to be 400 Hz.

Cutoff frequency $f_c = 100$ Hz,

$$\omega_c = \frac{2\pi f_c}{F_T} = \frac{2\pi \times 100}{400} = 0.5\pi$$

Step 1: The impulse response of an FIR lowpass filter of length 11 is obtained by truncating Eq. (6.2) as

$$h(n) = \frac{\sin(0.5\pi n)}{n\pi} \quad \text{for } -5 \leq n \leq 5$$

Table 6.4 Impulse responses $h(n)$ and $h_t(n)$

n	$h(n)$	$w_H(n)$	$h_t(n)$
0	0.2	1.0000	0.2
±1	0	0.8653	0
±2	−0.1871	0.5400	−0.1010
±3	0	0.2147	0
±4	0.1514	0.0800	0.0121

The filter coefficients are

$$h(0) = 0.5$$

$$h(1) = h(-1) = \frac{\sin{(0.5\pi)}}{\pi} = \frac{1}{\pi} = 0.3183$$

$$h(2) = h(-2) = \frac{\sin{(0.5\pi \times 2)}}{2\pi} = \frac{0}{2\pi} = 0$$

$$h(3) = h(-3) = \frac{\sin{(0.5\pi \times 3)}}{3\pi} = \frac{-1}{3\pi} = -0.1061$$

$$h(4) = h(-4) = \frac{\sin{(0.5\pi \times 4)}}{4\pi} = \frac{0}{4\pi} = 0$$

$$h(5) = h(-5) = \frac{\sin{(0.5\pi \times 5)}}{5\pi} = \frac{1}{5\pi} = 0.0637$$

Step 2: The Hamming window sequence for $N = 10$ is given by

$$w_H(n) = 0.54 + 0.46 \cos\left(\frac{\pi n}{5}\right) \quad \text{for } -5 \le n \le 5$$
$$= 0 \qquad\qquad\qquad\qquad \text{otherwise}$$

Hence,

$$w_H(0) = 1$$
$$w_H(-1) = w_H(1) = 0.9121$$
$$w_H(-2) = w_H(2) = 0.6821$$
$$w_H(-3) = w_H(3) = 0.3979$$
$$w_H(-4) = w_H(4) = 0.1679$$
$$w_H(-5) = w_H(5) = 0.0800$$

The filter coefficients using Hamming window are

$$h_t(n) = h(n)w_H(n) \quad \text{for } -5 \le n \le 5$$
$$= 0 \qquad\qquad\qquad \text{otherwise}$$

Thus,

$$h_t(0) = h(0)w_H(0) = 0.5000$$
$$h_t(1) = h_t(-1) = 0.2903$$
$$h_t(2) = h_t(-2) = 0$$
$$h_t(3) = h_t(-3) = -0.0422$$
$$h_t(4) = h_t(-4) = 0$$
$$h_t(5) = h_t(-5) = 0.0051$$

The impulse responses $h(n)$ and $h_t(n)$ are shown in Table 6.5.

Table 6.5 Impulse responses $h(n)$ and $h_t(n)$

n	$h(n)$	$w_H(n)$	$h_t(n)$
0	0.5	1.0000	0.5000
± 1	0.3183	0.9121	0.2903
± 2	0	0.6821	0
± 3	-0.1061	0.3979	-0.0422
± 4	0	0.1679	0
± 5	0.0637	0.0800	0.0051

Step 3: The transfer function of the filter is

$$H(z) = h_t(0) + \sum_{n=1}^{5} h_t(n)(z^n + z^{-n})$$

Delaying the above transfer function by $-N/2$, the realizable transfer function of the filter is

$$H(z) = z^{-5}\left[h_t(0) + \sum_{n=1}^{5} h_t(n)(z^n + z^{-n})\right]$$

6.3.5 Kaiser Window

As shown in Table 6.2, a trade-off has to be made between the main lobe width and the ripple ratio, since the ripple ratio decreases from window to window with increasing main lobe width. The main lobe width is inversely proportional to the filter order N. However, for a chosen window, the ripple ratio is approximately constant irrespective of the order N. To achieve the specified passband ripple and stopband attenuation, a designer has to select a window with an appropriate ripple ratio and then to choose N to obtain the specified transition width. In this design process, the designer has to settle for a window with low ripple ratio which results in a high main lobe width. Subsequently, to achieve the specified transition width, the filter order is to be increased to a high value unnecessarily. This problem can be overcome by using the Kaiser window, given by [3];

$$w_K(n) = \frac{I_0\left\{\beta\sqrt{1-(n/M)^2}\right\}}{I_0(\beta)}, \quad -M \le n \le M \qquad (6.65)$$

where $N = 2M$ is the order of the filter; β is an adjustable control parameter; and $I_0(x)$ is the modified zeroth-order Bessel function of the first kind given by

$$I_0(x) = 1 + \sum_{r=1}^{\infty} \left[\frac{(x/2)^r}{r!} \right]^2$$

$$= 1 + \frac{(0.25x^2)}{(1!)^2} + \frac{(0.25x^2)^2}{(2!)^2} + \frac{(0.25x^2)^3}{(3!)^2} + \cdots$$

(6.66)

which is positive for all real values of x. For most practical purposes, the summation up to the first 20 terms of Eq. (6.66) is sufficient to get a reasonably accurate value of $I_0(x)$.

The frequency response of the Kaiser window is given by

$$W_K(e^{j\omega}) = \frac{2}{I_0(\beta)} \frac{\sin\left[(N/2) \left\{ \omega^2 - (2\beta/N)^2 \right\}^{1/2} \right]}{\left\{ \omega^2 - (2\beta/N)^2 \right\}^{1/2}}$$

(6.67)

The minimum stopband attenuation α_s of the windowed filter response is controlled by the parameter β. For given α_s and normalized transition bandwidth $\Delta\omega$, the parameter and the filter order $N = 2 M$ can be computed by using the following empirical relations developed by Kaiser [3].

$$\beta = \begin{cases} 0.1102(\alpha_s - 8.7) & \text{for } \alpha_s > 50, \\ 0.5842(\alpha_s - 21)^{0.4} + 0.07886(\alpha_s - 21) & \text{for } 21 \le \alpha_s \le 50, \\ 0 & \text{for } \alpha_s < 21. \end{cases}$$

(6.68)

The filter order N is to be selected using the formula

$$N = \begin{cases} \dfrac{\alpha_s - 8}{2.285(\Delta\omega)} & \text{for } \alpha_s > 21 \\ \dfrac{5.797}{(\Delta\omega)} & \text{for } \alpha_s \le 21 \end{cases}$$

(6.69)

where $\Delta\omega = \omega_s - \omega_p$, ω_p and ω_s being the normalized angular passband and stopband edge frequencies, respectively, of the lowpass filter. From the above empirical relations, it should be noted that the Kaiser window has no independent control over the passband ripple δ_p. However, in practice, δ_p is approximately equal to δ_s.

6.3.6   Design Procedure for Linear Phase FIR Filter Using Kaiser Window

Step 1: Determine $h(n)$ for an ideal frequency response of the filter to be designed.

Step 2: Calculate stopband attenuation α_s in dB if the peak ripple value of the stopband is given in the specification instead of α_s.

Step 3: Determine the value of parameter β using Eq. (6.68).
Step 4: Determine the filter order using the formula given in Eq. (6.69), and choose the next higher even integer value for N.
Step 5: Compute the window sequence using Eq. (6.65).
Step 6: Determine the

$$h_t(n) = w_k(n)h(n) \tag{6.70}$$

Step 7: Formulate the realizable transfer function for the designed filter using $h_t(n)$

$$H_t(z) = z^{-M}\left[h_t(0) + \sum_{n=1}^{M} h_t(n)(z^n + z^{-n})\right] \tag{6.71}$$

The above procedure can be applied to the highpass, bandpass, and bandstop filters with the following specifications:

Highpass filter

$$\Delta\omega = \omega_p - \omega_s$$
$$H(e^{j\omega}) = 0 \quad \text{for } |\omega| < \omega_c \tag{6.72}$$
$$= 1 \quad \text{for } \omega_c \le |\omega| \le \frac{\omega_T}{2}$$

where

$$\omega_c = \frac{1}{2}[\omega_s + \omega_p]$$

Bandpass filter

$$\Delta\omega = \min[(\omega_{p1} - \omega_{s1}), (\omega_{s2} - \omega_{p2})]$$
$$H(e^{j\omega}) = 0 \quad \text{for } 0 \le |\omega| < \omega_{c1} \text{ and } \omega_{c2} \le |\omega| < \frac{\omega_T}{2} \tag{6.73}$$
$$= 1 \quad \text{for } \omega_{c1} \le |\omega| \le \omega_{c2}$$

where

$$\omega_{c1} = \omega_{p1} - \frac{\Delta\omega}{2}, \quad \omega_{c2} = \omega_{p1} + \frac{\Delta\omega}{2}$$

Bandstop filter

$$\Delta\omega = \min\left[\left(\omega_{s1} - \omega_{p1}\right), \left(\omega_{p2} - \omega_{s2}\right)\right]$$

$$H(e^{j\omega}) = 1 \quad \text{for } 0 \leq |\omega| < \omega_{c1} \text{ and } \omega_{c2} \leq |\omega| < \frac{\omega_T}{2} \tag{6.74}$$

$$= 0 \quad \text{for } \omega_{c1} \leq |\omega| \leq \omega_{c2}$$

where

$$\omega_{c1} = \omega_{p1} + \frac{\Delta\omega}{2}, \ \ \omega_{c2} = \omega_{p1} - \frac{\Delta\omega}{2}$$

Example 6.4 Design an FIR lowpass filter using Kaiser window with the following specifications:

Passband edge $\omega_p = 0.4\pi$, stopband edge $\omega_s = 0.6\pi$, and stopband attenuation ≥ 44 dB.

Solution

$$\text{cutoff frequency}(\omega_c) = \frac{\left(\omega_p + \omega_s\right)}{2} = \frac{(0.4\pi + 0.6\pi)}{2} = \frac{\pi}{2}\text{rad}$$

$$\text{Transition width } \Delta\omega = (\omega_s - \omega_p) = 0.6\pi - 0.4\pi = 0.2\pi$$

Step 1: Frequency response of the lowpass filter

$$h_{\text{LP}}(n) = \frac{\sin\left(\frac{\pi}{2}\right)n}{\pi n}, \quad -\infty \leq n \leq \infty$$

Step 2: From the given specifications, $\alpha_s = 44$ dB.

Step 3: From Eq. (6.68)

$$\beta = 0.5842(\alpha_s - 21)^{.4} + 0.07886(\alpha_s - 21)$$

$$= 0.5842(44 - 21)^{.4} + 0.07886(44 - 21) = 3.8614156$$

Step 4: The filter order

$$N = \frac{\alpha_s - 8}{2.285(\Delta\omega)} = \frac{44 - 8}{2.285(0.2\pi)} = \frac{36}{1.4357} = 25.075$$

We take the next higher even integer value of N, $N = 26$. Since $N = 2M$, $M = 13$.

Step 5: The window sequence

$$w_k(n) = \frac{I_0\left\{\beta\sqrt{1 - (n/M)^2}\right\}}{I_0(\beta)}, \quad -M \leq n \leq M$$

$$I_0(x) = 1 + \frac{(0.25x^2)}{(1!)^2} + \frac{(0.25x^2)^2}{(2!)^2} + \frac{(0.25x^2)^3}{(3!)^2} + \cdots$$

Substituting the value of β calculated in Step 3 and $M = 13$, $\omega(n)$ becomes

$$w_k(n) = \frac{I_0\left\{3.8614156\sqrt{1 - (n/13)^2}\right\}}{I_0(3.8614156)}, \quad -13 \leq n \leq 13$$

and

$$I_0(3.8614156) = 10.031.$$

Hence,

$$w_k(0) = \frac{I_0(\beta)}{I_0(\beta)} = 1$$

$$w_k(1) = w_k(-1) = \frac{I_0(3.84997)}{10.031} = \frac{9.93305}{10.031} = 0.99023$$

$$w_k(2) = w_k(-2) = \frac{I_0(3.8154)}{10.031} = \frac{9.643498}{10.031} = 0.961369$$

$$w_k(3) = w_k(-3) = \frac{I_0(3.7571)}{10.031} = \frac{9.175069}{10.031} = 0.91467$$

$$w_k(4) = w_k(-4) = \frac{I_0(3.674)}{10.031} = \frac{8.548}{10.031} = 0.85217$$

$$w_k(5) = w_k(-5) = \frac{I_0(3.56438)}{10.031} = \frac{7.7896}{10.031} = .77655$$

$$w_k(6) = w_k(-6) = \frac{I_0(3.4255)}{10.031} = \frac{6.9313}{10.031} = 0.690988$$

$$w_k(7) = w_k(-7) = \frac{I_0(3.2538)}{10.031} = \frac{6.008298}{10.031} = 0.598973$$

$$w_k(8) = w_k(-8) = \frac{I_0(3.04367)}{10.031} = \frac{5.05688}{10.031} = 0.504125$$

$$w_k(9) = w_k(-9) = \frac{I_0(2.7864)}{10.031} = \frac{4.1127}{10.031} = 0.409999$$

$$w_k(10) = w_k(-10) = \frac{I_0(2.4673)}{10.031} = \frac{3.20883}{10.031} = 0.31989$$

$$w_k(11) = w_k(-11) = \frac{I_0(2.057897)}{10.031} = \frac{2.3742}{10.031} = 0.236687$$

$$w_k(12) = w_k(-12) = \frac{I_0(1.4851)}{10.031} = \frac{1.6322}{10.031} = 0.16272$$

$$w_k(13) = w_k(-13) = \frac{I_0(0)}{10.031} = \frac{1}{10.031} = 0.09969$$

Step 6: Compute the truncated impulse response using

$$h_t(n) = h(n)w_k(n)$$

The impulse responses $h_t(n)$ and $h(n)$ are given in Table 6.6.
Step 7: The transfer function is given by

$$H_t(z) = z^{-13}\left[h_t(0) + \sum_{n=1}^{13} h_t(n)(z^n + z^{-n}) \right]$$

where the values of $h_t(n)$ are given in Table 6.6.

Table 6.6 Impulse responses $h(n)$ and $h_t(n)$

n	$h(n)$	$h_t(n) = h(n)w_k(n)$
0	0.5	0.5
±1	3.183	3.152
±2	0	0
±3	−1.061	−.097049
±4	0	0
±5	0.06366	.0494369
±6	0	0
±7	−4.547	−.027237
±8	0	0
±9	3.53677	.0145
±10	0	0
±11	−2.8937	−.006849
±12	0	0
±13	2.448537	.00244

Example 6.5 Design a lowpass FIR linear phase filter using the Kaiser window method such that the stopband ripple and passband ripple are 0.00056 and the transition width is 0.09 π. If input to the structure is a speech signal sampled at 44.1 kHz, will you be able to implement the filter on a DSP chip that does 20MIPS or 20 instructions per μs? One instruction includes one multiplication and one addition.

Solution $\alpha_s = -20\log_{10}(0.00056) = 65$

$N = \frac{65-8}{(2.285)(0.09\pi)} = 88.2259;$ $T = \text{sampling}$ time $\text{period} = 1/(44.1)(10^3) =$ 22.6 μs.

The next higher even integer value 90 is chosen as N. In one sampling time period, the DSP does $22.6 \times 20 = 452$ instructions. The filter requires $N/2$ multiplications and $(N-1)$ additions for implementation. Thus, the filter can be implemented on the DSP chip.

Example 6.6 Design an FIR highpass filter using Kaiser window with the following specifications:

Passband edge $\omega_p = 20$ rad/s, stopband edge $\omega_s = 15$ rad/s, sampling frequency 100 rad/s, and stopband ripple = 0.02.

Solution Sampling frequency $\omega_T = 100$ rad/s

$$2\pi F_T = 100, \quad F_T = \frac{100}{2\pi}$$

Sampling period $(T) = \frac{2\pi}{100}$

Passband edge frequency in radians $(\omega_p) = 20 \times T = 20 \times \frac{2\pi}{100} = 0.4\,\pi$ rad

Stopband edge frequency in radians $(\omega_s) = 15 \times T = 15 \times \frac{2\pi}{100} = 0.3\,\pi$ rad

Cutoff frequency $(\omega_c) = \frac{(\omega_p + \omega_s)}{2} = \frac{(0.4\pi + 0.3\pi)}{2} = 0.35\pi$ rad

Transition width $\Delta\omega = (\omega_p - \omega_s) = 0.4\pi - 0.3\pi = 0.1\pi$

Step 1: Frequency response of the highpass filter

$$H_{\text{HP}}(n) = -\frac{\sin(0.35\pi)n}{\pi n}, \quad -\infty \le n \le \infty$$

Step 2: The stopband attenuation,
If the stopband ripple (δ_s) is 0.02, then

$$\alpha_s = -20\log_{10}(0.02) = 33.9794 \text{ dB}$$

Step 3: From Eq. (6.68),

$$\beta = 0.5842(\alpha_s - 21)^{0.4} + 0.07886(\alpha_s - 21)$$
$$= 0.5842(33.9794 - 21)^{0.4} + 0.07886(33.9794 - 21)$$
$$= 2.652339$$

Step 4: The filter order

$$N = \frac{\alpha_s - 8}{2.285(\Delta\omega)} = \frac{33.9794 - 8}{2.28(.1\pi)} = \frac{25.9794}{0.71785} = 36.19$$

We take the next higher even integer value 38 as the order of the filter. Now, the filter is designed as Type 1 highpass filter.

$$N = 2M, \ M = 19.$$

Step 5: The window sequence

$$w_k(n) = \frac{I_0\left\{\beta\sqrt{1 - (n/M)^2}\right\}}{I_0(\beta)}, \quad -M \le n \le M$$

$$I_0(x) = 1 + \frac{(0.25x^2)}{(1!)^2} + \frac{(0.25x^2)^2}{(2!)^2} + \frac{(0.25x^2)^3}{(3!)^2} + \cdots$$

Substituting the value of β calculated in Step 3 and $M = 19$, $\omega(n)$ becomes

$$w_k(n) = \frac{I_0\left\{2.652339\sqrt{1 - (n/19)^2}\right\}}{I_0(2.652339)}, \quad -19 \le n \le 19$$

Also,

$$I_0(2.652339) = 3.70095$$

Hence,

$$w_k(0) = \frac{I_0(\beta)}{I_0(\beta)} = 1$$

$$w_k(1) = w_k(-1) = \frac{I_0(2.64866)}{3.70095} = \frac{3.69}{3.70095} = 0.9971356$$

$$w_k(2) = w_k(-2) = \frac{I_0(2.6376)}{3.70095} = \frac{3.65866}{3.70095} = 0.98857$$

$$w_k(3) = w_k(-3) = \frac{I_0(2.619)}{3.70095} = \frac{3.6063}{3.70095} = 0.9744158$$

$$w_k(4) = w_k(-4) = \frac{I_0(2.593)}{3.70095} = \frac{3.5338}{3.70095} = 0.95482389$$

$$w_k(5) = w_k(-5) = \frac{I_0(2.55885)}{3.70095} = \frac{3.441974}{3.70095} = 0.93$$

$$w_k(6) = w_k(-6) = \frac{I_0(2.51662)}{3.70095} = \frac{3.331976}{3.70095} = 0.90$$

$$w_k(7) = w_k(-7) = \frac{I_0(2.46577)}{3.70095} = \frac{3.205}{3.70095} = 0.86599885$$

$$w_k(8) = w_k(-8) = \frac{I_0(2.4058)}{3.70095} = \frac{3.06255}{3.70095} = 0.8275$$

$$w_k(9) = w_k(-9) = \frac{I_0(2.3359)}{3.70095} = \frac{2.9061675}{3.70095} = 0.785249$$

$$w_k(10) = w_k(-10) = \frac{I_0(2.2553)}{3.70095} = \frac{2.7376}{3.70095} = 0.7397$$

$$w_k(11) = w_k(-11) = \frac{I_0(2.1626)}{3.70095} = \frac{2.5588}{3.70095} = 0.6913948$$

$$w_k(12) = w_k(-12) = \frac{I_0(2.056)}{3.70095} = \frac{2.3716775}{3.70095} = 0.640829$$

$$w_k(13) = w_k(-13) = \frac{I_0(1.9343)}{3.70095} = \frac{2.17824}{3.70095} = 0.588563$$

$$w_k(14) = w_k(-14) = \frac{I_0(1.7932)}{3.70095} = \frac{1.98057}{3.70095} = 0.53515$$

$$w_k(15) = w_k(-15) = \frac{I_0(1.628)}{3.70095} = \frac{1.780}{3.70095} = 0.481157$$

$$w_k(16) = w_k(-16) = \frac{I_0(1.43)}{3.70095} = \frac{1.5807987}{3.70095} = 0.42713$$

$$w_k(17) = w_k(-17) = \frac{I_0(1.1845)}{3.70095} = \frac{1.382756}{3.70095} = 0.3736$$

$$w_k(18) = w_k(-18) = \frac{I_0(0.849)}{3.70095} = \frac{1.1885}{3.70095} = 0.32114589$$

$$w_k(19) = w_k(-19) = \frac{I_0(0)}{3.70095} = \frac{1.0000}{3.70095} = 0.2702$$

Step 6: Compute the truncated impulse response using

$$h_t(n) = h(n)w_k(n)$$

The impulse responses $h_t(n)$ and $h(n)$ are given in Table 6.7.
Step 7: The transfer function is given by

$$H_t(z) = z^{-19}\left[h_t(0) + \sum_{n=1}^{19} h_t(n)(z^n + z^{-n})\right]$$

where the values of $h_t(n)$ are given in Table 6.7.

Table 6.7 Impulse responses $h(n)$ and $h_t(n)$

n	$h(n)$	$h_t(n) = h(n)w_k(n)$
0	0.65	0.65
±1	−0.283616	−0.2828
±2	−0.128759	−0.12728789
±3	0.016598	0.01617356
±4	0.07568	0.0722636
±5	0.045	0.0418658
±6	−0.016393	−0.014759
±7	−0.0449	−0.03889
±8	−0.023387	−0.0193529966
±9	0.0160566	0.01260845
±10	0.03183	0.0235457
±11	0.013137	0.009083
±12	−0.01559	−0.00999148
±13	−0.0241839	−0.01423375
±14	−0.007025	−0.0037599
±15	0.015	0.007219895
±16	0.01892	0.0080816
±17	0.002929	0.00109437
±18	−0.0143	−0.00459449
±19	−0.014927	−0.00403333

Example 6.7 Design an FIR bandpass filter using Kaiser window with the following specifications:

Passband: 20–30 kHz, lower stopband edge: 10 kHz, upper stopband edge: 40 kHz, sampling frequency: 100 kHz, passband ripple value: 0.5 dB, and stopband attenuation: 30 dB.

Solution The passband edge frequencies are

$$\omega_{p1} = \frac{2\pi f_{p1}}{F_T} = \frac{2\pi \times 20 \times 10^3}{100 \times 10^3} = \frac{40\pi}{100} = 0.4\pi$$

$$\omega_{p2} = \frac{2\pi f_{p2}}{F_T} = \frac{2\pi \times 30 \times 10^3}{100 \times 10^3} = \frac{60\pi}{100} = 0.6\pi$$

The stopband edge frequencies are

$$\omega_{s1} = \frac{2\pi f_{s1}}{F_T} = \frac{2\pi \times 10 \times 10^3}{100 \times 10^3} = \frac{20\pi}{100} = 0.2\pi$$

$$\omega_{s2} = \frac{2\pi f_{s2}}{F_T} = \frac{2\pi \times 40 \times 10^3}{100 \times 10^3} = \frac{80\pi}{100} = 0.8\pi$$

$$\Delta\omega = \min\left[(\omega_{p1} - \omega_{s1}), (\omega_{s2} - \omega_{p2})\right]$$

$$\Delta\omega = [0.2\pi, 0.2\pi)] = 0.2\pi$$

Cutoff frequencies $\omega_{c1} = \omega_{p1} - \dfrac{\Delta\omega}{2} = 0.4\pi - \dfrac{0.2\pi}{2} = 0.3\pi$

$$\omega_{c2} = \omega_{p2} + \frac{\Delta\omega}{2} = 0.6\pi + \frac{0.2\pi}{2} = 0.7\pi$$

Step 1: The frequency response of the bandpass filter is

$$h_{\mathrm{BP}}(n) = \frac{\sin\omega_{c2}n}{\pi n} - \frac{\sin\omega_{c1}n}{\pi n}, \quad -\infty \leq n \leq \infty$$

$$h_{\mathrm{BP}}(n) = \frac{\sin(0.7\pi)n}{\pi n} - \frac{\sin(0.3\pi)n}{\pi n}, \quad -\infty \leq n \leq \infty$$

Step 2: From the given specifications, $\alpha_s = 30$ dB.

Step 3: From Eq. (6.68),

$$\beta = 0.5842(\alpha_s - 21)^{0.4} + 0.07886(\alpha_s - 21)$$

$$= 0.5842(30 - 21)^{0.4} + 0.07886(30 - 21) = 2.116624$$

Step 4: The filter order is

$$N = \frac{\alpha_s - 8}{2.285(\Delta\omega)} = \frac{30 - 8}{2.285(0.2\pi)} = 15.32345$$

We take the next higher even integer 16 as the filter order

$$N = 2M, \quad M = 8.$$

Step 5: The window sequence

$$w_k(n) = \frac{I_0\left\{\beta\sqrt{1 - (n/M)^2}\right\}}{I_0(\beta)}, \quad -M \leq n \leq M$$

$$I_0(x) = 1 + \frac{(0.25x^2)}{(1!)^2} + \frac{(0.25x^2)^2}{(2!)^2} + \frac{(0.25x^2)^3}{(3!)^2} + \cdots$$

Substituting the value of β calculated in the Step 3 and $M = 8$, $\omega(n)$ becomes

$$w_k(n) = \frac{I_0\left\{2.116624\sqrt{1 - (n/8)^2}\right\}}{I_0(2.116624)}, \quad -8 \le n \le 8$$

Also,

$$I_0(2.116624) = 2.4755$$

Thus,

$$w_k(0) = \frac{I_0(\beta)}{I_0(\beta)} = 1$$

$$w_k(1) = w_k(-1) = \frac{I_0(2.1)}{2.4755} = \frac{2.4463}{2.4755} = 0.988213$$

$$w_k(2) = w_k(-2) = \frac{I_0(2.0494)}{2.4755} = \frac{2.36}{2.4755} = 0.95335$$

$$w_k(3) = w_k(-3) = \frac{I_0(1.96216)}{2.4755} = \frac{2.22045}{2.4755} = 0.89697$$

$$w_k(4) = w_k(-4) = \frac{I_0(1.833)}{2.4755} = \frac{2.033785}{2.4755} = 0.821565$$

$$w_k(5) = w_k(-5) = \frac{I_0(1.65228)}{2.4755} = \frac{1.80819}{2.4755} = 0.730434$$

$$w_k(6) = w_k(-6) = \frac{I_0(1.4)}{2.4755} = \frac{1.5534}{2.4755} = 0.627513$$

$$w_k(7) = w_k(-7) = \frac{I_0(1.0247)}{2.4755} = \frac{1.28}{2.4755} = 0.517165$$

$$w_k(8) = w_k(-8) = \frac{I_0(0)}{2.4755} = \frac{1.00}{2.4755} = 0.4039587$$

Step 6: Compute the truncated impulse response using

$$h_t(n) = h(n)w_k(n)$$

The impulse responses $h_t(n)$ and $h(n)$ are given in Table 6.8.
Step 7: The transfer function is given by

$$H_t(z) = z^{-8}\left[h_t(0) + \sum_{n=1}^{8} h_t(n)(z^n + z^{-n})\right]$$

where the values of $h_t(n)$ are given in Table 6.8.

Table 6.8 Impulse responses $h(n)$ and $h_t(n)$

n	$h(n)$	$h_t(n) = h(n)w_k(n)$
0	0.4	0.4
± 1	0	0
± 2	−0.30273	−0.2886
± 3	0	0
± 4	0.0935489	0.0768565
± 5	0	0
± 6	0.0623659	0.039135
± 7	0	0
± 8	−0.07568267	−0.03057268

6.4 FIR Differentiator Design

The frequency response of an ideal differentiator is shown in Fig. 6.14. It can be expressed as

$$H\left(e^{j\omega}\right) = j\omega \quad -\pi \le \omega \le \pi \tag{6.75}$$

The impulse response of the ideal differentiator is computed using the following:

$$h(n) = \int_{-\pi}^{\pi} H\left(e^{j\omega}\right)e^{j\omega n}d\omega$$

Thus,

$$h(0) = 0$$

and

$$\begin{aligned} h(n) &= \frac{2\cos \pi n}{2\pi n} - \frac{j}{\pi n^2}\sin \pi n \quad \text{for} \quad n \ne 0 \\ &= \frac{\cos \pi n}{n}, \text{ since } \sin \pi n = 0 \text{ for all integer values of } n. \end{aligned} \tag{6.76}$$

Fig. 6.14 Frequency response of an ideal differentiator

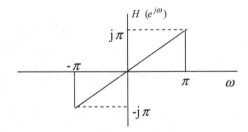

Hence,

$$h(n) = \begin{cases} 0 & \text{for } n = 0 \\ \frac{\cos \pi n}{n} & \text{for } |n| > 0 \end{cases}$$

A causal ideal differentiator frequency response can be represented as

$$H(e^{j\omega}) = j\omega e^{\frac{-j\omega N}{2}} \qquad -\pi \leq \omega \leq \pi \tag{6.77}$$

The corresponding ideal impulse response is

$$h(n) = \frac{\cos \pi \left(n - \frac{N}{2}\right)}{\left(n - \frac{N}{2}\right)} - \frac{\sin \pi \left(n - \frac{N}{2}\right)}{\pi \left(n - \frac{N}{2}\right)^2} \qquad \text{for } -\infty < n < \infty \tag{6.78}$$

It can be observed from Eq. (6.78) that an ideal differentiator is characterized by an antisymmetric impulse response. Hence, it can be realized by using either a Type 3 or Type 4 FIR filters. However, Eq. (6.75) implies that amplitude response $|H(\pi)| = \pi$ for an ideal differentiator. Hence, Type 3 FIR filter cannot be used as its transfer function has a zero at $z = -1$ that forces the amplitude response to be zero at $\omega = \pi$. Thus, only a Type 4 FIR filter can be used for the design of a differentiator. Since signals of interest are in a frequency range $0 \leq \omega \leq \omega_p$ for most practical applications, a differentiator with a band-limited frequency response

$$H_{\text{DIF}}(e^{j\omega}) = \begin{cases} j\omega & 0 \leq |\omega| \leq \omega_p \\ 0, & \omega_s \leq |\omega| \leq \pi \end{cases} \tag{6.79}$$

is desired. Now, it is possible to design a differentiator using both the Type 3 and Type 4 FIR filters with the frequency ω_p as its bandwidth.

Example 6.8 Design an ideal differentiator with frequency response

$$H(e^{j\omega}) = j\omega \, e^{\frac{-j\omega N}{2}} \qquad -\pi \leq \omega \leq \pi$$

Using Hamming window with $N = 11$.

Solution The impulse response of the ideal differentiator is given by

$$h(n) = \frac{\cos \pi \left(n - \frac{N}{2}\right)}{\left(n - \frac{N}{2}\right)} - \frac{\sin \pi \left(n - \frac{N}{2}\right)}{\pi \left(n - \frac{N}{2}\right)^2}$$

We find that the filter coefficients are antisymmetrical. Since N is odd, it is possible to design the differentiator as Type 4 linear phase system.

$$h(0) = -h(11) = -0.01052$$
$$h(1) = -h(10) = 0.015719$$
$$h(2) = -h(9) = -0.025984$$
$$h(3) = -h(8) = 0.05093$$
$$h(4) = -h(7) = -0.14147$$
$$h(5) = -h(6) = 1.273$$

The Hamming window sequence for $N = 11$ is given by

$$\omega_H(n) = 0.54 + 0.46 \cos\frac{\pi n}{5} \quad \text{for } 0 \le n \le 11$$
$$= 0 \qquad\qquad\qquad\qquad \text{otherwise}$$

Hence,

$$w_H(0) = w_H(11) = 1$$
$$w_H(1) = w_H(10) = 0.926977$$
$$w_H(2) = w_H(9) = 0.7311$$
$$w_H(3) = w_H(8) = 0.4745$$
$$w_H(4) = w_H(7) = 0.2388$$
$$w_H(5) = w_H(6) = 0.09863$$

The filter coefficients of the differentiator using Hamming window are

$$h_{\text{diff}}(n) = h(n)w_H(n) \quad \text{for } 0 \le n \le 11$$
$$= 0 \qquad\qquad\qquad \text{otherwise}$$

Thus,

$$h_{\text{diff}}(0) = -h_{\text{diff}}(11) = -0.01052$$
$$h_{\text{diff}}(1) = -h_{\text{diff}}(10) = -0.01457$$
$$h_{\text{diff}}(2) = -h_{\text{diff}}(9) = -0.018997$$
$$h_{\text{diff}}(3) = -h_{\text{diff}}(8) = -0.02417$$
$$h_{\text{diff}}(4) = -h_{\text{diff}}(7) = -0.033778$$
$$h_{\text{diff}}(5) = -h_{\text{diff}}(6) = -0.12558$$

The transfer function of the differentiator is

$$H(z) = \sum_{n=0}^{11} h_{\text{diff}}(n) z^{-n}$$

6.5 Hilbert Transformer

An ideal Hilbert transformer has a frequency response given by

$$\begin{aligned} H(e^{j\omega}) &= j \quad -\pi \le \omega \le 0 \\ &= -j \quad 0 \le \omega \le \pi \end{aligned} \tag{6.80}$$

The ideal frequency response is shown in Fig. 6.15.
The impulse response of an ideal Hilbert transformer is computed using

$$\begin{aligned} h_d(n) &= \frac{1}{2\pi} \int_{-\pi}^{\pi} H(e^{j\omega}) e^{j\omega n} \, d\omega \\ &= \frac{1 - \cos \pi n}{\pi n} \end{aligned} \tag{6.81}$$

As shown in Eq. (6.81), the ideal Hilbert transformer has an antisymmetric impulse response implying that it can be realized using either a Type 3 or a Type 4 FIR filter. It is evident from Eq. (6.80) that an ideal Hilbert transformer has unity magnitude response for all ω. This is not satisfied either by Type 3 FIR or by Type 4 FIR, since a Type 3 FIR filter has zero magnitude response at $\omega = 0$ and Type 4 FIR filter has zero magnitude response at $\omega = 0$ and $\omega = \pi$. However, in practice, $\omega_L \le |\omega| \le \omega_H$ is the finite frequency range of bandpass signals of interest. Consequently, the Hilbert transformer can be designed with a bandpass amplitude. From Eq. (6.81), we see that the impulse response of an ideal Hilbert transformer satisfies the condition that $h(n) = 0$ for n even. This property can be maintained by a Type 3 linear phase FIR filter if the desired amplitude response is symmetrical with respect to $\frac{\pi}{2}$.

Fig. 6.15 Frequency response of ideal Hilbert transformer

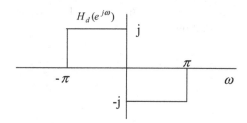

6.6   Kaiser Window-Based Linear Phase FIR Filter Design Using MAT LAB

Example 6.9 Design an FIR lowpass filter using a Kaiser window with the following specifications:

Passband edge $\omega_p = 0.4\pi$, stopband edge $\omega_s = 0.6\pi$, and stopband attenuation ≥ 44 dB.

Solution The following Program 6.3 is used to design the filter satisfying the specifications.

Program 6.3

```
clear;clc;
fedge=[0.4 0.6];%passband and stopband edges
mval=[1 0];% desired magnitudes in the passband and stopband
dev=[0.00630957344 0.00630957344];%desired ripples in the passband and
stopband
[N,Wc,beta,ftype]=kaiserord(fedge,mval,dev);
h=fir1(N,Wc,kaiser(N+1,beta))
[H, w]=freqz(h, 1, 256);
plot(w/(pi),20*log10(abs(H)), '-');grid;xlabel('\omega/\pi'); ylabel('Gain, dB')
```

The impulse response coefficients of the desired filter are

$h(0) = 4.995560142469730e - 001$ $h(7) = -2.721279075330305e - 002 = h(7)$
$h(1) = 3.149215182338651e - 001 = h(-1)$ $h(8) = -9.817111394490758e - 018 = h(-8)$
$h(2) = 1.872126164428770e - 017 = h(-2)$ $h(9) = 1.448785732452065e - 002 = h(-9)$
$h(3) = -9.696337108998725e - 002 = h(-3)$ $h(10) = 6.229417610845594e - 018 = h(-10)$
$h(4) = -1.659481196559194e - 017 = h(-4)$ $h(11) = -6.842987989798997e - 003 = h(-11)$
$h(5) = 4.939296671483609e - 002 = h(-5)$ $h(12) = -3.168771127001458e - 018 = h(-12)$
$h(6) = 1.345599160377898e - 017 = h(-6)$ $h(13) = 2.438800436380994e - 003 = h(-13)$

The magnitude response of the filter obtained from Program 6.3 is shown in Fig. 6.16.

Example 6.10 Design an FIR highpass filter using a Kaiser window with the following specifications:

Passband edge $\omega_p = 20$ rad/s, stopband edge $\omega_s = 15$ rad/s, sampling frequency = 100 rad/s, and stopband ripple = 0.02.

Solution Program 6.4 is used to design the desired filter with the above specifications.

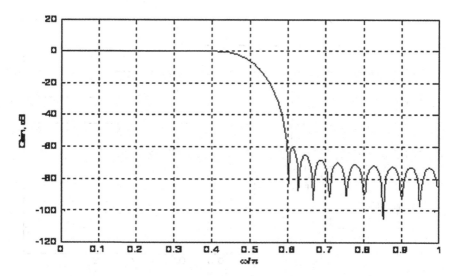

Fig. 6.16 Magnitude response of the FIR lowpass filter using Kaiser window

%Program 6.4

clear;clc;
fedge=[0.3 0.4];% stop band and pass band edges
mval=[0 1];% desired magnitudes in the stop band and pass band
dev=[0.02 0.02];%desired ripples in the stop band and pass band
[N,Wc,beta,ftype]=kaiserord(fedge,mval,dev);
h=fir1(N,Wc,ftype,kaiser(N+1,beta))
[H, w]=freqz(h, 1, 256);
plot(w/(pi),20*log10(abs(H)), '-');grid; xlabel('\omega/\pi'); ylabel('Gain, dB')

The magnitude response of the filter obtained from Program 6.4 is shown in Fig. 6.17.

Example 6.11 Design an FIR bandpass filter using a Kaiser window with the following specifications:

Passband frequency edges: 20 and 30 kHz, lower stopband frequency edge: 10 kHz, upper stopband frequency edge: 40 kHz, sampling frequency: 100 kHz, and stopband attenuation: 30 dB.

Solution The following Program 6.5 is used to design the filter satisfying the given specifications.

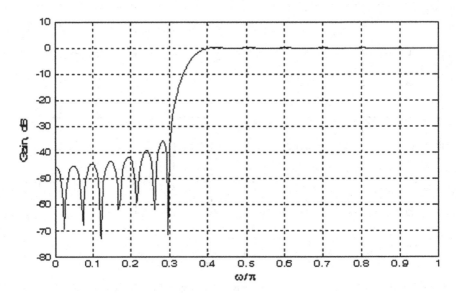

Fig. 6.17 Magnitude response of FIR highpass filter using Kaiser window

Program 6.5 Design of FIR bandpass filter using Kaiser window

```
clear;clc;
FT=100000; %sampling frequency
as=30;% stop band attenuation in dB
mval=[0 1 0];%desired magnitudes in the lower stopband,passband, and upper stop
band
fedge=[10000 20000 30000 40000];%lower stobandedge,passband edges, upper
stop band edge
ds=10^(-as/20);% peak ripple value in the stop bands
dp=ds;
dev=[ds,dp,ds];%desired ripples in the lower stopband, passband, and
upperstopband
[N,Wc,beta,ftype]=kaiserord(fedge,mval,dev,FT);
h=fir1 (N,Wc,ftype,kaiser(N+1,beta))
[H, w]=freqz(h, 1, 256);
plot (w/(pi),20*log10(abs(H)), '-');grid xlabel('\omega/\pi'); ylabel('Gain, dB')
```

The magnitude response of the filter designed using Program 6.5 is shown in
Fig. 6.18.

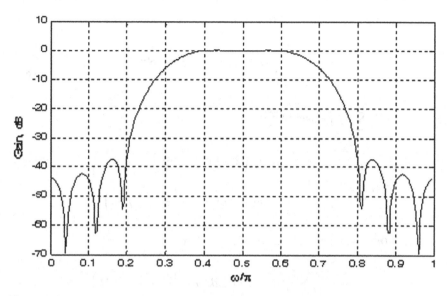

Fig. 6.18 Magnitude response of FIR bandpass filter using Kaiser window

6.7 Design of Linear Phase FIR Filters Using the Frequency Sampling Method

If $H(e^{j\omega})$ is the desired frequency response of the filter to be designed, by sampling it at discrete instants of frequency ω, we get M frequency samples $H(k)$ to be

$$H(k) = H(e^{j\omega})\big|_{\omega=\frac{2\pi k}{M}} \quad k = 0, 1, \ldots, M \tag{6.82}$$

Then, $H(k)$ can be expressed in polar form as

$$H(k) = |H(k)|e^{j\theta(k)} \quad k = 0, 1, \ldots, M - 1 \tag{6.83}$$

For linear phase,

$$\theta(k) = -\frac{M - 1}{M}\pi k \quad k = 0, 1, \ldots, M - 1 \tag{6.84}$$

By computing IDFT of $H(k)$, the filter coefficients $h(n)$ can be obtained, i.e.,

$$h(n) = \frac{1}{M}\sum_{k=0}^{M-1} H(k)e^{j2\pi kn/M} \tag{6.85}$$

Substituting Eqs. (6.83) and (6.84) in the above equation, we obtain

$$h(n) = \frac{1}{M}\left[\sum_{k=0}^{M-1}|H(k)|\{e^{-j\pi k(M-1)/M}\}\{e^{j2\pi kn/M}\}\right] \qquad (6.86)$$

$$= \frac{1}{M}\sum_{k=0}^{M-1}|H(k)|e^{j2\pi k\left(n-\frac{(M-1)}{2}\right)/M}$$

$$= \frac{1}{M}\sum_{k=0}^{M-1}|H(k)|\left(\begin{array}{c}\cos\left(2\pi k\left(n-\dfrac{(M-1)}{2}\right)\Big/M\right)\\[2mm]+j\sin\left(2\pi k\left(n-\dfrac{(M-1)}{2}\right)\Big/M\right)\end{array}\right) \qquad (6.87)$$

Since $h(n)$ is entirely real, Eq. (6.87) can be rewritten as

$$h(n) = \frac{1}{M}\sum_{k=0}^{M-1}|H(k)|\cos\left(2\pi k\left(n-\frac{(M-1)}{2}\right)\Big/M\right) \qquad (6.88)$$

The impulse response $h(n)$ must be symmetrical for the filter to have linear phase, and thus, we can rewrite Eq. (6.88) for even and odd M as follows:

$$\text{For even } M: \quad h(n) = \frac{1}{M}\left[H(0) + \sum_{k=1}^{\frac{M}{2}-1}2|H(k)|\cos\left(2\pi k\left(n-\tfrac{(M-1)}{2}\right)\Big/M\right)\right]$$
$$(6.89a)$$

$$\text{For odd } M: \quad h(n) = \frac{1}{M}\left[H(0) + \sum_{k=1}^{\frac{M-1}{2}}2|H(k)|\cos\left(2\pi k\left(n-\tfrac{(M-1)}{2}\right)\Big/M\right)\right]$$
$$(6.89b)$$

Example 6.12 Design an FIR bandpass filter of length 9 using the frequency sampling method for the following ideal characteristics.

$$\begin{aligned}H(e^{j\omega}) &= 0 \quad \text{for } 0 \le |\omega| \le 0.25\pi\\ &= 1 \quad \text{for } 0.25\pi \le |\omega| \le 0.75\pi\\ &= 0 \quad \text{for } 0.75\pi \le |\omega| \le \pi\end{aligned}$$

Solution

$$H(k) = H(e^{j\omega})\big|_{\omega=\frac{2\pi k}{9}} \quad k = 0, 1, \ldots, 8$$

$$|H(k)| = 0 \quad \text{for } k = 0, 1, 4$$
$$= 1 \quad \text{for } k = 2, 3$$

Hence,

$$h(n) = \frac{1}{9} \left[H(0) + \sum_{k=1}^{4} 2|H(k)| \cos\left(\frac{2\pi k}{9}(n-4)\right) \right]$$

$$h(n) = \frac{2}{9} \left[\sum_{k=2}^{3} |H(k)| \cos\left(\frac{2\pi k}{9}(n-4)\right) \right]$$

$$= \frac{2}{9} \left[\cos\frac{4\pi}{9}(n-4) + \cos\frac{6\pi}{9}(n-4) \right]$$

Therefore,

$$h(0) = h(8) = 0.0591; h(1) = h(7) = 0.1111;$$
$$h(2) = h(6) = -0.3199; h(3) = h(5) = -0.0725; h(4) = 0.4444.$$

6.8 Design of Optimal Linear Phase FIR Filters

The windowing method discussed in the preceding sections has an advantage that the filter responses can be obtained simply from the ideal filter response using closed-form expressions. However, the filter designs are suboptimal. One of the optimal techniques for the design of FIR filters is the equiripple design technique based on Chebyshev approximation.

From Eqs. (6.25), (6.29), (6.33), and (6.37), we see that the frequency response of a linear phase FIR filter can be written as

$$H(e^{j\omega}) = e^{j\beta} \; e^{-\frac{j\omega N}{2}} H_1(\omega) \tag{6.90}$$

where $\beta = 0$ or π for Types 1 and 2 filters, and $\beta = \frac{\pi}{2}$ or $\frac{3\pi}{2}$ for Types 3 and 4 filters. Furthermore, from (6.26) it is seen that for Type 1 filter, the amplitude response is given by

$$H_1(\omega) = \sum_{n=0}^{N/2} a(n)\cos(\omega n) \tag{6.91a}$$

where

$$a(0) = h\left(\frac{N}{2}\right), \quad a(n) = 2h\left(\frac{N}{2} - n\right) \quad 1 \le n \le \frac{N}{2} \tag{6.91b}$$

For Type 2, the amplitude response given by Eq. (6.30) can be rewritten as

$$H_1(\omega) = \sum_{n=1}^{(N+1)/2} b(n)\cos(\omega n) \tag{6.92a}$$

where

$$b(n) = 2h\left(\frac{N+1}{2} - n\right) \quad 1 \le n \le \frac{N+1}{2} \tag{6.92b}$$

The above equation can be expressed as

$$H_1(\omega) = \cos\left(\frac{\omega}{2}\right) \sum_{n=0}^{(N-1)/2} \tilde{b}(n)\cos(\omega n) \tag{6.93a}$$

where

$$
\begin{aligned}
b(1) &= \frac{1}{2}\left(\tilde{b}(1) + 2\tilde{b}(0)\right), \\
b(n) &= \frac{1}{2}\left(\tilde{b}(n) + 2\tilde{b}(n-1)\right), \quad 2 \le n \le \frac{N-1}{2}, \\
b\left(\frac{N+1}{2}\right) &= \frac{1}{2}\tilde{b}\left(\frac{N-1}{2}\right)
\end{aligned}
\tag{6.93b}
$$

For Type 3, the amplitude response given by Eq. (6.34) can be rewritten as

$$H_1(\omega) = \sum_{n=0}^{N/2} c(n)\sin(\omega n) \tag{6.94a}$$

where

$$c(n) = 2h\left(\frac{N}{2} - n\right) \quad 1 \le n \le \frac{N}{2} \tag{6.94b}$$

The above expression can be expressed as

$$H_1(\omega) = \sin\omega \sum_{n=0}^{\left(\frac{N}{2}\right)-1} \tilde{c}(n)\cos(\omega n) \tag{6.95a}$$

where

$$c(1) = \tilde{c}(0) - \frac{1}{2}\tilde{c}(1)$$

$$c(n) = \frac{1}{2}(\tilde{c}(n-1) - \tilde{c}(n)), \quad 2 \le n \le \frac{N}{2} - 1, \quad \quad (6.95b)$$

$$c\left(\frac{N}{2}\right) = \frac{1}{2}\tilde{c}\left(\frac{N}{2} - 1\right)$$

For Type 4, the amplitude response given by Eq. (6.38) can be rewritten in the following form

$$H_1(\omega) = \sum_{n=1}^{(N+1)/2} d(n)\sin\omega\left(n - \frac{1}{2}\right) . \quad \quad (6.96a)$$

where

$$d(n) = 2h\left(\frac{N+1}{2} - n\right) \quad 1 \le n \le \frac{N+1}{2} \quad \quad (6.96b)$$

The above equation can be expressed as

$$H_1(\omega) = \sin\left(\frac{\omega}{2}\right) \sum_{n=0}^{(N-1)/2} \tilde{d}(n)\cos(\omega n) \quad \quad (6.97a)$$

where

$$d(1) = \tilde{d}(0) - \frac{1}{2}\tilde{d}(1)$$

$$d(n) = \frac{1}{2}\left(\tilde{d}(n-1) - \tilde{d}(n)\right), \quad 2 \le n \le \frac{N-1}{2} \quad \quad (6.97b)$$

$$d\left(\frac{N+1}{2}\right) = \tilde{d}\left(\frac{N-1}{2}\right)$$

From Eqs. (6.91), (6.93a), (6.95a), and (6.97a), it is seen that the expression for $H_1(\omega)$ can be expressed as a product of a fixed function of $Q(\omega)$ and a function $A(\omega)$ that is a sum of cosines in the form

$$H_1(\omega) = Q(\omega)A(\omega) \quad \quad (6.98)$$

where

$$A(\omega) = \sum_{n=0}^{L} a(n) \cos \omega n \qquad (6.99)$$

$$a(n) = \begin{cases} a(n) & \text{for Type 1} \\ \tilde{b}(n) & \text{for Type 2} \\ \tilde{c}(n) & \text{for Type 3} \\ \tilde{d}(n) & \text{for Type 4} \end{cases} \qquad (6.100)$$

$$Q(\omega) = \begin{cases} 1 & \text{for Type 1} \\ \cos\left(\frac{\omega}{2}\right) & \text{for Type 2} \\ \sin(\omega) & \text{for Type 3} \\ \sin\left(\frac{\omega}{2}\right) & \text{for Type 4} \end{cases} \qquad (6.101)$$

$$L = \begin{cases} \frac{N}{2} & \text{for Type 1} \\ \frac{N-1}{2} & \text{for Type 2} \\ \frac{N}{2} - 1 & \text{for Type 3} \\ \frac{N-1}{2} & \text{for Type 4} \end{cases} \qquad (6.102)$$

Now, let $H_d(\omega)$, the desired frequency response, be given as a piecewise linear function of ω. Consider the difference between $H_1(\omega)$ and $H_d(\omega)$ specified as a weighted error function $\varepsilon(\omega)$ given by

$$\varepsilon(\omega) = W(\omega)[H_1(\omega) - H_d(\omega)] \qquad (6.103)$$

where $W(\omega)$ is a positive weighing function that can be chosen as the stopband:

$$W(\omega) = \begin{cases} \frac{\delta_s}{\delta_p} & \text{in the passband} \\ 1 & \text{in the stopband} \end{cases} \qquad (6.104)$$

where δ_s and δ_p are the peak ripple values in the stopband and passband, respectively.

Substituting Eq. (6.98) in Eq. (6.103), we get

$$\varepsilon(\omega) = W(\omega)[Q(\omega)A(\omega) - H_d(\omega)]$$

The above equation can be rewritten as

$$\varepsilon(\omega) = W(\omega)Q(\omega)\left[A(\omega) - \frac{H_d(\omega)}{Q(\omega)}\right] \qquad (6.105)$$

Using the notations $\tilde{W}(\omega) = W(w)Q(\omega)$ and $\tilde{H}_d(\omega) = H_d(\omega)/Q(\omega)$, we can rewrite the above equation as

$$\varepsilon(\omega) = \widetilde{W}(\omega)[A(\omega) - \widetilde{H}_d(\omega)] \tag{6.106}$$

A commonly used approximation measure, called the Chebyshev or minimax criterion, is to minimize the peak absolute value of the weighted error $\varepsilon(\omega)$,

$$\varepsilon = \max_{\omega \in S} |\varepsilon(\omega)| \tag{6.107}$$

The optimization problem now is to determine the coefficients $(n), 0 \le n \le L$, so that the weighted approximation error $\varepsilon(\omega)$ of Eq. (6.106) is minimum for all values of ω over closed subintervals of $0 \le \omega \le \pi$. Knowing the type of filter being designed, the filter coefficients are obtained using Eq. (6.100). For instance, if the filter being designed is of Type 3, it can be observed from Eq. (6.100) that $\widetilde{c}(n) = a(n)$, and from Eq. (6.102) that $N = 2(L + 1)$. Next, $c(n)$ is determined using Eq. (6.95b). Finally, the filter coefficients $h(n)$ are obtained by substituting $c(n)$ in Eq. (6.94b). Similarly, the other three types of FIR filter coefficients can be computed.

To solve the above optimization problem, Parks and McClellan applied the following theorem called Alternation theorem from the theory of Chebyshev approximation [4].

Alternation Theorem:

Let S be any closed subset of the closed interval $0 \le \omega \le \pi$. The amplitude function $A(\omega)$ is the best unique approximation of the desired amplitude response obtained by minimizing the peak absolute value ε of $\varepsilon(\omega)$ given by Eq. (6.106), if and only if the error function $\varepsilon(\omega)$ exhibits at least $(L + 2)$ 'alternations' or external frequencies in S such that $\omega_1 < \omega_2 \ldots < \omega_{L+2}$ and $\varepsilon(\omega_i) = -\varepsilon(\omega_{i+1})$ with $|\varepsilon(\omega_i)| = \varepsilon$ for all i in the range $1 \le i \le L+2$.

To obtain the optimum solution, the following set of equations are to be solved:

$$\widetilde{W}(\omega_i)[\widetilde{H}_d(\omega_i) - A(\omega_i)] = (-1)^{i+1}\varepsilon, \quad i = 1, 2, \ldots, L+2 \tag{6.108}$$

The above equation in matrix form can be written as

$$
\begin{bmatrix}
1 & \cos \omega_1 & \cos 2\omega_1 & \cdots & \cos L\omega_1 & \frac{1}{\widetilde{W}_{(\omega_1)}} \\
1 & \cos \omega_2 & \cos 2\omega_2 & \cdots & \cos L\omega_2 & \frac{-1}{\widetilde{W}_{(\omega_1)}} \\
1 & \cos \omega_3 & \cos 2\omega_3 & \cdots & \cos L\omega_3 & \frac{1}{\widetilde{W}_{(\omega_1)}} \\
\vdots & \vdots & \vdots & \cdots & \vdots & \vdots \\
1 & \cos \omega_{L+2} & \cos 2\omega_{L+2} & \cdots & \cos L\omega_{L+2} & \frac{-(-1)^{L+2}}{\widetilde{W}_{(\omega_{L+2})}}
\end{bmatrix}
\begin{bmatrix}
a(0) \\
a(1) \\
a(2) \\
\vdots \\
\varepsilon
\end{bmatrix}
=
\begin{bmatrix}
\widetilde{H}_d(\omega_1) \\
\widetilde{H}_d(\omega_2) \\
\widetilde{H}_d(\omega_3) \\
\vdots \\
\widetilde{H}_d(\omega_{L+2})
\end{bmatrix}
$$

$$\tag{6.109}$$

The above set of equations can be solved for unknowns a and ε iteratively starting with an initial guess of ω_i for $i = 1, 2, \ldots, L+2$ and to be continued with new set of extremal frequencies until the necessary and sufficient condition for optimal solution $|\varepsilon(\omega_i)| \leq \varepsilon$ is satisfied for all frequencies.

However, the Remez exchange algorithm, a highly efficient iterative procedure, is an alternative to find the desired sets of $(L + 2)$ extremal points. It consists of the following steps at each iteration stage.

1. Select an initial set of the $(L + 2)$ extremal points $\{\omega_n\}$ $n = 1, 2, \ldots, L + 2$.
2. Calculate the deviation associated with this set by using the following formula.

$$\varepsilon = \frac{\sum_{n=1}^{L+2} \gamma_n \widetilde{H}_d(\omega_n)}{\sum_{n=1}^{L+2} \frac{\gamma_n (-1)^{n+1}}{W(\omega_n)}} \tag{6.110}$$

where

$$\gamma_n = \prod_{\substack{i = 1 \\ i \neq n}}^{L+2} \frac{1}{(\cos \omega_k - \cos \omega_i)} \tag{6.111}$$

3. Compute $A(\omega)$ using the following Lagrange interpolation formula

$$A(\omega) = \frac{\sum_{i=1}^{L+1} \frac{d_i c_i}{(\cos\omega - \cos\omega_i)}}{\sum_{i=1}^{L+1} \frac{d_i}{(\cos\omega - \cos\omega_i)}} \tag{6.112}$$

where

$$c_i = \widetilde{H}_d(\omega_i) - \frac{(-1)^{i+1} \varepsilon}{\widetilde{W}(\omega_i)}$$

and

$$d_i = \gamma_i (\cos\omega_i - \cos\omega_{L+2})$$

4. Compute $\varepsilon(\omega)$
5. If $|\varepsilon(\omega)| \leq \varepsilon$ for all ω in the passband and stopband, the extrema are the same as the set used in Step 1, stop and calculate the impulse response corresponding to the frequency response calculated in Step 3. Otherwise, find a new set of extrema and repeat Steps 2–4.

Example 6.13 Design an optimal FIR lowpass filter of length 3 to meet the following specifications:

Passband edge frequency $= f_p = 750\,\text{Hz}$
Stopband edge frequency $= f_s = 1000\,\text{Hz}$.
Sampling frequency $= 5000\,\text{Hz}$
Tolerance ratio $= (\delta_p/\delta_s) = 2$

Solution

$$\omega_p = \frac{2\pi(750)}{5000} = 0.3\pi$$

$$\omega_s = \frac{2\pi(1000)}{5000} = 0.4\pi$$

Since the length of the filter is 3, the order is 2 and $L = 1$. Hence, we have to choose $L + 2$ extremal points; two of them will be the edge frequencies, and the third one can be chosen arbitrarily. Let us choose the extremal points to be $\omega_1 = 0.3\pi$, $\omega_2 = 0.4\pi$ and $\omega_3 = \pi$, as shown in Fig. 6.19.

The desired characteristics are

$$\tilde{H}_d(\omega_1) = 1; \quad \tilde{H}_d(\omega_2) = 0; \quad \tilde{H}_d(\omega_3) = 0;$$

The weighting functions are:

$$\tilde{W}(\omega_1) = \frac{\delta_s}{\delta_p} = 1/2, 0 < \omega < \omega_p; \quad \tilde{W}(\omega_2) = 1, \tilde{W}(\omega_3) = 1, \omega_s < \omega < \pi.$$

Fig. 6.19 $A(\omega)$ response with assumed peaks at extremal frequencies at the point

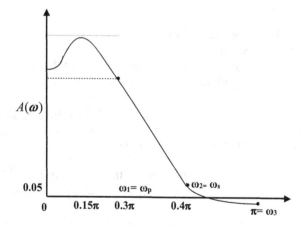

Now, Eq. (6.109) can be written as

$$
\begin{bmatrix}
1 & \cos\omega_1 & \frac{1}{\widetilde{W}(\omega_1)} \\
1 & \cos\omega_2 & \frac{-1}{\widetilde{W}(\omega_2)} \\
1 & \cos\omega_3 & \frac{1}{\widetilde{W}(\omega_3)}
\end{bmatrix}
\begin{bmatrix}
a(0) \\
a(1) \\
\varepsilon
\end{bmatrix}
=
\begin{bmatrix}
\widetilde{H}_d(\omega_1) \\
\widetilde{H}_d(\omega_2) \\
\widetilde{H}_d(\omega_3)
\end{bmatrix}
$$

This leads to

$$
\begin{bmatrix}
1 & 0.5878 & 2 \\
1 & 0.3090 & -1 \\
1 & -1.0000 & 1
\end{bmatrix}
\begin{bmatrix}
a(0) \\
a(1) \\
\varepsilon
\end{bmatrix}
=
\begin{bmatrix}
1 \\
0 \\
0
\end{bmatrix}
$$

Solving for $a(0)$, $a(1)$, and ε, we obtain $a(0)= 0.1541$, $a(1)= 0.4460$, and $\varepsilon = 0.2919$.

Hence,

$$A(\omega) = 0.1541 + 0.4460\cos\omega = h(0) + 2h(1)\cos\omega$$

The weighted approximation error is

$$\varepsilon(\omega) = \widetilde{W}(\omega)\big(\widetilde{H}_d(\omega) - A(\omega)\big).$$

and its values over the interval $0 \leq \omega \leq \pi$ are tabulated below:

ω	0	0.15π	0.3π	0.4π	π
$\varepsilon(\omega)$	0.2	0.2243	0.2919	-0.2919	0.2919

For an optimal solution, the necessary and sufficient condition is that $\varepsilon(\omega) \leq \varepsilon$ for all frequencies.

From the above table, it can be observed that $\varepsilon(\omega) \leq \varepsilon$ for all frequencies, and hence, the optimal solution is achieved. Thus,

$$A(z) = h(-1)z + h(0) + h(1)z^{-1}$$

The causal transfer function is

$$
\begin{aligned}
H(z) &= z^{-1}\big(0.223z + 0.1541 + 0.223z^{-1}\big) \\
&= 0.223 + 0.1541z^{-1} + 0.223z^{-2}
\end{aligned}
$$

Example 6.14 Design an optimal FIR highpass filter of length 3 to meet the following specifications:

$$\text{Passband edge frequency} = f_p = 1000\text{Hz}$$
$$\text{Stopband edge frequency} = f_s = 750\text{Hz}$$
$$\text{Sampling frequency} = 5000\text{Hz}$$
$$\text{Tolerance ratio} = (\delta_p/\delta_s) = 2$$

Solution

$$\omega_s = \frac{2\pi(750)}{5000} = 0.3\pi$$
$$\omega_p = \frac{2\pi(1000)}{5000} = 0.4\pi$$

Since the length of the filter is 3, the order is 2 and $L = 1$. Hence, we have to choose $L + 2$ extremal points; two of them will be the edge frequencies, and the third one can be chosen arbitrarily. Let us choose the extremal points to be $\omega_1 = 0.3\pi$, $\omega_2 = 0.4\pi$ and $\omega_3 = \pi$, as shown in Fig. 6.20.
The desired characteristics are

$$\tilde{H}_d(\omega_1) = 0; \quad \tilde{H}_d(\omega_2) = 1; \quad \tilde{H}_d(\omega_3) = 1;$$

The weighting functions are:

$$\tilde{W}(\omega_1) = \frac{\delta_s}{\delta_p} = 1, 0 < \omega < \omega_s; \quad \tilde{W}(\omega_2) = 1/2, \tilde{W}(\omega_3) = 1/2, \quad \omega_p < \omega < \pi.$$

Fig. 6.20 $A(\omega)$ response with assumed peaks at extremal frequencies at the point •

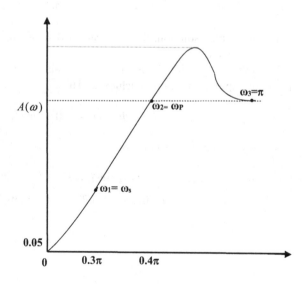

Now, Eq. (6.109) can be written as

$$
\begin{bmatrix}
1 & \cos \omega_1 & \frac{1}{\widetilde{W}(\omega_1)} \\
1 & \cos \omega_2 & \frac{-1}{\widetilde{W}(\omega_2)} \\
1 & \cos \omega_3 & \frac{1}{\widetilde{W}(\omega_3)}
\end{bmatrix}
\begin{bmatrix}
a(0) \\
a(1) \\
\varepsilon
\end{bmatrix}
=
\begin{bmatrix}
\widetilde{H}_d(\omega_1) \\
\widetilde{H}_d(\omega_2) \\
\widetilde{H}_d(\omega_3)
\end{bmatrix}
$$

This leads to

$$
\begin{bmatrix}
1 & 0.5878 & 1 \\
1 & 0.3090 & -2 \\
1 & -1.0000 & 2
\end{bmatrix}
\begin{bmatrix}
a(0) \\
a(1) \\
\varepsilon
\end{bmatrix}
=
\begin{bmatrix}
0 \\
1 \\
1
\end{bmatrix}
$$

Solving   for $a(0)$, $a(1)$,   and ε,   we   obtain $a(0)= 0.7259$, $a(1)= -0.7933$, and $\varepsilon = -0.2596$.
Hence,

$$
A(\omega) = 0.7259 - 0.7933 \cos \omega = h(0) + 2h(1) \cos \omega
$$

The weighted approximation error is

$$
\varepsilon(\omega) = \widetilde{W}(\omega)(\widetilde{H}_d(\omega) - A(\omega)).
$$

and its values over the interval $0 \le \omega \le \pi$ are tabulated below:

ω	0	0.15π	0.3π	0.4π	π
$\varepsilon(\omega)$	0.0674	-0.0191	-0.2596	0.2596	-0.2596

For an optimal solution, the necessary and sufficient condition is that $\varepsilon(\omega) \le \varepsilon$ for all frequencies.

From the above table, it can be observed that $\varepsilon(\omega) \le \varepsilon$ for all frequencies, and hence, the optimal solution is achieved. Thus,

$$
A(z) = h(-1)z + h(0) + h(1)z^{-1}
$$

The causal transfer function is

$$
\begin{aligned}
H(z) &= z^{-1}(-0.3967z + 0.7259 - 0.3967z^{-1}) \\
&= -0.3967 + 0.7259z^{-1} - 0.3967z^{-2}
\end{aligned}
$$

6.8.1 Optimal (Equiripple) Linear Phase FIR Filter Design Using MATLAB

The MATLAB command [N,fpts,mag,wt]=firpmord(fedge, mval, dev, FT) finds the approximate order N, the normalized frequency bandedges, frequency band amplitudes, and weights that meet input specifications fedge, mval, and dev.

fedge is a vector of frequency bandedges (between 0 and FT/2, where FT is the sampling frequency), and mval is a vector specifying the desired amplitude on the bands defined by fedge.

dev is a vector of the same size as mval that specifies the maximum allowable deviation of ripples between the frequency response and the desired amplitude of the output for each band.

firgr is used along with the resulting order N, frequency vector fpts, amplitude response mag, and weights wt to design the filter h which approximately meets the specifications given by firpmord input parameters fedge, mval, dev.

$$h=\text{firgr}(N, \text{fpts}, \text{mag}, \text{wt});$$

h is a row vector containing the N+1 coefficients of FIR filter of order N. The vector fpts must be in the range between 0 and 1, with the first element 0 and the last element 1, and sampling frequency being equal to 2. With the elements given in equal-valued pairs, the vector mag gives the desired magnitudes of the FIR filter frequency response at the specified bandedges.

The argument wt included in the function firgr is the weight vector. The desired magnitude responses in the passband and stopband can be weighted by the weight vector wt. The function can be used to design equiripple linear phase FIR filters of Types 1, 2, 3, and 4. Types 1 and 2 are the default designs for N even and odd, respectively. To design 'Hilbert transformer' and 'differentiator,' Type 3 (N even) and Type 4 (N odd) are used with the flags 'hilbert' and 'differentiator' for 'ftype' in the firgr function as given below.

$$h=\text{firgr}(N, \text{fpts}, \text{mag},' \text{ftype}');$$
$$h=\text{firgr}(N, \text{fpts}, \text{mag}, \text{wt},' \text{ftype}');$$

Example 6.14 A signal has its energy concentrated in the frequency range 0–3 kHz. However, it is contaminated with high-frequency noise. It is desired to enhance the signal preserving frequency components in the frequency band 0–2.5 kHz. To suppress the noise, design a lowpass FIR filter using the Remez exchange algorithm assuming the stopband edge frequency as 2.85 kHz, passband ripple as 1 dB, and stopband attenuation as 40 dB.

Solution From the specifications, the desired ripple in the passband and stopband is calculated as

$$\delta_p = 1 - 10^{-\alpha_p/20} = 1 - 10^{-0.05} = 0.010875$$
$$\delta_s = 10^{-\alpha_s/20} = 10^{-2} = 0.01$$

The sampling frequency (FT) is 10000 Hz [4 times 2.5 kHz)] since the desired frequency band is 0–2.5 kHz. The following MATLAB Program 6.6 is used to determine the order of the filter (N) and the filter coefficients (h).

Program 6.6 Program to design a lowpass FIR filter using Remez algorithm for Example 6.12.

```
clear;clf;
ap=1;%pass band ripple in dB
as=40;% stop band attenuation in dB
dp=(1-10^(-ap/20));%peak ripple value in passband,
ds=10^(-as/20);% peak ripple value in the stopbands
fedge=[2500 2850];%passband and stopband edge frequencies
mval=[1 0];% desired magnitudes in the passband and stopband
dev=[dp ds];% desired ripple values in the passband and stopband
FT=10000;%sampling frequency
[N,fpts,mag,wt]=firpmord(fedge,mval,dev,FT);
h=firgr(N,fpts,mag,wt);
[H,omega]=freqz(h,1,256);
plot(omega/(2*pi),20*log10(abs(H)));
grid;xlabel('\omega/2\pi');ylabel('Gain,dB')
```

The order of the filter (N) is found as 35. The gain response of the designed filter with $N = 35$ is shown in Fig. 6.21.

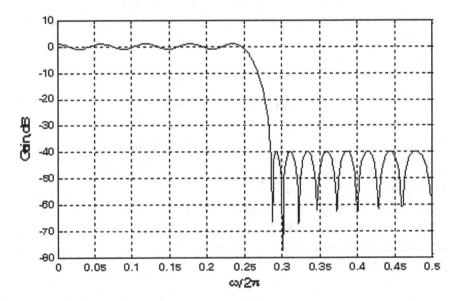

Fig. 6.21 Gain response of the FIR lowpass filter of order 35

From the frequency response of the designed filter, it is seen that the required specifications are not satisfied since the maximum passband amplitude is 1.1114 (maximum value of H in the passband) and the required stopband attenuation is not obtained. Thus, the filter order is to be increased until the specifications are satisfied.

Hence, the order N is increased to 41. The gain response of the filter with $N = 41$ is shown in Fig. 6.22. It may be observed that the filter with $N = 41$ has met the specifications. With $N = 41$, the maximum passband amplitude = 1.0818. Thus, the ripple in the passband for the designed filter is 0.0818 which equals $-20\log10$ $(1 - 0.0818) = 0.741$ dB.

Example 6.15 Design an FIR lowpass filter using the Remez exchange algorithm to meet the following specifications: passband edge 0.2π, stopband edge 0.3π, passband ripple 0.01, and stopband ripple 0.01.

Solution The fedge=[0.2 0.3]; dev=[0.01 0.01]; mval=[1 0];

With these values, the order of the filter is obtained using the following program statement

$$[N, \text{fpts}, \text{mag}, \text{wt}]=\text{firpmord}(\text{fedge}, \text{mval}, \text{dev});$$

The filter order obtained is $N = 39$. The log magnitude response of the designed filter with $N = 39$ is shown in Fig. 6.23.

From the frequency response of the designed filter, it is seen that the required specifications are not satisfied since the maximum passband amplitude is 1.0112 (maximum value of H in the passband) and the required stopband attenuation is not

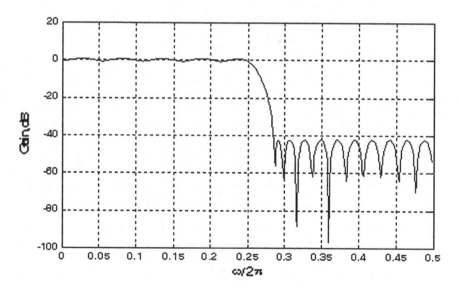

Fig. 6.22 Gain response of the FIR lowpass filter of order 41

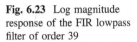

Fig. 6.23 Log magnitude
response of the FIR lowpass
filter of order 39

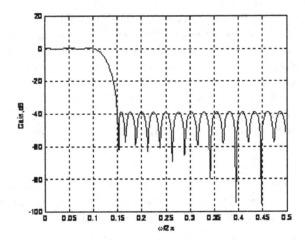

obtained. Thus, the filter order is to be increased until the specifications are satisfied. Hence, the order N is increased to 42. The gain response of the filter with
$N = 42$ is shown in Fig. 6.24.

It is observed that the filter with $N = 42$ has satisfied the specifications. With
$N = 42$, the maximum passband amplitude = 1.00982. Thus, the ripple in the
passband for the designed filter is 0.00982 which equals $-20\log10$
$(1 - 0.00982) = 0.0857$ dB.

Example 6.16 A linear phase FIR bandpass filter is required to meet the following
specifications:

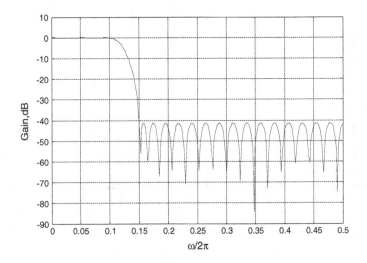

Fig. 6.24 Log magnitude response of the FIR lowpass filter of order 42

Pass band:	12 – 16 kHz
Transition width:	2 kHz
Pass band ripple:	1 db
Stop band attenuation:	40 db
Sampling frequency:	50 kHz
Lower stop band:	0 – 10 kHz
Upper stop band:	18 – 25 kHz

Solution The MATLAB Program 6.7 given below is used to design the bandpass filter for the given specifications.

Program 6.7

```
clear;clc;
FT=50000; %sampling frequency
ap=1;%pass band ripple in dB
as=40;% stop band attenuation in dB
mval=[0 1 0];%desired magnitudes in the lower stopband,passband, and upper stop
band
fedge=[10000 12000 16000 18000];%lower stop band edge,passband edges, upper
stop band edge
dp=(1-10^(-ap/20));%peak ripple value in pass band,
ds=10^(-as/20);% peak ripple value in the stop bands
dev=[ds,dp,ds];%desired ripples in the lower stopband,passband,and upperstopband
[N,fpts,mag,wt]=firpmord(fedge,mval,dev,FT);
h=firgr(N,fpts,mag,wt);
[H,omega]=freqz(h,1,1024);
plot(omega/(2*pi),20*log10(abs(H)));
grid;xlabel('\omega/2\pi');ylabel('Gain,dB')
```

The filter order obtained is $N = 31$. The log magnitude response of the designed filter with $N = 31$ is shown in Fig. 6.25.

From the frequency response of the designed filter, it is seen that the required specifications are not satisfied. Thus, the filter order is to be increased until the specifications are satisfied. When the filter order is increased to 35, the frequency response of the designed filter satisfying the specifications is as shown in Fig. 6.26.

6.9   Design of Minimum-Phase FIR Filters

A minimum-phase FIR filter yields a transfer function with a smaller group delay as compared to that of its equivalent linear phase filter. Thus, minimum-phase FIR filters are very useful in applications, where the linear phase is not a constraint. A method to design a minimum-phase FIR filter [5] is outlined below.

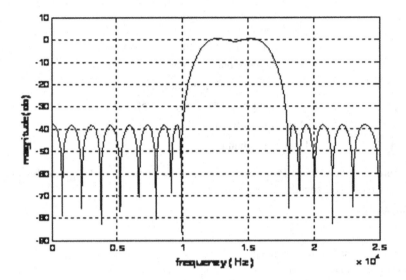

Fig. 6.25 Log magnitude response of the linear phase FIR bandpass filter of order 31

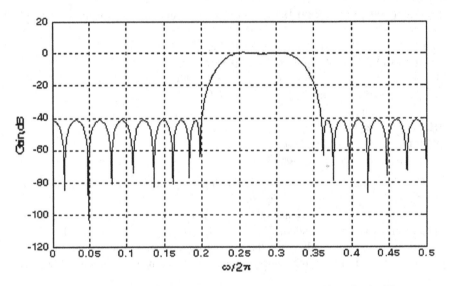

Fig. 6.26 Log magnitude response of the linear phase FIR bandpass filter of order 35

Let $H(z)$ be an arbitrary FIR transfer function of degree N and given by

$$H(z) = \sum_{n=0}^{N} h(n)z^{-n} = h(0) \prod_{k=1}^{N} (1 - a_k z^{-1}) \qquad (6.113)$$

Then, the mirror-image $\hat{H}(z)$ of $H(z)$ can be written as

$$\hat{H}(z) = z^{-N} H(z^{-1})$$
$$= \sum_{n=0}^{N} h(N - n)z^{-n} = h(N) \prod_{k=1}^{N} (1 - z^{-1}/a_k) \qquad (6.114)$$

The zeros of $\hat{H}(z)$ are reciprocals of the zeros of $H(z)$ at $z = a_k$. From Eqs. (6.113) and (6.114),
$H(z)\hat{H}(z)$ can be written as

$$G(z) = H(z)\hat{H}(z) = z^{-N} H(z)H(z^{-1}) \qquad (6.115)$$

Thus, $G(z)$ is a Type 1 linear phase transfer function of order $2N$ and has zeros exhibiting mirror-image symmetry in the z-plane. Also, the zeros on the unit circle of $G(z)$ with real coefficients are of order 2, since if $H(z)$ has a zero on the unit circle, $\hat{H}(z)$ will also have a zero on the unit circle at the conjugate reciprocal position. On the unit circle, Eq. (6.115) becomes

$$G(e^{jw}) = |H(e^{jw})|^2 \qquad (6.116)$$

Hence, the amplitude response $G(e^{jw})$ is positive, and it has double zeros in the frequency range $[0, \omega]$. The steps involved in the design of minimum-phase FIR filter are given below.

Step 1: From given specifications $\omega_p, \omega_s, \delta_p, \delta_s$ of the desired minimum-phase FIR filter, design a Type 1 linear phase FIR filter $F(z)$ of degree $2N$ with the specifications $\omega_p, \omega_s, \delta_{pF}, \delta_{sF}$, where

$$\delta_{sF} = \delta_s^2 / (2 - \delta_s^2) \qquad (6.117)$$

$$\delta_{pF} = (1 + \delta_{sF}) \left((\delta_p + 1)^2 - 1 \right) \qquad (6.118)$$

Step 2: Obtain the linear phase transfer function

$$G(z) = \delta_{sF} z^{-N} + F(z) \qquad (6.119)$$

Now, $G(z)$ can be expressed in the form of Eq. (6.115) as

$$G(z) = z^{-N} H_m(z) H_m(z^{-1}) \tag{6.120}$$

where $H_m(z)$ is a minimum-phase FIR transfer function and has for its zeros the zeros of $G(z)$ that are inside the unit circle and one each of the double zeros of $G(z)$ on the unit circle.

Step 3: Determine $H_m(z)$ from $G(z)$ by applying spectral factorization.

6.9.1 Design of Minimum-Phase FIR Filters Using MATLAB

The M-file *firminphase* can be used to design a minimum-phase FIR filter. The following example illustrates its use in the design.

Example 6.17 Design a minimum-phase lowpass FIR filter with the passband edge at $\omega_p = 0.3\pi$, stopband edge at $\omega_s = 0.4\pi$, passband ripple of $\alpha_p = 1$ dB, and a minimum stopband attenuation of $\alpha_s = 40$ dB.

Solution For the given specifications, the following MATLAB Program 6.8 can be used to design the minimum-phase lowpass FIR filter.

Program 6.8 Design of a minimum-phase lowpass FIR filter

```
% Design of a minimum-phase lowpass FIR filter
clear all;clc;
Wp=0.3; %passband edge frequency
Ws=0.4; %stopband edge frequency
Rp=1; %passband ripple?_p in dB
Rs=40; %stopband attenuation?_s in dB
dp=1-10^(-Rp/20);%passband ripple value for the minimum phase filter
ds=10^(-Rs/20);%stopband ripple value for the minimum phase filter
Ds=(ds*ds)/(2 - ds*ds);%stopband ripple value for the linear phase filter F(z)
Dp=(1+Ds)*((dp+1)*(dp+1) - 1);%passband ripple value for the linear
%phase filter F(z)
[N,fpts,mag,wt]=firpmord([Wp Ws], [1 0], [Dp Ds]);% Estimate filter
%order of F(z)
% Design the prototype linear phase filter F(z)
[hf,err,res]=firgr(N, fpts, mag, wt);
figure (1),zplane(hf) % Plots the zeros of F(z)
hmin=firminphase(hf);
figure (2), zplane(hmin) % Plots the zeros of the minimum-phase filter
[Hmin,w]=freqz(hmin,1, 512);
% Plot the gain response of the minimum-phase filter
figure (3), plot(w/pi, 20*log10(abs(Hmin)));grid
xlabel('\omega/\pi');ylabel('Gain, dB');
```

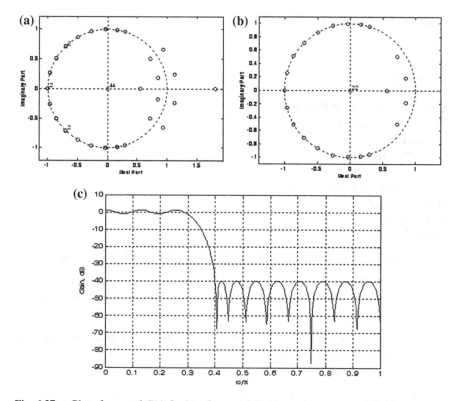

Fig. 6.27 a Plot of zeros of $G(z)$, **b** plot of zeros of $H_m(z)$, **c** gain response of $H_m(z)$

The zeros of $F(z)$ and $H_m(z)$ generated by the above program are shown in Fig. 6.27a and b, respectively. The gain response of the designed minimum-phase filter $H_m(z)$ is shown in Fig. 6.27c.

6.10 Design of FIR Differentiator and Hilbert Transformer Using MATLAB

Example 6.18 Design a Type 4 FIR differentiator of order 51 with the following specifications: passband edge $\omega_p = 0.3\pi$, stopband edge $\omega_s = 0.35\pi$.

Solution The MATLAB Program 6.9 is used to design the desired differentiator.

Program 6.9 To design a differentiator

```
clear all;clc;
N=51;
fpts=[0 0.3 0.35 1];
mag=[0 0.3*pi 0 0];
h=firpm(N,fpts,mag,'differentiator')
[Hz,w]=freqz(h,1,256);
h1=abs(Hz);
M1=20*log10(h1);
figure,plot(w/(pi),h1,'-');grid;
%figure,plot(w/(pi),M1,'-');grid;
xlabel('\omega/\pi');
%ylabel('gain,dB');
ylabel('Magnitude');
```

Figure 6.28 shows the magnitude response of the Type 4 FIR differentiator.

Example 6.19 Design a tenth-order linear phase bandpass FIR Hilbert transformer using Remez function. The passband is from $0.1\,\pi$ to $0.9\,\pi$ with a magnitude of unity in the passband.

Solution The MATLAB Program 6.10 given below is used to design the required Hilbert transformer satisfying the given specifications.

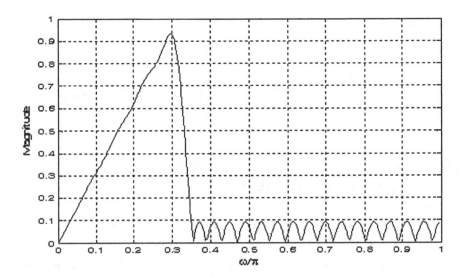

Fig. 6.28 Magnitude response of Type 4 FIR differentiator

Program 6.10 To design linear phase FIR Hilbert Transformer

```
clear all;clc;
N=10;% order of Hilbert Transformer
fpts=[ 0.1 0.9];% passband edges
mag=[ 1 1]% desired magnitude in the passband
h=firpm(N,fpts,mag,'Hilbert')
[Hz,w]=freqz(h,1,256);
h1=abs(Hz);
figure, plot(w/(pi),h1,'-');grid;
xlabel('\omega/\pi');
ylabel('Magnitude');
```

The magnitude response of the Type 3 FIR bandpass Hilbert transformer obtained is shown in Fig. 6.29. The impulse response coefficients of the designed Hilbert transformer are:

$$h(0) = 0$$
$$h(1) = 0.6264 \qquad h(-1) = -0.6264$$
$$h(2) = 0 \qquad h(-2) = 0$$
$$h(3) = 0.183 \qquad h(-3) = -0.183$$
$$h(4) = 0 \qquad h(-4) = 0$$
$$h(5) = 0.1075 \qquad h(-5) = -0.1075$$

We observe that the impulse response coefficients for n even are zero.

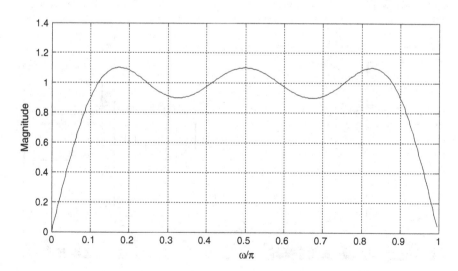

Fig. 6.29 Magnitude response of linear phase FIR Hilbert transformer

6.11 Linear Phase FIR Filter Design Using MATLAB GUI Filter Designer SPTOOL

With the aid of MATLAB GUI filter designer SPTOOL, an FIR filter for given specifications can be designed using the same stepwise procedure outlined in Chap. 5 for IIR filters design using GUI SPTOOL.

Example 6.20 Design an FIR lowpass filter satisfying the specifications of Example 6.12 using MAT LAB GUI SPTOOL.

Solution It can be designed following the stepwise procedure outlined in Chap. 5. After execution of Steps 1 and 2, the SPTOOL window displays the filter characteristics as shown in Fig. 6.30.

From Fig. 6.30, the filter order obtained with MATLAB GUI filter designer SPTOOL is 41. Thus, the ripple in the passband for the designed filter is 0. 0578 which equals $-20\log10 (1 - 0.0578) = 0.5171$ dB.

In Step 3, the name is changed as lpf2500. In Step 4, the lpf2500 is exported to the MATLAB workspace.

In Step 5, the execution of the command $\gg$lpf2500.tf.num displays the 42 coefficients of the designed filter, as given below.

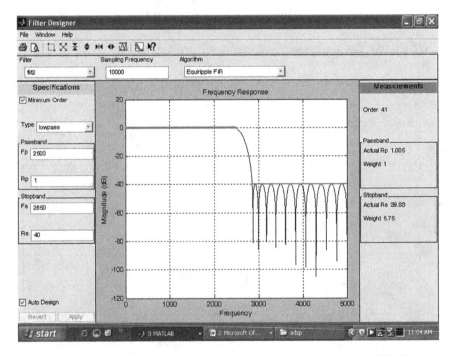

Fig. 6.30 Characteristics of equiripple FIR lowpass filter using MATLAB GUI SPTOOL

$h(1) = 1.162994100164804e-002 = h(42)$

$h(2) = 3.309103309154959e-003 = h(41)$

$h(3) = -1.678966041006541e-002 = h(40)$

$h(4) = -1.759177730507516e-002 = h(39)$

$h(5) = 3.573758441551627e-003 = h(38)$

$h(6) = 4.839870629729102e-003 = h(37)$

$h(7) = -1.574682217783095e-002 = h(36)$

$h(8) = -9.466471803432099e-003 = h(35)$

$h(9) = 1.712212641288460e-002 = h(34)$

$h(10) = 5.095143721625126e-003 = h(33)$

$h(11) = -2.539075267886383e-002 = h(32)$

$h(12) = -2.910221519433109e-003 = h(31)$

$h(13) = 3.293779459779279e-002 = h(30)$

$h(14) = -3.876877148168374e-003 = h(29)$

$h(15) = -4.500050880043559e-002 = h(28)$

$h(16) = 1.505788705350031e-002 = h(27)$

$h(17) = 6.370174039208193e-002 = h(26)$

$h(18) = -3.983455373294787e-002 = h(25)$

$h(19) = -1.071625205542393e-001 = h(24)$

$h(20) = 1.274212094792063e-001 = h(23)$

$h(21) = 4.702377654938166e-001 = h(22)$

In Step 5, the execution of the command $\gg$round (lpf2500.tf.num*2^15) displays the coded coefficients of the designed file listed below.

$h(1) = 381 = h(42)$ $h(7) = -516 = h(36)$

$h(2) = 108 = h(41)$ $h(8) = -310 = h(35)$

$h(3) = -550 = h(40)$ $h(9) = 561 = h(34)$

$h(4) = -576 = h(39)$ $h(10) = 167 = h(33)$

$h(5) = 117 = h(38)$ $h(11) = -832 = h(32)$

$h(6) = 159 = h(37)$

$h(12) = -95 = h(31)$ $h(17) = 2087 = h(26)$

$h(13) = 1079 = h(30)$ $h(18) = -1305 = h(25)$

$h(14) = -127 = h(29)$ $h(19) = -3512 = h(24)$

$h(15) = -1475 = h(28)$ $h(20) = 4175 = h(23)$

$h(16) = 493 = h(27)$ $h(21) = 15409 = h(22)$

Example 6.21 Design an FIR bandpass filter satisfying the specifications of Example 6.14 using MATLAB GUI SPTOOL.

Solution After execution of Steps 1 and 2 with type of filter as bandpass filter and with its specifications, the GUI SPTOOL window displays the filter characteristics as shown in Fig. 6.31.

From Fig. 6.31, the filter order obtained with MATLAB GUI filter designer SPTOOL is 35. The execution of Steps 3, 4, and 5 will display the designed filter coefficients and the coded coefficients.

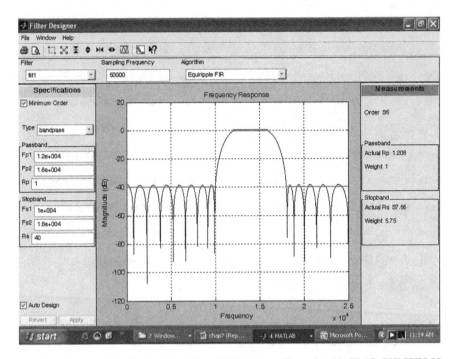

Fig. 6.31 Characteristics designed equiripple FIR bandpass filter using MATLAB GUI SPTOOL

6.12 Effect of FIR Filtering on Voice Signals

Example 6.22 Consider the voice signal from the sound file 'dontwory.wav' [6] as the input and contaminate it with noise, and illustrate the effect of a lowpass filter on the corrupted voice signal.

Solution To read the .wav file, use the following MATLAB function

$$[x, F_T, \text{bps}] = \text{wavread}\,('\text{dontwory.wav}'),$$

where x is the data, F_T is sampling frequency, and bps is bits per second.

The voice signal is converted into a digital signal (x) at a sampling rate of 22050 Hz and is shown in Fig. 6.32.

The digital signal is contaminated with the random noise using the MATLAB command

$$x = x + 0.1 * \text{randn}(\text{size}(x))$$

The contaminated voice signal is shown in Fig. 6.33.

The .wav file corresponding to the noisy signal can be created using the following MATLAB function.

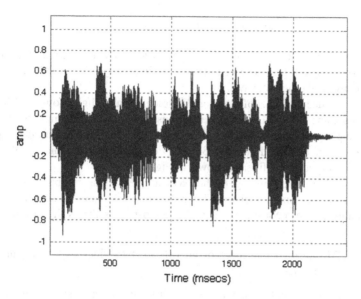

Fig. 6.32 Original voice signal

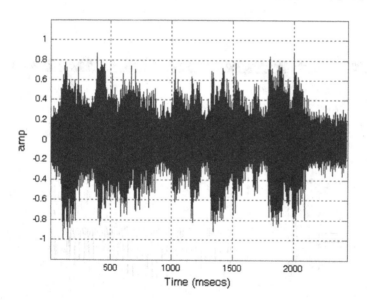

Fig. 6.33 Voice signal contaminated with noise

$$\text{wavwrite } (x, f_s' \text{ dontworynoise.wav}');$$

The noisy .wav file is connected to a speaker to verify the effect of noise on the original voice signal. It is observed that the audio quality is degraded with a lot of hissing noise in the noisy .wav file.

Now, to reduce the effect of noise on the voice signal, a digital lowpass FIR filter is designed using SPTOOL with specifications mentioned in Example 6.12 except for the sampling frequency, which is now 22050 Hz. The filter magnitude response and the order are displayed in Fig. 6.34. The designed filter is applied on the noisy signal. The output of the filter is shown in Fig. 6.35.

The .wav file corresponding to the reconstructed signal is obtained and connected to a speaker, and it is observed that the noise is suppressed significantly and audio quality of the reconstructed signal is almost the same as the original voice signal.

Example 6.23 Illustrate the effect of an FIR lowpass filter with different cutoff frequencies on a voice signal.

Solution In this example, the voice signal from the sound file 'theforce.wav' is sampled at 22050 Hz and shown in Fig. 6.36.

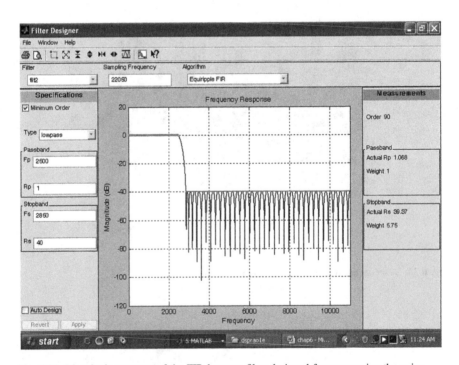

Fig. 6.34 Magnitude response of the FIR lowpass filter designed for suppressing the noise

Fig. 6.35 Reconstruction of
the voice signal after filtering

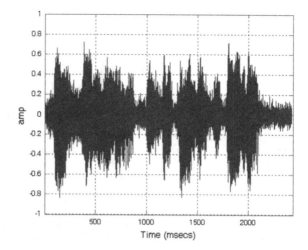

The voice signal shown in Fig. 6.36 is passed through a lowpass filter with different cutoff frequencies 600, 1500, and 3000 Hz. The lowpass filter is designed using the following MATLAB command using Hamming window.

$$h = \mathrm{fir1}(N, Wn),$$

the vector h of length $N+1$ containing the impulse response coefficients of a lowpass FIR filter of order N with a normalized cutoff frequency of Wn between 0 and 1.

First, the lowpass FIR filter with cutoff frequency 600 Hz is applied on the original voice signal shown in Fig. 6.36 and the filtered voice signal is shown in Fig. 6.37.

The corresponding .wav file is created and connected to a speaker. It is observed that it resulted in low clarity and intensity of the voice with the suppression of

Fig. 6.36 Original voice
signal

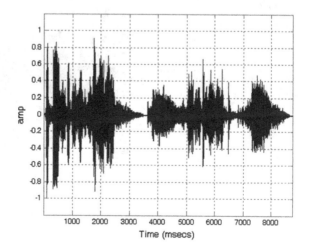

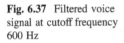

Fig. 6.37 Filtered voice signal at cutoff frequency 600 Hz

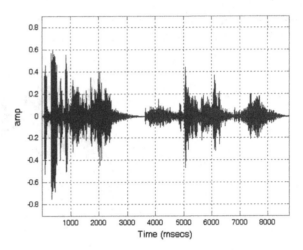

frequency components of the voice signal above 600 Hz. Next, the lowpass FIR filter with cutoff frequency 1500 Hz is applied on the original voice signal and the filtered voice signal is shown in Fig. 6.38. When the wav file of the filtered voice signal is connected to a speaker, it results in improved clarity and intensity of the voice as compared to the voice signal filtered with cutoff frequency 600 Hz.

Finally, the lowpass FIR filter with cutoff frequency 3000 Hz is applied on the original voice signal and the filtered voice signal is shown in Fig. 6.39. When the wav file of the filtered voice signal is connected to a speaker, it results in clarity and intensity of the voice similar to that of the original voice, since the energy of the voice signal is concentrated in the frequency band 0–3000 Hz.

Fig. 6.38 Filtered voice signal at cutoff frequency 1500 Hz

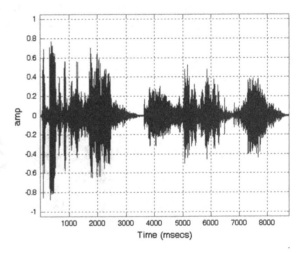

Fig. 6.39 Filtered voice signal at cutoff frequency 3000 Hz

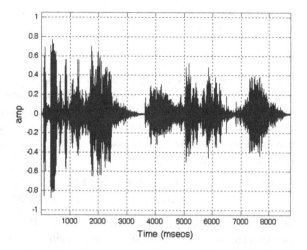

6.12.1   FIR Filters to Recover Voice Signals Contaminated with Sinusoidal Interference

FIR Null Filter

Let the interference signal $j(n)$ be associated with a shift-invariant FIR model such that, for any $n \geq m$,

$$h_j[j(n)] = \sum_{k=0}^{m} h_k j(n - k) = 0, h_j[x(n)] \neq 0, h_m = 1 \qquad (6.121)$$

where the operator $h_j[.]$ is a null filter [7, 8] which annihilates the signal $j(n)$. To reduce the transient that occurs over the interval of m samples, it is desirable to explore the FIR excision filter of the lowest order possible; in the present case, it means that $m = 2$ (a three-coefficient filter).

Consider a narrowband sinusoidal jammer of the form $j(n) = A \sin(\omega_0 n)$, where A is the jammer amplitude. If a three-coefficient filter with coefficients h_0, h_1 and 1 is considered, the jammer at the filter output $j_0(n)$ becomes:

$$j_0(n) = h_0 A \sin(\omega_0 n) + h_1 A \sin(\omega_0 (n - 1)) + A \sin(\omega_0(n - 2))$$

In the above equation, $j_0(n) = 0$, if $a_0 = 1$, and $a_1 = -2\cos(\omega_0)$.

As such, the FIR null filter has impulse response of three coefficients as shown in the following vector

$$h = [1 - 2\cos(\omega_0) 1]$$

where ω_0 is the interference frequency. When the three-coefficient FIR null filter is applied on the signal corrupted with sinusoidal interference, the interference can be suppressed. However, the recovered signal will have self-noise introduced by the null filter itself [7].

FIR Notch Filter

The FIR notch filter is an FIR bandstop filter with a very narrow transition width. The narrowband bandstop filter can be designed using MATLAB SPTOOL. If a narrowband bandstop filter is designed with the notch frequency as the frequency of the sinusoidal interference and applied on the signal contaminated with interference, the interference can be suppressed. However, the order of the FIR narrowband bandstop filter will be high.

Example 6.24 Design schemes using FIR null filters and FIR notch filters to recover a voice signal corrupted by two sinusoidal interferences.

Solution The voice signal from the sound file 'theforce.wav' [9] is considered here as the input signal. The voice signal is digitized with a sampling frequency of 22,050 Hz. The digital voice signal is contaminated with two sinusoidal signals of 900, and 2700 Hz, respectively. The spectrums of the original voice signal and the contaminated signal are shown in Figs. 6.40 and 6.41, respectively.

When the .wav file corresponding to the contaminated voice is connected to a speaker, it is observed that the voice signal is completely jammed. To suppress the interference, two FIR null filters are designed corresponding to the frequencies of the two sinusoidal interferences. The two null filters are connected in cascade as shown in Fig. 6.42.

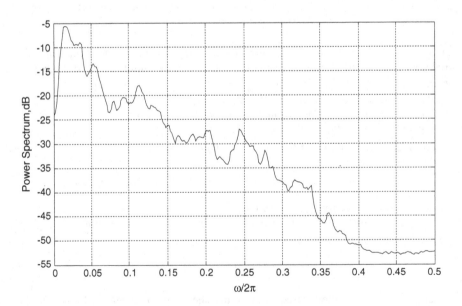

Fig. 6.40 Spectrum of the original voice signal from the sound file 'theforce.wav'

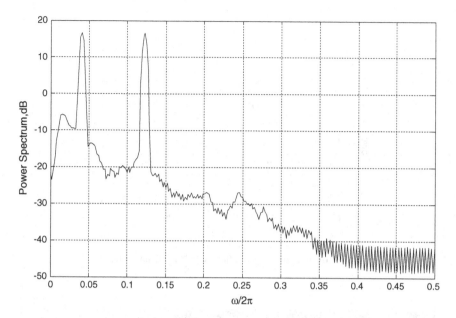

Fig. 6.41 Spectrum of the voice signal contaminated by two sinusoidal interferences

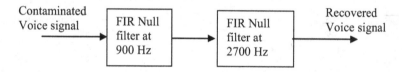

Fig. 6.42 Block diagram for cascade null filter

The contaminated voice signal is fed as the input to the cascade scheme. The output of the first null filter is used as the input to the second null filter. The output of the second null filter is the recovered voice signal.

The spectrum of the recovered signal is shown in Fig. 6.43. From Fig. 6.43, it can be observed that the sinusoidal interferences are suppressed. However, the spectrum of the recovered signal is not exactly the same as the original voice spectrum shown in Fig. 6.40. This may be attributed to the fact that there is self-noise in the recovered signal introduced by the null filters. When the .wav file corresponding to the recovered signal is connected to a speaker, the audio quality of the voice is good, but with some noise.

Now, to see the effect of FIR notch filters on the corrupted voice, two FIR notch filters with the sinusoidal interference frequencies as notch frequencies are designed using MATLAB SPTOOL. The magnitude responses and filter orders are displayed in Figs. 6.44 and 6.45 for the two notch filters with notch frequencies 900 and 2700 Hz, respectively.

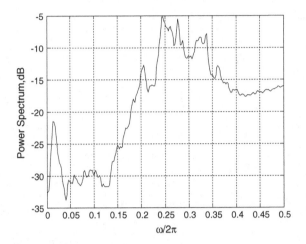

Fig. 6.43 Spectrum of recovered voice signal after cascaded FIR null filters

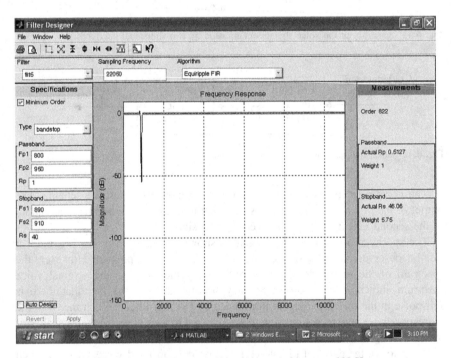

Fig. 6.44 Magnitude response of FIR notch filter with notch frequency 900 Hz

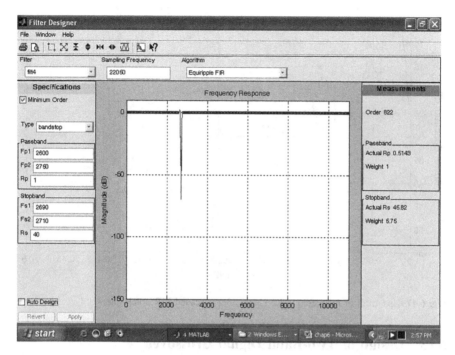

Fig. 6.45 Magnitude response of FIR notch filter with notch frequency 2700 Hz

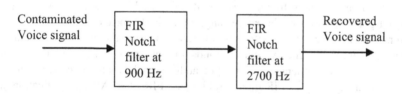

Fig. 6.46 Block diagram for cascaded notch filter

The designed FIR notch filters are connected in cascade as shown in Fig. 6.46.
The spectrum of the recovered voice signal from the cascaded FIR notch filters is
shown in Fig. 6.47. From Fig. 6.47, it can be observed that the sinusoidal interferences are suppressed and further the spectrum of the recovered voice signal is
similar to the original voice signal spectrum. When the .wav file corresponding to
the recovered signal from the notch filters is connected to a speaker, the audio
quality is close to the original voice without any noise. However, from Figs. 6.44
and 6.45, one can observe that the filter orders are high, namely 822.

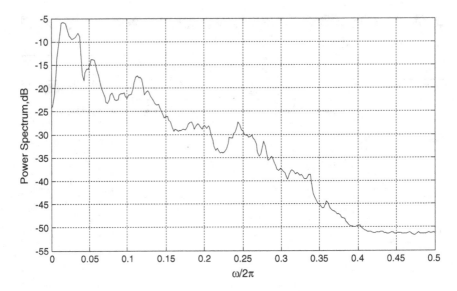

Fig. 6.47 Spectrum of recovered voice signal after cascaded FIR notch filters

6.13 Design of Two-Band Digital Crossover Using FIR Filters

The entire range of audio frequencies are often required in audio systems. However, it is not possible to cover the entire range by a single speaker driver. Hence, two drivers called woofer and tweeter are combined to cover the entire audio range as shown in Fig. 6.48. The woofer responds to low frequencies, and the tweeter responds to high frequencies. The input audio signal is split into two bands in parallel by using lowpass FIR filter and highpass FIR filter. After amplification, the separated low frequencies and high frequencies are sent to the woofer and tweeter,

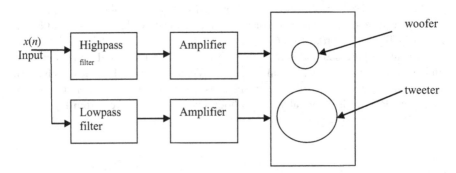

Fig. 6.48 Two-band digital crossover

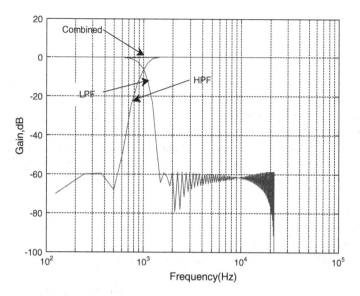

Fig. 6.49 Magnitude frequency response for the digital audio crossover system

respectively. The design of a two-band digital crossover objective is to design the lowpass FIR filter and the highpass FIR filter with sharp transitions such that the combined frequency response is flat preventing distortion in the transition range.

Example 6.25 Design a two-band digital crossover with crossover frequency of 1 kHz for an audio signal sampled at 44.1 kHz

Solution A lowpass filter and a highpass filter with the following specifications are used to design two-band digital crossover with crossover frequency of 1 kHz.

Lowpass filter specifications:
Passband: 0–600 Hz
Stopband edge frequency = 1.4 kHz
Passband ripple: 0.01 dB
Stopband ripple: 60 dB

Highpass filter specifications:
Passband: 1.4–44.1 kHz
Stopband edge frequency = 900 Hz
Passband ripple: 0.01 dB
Stopband ripple: 60 dB

The optimal FIR filter design method is used for both the lowpass and highpass filters. The order of the filter is 178 for both the filters. The magnitude frequency response of the two-digital crossover system is shown in Fig. 6.49. The MATLAB program for the design of the two-band digital crossover system is listed in Program 6.11.

Program 6.11

```
% MATLAB program for two-band digital crossover
clear;clf;
ap=0.01;%pass band ripple in dB
as=60;% stop band attenuation in dB
dp=(1-10^(-ap/20));%peak ripple value in passband,
ds=10^(-as/20);% peak ripple value in the stopbands
fedge=[600 1400];%passband and stopband edge frequencies
mval=[0 1];% desired magnitudes in the passband and stopband
dev=[ds dp];% desired ripple values in the passband and stopband
FT=44100;%sampling frequency
[N1,fpts,mag,wt]=firpmord(fedge,mval,dev,FT);
h=firgr(N1,fpts,mag,wt);
[H,F]=FREQZ(h,1,N1,FT);
semilogx(F,20*log10(abs(H)));
grid;xlabel('Frequency(Hz)');ylabel('Gain,dB')
hold on
mval=[1 0];% desired magnitudes in the passband and stopband
dev=[dp ds];% desired ripple values in the passband and stopband
[N2,fpts,mag,wt]=firpmord(fedge,mval,dev,FT);
h=firgr(N1,fpts,mag,wt);
[H,F]=FREQZ(h,1,N2,FT);
semilogx(F,20*log10(abs(H)));
```

6.14   Comparison Between FIR and IIR Filters

The choice between an IIR filter and an FIR filter depends on the importance attached to the design problem under consideration. A comparison of the advantages of IIR and FIR filters is given in Table 6.9.

6.15   Problems

1. Design an FIR linear phase digital filter using Hanning window using length of 11. The ideal frequency response is

$$H_d(\omega) = \begin{cases} 1 & \text{for } |\omega| \leq \pi/6 \\ 0 & \text{for } \pi/6 \leq |\omega| \leq \pi \end{cases}$$

Table 6.9 Comparison between IIR and FIR filters

	FIR filters	IIR filters
1.	Linear phase designs can be easily achieved	Exact linear phase designs are not possible
2.	Arbitrary frequency response can be readily be approximated arbitrarily and closely for large filter lengths	Arbitrary magnitude response can readily be approximated for small filter lengths
3.	Always stable	Stability is not guaranteed
4.	The number of filter coefficients required for sharp cutoff filters is generally quite large	Designs are very efficient with a small number of poles and zeros, especially for sharp cutoff filters
5.	Filter coefficients can be rounded to reasonable word lengths for most practical designs, and there are no limit cycles	Finite word length effects and limit cycles are to be controlled
6.	Most of the design techniques are iterative and require a personal computer or workstation	Feasible to design by manual computation

2. Design a linear phase bandpass FIR digital filter of appropriate length to meet the following specifications:

   Passband: 1000–2000 Hz
   Sampling frequency: 10 kHz

   Use Bartlett window.

3. Design an FIR lowpass discrete time filter for the following specifications:

$$H_d(\omega) = \begin{cases} 1 & \text{for } 0 \leq f \leq 5 \\ 0 & \text{otherwise} \end{cases}$$

   The sampling frequency is 20 samples/s, and the impulse response is to have duration of 1 s. Use Hamming window and determine the impulse response.

4. Design an FIR lowpass filter using Kaiser window with the following specifications:

   passband edge $\omega_p = 1.5$ rad/s, stopband edge $\omega_s = 2.5$ rad/s, stopband ripple ≥ 40 dB, and sampling frequency is 10 rad/s.

5. Design an FIR highpass filter using Kaiser window with the following specifications:

   Passband edge $\omega_p = 3$ rad/s; stopband edge $\omega_s = 2$ rad/s, Sampling frequency 10 rad/s, and stopband ripple = 0.00562

6. Design an FIR bandpass filter using Kaiser window with the following specifications:

passband: 40–60 rad/s, lower stopband edge: 20 rad/s,
upper stopband edge: 80 rad/s, sampling frequency: 200 rad/s, and
and stopband ripple: 35 dB.

7. Design an FIR bandstop filter using Kaiser window with the following
   specifications:

   passband: 1000–4000 rad/s, lower stopband edge: 2000 rad/s,
   upper stopband edge: 3000 rad/s, sampling frequency: 10,000 rad/s, and
   and stopband ripple: 40 dB.

8. Design a linear phase FIR lowpass filter of order 14 using frequency sampling
   method with cutoff frequency at 0.25π.

9. Design an optimal linear phase FIR lowpass filter of length 3 to meet the
   following specifications:

   Passband edge frequency $= f_p = 500\,\text{Hz}$
   Stopband edge frequency $= f_s = 1500\text{Hz}$
   Tolerance ratio $= (\delta_p/\delta_s) = 3$

   Assume a suitable sampling frequency

10. Design an optimal linear phase FIR lowpass filter of length 5 to meet the
    following specifications:

    Passband edge frequency $= f_p = 300\text{Hz}$
    Stopband edge frequency $= f_s = 650\text{Hz}$
    Tolerance ratio $= (\delta_p/\delta_s) = 2$

    Assume a suitable sampling frequency

6.16 MATLAB Exercises

1. Design a linear phase FIR differentiator of order 41 with passband edge frequency 0.2π, stopband edge frequency 0.3π, passband ripple value 0.01, and
   stopband ripple 40 dB.
2. It is desired to design a highpass filter that has a cutoff frequency of 750 Hz. It is
   known that there is no frequency in the filter input signal above 1000 Hz, and so
   the sampling frequency for the signal is selected as 4000 Hz. Plot the frequency
   response curve (magnitude only) for the filter having 51 coefficients with and
   without a Hamming window.
3. Design a linear phase bandpass FIR Hilbert transformer of order 42. The
   passband is from 0.2π to 0.8π with magnitude of unity.
4. Design a lowpass FIR filter with the following specifications:

   Passband edge: 0.75π
   Stopband edge: 0.85π

Stopband ripple: 40 dB

Use Remez algorithm.

5. Design a lowpass FIR filter with the aid of MATLAB using Kaiser window to meet the following specifications:

   Passband edge frequency: 100 Hz
   Stopband edge frequency: 110 Hz
   Sampling frequency: 1 kHz
   Stopband ripple: 0.01

6. Design an FIR bandpass filter with the aid of MATLAB using Remez algorithm to meet the following specifications:

   Passband edge frequencies: 900–1100 Hz
   Lower stopband edge frequency: 450 Hz
   Upper stopband edge frequency: 1550 Hz
   Sampling frequency: 15 kHz
   Stopband ripple: 0.01

7. Design a minimum-phase lowpass FIR filter with the aid of MATLAB for the following specifications:

   Passband edge at $\omega_p = 0.4\pi$
   Stopband edge at $\omega_s = 0.6\pi$
   Passband of ripple $R_p = 1.5$ dB
   Minimum stopband ripple $R_s = 45$ dB

8. Design an FIR highpass filter with the aid of MATLAB using Kaiser window to meet the following specifications:

   Passband edge at $\omega_p = 0.5\pi$
   Stopband edge at $\omega_s = 0.45\pi$
   Stopband ripple $\delta_s = 0.01$

9. Using MATLAB, design a linear phase FIR bandpass filter using Kaiser window with the following specifications:

   Passband edges: 0.6π and 0.8π
   Stopband edges: 0.5π and 0.7π
   Stopband ripple: 0.001

10. Design an FIR null filter to recover a voice signal corrupted by a sinusoidal signal interference of 1500 Hz. Connect the original voice signal, corrupted voice signal, and recovered voice signal to a speaker and comment on the audio quality of the voice.

11. Design a cascade of FIR notch filters to recover a voice signal corrupted by sinusoidal interferences of 600 and 2100 Hz. Connect the original voice signal, corrupted voice signal, and recovered voice signal to a speaker and comment on the audio quality of the voice.

References

1. T.W. Parks, C.S. Burrus, *Digital Filter Design* (Wiley, New York, NY, 1987)
2. A. Antoniou, *Digital Filters: Analysis and Design* (McGraw Hill Book Co., 1979)
3. J.F.Kaiser. Nonrecursive Digital Filter Design Using the l_o-*sinh Window Function*, in *Proceedings of 1974 IEEE International Symposium on Circuits and Systems*, San Francisco, CA, April 1974, pp. 20–23
4. T.W. Parks, J.H. McClellan, Chebyshev approximation for nonrecursive digital filters with linear phase. IEEE Trans. Circuit Theory, **19**, 189–194 (1972)
5. O. Herrmann, H.W. Schiissler, Design of nonrecursive digital filters with minimum phase. Electron. Lett. **6**, 329–330 (1970)
6. http://www.caselab.okstate.edu/cole/wavs/. Accessed 31 Oct 2008
7. K. Deergha Rao, M.N.S. Swamy, E.I.Plotkin, A nonlinear adaptive filter for narrowband interference mitigation in spread spectrum systems. Sig. Process. **85**, 625–635 (2005)
8. K. Deergha Rao, M.N.S. Swamy, New approach for suppression of FM jamming in GPS receivers. IEEE Trans. Aerosp. Electron. Syst. **42**(4), 1464–1474 (2006)
9. mwavs.com/0038475992/WAVS/Movies/Star_Wars/theforce.wav

Chapter 7
Structures for Digital Filter Realization and Effects of Finite Word Length

Various forms of structures can be developed for the same relationship between the input sequence $x(n)$ and the output sequence $y(n)$. In this chapter, we derive direct form-I, direct form-II, cascade, parallel, and lattice structures for IIR and FIR digital systems. The structural representation provides the relationship among the various variables, as well as ways for implementation of the filter. Since registers are of finite length, the signal variables and the filter coefficients cannot be represented with infinite numerical precision in practical implementation. Although two structures may be equivalent with regard to their input–output characteristics, they may have different behavior for implementation with finite word lengths. Hence, the concern of this chapter is also to analyze the effect of finite word length in implementation of different equivalent structures.

7.1 Signal Flow Graphs and Block Diagram Representation

A signal flow graph [1] is a directed graph consisting of nodes and directed edges, wherein the nodes or vertices represent the variables denoted by $x_1, x_2, \ldots$, and the various directed edges or branches represent the relations between the variables. If $x_j = t_{ji}x_i$, then there exists a branch directed to x_j from x_i; the quantity t_{ji} is called the transmittance of that branch and is labeled as such. If the transmittance is unity, then such a branch will remain unlabeled. If a number of branches are incident upon x_j and originate from nodes $x_1, x_2, \ldots, x_q$, then $x_j = \sum_{i=1}^{q} t_{ji}x_i$. If a node has only incoming branches, then it is called a source (input) node, while if it has only outgoing branches, it is called a sink (output) node. Signal flow graphs can be used to represent linear algebraic equations. We will now see how a difference equation or its z-transformed version can be represented by a signal flow graph (SFG).

© Springer Nature Singapore Pte Ltd. 2018
K. D. Rao and M. N. S. Swamy, *Digital Signal Processing*,
https://doi.org/10.1007/978-981-10-8081-4_7

Consider the difference equation

$$y(n) + a_1 y(n-1) = b_0 x(n) + b_1 x(n-1) \qquad (7.1)$$

This can be represented by the SFG shown in Fig. 7.1a. The same SFG represents the relationship between $Y(z)$ and $X(z)$, the z-transformed variables of $y(n)$ and $x(n)$, where now the nodes 1, 3, and 6 represent the variables $X(z)$, $W(z)$, and $Y(z)$, respectively. It is quite easy to verify that

$$\left(1 + a_1 z^{-1}\right) Y(z) = \left(b_0 + b_1 z^{-1}\right) X(z) \qquad (7.2a)$$

or

$$H(z) = \frac{Y(z)}{X(z)} = \frac{\left(b_0 + b_1 z^{-1}\right)}{\left(1 + a_1 z^{-1}\right)} \qquad (7.2b)$$

The quantity $H(z)$ is called the gain of the SFG; in fact, it is the transfer function of the system given by Eq. (7.1). If we now arrange Eq. (7.2b) as

$$W(z) = \left(b_0 + b_1 z^{-1}\right) X(z)$$
$$Y(z) = W(z) - a_1 z^{-1} Y(z)$$

then we can realize (7.2b) by the block diagram shown in Fig. 7.1b.

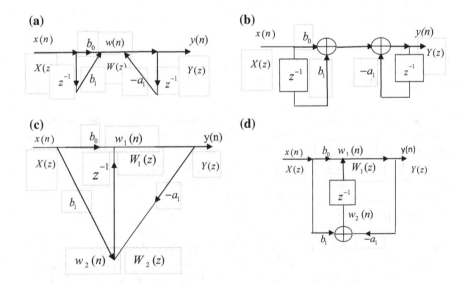

Fig. 7.1 a Flow graph representation of a first-order IIR digital filter, **b** block diagram representation of a first-order IIR digital filter, **c** signal flow graph, **d** block diagram representation

We can also rearrange the difference Eq. (7.1) as

$$
\begin{aligned}
w_1(n) &= w_2(n-1) + b_0 x(n) \\
w_2(n) &= b_1 x(n) - a_1 w_1(n) \\
y(n) &= w_1(n)
\end{aligned}
\tag{7.3}
$$

and obtain the SFG shown in Fig. 7.1c. Using the z-transformed variables, we see that the gain of the SFG of Fig. 7.1c is obtained as

$$
\begin{aligned}
W_1(z) &= b_0 X(z) + b_1 z^{-1} X(z) - a_1 z^{-1} Y(z) \\
Y(z) &= W_1(z)
\end{aligned}
\tag{7.4a}
$$

or

$$
\frac{Y(z)}{X(z)} = \frac{(b_0 + b_1 z^{-1})}{(1 + a_1 z^{-1})}
\tag{7.4b}
$$

Thus, the gain (transfer function) realized by the SFG of Fig. 7.1c is the same as that realized by the SFG of Fig. 7.1a. The corresponding block diagram representation is shown in Fig. 7.1d. Hence, the same transfer function has been realized by two different SFGs or block diagrams.

If we compare Fig. 7.1a, b, or c, d, then we see that there is a direct correspondence between the branches in the SFG and those in the block diagram. The main difference is that the nodes in a SFG can represent adders as well as branching points, while in a block diagram, a special symbol is used to denote adders. Also, in a SFG, a branching point has only one incoming branch, but may have many outgoing ones. Thus, it is quite easy to covert a SFG into a block diagram representation, or vice versa. It should be mentioned that often the difference equations of a SFG are difficult to manipulate when dealing with time variables; however, it is always possible to use the z-transform representation, where all the branch transmittances are simple gains (constants or z^{-1}).

7.1.1 Transposition

When the directions of the arrows in the branches are all reversed in a SFG, but with the transmittance values of the branches unchanged, then the resulting SFG is called the transposed graph. It is clear that the source and sink nodes in the original graph become the sink and source nodes, respectively, in the transposed graph; also, a branching node becomes a summing node and vice versa. Let us now consider the SFG of Fig. 7.2a and find its transpose as well as the gains of these two graphs. For the SFG of Fig. 7.2a, we have

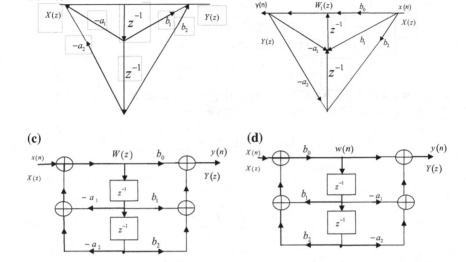

Fig. 7.2 **a** A signal flow graph, **b** transpose of signal flow graph of (**a**), **c** block diagram representation of the flow graph of (**a**), **d** block diagram representation of the flow graph of (**b**)

$$W(z) = X(z) - a_1 z^{-1} W(z) - a_2 z^{-2} W(z)$$

and

$$Y(z) = b_0 W(z) + b_1 z^{-1} W(z) + b_2 z^{-2} W(z)$$

Hence,

$$\frac{Y(z)}{X(z)} = \frac{b_0 + b_1 z^{-1} + b_2 z^{-2}}{1 + a_1 z^{-1} + a_2 z^{-2}} \qquad (7.5a)$$

The corresponding difference equations are

$$
\begin{aligned}
w(n) &= -a_1 w(n-1) - a_2 w(n-2) + x(n) \\
y(n) &= b_0 w(n) + b_1 w(n-1) + b_2 w(n-2)
\end{aligned}
\qquad (7.5b)
$$

The transposed SFG of Fig. 7.2a is shown in Fig. 7.2b. For this SFG, we can write

$$W_3(z) = b_2 X(z) - a_2 Y(z)$$
$$W_2(z) = z^{-1} W_3(z) - a_1 Y(z) + b_1 X(z)$$
$$W_1(z) = b_0 X(z) + z^{-1} W_2(z)$$
$$Y(z) = W_1(z)$$

Hence,

$$\frac{Y(z)}{X(z)} = \frac{b_0 + b_1 z^{-1} + b_2 z^{-2}}{1 + a_1 z^{-1} + a_2 z^{-2}} \tag{7.6a}$$

Also, the corresponding difference equations are

$$
\begin{aligned}
w_1(n) &= b_0 x(n) + w_2(n-1) \\
y(n) &= w_1(n) \\
w_2(n) &= -a_1 y(n) + b_1 x(n) + w_3(n-1) \\
w_3(n) &= -a_2 y(n) + b_2 x(n)
\end{aligned}
\tag{7.6b}
$$

Equations (7.5b) and (7.6b) are two different ways of arranging the computation of the samples $y(n)$ from $x(n)$. However, it is not directly evident that they are equivalent, even though we know that they are, since the corresponding gains in both the cases are the same, namely as given by (7.5a) and (7.6a). In fact, the set of Eqs. (7.5b) and (7.6b) is both equivalent to the second-order difference equation

$$y(n) + a_1 y(n-1) + a_2 y(n-2) = b_0 x(n) + b_1 x(n-1) + b_2 x(n-2) \tag{7.7}$$

The result that we have obtained, namely that the gains of the original SFG and its transpose are the same, can be shown to be true for any SFG with one input and one output, using Mason's gain formula for SFGs [1]. Since one can easily obtain the block diagram representation from a SFG or vice versa, we see that a transposed block diagram structure can be easily obtained from a given digital structure by simply reversing the direction of all the arrows in the various branches; the branching nodes in the original will now, of course, become adders in the transposed structure, while the adders in the original will become branching nodes in the transposed structure. Thus, an alternate structure can always be obtained for a given digital structure using the operation of transposition. The block diagram representations of SFGs of Fig. 7.2a and its transpose are shown in Fig. 7.2c, d, respectively.

7.2   IIR Filter Structures

7.2.1   Direct Form I Structure

The transfer function of an Nth-order IIR filter is of the form

$$H(z) = \frac{Y(z)}{X(z)} = \frac{b_0 + b_1 z^{-1} + \cdots + b_N z^{-N}}{1 + a_1 z^{-1} + \cdots + a_N z^{-N}} \tag{7.8}$$

$$= H_b(z) H_a(z), \quad \text{(say)} \tag{7.9a}$$

where

$$H_a(z) = \frac{Y(z)}{W(z)} = \frac{1}{1 + a_1 z^{-1} + \cdots + a_N z^{-N}} \tag{7.9b}$$

$$H_b(z) = \frac{W(z)}{X(z)} = b_0 + b_1 z^{-1} + \cdots + b_N z^{-N} \tag{7.9c}$$

The functions $H_a(z)$ and $H_b(z)$ can easily be realized and then cascaded to obtain $H(z)$; the resulting structure is shown in Fig. 7.3. This is called *direct form-I structure* and has $2N$ delay elements. We can easily obtain its transpose to get an alternate structure. Also, from Eqs. (7.9b) and (7.9c), we have the following corresponding difference equations:

$$w(n) = b_0 x(n) + b_1 x(n-1) + \cdots + b_N x(n-N)$$
$$y(n) = w(n) - a_1 y(n-1) + \cdots + a_N y(n-N)$$

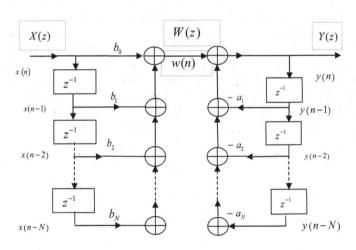

Fig. 7.3 Direct form-I IIR filter structure

7.2.2 Direct Form II Structure

Instead of writing Eq. (7.8) in the form of (7.9a–7.9c), suppose we write it as

$$H(z) = \frac{Y(z)}{X(z)} = H_a(z)H_b(z) \tag{7.10a}$$

where

$$H_a(z) = \frac{W(z)}{X(z)} = \frac{1}{1 + a_1 z^{-1} + \cdots + a_N z^{-N}} \tag{7.10b}$$

$$H_b(z) = \frac{Y(z)}{W(z)} = b_0 + b_1 z^{-1} + \cdots + b_N z^{-N} \tag{7.10c}$$

Then,

$$W(z) = X(z) - \left(a_1 z^{-1} + \cdots + a_N z^{-N}\right) W(z) \tag{7.11a}$$

and

$$Y(z) = \left(b_0 + b_1 z^{-1} + \cdots + b_N z^{-N}\right) W(z) \tag{7.11b}$$

Equations (7.11a, 7.11b) can be realized by the structure as shown in Fig. 7.4a. This is called *direct form-II structure* and utilizes only N delay elements, and hence is a canonic structure. The difference equations corresponding to Eqs. (7.11a, 7.11b) are

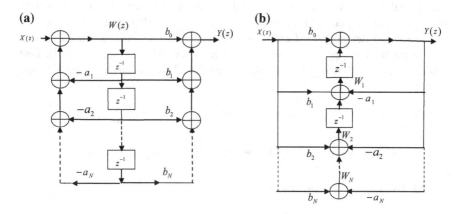

Fig. 7.4 a Direct form-II IIR filter structure and **b** transpose of the direct form-II IIR filter structure

$$w(n) = x(n) - a_1 w(n-1) - \cdots - a_N w(n-N) \qquad (7.12a)$$

$$y(n) = b_0 w(n) + b_1 w(n-1) + \cdots + b_N w(n-N) \qquad (7.12b)$$

The transpose of the direct form-II structure is shown in Fig. 7.4b, and of course, is also canonic. The corresponding difference equations are

$$
\begin{aligned}
y(n) &= w_1(n-1) + b_0 x(n) \\
w_i(n) &= w_{i+1}(n-1) - a_i y(n) + b_i x(n), \quad i = 1, 2, \ldots, N-1 \qquad (7.13) \\
w_N(n) &= b_N x(n) - a_N y(n)
\end{aligned}
$$

7.2.3   Cascade Structure

The transfer function given by Eq. (7.8) can be expressed as

$$H(z) = H_1(z) H_2(z) \ldots H_r(z) \qquad (7.14)$$

where $H_i(z)$ is a first-order or a second-order transfer function. The overall transfer function can be realized as a cascade of the individual transfer functions, as shown in Fig. 7.5.

The direct form-II can be implemented for each of the sections. As an example, the realization of the fourth-order IIR filter transfer function

$$H(z) = \frac{(b_{01} + b_{11}z^{-1} + b_{21}z^{-2})(b_{02} + b_{12}z^{-1} + b_{22}z^{-2})}{(1 + a_{11}z^{-1} + a_{21}z^{-2})(1 + a_{12}z^{-1} + a_{22}z^{-2})}$$

using two second-order sections in cascade is shown in Fig. 7.6. In Fig. 7.6, each of the second-order sections has been realized using the direct-II form shown in Fig. 7.2c. The output result is not very much influenced by the ordering of the numerator and denominator factors. On the other hand, quantization noise can be minimized with an appropriate ordering of each second-order section [2, 3]. One could have used the transposed structure of Fig. 7.2d for either or both of the sections.

Fig. 7.5 Cascade form IIR filter structure

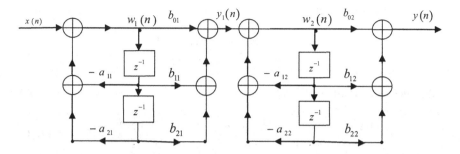

Fig. 7.6 Fourth-order IIR filter with two direct form II sections in cascade

Example 7.1 Obtain the cascade realization of the following third-order IIR transfer function:

$$H(z) = 0.05634 \frac{1+z^{-1}}{1-0.683z^{-1}} \frac{1-1.0166z^{-1}+z^{-2}}{1-1.4461z^{-1}+0.7957z^{-2}}$$

Solution The transfer function $H(z)$ can be written as

$$H(z) = KH_1(z)H_2(z)$$

where

$$K = \frac{X_1(z)}{X(z)}, \quad H_1(z) = \frac{Y_1(z)}{X_1(z)} = \frac{1+z^{-1}}{1-0.683z^{-1}} \quad \text{and}$$

$$H_2(z) = \frac{Y(z)}{Y_1(z)} = \frac{1-1.0166z^{-1}+z^{-2}}{1-1.4461z^{-1}+0.7957z^{-2}}$$

A possible cascade realization using the direct-II form structure of Fig. 7.4a for $H_1(z)$ and $H_2(z)$ is shown in Fig. 7.7. Using Eqs. (7.12a) and (7.12b), we see that the difference equations for the first-order and second-order sections are as follows.

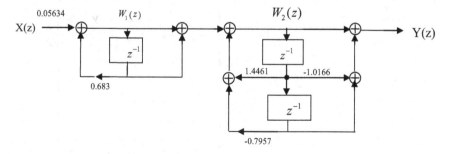

Fig. 7.7 Cascade realization of the third-order IIR filter of Example 7.1

For the first-order section:

$$w_1(n) = x(n) + 0.683w_1(n-1)$$
$$y_1(n) = w_1(n) + w_1(n-1)$$

(7.15)

For the second-order section:

$$w_2(n) = y_1(n) + 1.4461w_2(n-1) - 0.7957w_2(n-2)$$
$$y(n) = w_2(n) - 1.0166w_2(n-1) + w_2(n-2)$$

(7.16)

7.2.4   Parallel Form Structure

Equation (7.8) can be expanded by partial fractions in the form

$$H(z) = C + H_1(z) + H_2(z) + \cdots + H_r(z)$$

(7.17a)

where C is a constant and $H_i(z)$ is a first- or second-order function of the form $z/(1 + a_1 z^{-1})$ or $(b_0 + b_1 z^{-1} + b_2 z^{-2})/(1 + a_1 z^{-1} + a_2 z^{-2})$. This is known as the *parallel form* realization of $H(z)$ and is shown in Fig. 7.8. Each of the functions $H_i(z)$ may be realized by using the direct form-II structure. The output is given by

$$y(n) = Cx(n) + \sum_{i=1}^{r} y_i(n)$$

(7.17b)

Fig. 7.8 Parallel form IIR filter structure

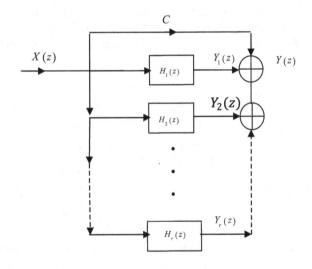

Example 7.2 A causal LTI system has system function

$$H(z) = \frac{1 - 2.3z^{-1} - 1.7z^{-2}}{(1 - 0.5z^{-1})(1 + 0.5z^{-1})(1 + 0.8z^{-1})}$$

Obtain an implementation of the system in parallel form using first- and second-order direct form-II sections. Draw the corresponding SFG and its transpose. Find the gain (or the transfer function) of the transposed SFG.

Solution The transfer function can be written in partial fraction form as

$$H(z) = \frac{-1 - 1.5z^{-1}}{(1 - 0.25z^{-2})} + \frac{2}{(1 + 0.8z^{-1})}$$

The corresponding realization using direct form-II structure is shown in Fig. 7.9a. The corresponding SFG can be directly drawn and is shown Fig. 7.9b. The transpose of this SFG is shown in Fig. 7.9c, which is obtained by simply reversing the directions of the arrows in Fig. 7.9b. It is straightforward to show that

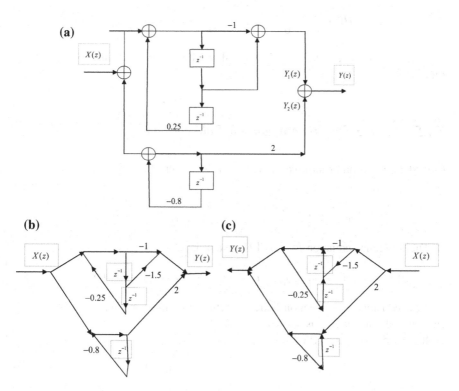

Fig. 7.9 **a** Parallel form realization of the IIR filter of Example 7.2, **b** SFG corresponding to block diagram of (**a**), **c** transpose of SFG of (**a**)

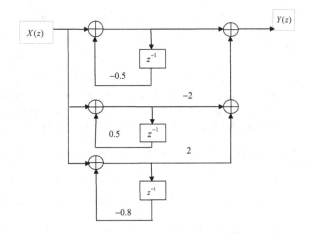

Fig. 7.10 An alternate parallel form realization of the IIR filter of Example 7.2

the gain $Y(z)/X(z)$ of the SFG of Fig. 7.9c is the same as that of the original graph, namely $H(z)$.

It is noted that we could have expressed $H(z)$ as

$$H(z) = \frac{1}{1+0.5z^{-1}} + \frac{-2}{1-0.5z^{-1}} + \frac{2}{1+0.8z^{-1}}$$

and realized it in parallel form using three first-order sections, as shown in Fig. 7.10.

7.2.5 Lattice Structure of an All-Pole System

Consider an Nth-order all-pole system with the system function

$$H_N(Z) = \frac{1}{1+a_{N1}z^{-1}+a_{N2}z^{-2}+\cdots+a_{N(N-1)}z^{-(N-1)}+a_{NN}z^{-N}} \quad (7.18)$$

The difference equation for this IIR system is

$$y(n) = -a_{N1}y(n-1) - a_{N2}y(n-2) - \cdots - a_{NN}y(n-N) + x(n) \quad (7.19)$$

Now consider the N-section lattice structure, as shown in Fig. 7.11a. A typical section, the mth section, is shown in Fig. 7.12b.

It is seen from Fig. 7.11 that

$$w_{m-1}(n) = w_m(n) - k_m s_{m-1}(n-1) \quad (7.20)$$

(a)

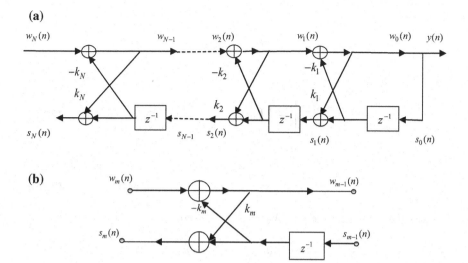

(b)

Fig. 7.11 **a** An N-section lattice structure realizing an all-pole filter and **b** the mth section of the lattice filter

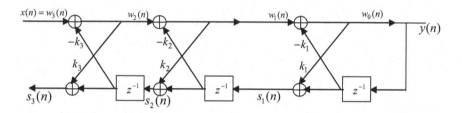

Fig. 7.12 A three-section lattice structure realizing a third-order all-pole filter

$$s_m(n) = k_m w_{m-1}(n) + s_{m-1}(n-1) \qquad (7.21)$$

with

$$w_0(n) = s_0(n) = y(n) \qquad (7.22)$$

$$w_N(n) = x(n) \qquad (7.23)$$

Taking z-transforms of the above equations, we get

$$\begin{bmatrix} W_m(z) \\ S_m(z) \end{bmatrix} = \begin{bmatrix} 1 & k_m z^{-1} \\ k_m & z^{-1} \end{bmatrix} \begin{bmatrix} W_{m-1}(z) \\ S_{m-1}(z) \end{bmatrix}, \quad m = 1, 2, \ldots, N \qquad (7.24a)$$

with

$$W_0(z) = S_0(z) = Y(z), \quad \text{and} \quad W_N(z) = X(z) \tag{7.24b}$$

Thus,

$$\begin{bmatrix} W_N(z) \\ S_N(z) \end{bmatrix} = \prod_{m=1}^{N} \begin{bmatrix} 1 & k_m z^{-1} \\ k_m & z^{-1} \end{bmatrix} \begin{bmatrix} W_0(z) \\ S_0(z) \end{bmatrix} \tag{7.25a}$$

with

$$W_0(z) = S_0(z) = Y(z), \quad \text{and} \quad W_N(z) = X(z) \tag{7.25b}$$

From the above, we see that $W_N(z)$ is of the form

$$W_N(z) = \left(a_{N0} + a_{N1}z^{-1} + \cdots + a_{NN}z^{-N} \right) W_0(z)$$

or

$$X(z) = \left(a_{N0} + a_{N1}z^{-1} + \cdots + a_{NN}z^{-N} \right) Y(z) \tag{7.26}$$

It is clear from Fig. 7.11 that $a_{N0} = 1$ and $a_{NN} = k_N$. Hence, from Eq. (7.26) we get

$$H_N(z) = \frac{Y(z)}{X(z)} = \frac{1}{1 + a_{N1}z^{-1} + \cdots + a_{NN}z^{-N}} = \frac{1}{A_N(z)} \tag{7.27}$$

where $a_{NN} = k_N$, thus resulting in the all-pole transfer function (7.18). We now need to find the values of the lattice coefficients k_i's in terms of a_{Nj}, $j = 1, 2, \ldots, N$. Let us denote by $H_m(z)$ and $\widetilde{H}_m(z)$, the transfer functions $W_0(z)/W_m(z)$ and $S_m(z)/S_0(z)$ for an m-section lattice; that is, let

$$H_m(z) = \frac{1}{A_m(z)} = \frac{W_0(z)}{W_m(z)} = \frac{1}{1 + \sum_{i=1}^{m} a_{mi}z^{-i}} \tag{7.28}$$

and

$$\widetilde{H}_m(z) = \widetilde{A}_m(z) = \frac{S_m(z)}{S_0(z)} \tag{7.29}$$

Then, from Eqs. (7.24a, 7.24b), we get

$$\begin{bmatrix} A_m(z) \\ \tilde{A}_m(z) \end{bmatrix} = \begin{bmatrix} 1 & k_m z^{-1} \\ k_m & z^{-1} \end{bmatrix} \begin{bmatrix} A_{m-1}(z) \\ \tilde{A}_{m-1}(z) \end{bmatrix}, \quad m = 1, 2, \ldots, N \tag{7.30a}$$

with

$$A_0(z) = \tilde{A}_0(z) = 1 \tag{7.30b}$$

It can be shown by mathematical induction that

$$A_m(z) = A_{m-1}(z) + k_m z^{-m} A_{m-1}(z^{-1}) \tag{7.31}$$

and

$$\tilde{A}_m(z) = z^{-m} A_m(z^{-1}) \tag{7.32}$$

Now substituting the expressions for $A_m(z)$ and $\tilde{A}_m(z)$ in Eq. (7.31), we get

$$1 + \sum_{i=1}^{m-1} a_{mi} z^{-i} + a_{mm} z^{-m} = \left[1 + \sum_{i=1}^{m-1} a_{m-1,i} z^{-i} \right] + k_m z^{-m} \left[1 + \sum_{i=1}^{m-1} a_{m-1,i} z^{i} \right] \tag{7.33}$$

Comparing the coefficients of z^{-i} on both sides of Eq. (7.33), we have

$$k_m = a_{mm} \tag{7.34}$$

$$a_{mi} = a_{m-1,i} + a_{m-1,m-i} a_{mm} \tag{7.35}$$

Also, the coefficients of z^{m-i} give

$$a_{m,m-i} = a_{m-1,m-i} + a_{m-1,i} a_{mm} \tag{7.36}$$

Solving Eqs. (7.35) and (7.36), we get

$$a_{m-1,i} = \frac{a_{mi} - a_{m,m-i} a_{mm}}{1 - a_{mm}^2}, \quad m = N, N-1, \ldots, 2 \text{ and } i = 1, 2, \ldots, N-1. \tag{7.37}$$

Using (7.34) and (7.37), we can successively calculate the lattice coefficients $k_N, k_{N-1}, \ldots, k_1$, which are also known as the reflection coefficients. Thus, for example, for a third-order all-pole function realized by the lattice structure, we have

$$k_3 = a_{33}$$

$$k_2 = a_{22} = \frac{a_{32} - a_{31}a_{33}}{1 - a_{33}^2}, \quad a_{21} = \frac{a_{31} - a_{32}a_{33}}{1 - a_{33}^2} \tag{7.38}$$

$$k_1 = a_{11} = \frac{a_{21} - a_{21}a_{22}}{1 - a_{22}^2}$$

The following are some important properties of the lattice structure.

(i)   If $A_N(z)$ has all its zeros inside the unit circle, then the lattice structure will have all its coefficients k_i to have their magnitudes less than unity. Otherwise, the system would be unstable. Thus, the stability of the given all-pole function is automatically checked for when deriving the lattice structure.

(ii) Since from (7.32), $\widetilde{A}_N(z) = z^{-N}A_N(z^{-1})$, we have

$$\widetilde{A}_N(z) = a_{NN} + a_{N,N-1}z^{-1} + \cdots + z^{-N} \tag{7.39}$$

Hence, the zeros of $\widetilde{A}_N(z)$ are reciprocals of those of $A_N(z)$.

(iii) From Eq. (7.29), it is known that the transfer function

$$\widetilde{H}_N(z) = \frac{S_N(z)}{S_0(z)} = \widetilde{A}_N(z) \tag{7.40}$$

and thus $\widetilde{H}_N(z)$ realizes the FIR transfer function given by Eq. (7.39).

(iv) From Eqs. (7.28) and (7.29), we have

$$H_N(z)\widetilde{H}_N(z) = \frac{W_0(z)}{W_N(z)}\frac{S_N(z)}{S_0(z)} = \frac{S_N(z)}{X(z)} = \frac{1}{\widetilde{A}_N(z)}A_N(z)$$

$$= \frac{a_{NN} + a_{N,N-1}z^{-1} + \cdots + z^{-N}}{1 + a_{N1}z^{-1} + \cdots + a_{NN}z^{-N}}$$

Hence, the lattice structure of Fig. 7.11 can be used to realize the all pass transfer function

$$H_{AP}(z) = \frac{S_N(z)}{X(z)} = \frac{a_{NN} + a_{N,N-1}z^{-1} + \cdots + z^{-N}}{1 + a_{N1}z^{-1} + \cdots + a_{NN}z^{-N}} \tag{7.41}$$

Example 7.3 Obtain the cascade lattice structure for the following third-order all-pole IIR transfer function

$$H(z) = \frac{1}{1 + 0.54167z^{-1} + 0.625z^{-2} + 0.333z^{-3}}$$

Solution Using Eq. (7.38), we realize $H(z)$ in the form of a three-section lattice structure, as shown in Fig. 7.12 with the following lattice coefficients:

$$k_3 = a_{33} = 0.333$$
$$k_2 = a_{22} = \frac{a_{32} - a_{31}a_{33}}{1 - a_{33}^2} = 0.50008, \quad a_{21} = \frac{a_{31} - a_{32}a_{33}}{1 - a_{33}^2} = 0.37501$$
$$k_1 = a_{11} = \frac{a_{21} - a_{21}a_{22}}{1 - a_{22}^2} = 0.25008.$$

7.2.6 Gray–Markel's Lattice Structure for a General IIR Filter

Suppose we are given an IIR filter of the form

$$H_N(z) = \frac{B_N(z)}{A_N(z)} = \frac{b_{N0} + b_{N1}z^{-1} + \cdots + b_{NN}z^{-N}}{1 + a_{N1}z^{-1} + \cdots + a_{NN}z^{-N}} \tag{7.42}$$

We first obtain the all-pole filter of Fig. 7.11 corresponding to the transfer function $1/A_N(z)$. Then, we know that the transfer function of the m-section lattice is given by

$$\begin{aligned} \frac{S_m(z)}{X(z)} &= \frac{S_m(z)}{S_0(z)}\frac{S_0(z)}{X(z)} = \frac{S_m(z)}{S_0(z)}\frac{W_0(z)}{X(z)} \\ &= \frac{\widetilde{A}_m(z)}{A_N(z)}, \text{ using Eqs.(7.40) and (7.28)} \end{aligned} \tag{7.43}$$

Substituting Eqs. (7.32) in (7.43), we get

$$S_m(z) = \frac{z^{-m}A_m(z^{-1})}{A_N(z)}X(z) \tag{7.44}$$

We can now take the outputs $S_0(z), S_1(z), \ldots, S_N(z)$ and multiply them by the feedforward multipliers $\alpha_0, \alpha_1, \ldots, \alpha_N$, respectively, and add all the resulting quantities, as shown in Fig. 7.13.

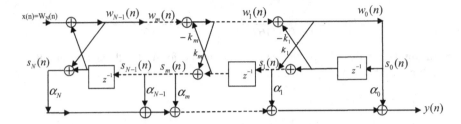

Fig. 7.13 Gray–Markel's lattice structure to realize a general IIR filter of order N

Then the output $Y(z)$ is related to $X(z)$ by

$$\frac{Y(z)}{X(z)} = \frac{\sum_{m=0}^{N} \alpha_m z^{-m} A_m(z^{-1})}{A_N(z)} \tag{7.45}$$

It is seen that the numerator of Eq. (7.45) is an Nth degree polynomial in z^{-1}. If we now equate (7.42) to (7.45), we have

$$H_N(z) = \frac{Y(z)}{X(z)} = \frac{\sum_{m=0}^{N} \alpha_m z^{-m} A_m(z^{-1})}{A_N(z)} = \frac{B_N(z)}{A_N(z)} \tag{7.46}$$

Thus,

$$B_N(z) = \sum_{m=0}^{N} \alpha_m z^{-m} A_m(z^{-1}) \tag{7.47}$$

or

$$\sum_{m=0}^{N} b_{nm} z^{-m} = \sum_{m=0}^{N} \alpha_m \left[a_{mm} + a_{m,m-1} z^{-1} + \cdots + a_{m1} z^{-(m-1)} + z^{-m} \right] \tag{7.48}$$

Hence,

$$\alpha_N = b_{NN}$$
$$\alpha_m = b_{Nm} - \sum_{i=m+1}^{N} \alpha_i a_{i,i-m}, \quad m = N - 1, N - 2, \ldots, 0 \tag{7.49}$$

Thus, the feedforward multiplier values can be determined from the given IIR transfer function coefficients. For example, for a third-order IIR transfer function given by

$$H_3(z) = \frac{B_3(z)}{A_3(z)} = \frac{b_{30} + b_{31}z^{-1} + b_{32}z^{-2} + b_{33}z^{-3}}{1 + a_{31}z^{-1} + a_{32}z^{-2} + a_{33}z^{-3}} \tag{7.50}$$

we have

$$
\begin{aligned}
\alpha_3 &= b_{33} \\
\alpha_2 &= b_{32} - \alpha_3 a_{31} \\
\alpha_1 &= b_{31} - \alpha_3 a_{32} - \alpha_2 a_{21} \\
\alpha_0 &= b_{30} - \alpha_3 a_{33} - \alpha_2 a_{22} - \alpha_1 a_{11}
\end{aligned}
\tag{7.51}
$$

Since the lattice structure requires more multiplication operations, it is computationally inefficient as compared to the direct form-II or cascade structures. However, the lattice structures are less sensitive to quantization errors [4–6].

Example 7.4 Obtain the Gray–Markel realization of the following third-order IIR transfer function

$$H(z) = 0.05634 \frac{1 + z^{-1}}{1 - 0.683z^{-1}} \frac{1 - 1.0166z^{-1} + z^{-2}}{1 - 1.4461z^{-1} + 0.7957z^{-2}} \tag{7.52}$$

Solution The transfer function $H(z)$ can be written in the form

$$H(z) = \frac{b_{30} + b_{31}z^{-1} + b_{32}z^{-2} + b_{33}z^{-3}}{1 + a_{31}z^{-1} + a_{32}z^{-2} + a_{33}z^{-3}}$$

where

$$b_{30} = 0.05634, \quad b_{31} = b_{32} = -(9.3524)10^{-4}, \quad b_{33} = 0.05634$$
$$a_{31} = -2.1291, \quad a_{32} = 1.7834, \quad a_{33} = -0.5435$$

Using Eq. (7.38), we can get the lattice reflection coefficients as

$$k_3 = a_{33} = -0.5435$$
$$k_2 = a_{22} = \frac{a_{32} - a_{31}a_{33}}{1 - a_{33}^2} = 0.8888, \quad a_{21} = \frac{a_{31} - a_{32}a_{33}}{1 - a_{33}^2} = -1.6461$$
$$k_1 = a_{11} = \frac{a_{21} - a_{21}a_{22}}{1 - a_{22}^2} = -0.8715.$$

x(n)

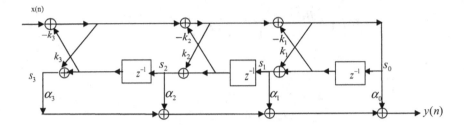

Fig. 7.14 A three-section Gray–Markel's structure

The values of the feedforward multipliers are calculated using Eq. (7.50) as

$$\alpha_3 = b_{33} = 0.05634,$$
$$\alpha_2 = b_{32} - \alpha_3 a_{31} = 0.1190,$$
$$\alpha_1 = b_{31} - \alpha_3 a_{32} - \alpha_2 a_{21} = 0.0945,$$
$$\text{and} \quad \alpha_0 = b_{30} - \alpha_3 a_{33} - \alpha_2 a_{22} - \alpha_1 a_{11} = 0.0635$$

The Gray–Markel realization of $H(z)$ is shown in Fig. 7.14 with the above values for the lattice coefficients and the feedforward multipliers.

7.3   Realization of IIR Structures Using MATLAB

Example 7.5 Obtain the cascade realization for the system described by the following difference equation:

$$y(n) = - 1.25y(n - 1) - y(n - 2) - 0.5625y(n - 3) - 0.125y(n - 4) \\ + x(n) - 0.75x(n - 2) + 0.75x(n - 3) + 0.5x(n - 4) \tag{7.53}$$

Solution Applying z-transform on both sides, we get

$$Y(z)(1 + 1.25z^{-1} + z^{-2} + 0.5625z^{-3} + 0.125z^{-4}) \\ = X(z)(1 - 0.75z^{-2} + 0.75z^{-3} + 0.5z^{-4})$$

$$\frac{Y(z)}{X(z)} = H(z) = \frac{(1 - 0.75z^{-2} + 0.75z^{-3} + 0.5z^{-4})}{(1 + 1.25z^{-1} + z^{-2} + 0.5625z^{-3} + 0.125z^{-4})}$$

For cascade realization, $H(z)$ is to be written in the following form:

$$H(z) = H_1(z)H_2(z)$$

Using the following MATLAB Program 7.1, it can be written as the product of second-order sections.

Program 7.1 Cascade Realization of an IIR Filter

```
clear; clc;
num=[1 0 -0.75 0.75 0.5];
den=[1 1.25 1 0.5625 0.125];
[z,p,k]=tf2zp(num,den);
sos=zp2sos(z,p,k);
sos =
    1.0000    1.5000    0.5000    1.0000    1.0000    0.2500
    1.0000   -1.5000    1.0000    1.0000    0.2500    0.5000
```

From the results of the program shown above, $H(z)$ can be written as the product of second-order sections as given below:

$$H(z) = \frac{(1 + 1.5z^{-1} + 0.5z^{-2})}{(1 + z^{-1} + 0.25z^{-2})} \frac{(1 - 1.5z^{-1} + z^{-2})}{(1 + 0.25z^{-1} + 0.5z^{-2})} \tag{7.54}$$

where

$$H_1(z) = \frac{(1 + 1.5z^{-1} + 0.5z^{-2})}{(1 + z^{-1} + 0.25z^{-2})}; \quad H_2(z) = \frac{(1 - 1.5z^{-1} + z^{-2})}{(1 + 0.25z^{-1} + 0.5z^{-2})};$$

The cascade realization of $H(z)$ is shown in Fig. 7.15.

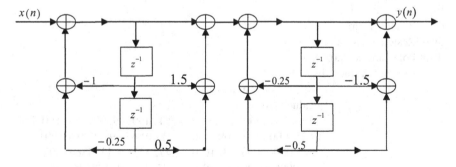

Fig. 7.15 Cascade realization of a system function $H(z)$

Example 7.6 Obtain the parallel form structure for the fourth-order IIR transfer function

$$
\begin{aligned}
6y(n) &= 12y(n-1) - 11y(n-2) + 5y(n-3) - y(n-4) + 9x(n) \\
&\quad + 33x(n-1) + 57x(n-2) + 33x(n-3) + 12x(n-4)
\end{aligned}
\tag{7.55}
$$

Solution Applying z-transform on both sides of the given difference equation, we get

$$
\begin{aligned}
Y(z)(6 - 12z^{-1} &+ 11z^{-2} - 5z^{-3} + z^{-4}) \\
&= X(z)(9 + 33z^{-1} + 57z^{-2} + 33z^{-3} + 12z^{-4})
\end{aligned}
$$

$$
\frac{Y(z)}{X(z)} = \frac{(9 + 33z^{-1} + 57z^{-2} + 33z^{-3} + 12z^{-4})}{(6 - 12z^{-1} + 11z^{-2} - 5z^{-3} + z^{-4})}
\tag{7.56}
$$

For parallel realization, $H(z)$ is to be written in the following form:

$$
H(z) = \text{constant} + H_1(z) + H_2(z)
$$

Using the MATLAB Program 7.2, it can be written in the above form.

Program 7.2 Realization of an IIR Filter in Parallel Form

```
clear;clc;
num=[9 33 57 33 12];
den=[6 -12 11 -5 1];
[r1,p1,k1]=residuez(num,den);
R1=[r1(1) r1(2)];
P1=[p1(1) p1(2)];
[b1,a1]=residuez(R1, P1, 0)
R2=[r1(3) r1(4)];
P2=[p1(3) p1(4)];
[b2,a2]=residuez(R2, P2, 0)
disp('Parallel form')
disp('Residues are');disp(r1);
disp('Poles are');disp(p1);
disp('Constant value');disp(k1);
```

                            Parallel form
                    Residues are
                    -3.525000000000111e+001 +7.124999999999898e+001i
                    -3.525000000000109e+001 -7.124999999999902e+001i
                    3.000000000000101e+001 -1.472243186433536e+002i
                    3.000000000000101e+001 +1.472243186433536e+002i
                    Poles are
                    5.000000000000013e-001 +5.000000000000007e-001i

> 5.000000000000013e-001 -5.000000000000007e-001i
> 4.999999999999986e-001 +2.886751345948124e-001i
> 4.999999999999986e-001 -2.886751345948124e-001i
> Constant value 12

b1 =

-70.5000 - 0.0000i -36.0000 + 0.0000i   0

    a1 =

            1.000000000000000e+000 -1.000000000000003e+000 5.000000
000000020e-001

b2 =

            6.000000000000203e+001 5.499999999999838e+001 0

    a2 =

            1.000000000000000e+000 -9.999999999999971e-001 3.333333
333333316e-001

From the results of the program as shown above, $H(z)$ as can be written as the sum of two second-order sections.

$$H(z) = 12 + \frac{-70.5 - 36z^{-1}}{(1 - z^{-1} + 0.5z^{-2})} + \frac{60 + 55z^{-1}}{(1 - z^{-1} + 0.3333z^{-2})}$$

where

$$H_1(z) = \frac{-70.5 - 36z^{-1}}{(1 - z^{-1} + 0.5z^{-2})}; \quad H_2(z) = \frac{60 + 55z^{-1}}{(1 - z^{-1} + 0.3333z^{-2})};$$

The parallel form realization of $H(z)$ using second-order direct form-II sections is shown in Fig. 7.16.

Example 7.7 Obtain the parallel form structure for the following system

$$y(n) = -0.375y(n-1) + 0.09375y(n-2) + 0.015625y(n-3)$$
$$+ x(n) + 3x(n-1) + 2x(n-1) \tag{7.57}$$

Solution Applying z-transform on both sides of the given difference equation, we get

$$Y(z)(1 + 0.375z^{-1} - 0.09375z^{-2} - 0.015625z^{-3})$$
$$= X(z)(1 + 3z^{-1} + 2z^{-2})$$

$$\frac{Y(z)}{X(z)} = H(z) = \frac{(1 + 3z^{-1} + 2z^{-2})}{(1 + 0.375z^{-1} - 0.09375z^{-2} - 0.015625z^{-3})} \tag{7.58}$$

For parallel realization, $H(z)$ is to be written in the following form.

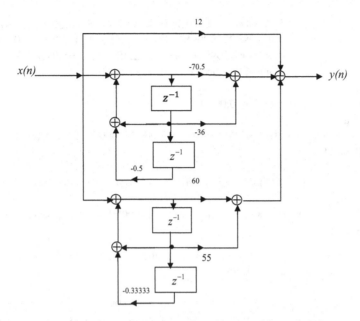

Fig. 7.16 Parallel form structure realizing the transfer function of Example 7.6

$$H(z) = H_1(z) + H_2(z) + H_3(z)$$

Using the following MATLAB Program 7.3, it can be written in the above form.

Program 7.3 Realization of an IIR filter in Parallel Form

```
clear; clc;
num=[1 3 2];
den=[1 0.375 -0.09375 -0.015625];
[r,p,k]=residuez(num,den);
disp('Parallel form')
disp('Residues are');disp(r);
disp('Poles are');disp(p);
disp('Constant value');disp(k);
```

                            Parallel form
            Residues are   Poles are
            2.666666666666661e+000   -5.000000000000002e-001
            9.999999999999984e+000   2.500000000000005e-001
            -1.166666666666664e+001 -1.249999999999999e-001

From the results of the program as shown above, $H(z)$ can be written as the sum of three second-order sections.

$$H(z) = \frac{2.667}{(1 + 0.5z^{-1})} + \frac{10}{(1 - 0.25z^{-1})} + \frac{11.667}{(1 + 0.125z^{-1})} \qquad (7.59)$$

where

$$H_1(z) = \frac{2.667}{(1 + 0.5z^{-1})}; \quad H_2(z) = \frac{10}{(1 - 0.25z^{-1})}; \quad H_3(z) = -\frac{11.667}{(1 + 0.125z^{-1})}$$

Thus, $H(z)$ can be realized in the parallel form using three first-order sections.

Example 7.8 Obtain the cascade lattice structure for the following all-pole IIR transfer function.

$$H(z) = \frac{1}{1 + 0.75z^{-1} + 0.5z^{-2} + 0.25z^3} \qquad (7.60)$$

Solution The MATLAB Program 7.4 is used to obtain lattice parameters of the given transfer function.

Program 7.4 Cascaded Lattice Structure of all-pole IIR Transfer Function

```
%k is the lattice parameter vector
num=1;%num is the numerator coefficient vector
den=[1 0.75 0.5 0.25];%den is the denominator coefficient vector
num=num/den(1);
den=den/den(1);
k=tf2latc(num,den);
disp('Lattice parameters are');disp(k');
```

The lattice parameters obtained from the above program are
$$5.0000\text{e-}001 \quad 3.3333\text{e-}001 \quad 2.5000\text{e-}001$$
Hence,

$$k_1 = 0.5, \quad k_2 = 0.3333 \quad \text{and} \quad k_3 = 0.25.$$

The lattice structure realizing (7.60) is as shown in Fig. 7.12 with the above values for the lattice coefficients k_1, k_2 and k_3.

Example 7.9 Obtain the Gray–Markel cascade lattice structure for the following IIR transfer function

$$H(z) = \frac{1 + 2z^{-1} + 2z^{-2} + z^{-3}}{1 + 0.54167z^{-1} + 0.625z^{-2} + 0.333z^{-3}} \qquad (7.61)$$

Solution The following MATLAB Program 7.5 is used to obtain the Gray–Markel cascade lattice structure

Program 7.5 Gray–Markel Cascaded Lattice Structure

%k is the lattice parameter vector
%alpha is the vector of feedforward multipliers
num=[1 2 2 1];%num is the numerator coefficient vector
den=[1 0.54167 0.625 0.333];%den is the denominator coefficient vector
num=num/den(1);den=den/den(1);
[k,alpha]=tf2latc(num,den);
disp('Lattice parameters are');disp(k');
disp('Feedforward multipliers are');disp(fliplr(alpha'));

The lattice parameters and feedforward multipliers obtained from the above program are as given below.

Lattice parameters are

2.500834329976927e-001　　5.000769195297325e-001
3.330000000000000e-001

Feedforward multipliers are

1.0e+000　　1.45833e+000　　8.279156878612457e-001
-2.693251715107814e-001

Thus,

$$k_1 = 0.25, \quad k_2 = 0.5, \quad k_3 = 0.333$$
$$\alpha_0 = -0.2695, \quad \alpha_1 = 0.8281, \quad \alpha_2 = 1.4583, \quad \alpha_3 = 1$$

The Gray–Markel lattice structure of $H(z)$ is shown in Fig. 7.14, where the values of the lattice and feedforward parameters are as given above.

7.4　FIR Filter Structures

7.4.1　Direct Form Realization

The system function of an FIR filter can be written as

$$H(z) = \sum_{n=0}^{N} h(n)z^{-n} = h(0) + h(1)z^{-1} + h(2)z^{-2} + \cdots + h(N)z^{-N} \quad (7.62)$$

$$Y(z) = h(0)X(z) + h(1)z^{-1}X(z) + h(2)z^{-2}X(z) + \cdots + h(N)z^{-N}X(z) \quad (7.63)$$

Fig. 7.17 Direct form realization of the FIR transfer function

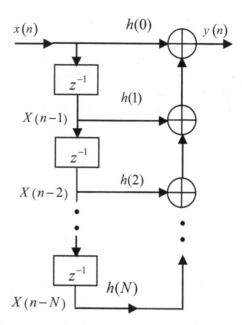

The FIR filter given by Eq. (7.63) can be realized as shown in Fig. 7.17.

Example 7.10 Obtain the direct form realization of the following system function.

$$H(z) = 1 + 2z^{-1} + 0.5z^{-2} - 0.5z^{-3} - 0.5z^{-4}$$

Solution For the given transfer function, we can write the output $Y(z)$ as

$$Y(z) = X(z) + 2z^{-1}X(z) + 0.5z^{-2}X(z) - 0.5z^{-3}X(z) - 0.5z^{-4}X(z)$$

Hence, the direct form realization of $H(z)$ is shown in Fig. 7.18.

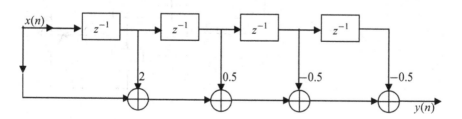

Fig. 7.18 Direct form realization of transfer function $H(z)$ of Example 7.10

7.4.2   *Cascade Realization*

The FIR transfer function given by Eq. (7.60) can be realized as a cascade of FIR sections with each section characterized by a second-order transfer function if N is even, or by one first-order section, the remaining sections being second-order ones, if N is odd. The FIR transfer function of Eq. (7.62) can be written as

$$H(z) = \prod_{k=1}^{\frac{N}{2}} \left(b_{k0} + b_{k1}z^{-1} + b_{k2}z^{-2} \right), \quad N \text{ even} \tag{7.64}$$

or as

$$H(z) = \left(1 + b_{10}z^{-1} \right) \prod_{k=2}^{\frac{N+1}{2}} \left(b_{k0} + b_{k1}z^{-1} + b_{k2}z^{-2} \right), \quad N \text{ odd} \tag{7.65}$$

Each of the sections can be realized in the direct form and then cascaded, as was done in the case of an IIR filter. For illustration, consider the following example.

Example 7.11 Obtain the cascade realization of the FIR filter

$$H(z) = 1 + 2.5z^{-1} + 2z^{-2} + 2z^{-3} \tag{7.66}$$

Solution The given FIR filter function can be written as

$$H(z) = \left(1 + 2z^{-1} \right)\left(1 + 0.5z^{-1} + z^{-2} \right)$$

The cascade form realization using the direct form for the two sections is shown in Fig. 7.19.

Fig. 7.19 Cascade form realization of FIR transfer function

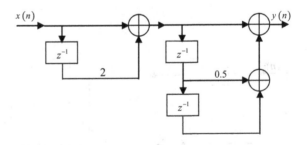

7.4.3   Linear Phase FIR Structures

The symmetry or the antisymmetric property of a linear phase FIR filter can be exploited to reduce the total number of multipliers to almost half of that required in the direct form implementation of the FIR filter. Consider the realization of a Type 1 linear phase FIR filter of order $2N$. It has a symmetric impulse response, and its transfer function is of the form

$$H(z) = h(0) + h(1)z^{-1} + h(2)z^{-2} + \cdots + h(N)z^{-N} + h(1)z^{-(2N-1)} + h(0)z^{-2N} \tag{7.67}$$

which can be rewritten in the form

$$H(z) = h(0)\left[1 + z^{-2N}\right] + h(1)\left[z^{-1} + z^{-(2N-1)}\right] + \cdots + h(N)z^{-N} \tag{7.68}$$

A direct realization of $H(z)$ based on the above decomposition is shown in Fig. 7.20.

Similarly, the transfer function of a Type 2 linear phase FIR filter of order $(2N + 1)$ given by

$$H(z) = h(0) + h(1)z^{-1} + h(2)z^{-2} + \cdots + h(1)z^{-2N} + h(0)z^{-(2N+1)} \tag{7.69}$$

can be rewritten as

$$H(z) = h(0)\left[1 + z^{-(2N+1)}\right] + h(1)\left[z^{-1} + z^{-2N}\right] + \cdots + h(N)\left[z^{-N} + z^{-(N+1)}\right] \tag{7.70}$$

This can be realized in direct form as shown in Fig. 7.21.

Example 7.12 Realize the following Type 3 linear phase FIR transfer function of length-7 with antisymmetric impulse response.

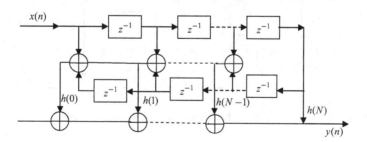

Fig. 7.20 Realization of Type 1 FIR linear phase FIR linear filter

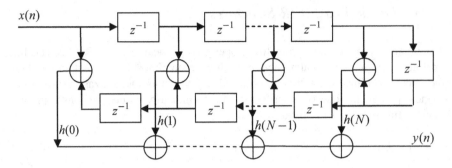

Fig. 7.21 Realization of Type 2 FIR linear FIR filter

$$H(z) = 1 + 2z^{-1} + 3z^{-2} + 4z^{-3} - 3z^{-4} - 2z^{-5} - z^{-6}$$

Solution The transfer function can be rewritten as

$$H(z) = \left(1 - z^{-6}\right) + 2(z^{-1} - z^{-5}) + 3(z^{-2} - z^{-4}) + 4z^{-3}$$

The above can be realized in direct form as shown in Fig. 7.22.

Example 7.13 Obtain a cascade realization for the following system function with minimum number of multipliers

$$H(z) = \left(1 + 2z^{-1}\right)\left(0.5 + z^{-1} + 0.5z^{-2}\right)\left(1 + 0.33z^{-1} + z^{-2}\right)$$

Solution The given transfer function $H(z)$ is a product of a first-order filter function and two second-order Type 1 linear phase FIR filters; each of these second-order sections can be realized using minimum number of multipliers. The cascade realization is shown in Fig. 7.23.

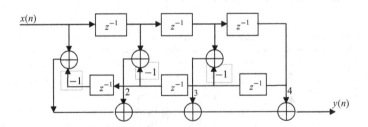

Fig. 7.22 Realization of Type 3 FIR linear phase transfer function of length-7

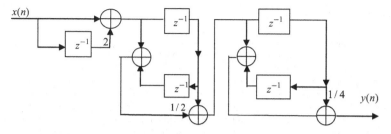

Fig. 7.23 Realization of $H(z)$

7.4.4   FIR Lattice Structures

Consider an Nth-order FIR system described by the following system function

$$H_N(Z) = 1 + b_{N1}z^{-1} + b_{N2}z^{-2} + \cdots + b_{N(N-1)}z^{-(N-1)} + b_{NN}z^{-N} \qquad (7.71)$$

The difference equation for this system is

$$y(n) = x(n) + b_{N1}x(n-1) + b_{N2}x(n-2) + \cdots + b_{NN}x(n-N) \qquad (7.72)$$

Now consider the N-section lattice structure shown in Fig. 7.24a, the mth section of which is shown in Fig. 7.24b.

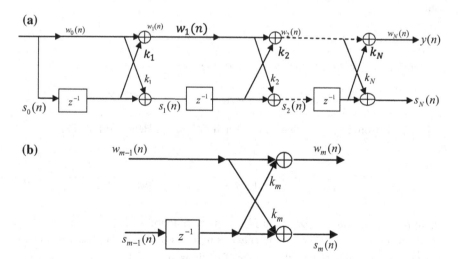

Fig. 7.24   a An N-section lattice structure realizing an FIR filter, **b** the mth section of the lattice structure

It is seen from Fig. 7.24 that

$$
\begin{bmatrix} w_m(n) \\ s_m(n) \end{bmatrix} = \begin{bmatrix} 1 & k_m \\ k_m & 1 \end{bmatrix} \begin{bmatrix} w_{m-1}(n) \\ s_{m-1}(n-1) \end{bmatrix}, \quad m = 1, 2, \ldots, N \qquad (7.73)
$$

with

$$
w_0(n) = s_0(n) = x(n) \qquad (7.74)
$$

$$
w_N(n) = y(n) \qquad (7.75)
$$

Taking z-transforms of the above equations, we get

$$
\begin{bmatrix} W_m(z) \\ S_m(z) \end{bmatrix} = \begin{bmatrix} 1 & k_m z^{-1} \\ k_m & z^{-1} \end{bmatrix} \begin{bmatrix} W_{m-1}(z) \\ S_{m-1}(z) \end{bmatrix}, \quad m = 1, 2, \ldots, N \qquad (7.76a)
$$

with

$$
W_0(z) = S_0(z) = X(z), \quad \text{and} \quad W_N(z) = Y(z) \qquad (7.76b)
$$

Thus,

$$
\begin{bmatrix} W_N(z) \\ S_N(z) \end{bmatrix} = \prod_{m=1}^{N} \begin{bmatrix} 1 & k_m z^{-1} \\ k_m & z^{-1} \end{bmatrix} \begin{bmatrix} W_0(z) \\ S_0(z) \end{bmatrix} \qquad (7.77a)
$$

with

$$
W_0(z) = S_0(z) = X(z), \quad \text{and} \quad W_N(z) = Y(z) \qquad (7.77b)
$$

From the above we see that $W_N(z)$ is of the form

$$
Y(z) = W_N(z) = \left(b_{N0} + b_{N1} z^{-1} + \cdots + b_{NN} z^{-N} \right) X(z) \qquad (7.78)
$$

It is clear from Fig. 7.24 that $b_{N0} = 1$ and $b_{NN} = k_N$. Hence, Eq. (7.78) reduces
to

$$
H_N(z) = \frac{Y(z)}{X(z)} = 1 + b_{N1} z^{-1} + \cdots + b_{NN} z^{-N} \qquad (7.79)
$$

where $b_{NN} = k_N$, thus realizing the FIR transfer function (7.71). We now need to
find the values of the lattice coefficients k_i's in terms of $b_{Nj}, j = 1, 2, \ldots, N$. Let us
denote by $H_m(z)$ and $\widetilde{H}_m(z)$, the transfer functions $W_m(z)/W_0(z)$ and $S_m(z)/S_0(z)$
for an m-section lattice; that is, let

$$H_m(z) = \frac{W_m(z)}{W_0(z)} = B_m(z) = 1 + \sum_{i=1}^{m} b_{mi} z^{-i} \tag{7.80}$$

and

$$\widetilde{H}_m(z) = \widetilde{B}_m(z) = \frac{S_m(z)}{S_0(z)} \tag{7.81}$$

Substituting these in Eqs. (7.76a, 7.76b), we see that

$$\begin{bmatrix} B_m(z) \\ \widetilde{B}_m(z) \end{bmatrix} = \begin{bmatrix} 1 & k_m z^{-1} \\ k_m & z^{-1} \end{bmatrix} \begin{bmatrix} B_{m-1}(z) \\ \widetilde{B}_{m-1}(z) \end{bmatrix}, \quad m = 1, 2, \ldots, N \tag{7.81a}$$

with

$$B_0(z) = \widetilde{B}_0(z) = 1 \tag{7.81b}$$

These are identical to Eqs. (7.30a, 7.30b), with $B_m(z)$ and $\widetilde{B}_m(z)$ replacing $A_m(z)$ and $\widetilde{A}_m(z)$, respectively. Hence, we can establish that

$$B_m(z) = B_{m-1}(z) + k_m z^{-m} B_{m-1}(z^{-1}) \tag{7.82}$$

$$\widetilde{B}_m(z) = z^{-m} B_m(z^{-1}) \tag{7.83}$$

$$k_m = b_{mm} \tag{7.84}$$

and

$$b_{m-1,i} = \frac{b_{mi} - b_{m,m-i} b_{mm}}{1 - b_{mm}^2}, \quad m = N, N-1, \ldots, 2 \quad \text{and} \quad i = 1, 2, \ldots, N-1. \tag{7.85}$$

Just as in the case of the IIR filter, using (7.84) and (7.85), we can successively calculate the lattice coefficients $k_N, k_{N-1}, \ldots, k_1$. Also, in view of (7.83), we see that the transfer function

$$\frac{S_N(z)}{X(z)} = \widetilde{B}_m(z) = z^{-m} B_m(z^{-1}) = b_{NN} + b_{N,N-1} z^{-1} + \cdots + z^{-N} \tag{7.86}$$

realizes an FIR filter function, whose zeros are the reciprocals of those of the FIR filter realized by $Y(z)/X(z)$.

Example 7.14 Obtain a lattice realization for the FIR system described by the following difference equation

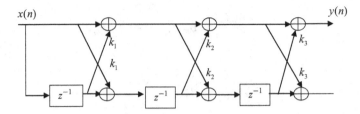

Fig. 7.25 Lattice realization of $H(z)$ of the FIR filter of Example 7.14

$$y(n) = x(n) + \frac{3}{4}x(n-1) + \frac{1}{2}x(n-2) + \frac{1}{4}x(n-3)$$

Solution Using (7.84) and (7.85), we have

$$k_3 = b_{33} = \frac{1}{4};$$

$$b_{21} = \frac{b_{31} - b_{33}b_{32}}{1 - b_{33}^2} = \frac{\frac{3}{4} - \frac{1}{4}\cdot\frac{1}{2}}{1 - \left(\frac{1}{4}\right)^2} = \frac{2}{3};$$

$$k_2 = b_{22} = \frac{b_{32} - b_{33}b_{31}}{1 - b_{33}^2} = \frac{\frac{1}{2} - \frac{1}{4}\cdot\frac{3}{4}}{1 - \left(\frac{1}{4}\right)^2} = \frac{1}{3}$$

$$k_1 = b_{11} = \frac{b_{21} - b_{22}b_{21}}{1 - b_{22}^2} = \frac{\frac{2}{3} - \frac{1}{3}\cdot\frac{2}{3}}{1 - \left(\frac{1}{3}\right)^2} = \frac{1}{2};$$

Hence, the FIR lattice coefficients are

$$k_1 = \frac{1}{2}, \quad k_2 = \frac{1}{3}, \quad k_3 = \frac{1}{4}.$$

The lattice realization of the given FIR filter is shown in Fig. 7.25.

7.4.5 Realization of FIR Lattice Structure Using MATLAB

The function tf2latc in MATLAB can be used to compute the lattice coefficients of the cascade lattice structure of Fig. 7.24. Program 7.7 given in the following example is used to illustrate the computation of lattice coefficients.

Example 7.15 Realize the following FIR transfer function using the lattice structure

$$H(z) = 1 + 0.75z^{-1} + 0.5z^{-2} + 0.25z^{-3}$$

Solution The following MATLAB Program 7.6 is used for cascade lattice structure realization.

Program 7.6 FIR Cascade Lattice Realization

```
clear;clc;
num=[1 0.75 0.5 0.25];
k=tf2latc(num);
disp('Lattice coefficients are');disp(fliplr(k)');
```

Lattice coefficients are

5.0000e-001   3.3333e-001   2.5000e-001

Hence,

$$k_1 = 0.5, \quad k_2 = 0.3333 \quad \text{and} \quad k_3 = 0.25$$

The lattice realization of the given FIR filter is shown in Fig. 7.25, with above values for the lattice coefficients k_1, k_2 and k_3.

7.5   Effect of Finite Word Length

7.5.1   Number Representation

Any number N can be represented with finite precision in the form

$$N = \sum_{i=-n}^{m} \alpha_i r^i, \quad 0 \le \alpha_i \le (r-1) \tag{7.87}$$

where α_i is the ith coefficient and r is the radix of the representation. For example, when $r = 10$, we use the symbols 0, 1, 2,..., 9 to represent the distinct $(r-1)$ admissible values of α_i, and we have the decimal system. When $r = 2$, the only distinct admissible values for α_i are 0 and 1, and we have the binary representation; the 0s and 1s are called binary digits or bits. For example,

$$N = 1 \times 2^4 + 0 \times 2^3 + 0 \times 2^2 + 1 \times 2^1 + 1 \times 2^0 + 0 \times 2^{-1} + 1 \times 2^{-2} \tag{7.88}$$

would represent the binary number $(N)_2 = 10011.01$, where the dot, called the binary dot or binary point, separates the two parts of the number, namely those for which the radix power $i \ge 0$ and those for which $i < 0$; thus, the integer part and the fractional part are separated by the binary point. In digital computers and signal processors, the numbers are represented in terms of the binary digits. From (7.88), it is seen that in decimal representation, $(N)_{10} = 19.01$. It is quite easy to convert a decimal number $(N)_{10}$ to binary form using the following three steps: (i) divide the integer part repeatedly by 2 and arrange the remainders in the reverse order,

(ii) multiply the fractional part repeatedly by 2, remove the integer parts at each step, and arrange them in the same order, and (iii) place the binary point between the results of (i) and (ii). For example, consider the number $(N)_{10} = 13.375$. The steps are shown below to get the equivalent binary representation.

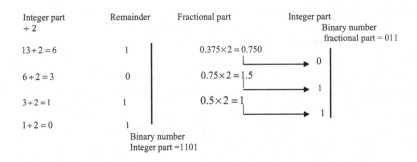

Therefore, the decimal number 13.375 can be represented in binary form as 1101.011; that is, $(13.375)_{10} = (1101.011)_2$. The binary number $(1101.011)_2$ uses seven bits, the left four bits of the binary point representing the integer part and the right three bits representing the fractional part. Hence, it is referred to as a seven-bit binary number, or a binary number with a word length of 7. In general, an $(m + n)$-bit binary number with m integer bits and n fractional bits is represented by

$$b_{m-1}b_{m-2}...b_1b_0 \cdot b_{-1}b_{-2}...b_{-n}$$

where b_i represents either a 0 or 1; the bit b_{m-1} is the most significant bit (MSB), and the bit b_{-n} is the least significant bit (LSB). Since a register has a finite word length, the digital representation of a number can assume only a finite range of values. For example, the set of all m-bit positive integers is confined to be in the range $0 \leq N \leq 2^m - 1$. This range is known as the dynamic range. As long as the arithmetic operations are such that the resulting number is within this range, the result is faithfully represented; otherwise, it cannot be and in such a case, we say that an overflow has occurred.

Basic Types of Number Representations

The two basic types of binary representations of numbers that are used in digital computers and other digital circuits are fixed-point and floating-point representations.

Fixed-Point Representation

The location of the binary point is held at a fixed position in the fixed-point representation for all of the arithmetic operations. In fixed-point representation, a sign bit is used at the MSB position of the register, which specifies the sign of the number whether it is an integer, a fraction, or a mixed number. For a positive number, the sign bit is 0. For example, the number $(01011.01)_2$ represents

$(+11.25)_{10}$. There are three methods that are most commonly used to represent a negative number, but in all the three methods the positive number is represented the same way, namely by assigning the sign bit as 0. The three methods for representing the negative numbers are the (i) signed magnitude, (ii) one's complement, and (iii) two's complement. In the signed magnitude method, the sign bit for a negative number is 1. Hence, the number $(11011.01)_2$ represents $(-11.25)_{10}$. In the one's complement method, the negative number is formed by taking one's complement of the corresponding positive number (i.e., the negative number is formed by changing all the 0s to 1s and vice versa in the corresponding positive number). For example, the decimal number -0.375 can be represented in the one's complement method as follows:

$$(0.375)_{10} = (0.011)_2$$

Hence,

$$(-0.375)_2 = \text{one's complement of } (0.011)_2 = (1.100)_2$$

In the two's complement method, the negative number is formed by taking the two's complement of the corresponding positive number. Hence, it can be formed by first taking the one's complement of the corresponding positive number, and then adding a 1 to the LSB. For example,

$$(0.375)_{10} = (0.011)_2$$
$$\text{one's complement of } (0.011)_2 = 1.100$$
$$\text{2's complement of } (0.011)_2 = 1.100 + 0.001$$

Hence, in the two-s complement method,

$$(-0.375)_{10} = (1.101)_2.$$

If two fixed-point numbers each of b-bits are added, the result may need more than b-bits to be represented. For example, consider $(0.875)_{10} + (0.375)_{10} = (1.25)_{10}$. If four bits including the sign bit are used to represent the two numbers as shown below,

$$(0.875)_{10} = 0.111_2$$
$$(0.375)_{10} = 0.011_2$$
$$\overline{1.010_2} = (-0.25)_{10}$$

then a carry is passed on to the sign bit position, thus giving the summation result incorrectly as $(-0.25)_{10}$. This condition is treated as an overflow, since the actual result $(1.25)_{10}$ cannot be represented in a four-bit format (including the sign bit). In general, fixed-point addition results in overflow.

If two numbers each of b-bits are multiplied, the result, in general, is of length $2b$-bits. The product of two integers is also an integer; however, the result may overflow. Similarly, the product of two fractional numbers is also a fractional number, but in this case, there is no overflow. In digital signal processing applications, it is essential to approximate the $2b$-bit product of two b-bit numbers by b-bits. Hence, in such applications it is assumed that only proper fractions are utilized, since the number of bits in a product can be simply reduced by *truncation* or *rounding*; it is not possible to reduce the number of bits by truncation or rounding in the case of a product of two integers. In truncation, all the bits after the bth bit is dropped; in the case of rounding to b-bits, if the $(b + 1)$th bit is a 0, then the bth bit is left unaltered, while if it is a 1, then we add a 1 to the bth bit. There is no loss of generality in this assumption, since all numbers can be suitably scaled so that they are all fractions. We illustrate truncation and rounding with the following example.

Example 7.16 Convert the number $(0.93)_{10}$ into binary notation having five bits including the sign bit.

$$
\begin{array}{ll}
 & \text{integer part} \\
0.93 \times 2 = 1.86 & 1 \\
0.86 \times 2 = 1.72 & 1 \\
0.72 \times 2 = 1.44 & 1 \\
0.44 \times 2 = 0.88 & 0 \\
0.88 \times 2 = 1.76 & 1
\end{array}
$$

Hence,

$$
\begin{aligned}
(0.93)_{10} &= (0.1110)_2 = (0.875)_{10} & \text{by truncation} \\
&= (0.1111)_2 = (0.9375)_{10} & \text{by rounding}
\end{aligned}
$$

It is seen that the absolute value of the error due to rounding is less than that due to truncation.

In most of the digital signal processing applications, the two's complement is used to represent negative numbers, since it can be easily implemented. We can represent a decimal fraction by a $(b + 1)$-bit binary number in two's complement form as follows:

$$
(N)_{10} = \alpha_0 + \sum_{i=1}^{b} \alpha_i 2^{-i} \tag{7.89}
$$

where α_0 is the sign bit and $\alpha_i = 0$ or 1. The numbers that can be represented by the above fall in the range $-1 \le (N)_2 \le 1 - 2^{-b}$.

It was seen in Example 7.16 that the absolute value of the error due to rounding is less than that due to truncation, and this is true in general; hence, it is preferable to use rounding than to use truncation to reduce the number of bits of a given number in the binary representation. Also, the absolute value of this round-off error decreased as the word length is increased.

Floating-Point Number Representation

The fixed-point system of representation has two main drawbacks. First, the range of numbers that can be represented as mentioned earlier is limited to the range $-1 \leq (N)_2 \leq 1 - 2^{-b}$. Second, as the magnitude of the number decreases, the error due to truncation or rounding increases. A partial solution to these is the floating-point number representation, where each number consists of three parts: a sign bit (s), an exponent (E), and a mantissa (M); a floating-point number is expressed as

$$N = (-1)^s M 2^E \tag{7.90}$$

where the sign bit is either 0 or 1, $1/2 \leq M < 1$, and E is a positive or a negative integer. Both M and E are expressed as fixed-point numbers.

In floating-point representation, the decimal numbers 5.5, 6.25, and 0.625 can be denoted as:

$$(5.5)_{10} = (0.6875)2^3 = (0.1011)2^{011}$$
$$(6.25)_{10} = (0.78125)2^3 = (0.11011)2^{011}$$
$$(0.625)_{10} = (0.625)2^0 = (0.101)2^{000}$$

respectively. To add two floating-point numbers, the exponent of the smaller number is adjusted such that it matches with the exponent of the larger one and addition is then performed. Later its mantissa is to be rescaled so that it lies between 0.5 and 1. For example, consider the addition of the numbers $(4)_{10}$ and $(0.625)_{10}$. Since $(4)_{10} = (0.100)2^{011}$ and $(0.625)_{10} = (0.101)2^{100}$, the number $(0.625)_{10}$ has to be rewritten as $(0.000101)2^{011}$ so that the exponents of both the numbers are equal. The sum of the two numbers is $(0.100)2^{011} + (0.000101)2^{011} = (0.100101)2^{011}$.

To multiply two floating-point numbers, the mantissa of the two numbers is multiplied using fixed-point multiplication and the two exponents added. The exponent is then adjusted so that the new mantissa satisfies $1/2 \leq M < 1$ corrected if the mantissa is less than 0.5 along with altering the exponent. For example, consider multiplication of $(1.75)_{10} = (0.111)2^{01}$ and $(2.5)_{10} = (0.101)2^{10}$. The product of the two numbers is given by

$$(0.111)(0.101)2^{01 + 10} = (0.100011)2^{11}$$

It is noted that truncation or rounding has to be performed after both addition and multiplication in floating-point representation, whereas it is necessary only after multiplication in fixed-point arithmetic. Since the dynamic range of numbers that can be represented in the floating-point system is rather large, there is no overflow

when addition is performed in floating-point arithmetic, whereas in the case of
fixed-point arithmetic, addition can result in overflow.

7.5.2 Effect of Quantization

In the implementation of digital filters, the numbers are stored in finite length
registers. Consequently, the filter coefficients and signals have to be quantized by
rounding or truncation. Quantization gives rise to the following three types of
errors.

(i)   Coefficient-quantization errors
(ii) Input-quantization errors
(iii) Product-quantization errors

As we have seen in Chap. 5, the coefficients of a filter transfer function are
evaluated with high accuracy during the approximation stage. For implementation,
these coefficients are quantized and represented by a finite number of bits. In such a
case, the frequency response of the resulting filter may differ appreciably from the
desired response or may even fail to satisfy the desired specifications. It may even
cause the filter to become unstable.

As discussed in Chap. 1, digital signal processing of analog signals consists of
sampling a continuous-time signal into a discrete sequence using an ADC, which
represents the sampled values with a finite number of bits. Thus, input-quantization
errors are inherent in the analog-to-digital converters.

Since a fixed length, say b-bits, is used for all the registers throughout the filter,
when a signal represented by b-bits is multiplied by a coefficient represented by b-
bits, the product is of length $2b$-bits. However, this number has to be truncated or
rounded to b-bits, the length of the registers. This results in the product-quantization
errors.

7.5.3 Fixed-Point Number Quantization

Let us assume that the register length in a fixed-point implementation is b-bits,
excluding the sign bit. Hence, any number that consists of B-bits (excluding the
sign bit) with $B > b$ has to be quantized, either by truncation or rounding. Let us
first consider the case of truncation. Let $x_q(n)$ represent the quantized value of a
number $x(n)$ and $e_T = x_q(n) - x(n)$ the quantization error due to truncation. For any
positive number, the truncation can only reduce its value, and hence $e_T < 0$; also,
the error is maximum when all the $(B - b)$ bits that are discarded are 1's. Hence,

$$-\left(2^{-b}-2^{-B}\right)\le e_T\le 0 \quad \text{when } x(n)\ge 0. \tag{7.91}$$

Let us assume that we are using two's complement for the negative numbers. In this case, since the negative number is obtained by subtracting the corresponding positive number from 2, the truncation increases the magnitude of the negative number. Hence

$$-\left(2^{-b}-2^{-B}\right)\le e_T\le 0 \quad \text{when } x(n)<0 \tag{7.92}$$

Thus, if two's complement is used for representing a number, then for all $x(n)$, the truncation error satisfies $-\left(2^{-b}-2^{-B}\right)\le \epsilon_T \le 0$. Since, in general, $B\gg b$, the range of truncation error e_T is given by

$$-2^{-b}\le e_T\le 0 \tag{7.93a}$$

or

$$-q\le e_T\le 0 \tag{7.93b}$$

where

$$q=2^{-b} \tag{7.94}$$

is the quantization step.

The characteristics for truncation using two's complement are shown in Fig. 7.26a.

Similarly, it can be shown that if one's complement or signed magnitude representation is used, then the range of e_T is given by

$$-2^{-b}\le e_T\le 2^{-b} \tag{7.95a}$$

or

$$-q\le e_T\le q \tag{7.95b}$$

The characteristics for truncation using one's complement or signed magnitude representation are shown in Fig. 7.26b.

Let us now consider the case of rounding. Let $x_q(n)$ represent the quantized value of a number $x(n)$ due to rounding and $e_R=x_q(n)-x(n)$ the round-off error. The round-off error is independent of the type of fixed-point representation. The maximum error that can occur is $(2^{-b}-2^{-B})/2$ and can be negative or positive. Hence, the range of the round-off error is

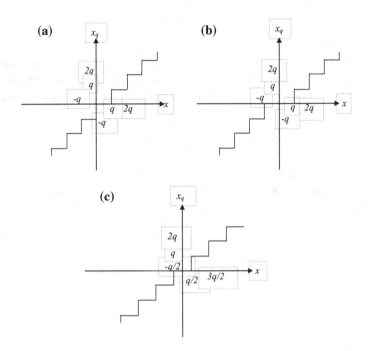

Fig. 7.26 Quantization characteristics: **a** for truncation using two's complement, **b** for truncation with one's complement or signed magnitude response, and **c** for rounding

$$-\frac{1}{2}\left(2^{-b} - 2^{-B}\right) \le e_R \le \frac{1}{2}\left(2^{-b} - 2^{-B}\right) \tag{7.96}$$

Again, since $B \gg b$, the above range becomes

$$-\frac{1}{2}\left(2^{-b}\right) \le e_R \le \frac{1}{2}\left(2^{-b}\right) \tag{7.97a}$$

or

$$-q/2 \le e_R \le q/2 \tag{7.97b}$$

The characteristics for rounding are shown in Fig. 7.26c.

We assume the number quantization to be a random process, and the quantization error has a uniform probability density function (PDF) $p(e)$ in the range given by (7.93a, 7.93b), (7.95a, 7.95b), or (7.97a, 7.97b), depending on whether the error is due truncation using two's complement, truncation using one's complement (or sign magnitude), or due to rounding. The PDFs for these three cases are shown in Fig. 7.27a–c, respectively.

The quantization error can be treated as an additive noise to the unquantized value and write it as

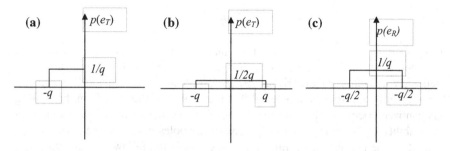

Fig. 7.27 Probability density functions: **a** for truncation error using two's complement, **b** for truncation error using one's complement or sign magnitude, and **c** for round-off error

$$x_q(n) = x(n) + e(n) \tag{7.98}$$

where $e = e_T$ or e_R. Now, the mean value m_e and σ_e^2, the variance or the power of the error signal $e(n)$ are given by (see Eqs. 2.143 and 2.145)

$$m_e = E[e(n)] = \int_{-\infty}^{\infty} e p(e) \mathrm{d}e \tag{7.99a}$$

and

$$\sigma_e^2 = E[e^2(n)] - E^2[e(n)] = \int_{-\infty}^{\infty} e^2 p(e) \mathrm{d}e - m_e^2 \tag{7.99b}$$

Using the above two expressions, it is easy to derive the following results.

(a) For truncation using two's complement representation:

$$m_{e_T} = -\frac{q}{2} = -\frac{2^{-b}}{2}, \quad \sigma_{e_T}^2 = \frac{q^2}{12} = \frac{2^{-2b}}{12} \tag{7.100}$$

(b) For truncation using one's complement or sign magnitude representation:

$$m_{e_T} = 0, \quad \sigma_{e_T}^2 = \frac{q^2}{3} = \frac{2^{-2b}}{3} \tag{7.101}$$

(c) For rounding:

$$m_{e_R} = 0, \quad \sigma_{e_R}^2 = \frac{q^2}{12} = \frac{2^{-2b}}{12} \tag{7.102}$$

Similar results can be established for quantization errors using floating-point representation of numbers. The readers are referred to [7–9].

7.5.4   Quantization of Filter Coefficients

It is important to analyze the effect of the quantization of the filter coefficients, since the coefficients used in the filter are not the exact ones derived from the given frequency response characteristics using appropriate approximation techniques. Consequently, the zeros and poles of the filter would be different from those using unquantized coefficients. In the case of an IIR filter, quantization may even affect the stability of the filter by moving some of the poles closest to the unit circle to outside of the unit circle. To illustrate this, consider the following example:

Example 7.17 Consider the following transfer function.

$$H(z) = \frac{1}{(1 - 0.943z^{-1})(1 - 0.902z^{-1})} \quad \text{(cascade form)}$$

$$= \frac{1}{1 - 1.845z^{-1} + 0.850586z^{-2}} \quad \text{(direct form)}$$

If the coefficients are quantized by truncation or rounding so that they can be expressed in six-bit binary form of which two bits are used to represent integers (including the sign bit) and four bits to represent fractions, find the pole positions for the cascade and direct forms with quantized coefficients.

Solution

$$(1.845)_{10} = (01.11011)_2 = \begin{cases} (01.1101)_2 & \text{after truncation} = (1.8125)_{10} \\ (01.1110)_2 & \text{after rounding} = (1.875)_{10} \end{cases}$$

$$(0.850586)_{10} = (00.11011)_2 = \begin{cases} (00.1101)_2 & \text{after truncation} = (0.8125)_{10} \\ (00.1110)_2 & \text{after rounding} = (0.875)_{10} \end{cases}$$

$$(0.943)_{10} = (00.11110)_2 = \begin{cases} (00.1111)_2 & \text{after truncation} = (0.9375)_{10} \\ (00.1111)_2 & \text{after rounding} = (0.9375)_{10} \end{cases}$$

$$(0.902)_{10} = (00.11100)_2 = \begin{cases} (00.1110)_2 & \text{after truncation} = (0.875)_{10} \\ (00.1110)_2 & \text{after rounding} = (0.875)_{10} \end{cases}$$

Hence, after truncation, we have

$$H_T(z) = \frac{1}{(1 - 0.9375z^{-1})(1 - 0.875z^{-1})} \quad \text{(cascade form)}$$

$$= \frac{1}{(1 - 0.8125z^{-1})(1 - z^{-1})} \quad \text{(direct form)}$$

Also, after rounding the transfer function becomes

$$H_R(z) = \frac{1}{(1 - 0.9375z^{-1})(1 - 0.875z^{-1})} \quad \text{(cascade form)}$$

$$= \frac{1}{(1 - 0.875z^{-1})(1 - z^{-1})} \quad \text{(direct form)}$$

Thus, it is seen from the above expressions that the movement of the poles is less in the case of the cascade form compared to that in the case of the direct form; in fact, in the latter case, the system is unstable, since one of the poles is on the unit circle. This example also shows that the poles are more sensitive to coefficient quantization as the order increases.

Pole Sensitivity to Coefficient Quantization

Let us consider the effect of quantization of the coefficients on the poles and zeros by studying the sensitivity of the poles and zeros to changes in the coefficients. Let us consider an IIR filter with system function

$$H(z) = \frac{\sum_{k=0}^{M} b_k z^{-k}}{1 + \sum_{k=1}^{N} a_k z^{-k}} \tag{7.103}$$

The coefficients a_k and b_k are the ideal infinite precision coefficients in the direct form realization of the IIR filter. The denominator of Eq. (7.103) can be expressed in the form

$$D(z) = 1 + \sum_{k=0}^{N} a_k z^{-k} = \prod_{i=1}^{N} (1 - p_i z^{-1}) \tag{7.104}$$

where we assume for simplicity that the poles p_i's of $H(z)$ are simple. If Δp_i is the total change in the pole location of p_i due to changes in a_k, then Δp_i can be expressed as

$$\Delta p_i = \sum_{k=1}^{N} \frac{\partial p_i}{\partial a_k} \Delta a_k \tag{7.105}$$

It can be shown that $\partial p_i / \partial a_k$, the sensitivity of the pole location p_i to the quantization of a_k, can be expressed as [10]

$$\frac{\partial p_i}{\partial a_k} = \frac{-p_i^{N-k}}{\prod_{\substack{i=1 \\ i \neq j}}^{N} (p_i - p_j)} \tag{7.106}$$

If $\hat{a}_k$'s are the quantized coefficients and $\hat{p}_i$ is the quantized ith pole, then the perturbation in the ith pole due to the quantization of the coefficients can be written as

$$\Delta p_i = \hat{p}_i - p_i = -\sum_{k=1}^{N} \frac{p_i^{N-k}}{\prod_{i=1}^{N}(p_i - p_j)} \Delta a_k \qquad (7.107)$$
$$i \neq j$$

Similar results can be obtained for $\partial z_i/\partial b_k$, the sensitivity of the ith zero, as well as for the total perturbation of the ith zero, due to the quantization of the coefficients b_k.

From the above equations, it can be noted that if the poles are closely spaced as in the case of a narrow-bandpass filter or a narrow band lowpass filter, the sensitivity increases highly for the direct form structure. Thus, small errors in the coefficients can cause large shifts in the location of the poles for direct form realizations.

In the cascade and parallel form system functions, the numerator and denominator polynomials are grouped into second-order direct form sections and quantized section by section. Since the complex conjugate poles of a second-order section are sufficiently spaced out, the changes in the pole positions due to coefficient quantization are minimized. Similarly, the cascade form improves the sensitivity of zeros, but the parallel form cannot since the zeros in a parallel form are globally distributed.

Example 7.18 Consider the transfer function of Example 7.17. Find the sensitivities of the poles with respect to the coefficients for the direct as well as the cascade forms.

Solution For direct form realization,

$$H(z) = \frac{1}{1 + a_1 z^{-1} + a_2 z^{-2}} = \frac{1}{[1 - (p_1 + p_2)z^{-1} + p_1 p_2 z^{-2}]}$$
$$= \frac{1}{(1 - 0.943 z^{-1})(1 - 0.902 z^{-1})}$$

Hence,

$$a_1 = -(p_1 + p_2), \quad a_2 = p_1 p_2$$

Using Eq. (7.106), we get

$$\frac{\partial p_2}{\partial a_2} = -\frac{\partial p_1}{\partial a_2} = \frac{1}{p_1 - p_2} = \frac{1}{0.041} = 24.39$$

$$\frac{\partial p_2}{\partial a_1} = \frac{p_2}{p_1 - p_2} = \frac{0.902}{0.041} = 22, \quad \frac{\partial p_1}{\partial a_1} = \frac{p_1}{p_1 - p_2} = \frac{0.943}{0.041} = 23$$

Now, for cascade form realization,

$$H(z) = \frac{1}{1 - p_1 z^{-1}} \frac{1}{1 - p_2 z^{-1}} = H_1(z)H_2(z) = \frac{1}{(1 - a_{11}z^{-1})} \frac{1}{(1 - a_{12}z^{-1})}$$

Thus,

$$p_1 = a_{11}, \quad p_2 = a_{12}$$

Therefore,

$$\frac{\partial p_1}{\partial a_{11}} = 1, \quad \frac{\partial p_1}{\partial a_{12}} = 0$$
$$\frac{\partial p_2}{\partial a_{11}} = 0, \quad \frac{\partial p_2}{\partial a_{12}} = 1$$

Thus, we see that the sensitivities of the poles with respect to coefficient quantization are large for the direct form realization, while they are very small for the cascade realization. These results are consistent with the observations made in Example 7.17.

A MATLAB function is provided below for coefficient quantization using truncation or rounding. In this program, flag=1 for truncation and flag=2 for rounding.

Program 7.7 Coefficient Quantization Using Truncation or Rounding

```
function beq=truncround(b,n,flag)
l=0;d=abs(b);
while fix(d)>0
    l=l+1
    d=abs(b)/(2^l);
end
if flag==1
beq=fix(d*2^n);
end
if flag==2
beq=fix(d*2^n+0.5);
end
beq=sign(b).*beq.*2^(l-n);
```

For illustration, let us consider a narrowband bandpass elliptic IIR filter with the specifications:

Lower passband edge $\omega_{p1} = 0.4\pi$ radians, upper passband edge $\omega_{p2} = 0.45\pi$ radians, lower stopband edge $\omega_{s1} = 0.35\pi$ radians, upper stopband edge $\omega_{s2} = 0.5\pi$ radians, and passband ripple = 0.5 dB, and stopband ripple = 40 dB.

The poles and zeros of the designed filter and its gain response without quantization are shown in Fig. 7.28a, b, respectively. Also, the poles and zeros of the designed filter and its gain response with quantization are shown in Fig. 7.29a, b,

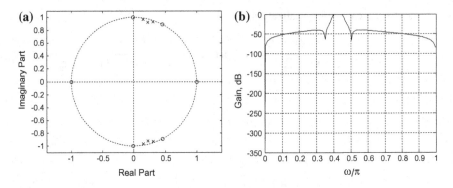

Fig. 7.28 **a** Poles and zeros without quantization, **b** magnitude response without quantization

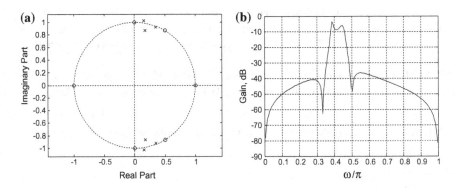

Fig. 7.29 **a** Poles and zeros with quantization (rounding) using direct form realization, **b** magnitude response with quantization (rounding) using direct form realization

respectively, for direct form realization when the filter coefficients are rounded to seven bits. Finally, the poles and zeros of the filter and the corresponding gain response are shown in Fig. 7.30a, b, respectively, for cascade realization when the filter coefficients are rounded to seven bits.

From Fig. 7.29a, b, we see that in the case of direct realization, some of the poles have moved outside of the unit circle making the filter unstable with a highly distorted gain response. Thus, the direct form realization is very sensitive to quantization errors in the coefficients. Comparing Figs. 7.28 and 7.30, we see that the poles, zeros, and gain response are in close proximity with the those of the filter without quantization. Hence, the cascade realization is less sensitive to quantization errors in the coefficients.

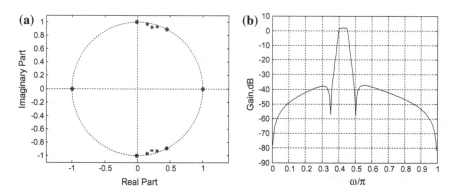

Fig. 7.30 a Poles and zeros with quantization (rounding) using cascade realization, **b** magnitude response with quantization (rounding) using cascade realization

Example 7.19 The specifications of a lowpass filter are:

passband edge frequency = 5000 Hz
stopband edge frequency = 5500 Hz
passband ripple value = 0.01 dB
minimum stopband attenuation = 60 dB
sampling frequency = 20000 Hz

Design an elliptic lowpass IIR filter for the above specifications. Use truncation and round-off to quantize the filter coefficients to six bits. Plot the gain responses as well as the pole-zero locations for the case of both the quantized and the unquantized coefficients of the filter implemented in direct form. Comment on the results.

Solution The MATLAB Program 7.8 given below is used to design the filter and to study the effect on the gain responses and pole-zero locations when the filter coefficients are truncated or rounded to six bits.

Program 7.8 Coefficient Quantization Effects on the Frequency Response of a Direct Form IIR Filter

```
clear all;close all;clc;
[N,wn]=ellipord(0.5,0.55,0.01,60)
[b,a]=ellip(N,0.01,60,wn);
[h,w]=freqz(b,a,512);
flag=input('enter 1 for truncation, 2 for rounding=');
bq=truncround(b,6,flag);aq=truncround(a,6,flag);
[hq,w]=freqz(bq,aq,512);
figure(1)
plot(w/pi,20*log10(abs(h)),'-',w/pi,20*log10(abs(hq)),'--');grid
axis([0 1 -80 5]);xlabel('\omega/\pi');ylabel('Gain,dB');
if flag==1
```

```
legend('without quantization','with truncation');
end
if flag==2
  legend('without quantization','with rounding');
end
figure(2)
zplane(bq,aq);
hold on
[zz,pp,kk]=tf2zp(b,a);
plot(real(zz),imag(zz),'*')
plot(real(pp),imag(pp),'+')
```

The above program is run with flag=1 for truncation. For direct form realization, the gain response and the pole-zero locations of the elliptic IIR lowpass filter with and without quantization are shown in Figs. 7.31a, b, respectively. It is seen from Fig. 7.31a that the effect of the coefficient quantization due to truncation is more around the bandedges with a higher passband ripple. From Fig. 7.31b, it is seen that the effect of the coefficient quantization due to truncation is to make the system unstable by moving some of the poles to outside of the unit circle; also, the quantization has made the minimum phase system into a non-minimum phase system by moving some of the zeros also outside the unit circle.

For quantization using rounding, the above Program 7.8 is run with flag=2. For direct form realization, the gain response and the pole-zero locations of the elliptic IIR lowpass filter with and without quantization are shown in Fig. 7.32a, b, respectively. It can be seen from Fig. 7.32a that the effect of the coefficient quantization due to rounding is more around the bandedges with a higher passband

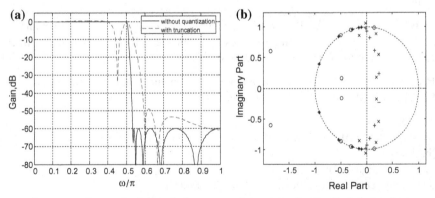

0 – represent the zeros for the quantized case   * - represent the zeros for the unquantized case
x – represent the poles for the quantized case   + - represent the poles for the unquantized case

Fig. 7.31 **a** Gain response of the elliptic IIR lowpass filter of Example 7.19 with quantization (truncation) and without quantization of the coefficients for direct form realization. **b** Pole-zero locations of the filter with truncation and without quantization of the coefficients for direct form realization

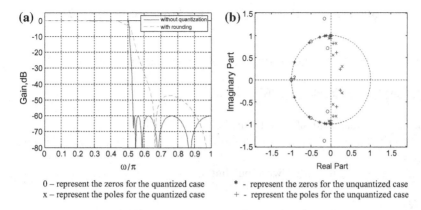

0 – represent the zeros for the quantized case
x – represent the poles for the quantized case

* - represent the zeros for the unquantized case
+ - represent the poles for the unquantized case

Fig. 7.32 **a** Gain response of the elliptic IIR lowpass filter of Example 7.19 with quantization (rounding) and without quantization of the coefficients for direct form realization. **b** Pole-zero locations of the filter with rounding and without quantization of the coefficients for direct form realization

ripple. It is observed from Fig. 7.32b that the effect of the coefficient quantization due to rounding is in making the minimum phase system into a non-minimum phase system by moving some of the zeros outside the unit circle. However, the quantization due to rounding has not affected the stability of the system.

Example 7.20 Design an elliptic lowpass IIR filter for the specifications given in Example 7.19. Use truncation and round-off to quantize the filter coefficients to six bits. Plot the gain responses as well as the pole-zero locations for the cases of quantized and unquantized coefficients of the filter implemented in cascade form. Comment on the results.

Solution The MATLAB Program 7.9 given below is used to design the filter and to study the effect on the gain responses and the pole-zero locations when the filter coefficients are truncated or rounded to six bits.

Program 7.9 Coefficient Quantization Effects on the Frequency Response of a Cascade IIR Filter

```
clear all;close all;clc;
[N,wn]=ellipord(0.5,0.55,0.01,60)
[z,p,k]=ellip(N,0.01,60,wn);
[b,a]=zp2tf(z,p,k);
[h,w]=freqz(b,a,512);
sos=zp2sos(z,p,k);
flag=input('enter 1 for truncation, 2 for rounding=');
sosq=truncround(sos,6,flag);
R1=sosq(1,:);R2=sosq(2,:);R3=sosq(3,:);R4=sosq(4,:);R5=sosq(5,:);
b11=conv(R1(1:3),R2(1:3));b12=conv(R3(1:3),R4(1:3));b1=conv(b11,b12);
bq=conv(R5(1:3),b1);
```

```
al1=conv(R1(4:6),R2(4:6));a12=conv(R3(4:6),R4(4:6));a1=conv(al1,a12);
aq=conv(R5(4:6),a1)
[hq,w]=freqz(bq,aq,512);
figure(1),plot(w/pi,20*log10(abs(h)),'-',w/pi,20*log10(abs(hq)),'--');grid
axis([0 1 -70 5]);
xlabel('\omega/\pi');ylabel('Gain,dB');
if flag==1
legend('without quantization','with truncation');
end
if flag==2
  legend('without quantization','with rounding');
end
figure(2)
zplane(bq,aq);
hold on
[zz,pp,kk]=tf2zp(b,a);plot(real(zz),imag(zz),'*')
plot(real(pp),imag(pp),'+')
```

Figure 7.33a shows the gain response with infinite precision coefficients (solid line) as well as that of the transfer function obtained with the coefficients truncated to six bits (dashed line), both being realized in the cascade form. The pole-zero locations for these two cases are shown in Fig. 7.33b. It can be seen from Fig. 7.33a that due to truncated coefficients, a flat loss has been added to the passband response with an increase in the passband ripple and one of the zeros has been moved outside the unit circle making the system non-minimum phase. However, the stability of the system is not affected. The overall effect on the stopband response is minimal.

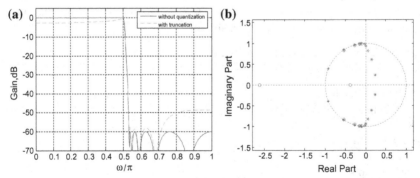

0 – represent the zeros for the quantized case   * - represent the zeros for the unquantized case
x – represent the poles for the quantized case   + - represent the poles for the unquantized case

Fig. 7.33 **a** Gain response of the elliptic IIR lowpass filter of Example 7.20 with truncation and without quantization of the coefficients for cascade realization. **b** Pole-zero locations of the filter with and without truncation of the coefficients for cascade realization

For quantization using rounding, the above Program 7.9 is run with bq=truncround(b,6,2);aq=truncround(a,6,2). Figure 7.34a shows the gain response with infinite precision coefficients (solid line) as well as that of the transfer function obtained with the coefficients rounded to six bits (dashed line), both being realized in the cascade form. The pole-zero locations for these two cases are shown in Fig. 7.34b.

It can be seen from Fig. 7.34 that there is almost no effect on the gain response and very little effect on the pole-zero locations of the filter due to rounding of the coefficients. But, there is a small effect on the stopband response. In general, quantization due to rounding has much less effect on the gain response as well as on the pole-zero locations compared to that of due to truncation of the coefficients.

Example 7.21 Design a lowpass FIR filter for the specifications given in Example 7.19. Use truncation and round-off to quantize the filter coefficients to six bits. Plot the gain responses as well as the pole-zero locations for the cases of quantized and unquantized coefficients of the filter implemented in direct form. Comment on the results.

Solution The MATLAB Program 7.10 given below is used to design the filter and to study the effect on the gain responses and pole-zero locations when the filter coefficients are truncated or rounded to six bits.

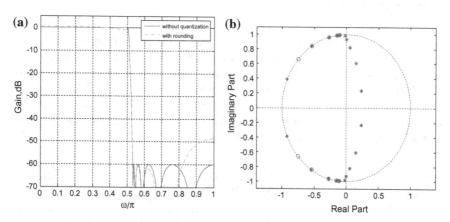

0 – represent the zeros for the quantized case   * - represent the zeros for the unquantized case

x – represent the poles for the quantized case   + - represent the poles for the unquantized case

Fig. 7.34 a Gain response of the elliptic IIR lowpass filter of Example 7.20 with rounding and without quantization of the coefficients for cascade realization. **b** Pole-zero locations of the filter with and without rounding of the coefficients for cascade realization

Program 7.10 Coefficient Quantization Effects on the Frequency Response of a Direct Form FIR Filter

```
Clear all;close all;clc;
f=[5000 5500];
a=[1 0];
dev=[0.00115062 0.001];
FT=20000;
[N,fo,ao,w] = firpmord(f,a,dev,FT);
b = firpm(N,fo,ao,w);
[h,w]=freqz(b,1,512);
flag=input('enter 1 for truncation, 2 for rounding=');
bq=truncround(b,6,flag);
[hq,w]=freqz(bq,1,512);
figure(1)
plot(w/pi,20*log10(abs(h)),'-',w/pi,20*log10(abs(hq)),'-');grid;
axis([0 1 -80 5]);
xlabel('\omega/\pi');ylabel('Gain,dB');
if flag==1
legend('without quantization','with truncation');
end
if flag==2
legend('without quantization','with rounding');
end
```

For quantization using truncation and rounding, the above Program 7.10 is run with flag=1 and flag=2, respectively. For direct form realization, the gain response of the lowpass FIR filter with and without truncation of the coefficients is shown in Fig. 7.35a, while that with and without rounding is shown in Fig. 7.35b.

It can be seen from Fig. 7.35a that the effect of the coefficient quantization due to truncation on an FIR filter implemented in direct form is to reduce the passband width, increase the passband ripple, increase the transition band, and reduce the

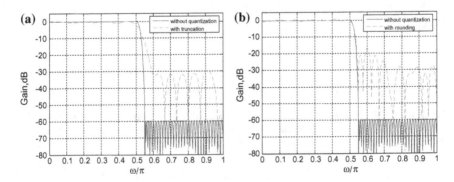

Fig. 7.35 **a** Gain response of the FIR equiripple lowpass filter with and without truncation of the coefficients implemented in direct form. **b** Gain response of the filter with and without rounding of the coefficients implemented in direct form

minimum stopband attenuation, whereas from Fig. 7.35b, it is observed that quantization due to rounding reduces the minimum stopband attenuation.

There is need to find the minimum word length for direct form or cascade realization satisfying the specifications. To find the optimum word length, the following iterative approach may be adopted.

7.5.5 Approach for Determining Optimum Coefficient Word Length

The error arising due to coefficient quantization may be described by examining the deviation of the desired frequency response from the actual frequency response as a consequence of using finite word length. Consider a digital filter characterized by its transfer function $H(z)$ and let

$$\left|H\left(e^{j\omega}\right)\right| = M(\omega) = \text{magnitude response without coefficient quantization}$$

$$\left|H_q\left(e^{j\omega}\right)\right| = M_q(\omega) = \text{magnitude response with coefficient quantization}$$

$$\left|H_I\left(e^{j\omega}\right)\right| = M_I(\omega) = \text{ideal magnitude response}$$

$$\delta_p = \text{passband tolerance}$$

$$\delta_s = \text{stopband tolerance}$$

The error ΔM due to coefficient quantization in the magnitude response is given by

$$\Delta M = \left|H\left(e^{j\omega}\right)\right| - \left|H_q\left(e^{j\omega}\right)\right| \tag{7.108}$$

Let us denote the difference between the ideal magnitude response and the desired one by ΔM_I, that is,

$$\Delta M_I = \left|H_I\left(e^{j\omega}\right)\right| - \left|H\left(e^{j\omega}\right)\right| \tag{7.109}$$

Then, the maximum permissible value of $|\Delta M|$ can be written as [11]

$$|\Delta M_{\max}(\omega)| = \begin{cases} \delta_p - |\Delta M_I| & \text{in the passband} \\ \delta_s - |\Delta M_I| & \text{in the stopband} \end{cases} \tag{7.110}$$

If

$$|\Delta M| \leq |\Delta M_{\max}(\omega)| \tag{7.111}$$

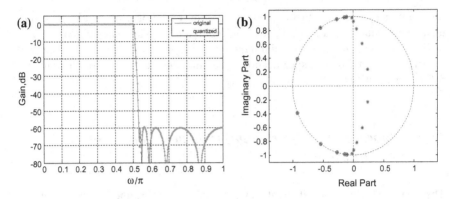

Fig. 7.36 a Direct form realization of the IIR lowpass filter with 16-bit word length for coefficients, **b** pole-zero locations for the direct form realization of the filter

in the passband as well as the stopband, then the desired specification will be satisfied. The optimum word length can be found by evaluating $|\Delta M|$ as a function of frequency by successively increasing the word length until Eq. (7.111) is satisfied.

Example 7.22 Find the optimum word lengths for cascade and direct form IIR and direct form FIR filter realizations satisfying the specifications given in Example 7.19.

Solution Adopting the procedure described above for determining the optimum word length using rounding for quantization, the optimum word lengths can be obtained; the word lengths so obtained are shown in Table 7.1. The corresponding gain response and the pole-zero locations for the direct form IIR realization are depicted in Fig. 7.36a, b, respectively, while those for the cascade realization are shown in Fig. 7.37a, b, respectively. The gain response of the direct form FIR realization is shown in Fig. 7.38.

7.5.6 Input-Quantization Errors

The input signal is quantized in all practical ADCs. This can be represented as shown in Fig. 7.39a, where Q is a quantizer, which rounds the sampled signal to the

Table 7.1 Optimum word lengths for direct form and cascade realizations of IIR and direct form FIR filters of Example 7.22

Filter type	Structure	Word length
IIR	Direct form	16
	Cascade form	10
FIR	Direct form	16

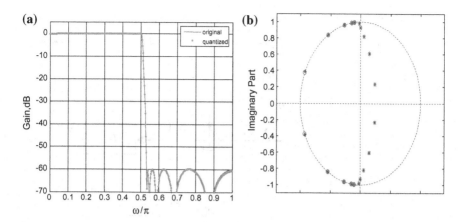

Fig. 7.37 **a** Cascade realization of the IIR lowpass filter with ten-bit word length for coefficients, **b** pole-zero locations for the cascade realization of the filter

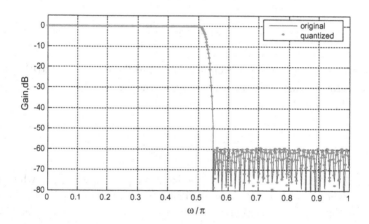

Fig. 7.38 Direct form realization of the FIR lowpass filter with 16-bit word length for the coefficients

nearest quantized level of the output. The quantization error is commonly treated as an additive noise signal, and the quantized input $x_q(n)$ is represented as a sum of the input signal $x(n)$ and the quantization noise $e(n)$ as shown in Fig. 7.39b, that is,

$$x_q(n) = x(n) + e(n) \qquad (7.112)$$

Extensive simulations have shown that the quantization error $e(n)$ due to rounding can be well approximated as a white noise that is uncorrelated with $x(n)$, and is uniformly distributed over the interval $(-q/2, q/2)$, where q is the

Fig. 7.39 Model for the
analysis of the
input-quantization effect

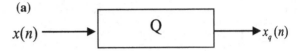

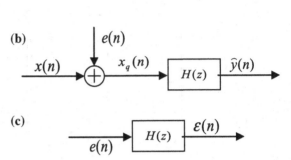

quantization step size. It can easily be shown that for such a white noise, the mean
and variance are given by

$$m_e = 0, \quad \sigma_e^2 = \frac{q^2}{12} \tag{7.113}$$

Assuming the signal to have been normalized to the range $-\frac{1}{2} \le x(n) \le \frac{1}{2}$, if b-
bits (including the sign bit) are used to represent the members of the quantized
sequence, then $q = 2^{-b}$, and hence

$$\sigma_e^2 = \frac{q^2}{12} = \frac{2^{-2b}}{12} \tag{7.114}$$

Thus, the effect of ADC can be modeled simply by adding a white noise
sequence, whose mean is zero and variance is given by (7.114), as shown in
Fig. 7.39b. Since the quantized input is processed by an LTI system, the output $\hat{y}(n)$
is given by

$$\hat{y}(n) = y(n) + \varepsilon(n) \tag{7.115}$$

where $\varepsilon(n)$ is the output noise of the LTI system due to the error $e(n)$ introduced by
input quantization, as shown in Fig. 7.39c.

Now, we define the peak power of the coder to be the power of a sinusoidal
signal that the coder can handle without clipping, and the coder dynamic range
(*DR*) to be the ratio of the peak power to the quantization noise power. Hence,

$$DR = 3(2^{2b-1}) \tag{7.116a}$$

or

$$DR = (6.02b + 1.76)\, \text{dB} \qquad (7.116b)$$

Thus, for example, the dynamic range of an eight-bit coder is about 50 dB. It is also seen from (7.116a, 7.116b) that each additional bit used in the ADC increases the dynamic range by 6 dB. If the signal amplitude is such that the dynamic range of the coder is exceeded, then the signal must be scaled before quantization is applied to reduce the amplitude range, and thus avoid clipping. In such a case, the quantizer model given by (7.112) is modified by introducing a scaling factor k:

$$x_q(n) = kx(n) + e(n), \quad 0 < k < 1 \qquad (7.117)$$

In such a case, the signal-to-noise ratio is given by

$$(\text{SNR})_q = 10 \log_{10} \left(\frac{k^2 \sigma_x^2}{\sigma_e^2} \right)$$

where $k^2 \sigma_x^2$ is the power of the scaled signal. It is found that clipping is negligible if $k = 1/\sigma_x$, in which case,

$$(SNR)_q = (6.02b - 1.25)\, \text{dB} \qquad (7.118)$$

Thus, for an eight-bit ADC, $(SNR)_q = 46.91$ dB, i.e., the noise power is about 47 dB below the signal level, and therefore, the noise due to the ADC is negligible.

Output Noise Power

The output noise $\varepsilon(n)$ can be obtained by convolution as

$$\varepsilon(n) = e(n) * h(n)$$
$$= \sum_{k=0}^{n} h(k)e(n-k) \qquad (7.119)$$

Since the input to the LTI system is a random white noise, using Eq. (2.169), the output noise power can be expressed as

$$\sigma_\varepsilon^2 = \frac{\sigma_e^2}{2\pi} \int_{-\pi}^{\pi} |H(e^{j\omega})|^2 d\omega \qquad (7.120)$$

Alternatively, the above equation can be written as

$$\sigma_\epsilon^2 = \frac{\sigma_e^2}{2\pi j} \oint H(z)H(z^{-1})z^{-1}dz \qquad (7.121)$$

where $\oint$ indicates the contour integral around $|z| = 1$. Using Parseval's theorem in Eq. (7.120), the output noise power can also be expressed as (see Eq. 2.170)

$$\sigma_\epsilon^2 = \sigma_e^2 \sum_{n=0}^{\infty} |h(n)|^2 \qquad (7.122)$$

Example 7.23 Determine the variance of the noise in the output due to quantization of the input for the first-order filter

$$y(n) = cy(n-1) + x(n), \quad 0 < |c| < 1 \qquad (7.123)$$

Solution Taking z-transform on both sides of Eq. (7.123), we get

$$H(z) = \frac{Y(z)}{X(z)} = \frac{1}{1 - cz^{-1}}$$

Hence,

$$h(n) = (-c)^n u(n)$$

Using (7.122), the output noise variance is given by

$$\sigma_\epsilon^2 = \sigma_e^2 \sum_{n=0}^{\infty} c^{2n} = \sigma_e^2 \frac{1}{1 - c^2}$$

Alternatively, we can use Eq. (7.121) to find the output noise variance:

$$\sigma_\epsilon^2 = \frac{\sigma_e^2}{2\pi j} \oint \frac{1}{1 - cz^{-1}} \frac{1}{1 - cz} z^{-1} dz$$

$$= \sigma_e^2 \left[\text{residue of} \left\{ \frac{1}{1 - cz^{-1}} \frac{1}{1 - cz} z^{-1} \right\} \text{ at } z = c \right]$$

since $z = c$ is the only pole of the integrand within the unit circle. Thus,

$$\sigma_\epsilon^2 = \sigma_e^2 \frac{1}{1 - c^2}$$

As $c^2 \to 1, \sigma_\epsilon^2 \to \infty$, while $\sigma_e^2 \to \sigma_\epsilon^2 = 2^{-2b}/12$; hence, we see that for this example the pole location has very considerable effect on the value of σ_ϵ^2.

7.5.7 Effect of Product Quantization

As mentioned earlier, in the case of fixed-point arithmetic, the product of two b-bit numbers results in a $2b$-bit number; hence, it is necessary to round it to a b-bit number. The output of finite word length multiplier can be expressed as

$$[au(n)]_q = [au(n)] + e(n) \tag{7.124}$$

or

$$\hat{v}(n) = [v(n)] + e(n) \tag{7.125}$$

where $v(n) = au(n)$ is the exact product, $e(n)$ is the quantization error due to product round-off, and $\hat{v}(n)$ is the actual product after quantization. Thus, the multiplier of a fixed-point arithmetic can be modeled as shown in Fig. 7.40. The error signal $e(n)$ can be considered as a random process having a uniform probability density function with mean and variance (or average power) given by

$$m_e = 0, \quad \sigma_e^2 = \frac{q^2}{12} = \frac{2^{-2b}}{12} \tag{7.126}$$

The effect of the multiplication rounding on the output of the filter can be studied by considering the response of the filter due to the noise signal $e(n)$. Let $G(z)$ be the transfer function from the output of the product quantizer to the filter output. Let the output noise due to error input $e(n)$ be $\epsilon(n)$. Then the output noise $\epsilon(n)$ can be obtained by convolution as

$$\begin{aligned} \epsilon(n) &= e(n) * g(n) \\ &= \sum_{k=0}^{n} g(k) e(n-k) \end{aligned} \tag{7.127}$$

Fig. 7.40 a Model for product quantization, **b** alternate way of depicting (a)

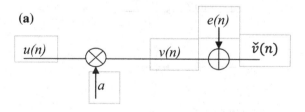

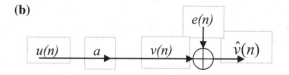

Since the input to the LTI system is a random white noise, using Eq. (2.169), the output noise power can be expressed as

$$\sigma_\epsilon^2 = \frac{\sigma_e^2}{2\pi} \int_{-\pi}^{\pi} \left| G\left(e^{j\omega}\right) \right|^2 d\omega \qquad (7.128)$$

Alternatively, the above equation can be written as

$$\sigma_\epsilon^2 = \frac{\sigma_e^2}{2\pi j} \oint G(z)G(z^{-1})z^{-1}dz \qquad (7.129)$$

where $\oint$ indicates the contour integral around $|z| = 1$. Using Parseval's theorem in Eq. (7.128), the output noise power can also be expressed as (see Eq. 2.170)

$$\sigma_\epsilon^2 = \sigma_e^2 \sum_{n=0}^{\infty} g^2(n) \qquad (7.130)$$

Let us now consider a filter, where several product quantizers are feeding the same adder as shown in Fig. 7.41. Let $G(z)$ be the transfer function from the noise source to the filter output.

Assuming that the product-quantization errors $e_i(n), i = 1, 2, \ldots, P$ are uncorrelated, the total noise variance at the output of the filter due to the quantization errors is given by

$$\sigma_\epsilon^2 = P\sigma_e^2 \sum_{n=0}^{\infty} g^2(n) \qquad (7.131a)$$

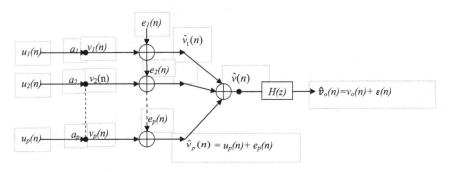

$v_o(n)$= Output due to unquantized signals $\sum_1^P u_i(n)$

$\varepsilon(n)$= Output due to the error signals $\sum_1^P e_i(n)$

Fig. 7.41 Model for the analysis of product round-off error with several multipliers feeding the same adder

or equivalently by

$$\sigma_\epsilon^2 = P \frac{\sigma_e^2}{2\pi j} \oint G(z)G(z^{-1})z^{-1}\mathrm{d}z \qquad (7.131\mathrm{b})$$

where

$$\sigma_e^2 = \frac{q^2}{12} = \frac{2^{-2b}}{12} \qquad (7.131\mathrm{c})$$

and $\oint$ indicates the contour integral around $|z| = 1$.

Since the product-quantization errors are dependent on the particular form of realization, we will consider some of these forms separately.

(a) **Direct Form I Structure**:

Consider a general IIR filter given by the transfer function

$$H(z) = \frac{Y(z)}{X(z)} = \frac{b_0 + b_1 z^{-1} + \cdots + b_N z^{-M}}{1 + a_1 z^{-1} + \cdots + a_N z^{-N}} \qquad (7.132)$$

From Sect. 7.2.1, it is known that $H(z)$ can be realized in direct form I using $(M + 1 + N)$ multipliers, as shown in Fig. 7.42. It is clear from the figure that all the $(M + 1 + N)$ quantization errors feed the same adder. Also, the transfer function $G(z)$ from the noise source to the filter output is given by

$$G(z) = \frac{\epsilon(z)}{W(z)} = \frac{1}{1 + a_1 z^{-1} + \cdots + a_N z^{-N}} \qquad (7.133)$$

where $\epsilon(z)$ is the z-transform of the output error $\epsilon(n)$ due to all the product-quantization errors. The total noise variance at the output of the filter due to the quantization errors can now be obtained using Eqs. (7.131a–7.131c) as

$$\sigma_\epsilon^2 = (M + N + 1)\sigma_e^2 \sum_{n=0}^{\infty} g^2(n) = (M + N + 1)\frac{\sigma_e^2}{2\pi j} \oint G(z)G(z^{-1})z^{-1}\mathrm{d}z \quad (7.134)$$

where σ_e^2 is given by Eq. (7.131c).

(b) **Direct Form II (Canonic) Structure**:

Again consider the IIR transfer function given by Eq. (7.132). From Sect. 7.2.2, we know that this can be realized in direct form II (canonic) structure, as shown in

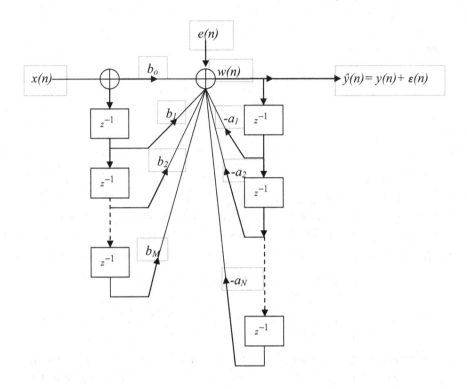

$e(n)$=round-off noise signal due to all the product quantizers feeding the adder

$\quad = \sum_{i=1}^{N} e_{ai}(n) + \sum_{i=0}^{M} e_{bi}(n)$

$\varepsilon(n)$= error at the output due to product quantization errors

$y(n)$=output due to input $x(n)$ without quantization errors

$\hat{y}(n)$= output with quantization errors included

Fig. 7.42 Model for product round-off errors in direct form I structure

Fig. 7.43. Let the product round-off errors due to the N multipliers feeding the first adder be $e_b(n)$ and that due to the $(M+1)$ multipliers feeding the second adder be $e_a(n)$.

It is easy to see that the transfer functions for the noise signals $e_a(n)$ and $e_b(n)$ are, respectively,

$$G_a(z) = H(z) \quad \text{and} \quad G_b(z) = 1$$

Hence, from (7.133), the variance of the corresponding noise outputs $\epsilon_a(n)$ and $\epsilon_b(n)$ is

Fig. 7.43 Model for product round-off errors in direct form II structure

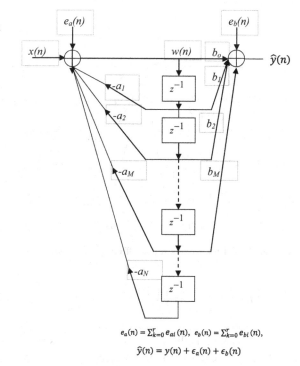

$$e_a(n) = \Sigma_{k=0}^r e_{ai}(n), \quad e_b(n) = \Sigma_{k=0}^r e_{bi}(n),$$

$$\hat{y}(n) = y(n) + \epsilon_a(n) + \epsilon_b(n)$$

$$\sigma_{\epsilon_a}^2 = N \frac{q^2}{12} \sum_{n=0}^{\infty} h^2(n) = N \frac{q^2}{12} \frac{1}{2\pi j} \oint H(z) H(z^{-1}) z^{-1} dz \qquad (7.135a)$$

and

$$\sigma_{\epsilon_b}^2 = (M+1)\sigma_{e_b}^2 = (M+1)\frac{q^2}{12} \qquad (7.135b)$$

Therefore, the variance of the output noise is given by

$$\sigma_{\epsilon}^2 = \sigma_{\epsilon_a}^2 + \sigma_{\epsilon_b}^2 \qquad (7.136)$$

(c) Cascade Structure:

As mentioned in Sect. 7.2.3, the IIR transfer function (7.133) can be expressed as a product of a number of first- and second-order transfer functions. We will consider here a function which is a product of r second-order sections; however, the same procedure can be used if the IIR transfer function contains some first-order functions. Let the transfer function be

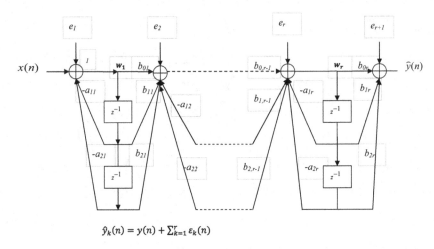

$$\hat{y}_k(n) = y(n) + \sum_{k=1}^{r} \varepsilon_k(n)$$

Fig. 7.44 Model for product round-off errors in a cascade structure

$$H(z) = \frac{Y(z)}{X(z)} = \prod_{k=1}^{r} \left[\frac{b_{0k} + b_{1k}z^{-1} + b_{2k}z^{-2}}{1 + a_{1k}z^{-1} + a_{2k}z^{-2}} \right] = \prod_{k=1}^{r} [H_k(z)] \qquad (7.137)$$

Then it can be realized as a cascade of r second-order sections as shown in Fig. 7.44. It is seen from Fig. 7.44 that two quantization errors feed adder 1, 3 for the last adder, and 5 for each of the remaining adders. It is also seen that the noise transfer functions for the noise source $e_k(n)$ is $\prod_{i=k}^{r} [H_i(z)]$. Let

$$G_k(z) = \prod_{i=k}^{r} [H_i(z)] \qquad (7.138)$$

and

$$\eta_k = \frac{1}{2\pi j} \oint G_k(z) G_k(z^{-1}) z^{-1} \mathrm{d}z \qquad (7.139)$$

Then, using Eqs. (7.131a–7.131c), we see that the variance of the noise output is given by

$$\sigma_\epsilon^2 = \frac{q^2}{12} \left[2\eta_1 + 5 \sum_{k=2}^{r} \eta_k + 3 \right] \qquad (7.140)$$

(d) **Parallel Structure**:

As mentioned in Sect. 7.2.4, the IIR transfer function (7.133) can be expanded by partial fractions in the form

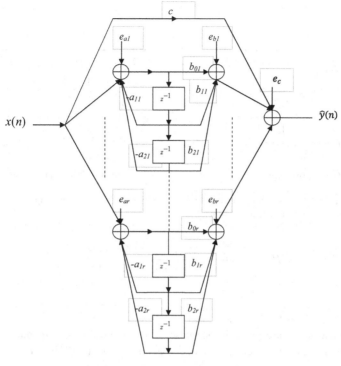

$$\hat{y}(n) = y(n) + \sum_{k=1}^{r}\epsilon_{ak}(n) + \sum_{k=0}^{r}\epsilon_{bk(n)} + \epsilon_{c}(n)$$

Fig. 7.45 Model for product round-off errors for a parallel structure

$$H(z) = C + H_1(z) + H_2(z) + \cdots + H_r(z) \qquad (7.141)$$

where C is a constant and $H_k(z)$ is a first- or second-order function. We will
consider here $H_k(z)$ to be a second-order section; however, the same procedure can
be used even if some of the sections are first-order ones. Let each of the
second-order sections be realized by direct-II form. Then the overall realization is as
shown in Fig. 7.45. It is seen from Fig. 7.45 that the noise transfer function for
$e_{ak}(n)$ is $H_k(z)$, while the transfer function for $e_{bk}(n)$ and $e_c(n)$ is unity. Hence,
using Eqs. (7.131a–7.131c) we see that the variance of the noise output is given by

$$\sigma_{\epsilon}^2 = \frac{q^2}{12}\left[2\frac{1}{2\pi j}\sum_{k=1}^{r}\oint H_k(z)H_k(z^{-1})z^{-1}dz\right] + (3r+1)\frac{q^2}{12} \qquad (7.142)$$

Example 7.24 Find the output noise power in the direct form I and II realizations of
the transfer function

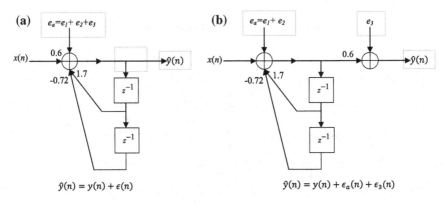

(a) $e_a = e_1 + e_2 + e_3$

$\hat{y}(n) = y(n) + \epsilon(n)$

(b) $e_a = e_1 + e_2$ e_3

$\hat{y}(n) = y(n) + \epsilon_a(n) + \epsilon_3(n)$

Fig. 7.46 **a** Output noise power in direct form I realization of $H(z)$ of Example 7.24, **b** output noise power in direct form II realization

$$H(z) = \frac{Y(z)}{X(z)} = \frac{0.6}{(1 - 0.9z^{-1})(1 - 0.8z^{-1})} \tag{7.143}$$

Solution

(a) For direct form I: The direct form I realization is shown in Fig. 7.46a, where $e_1(n)$ and $e_2(n)$ are the noise signals due to the product-quantization errors. The output noise power or variance is given by Eq. (7.131b) with $P = 3$ and

$$G(z) = \frac{1}{(1 - 0.9z^{-1})(1 - 0.8z^{-1})}$$

Hence,

$$\begin{aligned}
\sigma_\epsilon^2 &= 3\frac{q^2}{12}\frac{1}{2\pi j}\oint \frac{1}{(1 - 0.9z^{-1})(1 - 0.8z^{-1})}\frac{1}{(1 - 0.9z)(1 - 0.8z)}z^{-1}dz \\
&= \frac{q^2}{4}\frac{1}{2\pi j}\oint \frac{z}{(z - 0.9)(z - 0.8)(1 - 0.9z)(1 - 0.8z)}dz = \frac{q^2}{4}\frac{1}{2\pi j}\oint F(z)dz \\
&= \frac{q^2}{4}[\text{sum of the residues of } F(z) \text{ at the poles } z = 0.9 \text{ and } z = 0.8] \\
&= \frac{q^2}{4}[169.17 - 79.36] = 22.45q^2.
\end{aligned}$$
$$\tag{7.144a}$$

(b) For direct form II: The direct form II realization is shown in Fig. 7.46b, where $e_1(n)$, $e_2(n)$, and $e_3(n)$ are the noise signals. For $e_a(n) = e_1(n) + e_2(n)$, the noise transfer function is $G_a(z)$ and for $e_3(n)$, it is $G_3(z)$, where

$$G_a(z) = \frac{0.6}{(1 - 0.9z^{-1})(1 - 0.8z^{-1})} \quad \text{and} \quad G_3(z) = 1$$

The output variance σ_ϵ^2 is given by

$$\sigma_\epsilon^2 = \sigma_{\epsilon_a}^2 + \sigma_{\epsilon_3}^2$$

where $\sigma_{\epsilon_a}^2$ and $\sigma_{\epsilon_3}^2$ are the variances of the output noise due to $e_a(n)$ and $e_3(n)$, respectively. Now, from Eq. (7.131b), the output noise variances $\sigma_{\epsilon_a}^2$ and $\sigma_{\epsilon_3}^2$ are given by

$$\sigma_{\epsilon_a}^2 = 2\frac{q^2}{12}\frac{1}{2\pi j}\oint \frac{0.6}{(1 - 0.9z^{-1})(1 - 0.8z^{-1})}\frac{0.6}{(1 - 0.9z)(1 - 0.8z)}z^{-1}dz = 5.39q^2$$

and

$$\sigma_{\epsilon_3}^2 = \frac{q^2}{12}$$

Thus,

$$\sigma_\epsilon^2 = 5.47q^2. \tag{7.144b}$$

It is observed that, for this example, the output noise power of the canonic (direct form II) realization is drastically less than that of the direct form I.

Example 7.25 Find the output noise power in the cascade form realization of the transfer function of Example 7.24.

Solution The given transfer function can be written as

$$H(z) = \frac{0.6}{(1 - 0.9z^{-1})}\frac{1}{(1 - 0.8z^{-1})}$$

and each of the factors realized by a first-order section and connected in cascade. The first-order sections may themselves be realized using either direct form I or direct form II. We now show the output noise power can vary drastically depending on which form we use in the cascade realization. Let us first realize $H(z)$ in cascade form using direct form I first-order sections, as shown Fig. 7.47a.

For the noise sources $e_a(n) = e_1(n) + e_2(n)$ and $e_3(n)$, the noise transfer functions are given by $G_a(z)$ and $G_3(z)$, respectively, where

$$G_a(z) = \frac{1}{(1 - 0.9z^{-1})(1 - 0.8z^{-1})} \quad \text{and} \quad G_3(z) = \frac{1}{(1 - 0.8z^{-1})}$$

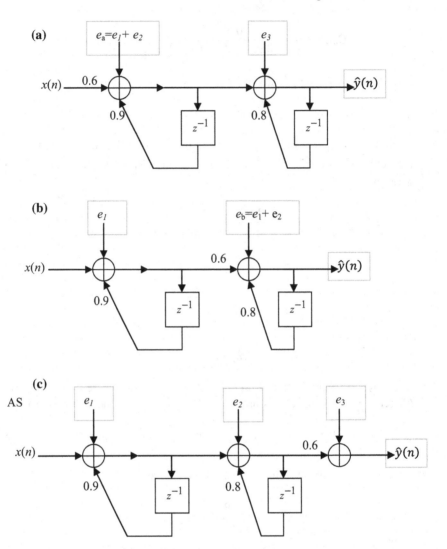

Fig. 7.47 Output noise of $H(z)$ of Example 7.25 realized in three different cascade forms shown in (a), (b), and (c)

Using the same procedure as used in Example 7.24, we can show that the output noise variances $\sigma_{\epsilon_a}^2$ and $\sigma_{\epsilon_3}^2$ due to $e_a(n)$ and $e_3(n)$, respectively, are given by

$$\sigma_{\epsilon_a}^2 = 2\frac{q^2}{12}(89.81) \quad \text{and} \quad \sigma_{\epsilon_3}^2 = \frac{q^2}{12}(2.78)$$

Thus,

$$\sigma_\epsilon^2 = \sigma_{\epsilon_a}^2 + \sigma_{\epsilon_3}^2 = 15.2q^2. \tag{7.145a}$$

Let the transfer function be now realized as a cascade of two first-order sections $\frac{1}{(1-0.9z^{-1})}$ and $\frac{0.6}{(1-0.8z^{-1})}$ realized using direct form I, as shown in Fig. 7.47c. In this case, for the noise sources $e_1(n), e_b(n) = e_2(n) + e_3(n)$, the noise transfer functions are given by $G_1(z)$, and $G_b(z)$, respectively, where

$$G_1(z) = \frac{0.6}{(1 - 0.9z^{-1})(1 - 0.8z^{-1})} \quad \text{and} \quad G_b(z) = \frac{1}{(1 - 0.8z^{-1})}$$

Using the above and Eq. (7.131b), we can obtain the total output noise power to be

$$\sigma_\epsilon^2 = \sigma_{\epsilon_1}^2 + \sigma_{\epsilon_b}^2 = \frac{q^2}{12}(32.33) + 2\frac{q^2}{12}(2.78) = 3.16q^2 \tag{7.145b}$$

Finally, let us obtain the given transfer function as a cascade of the two first-order sections $\frac{1}{(1-0.9z^{-1})}$ and $\frac{0.6}{(1-0.8z^{-1})}$ realized using direct form II structures, as shown in Fig. 7.47c. In this case, for the noise sources $e_1(n), e_2(n)$ and $e_3(n)$, the noise transfer functions are given by $G_1(z)$, $G_2(z)$, and $G_3(z)$, respectively, where

$$G_1(z) = \frac{0.6}{(1 - 0.9z^{-1})(1 - 0.8z^{-1})}, \quad G_2(z) = \frac{0.6}{(1 - 0.8z^{-1})} \text{ and } \quad G_3(z) = 1.$$

Using the above and Eq. (7.131b), we can obtain the total output noise power to be

$$\sigma_\epsilon^2 = \sigma_{\epsilon_1}^2 + \sigma_{\epsilon_2}^2 + \sigma_{\epsilon_3}^2 = \frac{q^2}{12}(32.33) + \frac{q^2}{12}(1) + \frac{q^2}{12} = 2.86q^2 \tag{7.145c}$$

It is seen that the cascade realization using direct form II structure has the least output noise power among the three cascade realization; further this example also shows that the output noise power for all the cascade realizations is very much reduced compared to that realized using direct form I structure.

Example 7.26 Find the output noise power in the parallel form realization of the transfer function of Example 7.24.

Solution The given transfer function can be written in partial fraction form as

$$H(z) = \frac{5.4}{(1 - 0.9z^{-1})} - \frac{4.8}{(1 - 0.8z^{-1})}$$

and realized as a parallel structure, as shown in Fig. 7.48.

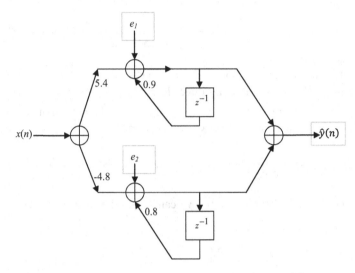

Fig. 7.48 Output noise power of $H(z)$ of Example 7.26 realized by parallel structure

The transfer functions for the two noise sources $e_1(n)$ and $e_2(n)$ are given by $G_1(z)$ and $G_2(z)$, where

$$G_1(z) = \frac{1}{(1 - 0.9z^{-1})}, \quad \text{and} \quad G_2(z) = \frac{1}{(1 - 0.8z^{-1})}$$

Again using (7.131b), we get the output noise power to be

$$\sigma_\epsilon^2 = \sigma_{\epsilon_1}^2 + \sigma_{\epsilon_2}^2 = 2\frac{q^2}{12}(5.26) + 2\frac{q^2}{12}(2.78) = 1.34q^2 \tag{7.146}$$

Thus, for this example, this parallel realization has the least amount of output noise than any of the direct or cascade realizations.

7.6 Scaling in Fixed-Point Realization of IIR Digital Filters

IIR digital filter realization using fixed-point arithmetic may cause overflow at certain internal nodes, such as inputs to multipliers, and this may give rise to large amplitude oscillations at the filter output. The overflow can be minimized significantly by scaling the input. The selection of the scaling factor is done in such a way that the overflow at a node is reduced or avoided. The numerator coefficients of the transfer function are multiplied by the scaling factor in order to have the overall filter gain unchanged. We assume that all fixed-point numbers are represented as fractions and the input sequence is bounded by unity. The objective of the scaling is to make sure that the sequence at any of the internal nodes is also bounded by unity.

An ith scale factor is defined by the L_p norm as

$$s_i = \|F_i\|_p \triangleq \left[\sum_{k=0}^{\infty} |f_i(k)|^p \right]^{1/p} = \left[\frac{1}{2\pi} \int_{-\pi}^{\pi} |F_i(e^{j\omega})|^p \, d\omega \right]^{1/p} \tag{7.147}$$

where $f_i(k)$ is the impulse response of the scaling transfer function $F_i(z)$ from the input to the ith node. The three common approaches used to determine the scaling factors of a filter are the L_1, L_2, and L_∞ norms. From the above definition, it follows that in the case of the L_1 norm

$$s_i = \|F_i\|_1 = \sum_{k=0}^{\infty} |f_i(k)| = \frac{1}{2\pi} \int_{-\pi}^{\pi} |F_i(e^{j\omega})| \, d\omega \tag{7.148}$$

whereas in the L_2 norm method, the scale factor is obtained by

$$s_i = \|F_i\|_2 = \left[\sum_{k=0}^{\infty} |f_i(k)|^2 \right]^{1/2} = \left[\frac{1}{2\pi} \int_{-\pi}^{\pi} |F_i(e^{j\omega})|^2 \, d\omega \right]^{1/2} \tag{7.149}$$

and in the case of L_∞ norm, the scale factor is given by

$$s_i = \max_{-\pi \leq \omega \leq \pi} |F_i(e^{j\omega})| \tag{7.150}$$

which is the peak absolute value of the frequency response of the scaling transfer function.

Let $w_i(n)$ be the output at ith node when the input sequence is $x(n)$ and let $f_i(n)$ be the impulse response at the ith node. Then, $w_i(n)$ can be expressed by convolution as

$$w_i(n) = \sum_{k=0}^{\infty} f_i(k) x(n - k) \tag{7.151}$$

Since it has been assumed that the input sequence $x(n)$ is bounded by unity,

$$|x(n)| < 1 \quad \text{for all } n. \tag{7.152}$$

Hence,

$$|w_i(n)| = \left| \sum_{k=0}^{\infty} f_i(k) x(n-k) \right| \leq \left| \sum_{k=0}^{\infty} f_i(k) \right| \leq \sum_{k=0}^{\infty} |f_i(k)|$$

The condition $|w_i(n)| \leq 1$ can be satisfied only if

$$\sum_{k=0}^{\infty} |f_i(k)| \leq 1 \quad \text{for all } i. \tag{7.153}$$

Thus, the input signal has to be scaled by a multiplier of value $1/(s_i)$ by

$$\frac{1}{s_i} = \frac{1}{\sum_{k=0}^{\infty} |f_i(k)|} \tag{7.154}$$

The above scaling is based on the L_1 norm. It is a harsh scaling, especially for narrowband input sequences such as the sinusoidal sequences. The frequency response of the system can be utilized for appropriate scaling for sinusoidal sequences. As such the L_∞ norm is more appropriate for such sequences.

Let us now consider L_2 norm for scaling purpose. By applying DTFT to the convolution sum (7.151), we get

$$W_i(e^{j\omega}) = F_i(e^{j\omega}) X(e^{j\omega}) \tag{7.155}$$

Taking the inverse transform, we have

$$w_i(n) = \frac{1}{2\pi} \int_{-\pi}^{\pi} F_i(e^{j\omega}) X(e^{j\omega}) e^{j\omega n} d\omega \tag{7.156}$$

Hence,

$$|w_i(n)|^2 = \left| \frac{1}{2\pi} \int_{-\pi}^{\pi} F_i(e^{j\omega}) X(e^{j\omega}) e^{j\omega n} d\omega \right|^2$$

Using Schwartz's inequality, we get

$$|w_i(n)|^2 \leq \left[\frac{1}{2\pi} \int_{-\pi}^{\pi} |F_i(e^{j\omega})|^2 d\omega \right] \left[\frac{1}{2\pi} \int_{-\pi}^{\pi} |X(e^{j\omega})|^2 d\omega \right]$$

Hence,

$$|w_i(n)| \le \|F_i\|_2 \|X\|_2 \qquad (7.157)$$

Thus, the L_2 norm sets an energy constraint on both the input and the transfer function. If the filter input has finite energy bounded by unity, then by scaling the filter coefficients so that $s_i = \|F_i\|_2$ (i.e., the L_2 norm) is bounded by unity for all i, we ensure that

$$|w_i(n)| \le 1 \quad \text{for all } i. \qquad (7.158)$$

Hence, in order to avoid the overflow at the internal nodes, we have to evaluate the scale factor $s_i = \|F_i\|_2$ for each of the internal nodes at the input of the multipliers and choose the maximum of s_i for scaling.

The scale factors using L_1, L_2, and L_∞ norms can be shown to satisfy the relation

$$\left[\sum_{k=0}^{\infty} |f_i(k)|^2 \right]^{1/2} \le \max_{-\pi \le \omega \le \pi} |F_i(e^{j\omega})| \le \sum_{k=0}^{\infty} |f_i(k)| \qquad (7.159)$$

7.6.1 Scaling for a Second-Order Filter

Consider a second-order system characterized by the transfer function

$$H(z) = \frac{Y(z)}{X(z)} = \frac{b_0 + b_1 z^{-1} + b_2 z^{-2}}{1 + a_1 z^{-1} + a_2 z^{-2}} \qquad (7.160)$$

The direct form II (canonic) realization is shown in Fig. 7.49a. Let us obtain the scaling factor for this realization using the L_2 norm.

Since w is the only internal node feeding the multipliers, we have to calculate only the L_2 norm corresponding to that node. The transfer function $F(z)$ from the input to this node is given by

(a)   **(b)**

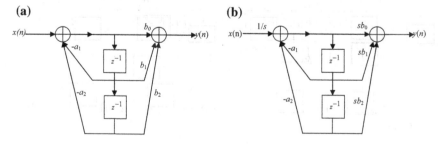

Fig. 7.49 **a** Second-order IIR filter realized by direct form II, **b** scaled version of (a)

$$F(z) = \frac{1}{1 + a_1 z^{-1} + a_2 z^{-2}} \tag{7.161}$$

Hence,

$$s^2 = \|F\|_2^2 \tag{7.162}$$

To avoid the overflow at the internal node w, we now scale the input sequence $x(n)$ by $(1/s)$. In order not to alter the given gain of $H(z)$, we multiply the numerator of $H(z)$ by s so that we have

$$H(z) = \frac{Y(z)}{X(z)} = \frac{X_1(z)}{X(z)} \frac{Y(z)}{X_1(z)} = \frac{1}{S}[SH(z)]$$

or

$$\frac{Y(z)}{X_1(z)} = \frac{s(b_0 + b_1 z^{-1} + b_2 z^{-2})}{1 + a_1 z^{-1} + a_2 z^{-2}} \tag{7.163}$$

The above can be very simply realized by replacing the multipliers b_0, b_1 and b_2 by sb_0, sb_1 and sb_2, respectively in Fig. 7.49a. The second-order IIR filter realizing $H(z)$ but with its input scaled by $(1/s)$ is shown in Fig. 7.49b.

We now find an explicit expression for the scale factor s. For this purpose, we let

$$1 + a_1 z^{-1} + a_2 z^{-2} = \left(1 - p_1 z^{-1}\right)\left(1 - p_2 z^{-1}\right) \tag{7.164}$$

Using (7.149), we have

$$
\begin{aligned}
s^2 &= \left[\frac{1}{2\pi} \int_{-\pi}^{\pi} |F(e^{j\omega})|^2 d\omega\right] \\
&= \frac{1}{2\pi j} \oint F(z)F(z^{-1})z^{-1} dz, \text{ the integral being taken over the unit circle.} \\
&= \frac{1}{2\pi j} \oint \frac{z}{(z - p_1)(z - p_2)(1 - p_1 z)(1 - p_2 z)} dz \\
&= \frac{p_1}{(p_1 - p_2)(1 - p_1 p_2)\left(1 - p_1^2\right)} + \frac{p_2}{(p_2 - p_1)(1 - p_1 p_2)\left(1 - p_2^2\right)} \\
&= \frac{1 + p_1 p_2}{1 - p_1 p_2} \frac{1}{\left(1 - p_1^2\right)\left(1 - p_2^2\right)}
\end{aligned}
$$

$$\tag{7.165}$$

From (7.164), we know that $p_1 + p_2 = -a_1$ and $p_1 p_2 = a_2$. Hence Eq. (7.165) reduces to

$$s^2 = \frac{1+a_2}{1-a_2} \frac{1}{(1+a_2)^2-a_1^2}$$ (7.166)

7.6.2   Scaling in a Parallel Structure

It has been shown in Sect. 7.2.4 that a general IIR transfer function given by

$$H(z) = \frac{Y(z)}{X(z)} = \frac{b_0+b_1z^{-1}+\cdots+b_Nz^{-N}}{1+a_1z^{-1}+\cdots+a_Nz^{-N}}$$

can be expanded by partial fractions in the form

$$H(z) = C + H_1(z) + H_2(z) + \cdots + H_R(z)$$ (7.167)

where C is a constant, and $H_i(z)$ is a first- or second-order function of the form $z/(1+a_1z^{-1})$ or $(b_0+b_1z^{-1}+b_2z^{-2})/(1+a_1z^{-1}+a_2z^{-2})$, and realized in parallel form as shown in Fig. 7.8. Let us assume that each of the functions $H_i(z)$ is realized by the direct form II structure. Then we can scale each of the realized first- or second-order sections employing the L_2 norm using the method detailed in Sect. 7.6.1. We now illustrate the method by the following example.

Example 7.27 Realize the following transfer function in parallel form with scaling using the L_2 norm.

$$H(z) = 1 + \frac{0.3243 - 0.4595z^{-1}}{1 - 0.5z^{-1} + 0.5z^2} + \frac{0.6757 - 0.3604z^{-1}}{1 + z^{-1} + 0.3333z^{-2}}$$

Solution $H(z)$ can be written in partial fraction expansion as

$$H(z) = 1 + H_1(z) + H_2(z)$$

$$= 1 + \frac{0.3243 - 0.4595z^{-1}}{1 - 0.5z^{-1} + 0.5z^2} + \frac{0.6757 - 0.3604z^{-1}}{1 + z^{-1} + 0.3333z^{-2}}$$

Using Eq. (7.166), we can determine the scaling factors s_1 and s_2 for $H_1(z)$ and $H_2(z)$ to be

$$s_1 = \sqrt{1.5} = 1.2247 \quad \text{and} \quad s_2 = \sqrt{2.5714} = 1.6035$$

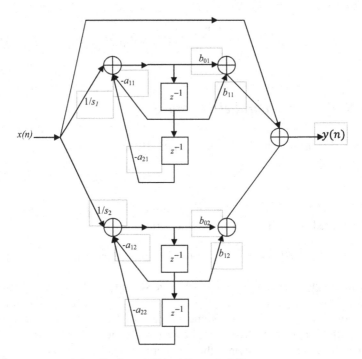

Fig. 7.50 L_2-norm scaled realization of $H(z)$ of Example 7.27

The required realization is shown in Fig. 7.50, where

$$a_{11} = -0.5, \quad a_{21} = 0.5, \quad b_{01} = 0.3243, \quad b_{11} = -0.4595$$

$$a_{12} = 1, \quad a_{22} = 0.3333, \quad b_{02} = 0.6757, \quad b_{12} = -0.3604$$

7.6.3 Scaling in a Cascade Structure

It has been shown in Sect. 7.2.3 that a general IIR transfer function given by

$$H(z) = \frac{Y(z)}{X(z)} = \frac{b_0 + b_1 z^{-1} + \cdots + b_N z^{-N}}{1 + a_1 z^{-1} + \cdots + a_N z^{-N}}$$

can be expanded in the form

$$H(z) = H_1(z)H_2(z)\ldots H_R(z) \tag{7.168}$$

$H_i(z)$ is a first- or second-order function. Without loss of generality, we assume that each of the functions $H_i(z)$ is a second-order one of the form

$$H_i(z) = \frac{b_{0i} + b_{1i}z^{-1} + b_{2i}z^{-2}}{1 + a_{1i}z^{-1} + a_{2i}z^{-2}} \tag{7.169}$$

and is realized by the direct form II structure. In the ith section, realizing $H_i(z)$, let w_i be the internal node which is the input node to all the multipliers in that section. Let the transfer function from the input node to the node w_i be $F_i(z)$. Then, it is clear that

$$F_i(z) = \left[\prod_{k=0}^{i-1} H_k(z)\right]\frac{1}{A_i(z)} \tag{7.170a}$$

where

$$A_i(z) = 1 + a_{1i}z^{-1} + a_{2i}z^{-2} \tag{7.170b}$$

As a consequence, the various L_2 norm scaling functions s_i are given by

$$s_i^2 = \|F_i\|_2^2 \triangleq \frac{1}{2\pi} \int_{-\pi}^{\pi} |F_i(e^{j\omega})|^2 d\omega \tag{7.171}$$

Hence, the scaling functions s_i may be written as

$$s_i^2 = \|F_i\|_2^2 = = \frac{1}{2\pi j} \oint F_i(z)F_i(z^{-1})z^{-1}dz \tag{7.172}$$

Just as in the case of scaling of a single second-order section, in order not to alter the gain of the given IIR transfer function $H(z)$, the multipliers b_{0i}, b_{1i} and b_{2i} are now replaced by

$$(s_i/s_{i+1})b_{0i}, \quad (s_i/s_{i+1})b_{1i}, \quad \text{and} \quad (s_i/s_{i+1})b_{1i} \quad \text{for } i = 1, 2, \ldots, R-1$$

and the multipliers b_{0R}, b_{1R} and b_{2R} by

$$s_R b_{0R}, \quad s_R b_{1R} \quad \text{and} \quad s_R b_{2R}$$

respectively. Thus, except for the multiplier $(1/s_1)$, all the other scaling multipliers are absorbed into the existing ones. For illustration, a scaled cascade realization of a sixth-order IIR filter using L_2 norm is shown in Fig. 7.51.

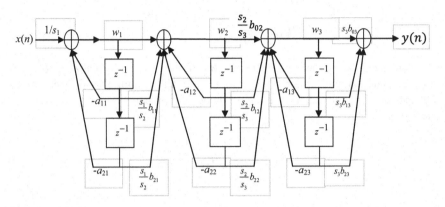

Fig. 7.51 L_2-norm scaled realization of a sixth-order IIR filter

Example 7.28 Realize the following transfer function in cascade form with scaling using the L_2 norm.

$$H(z) = \frac{(1+0.7z^{-1})}{(1+0.8z^{-1})} \frac{(1+1.6z^{-1}+0.6375z^{-2})}{(1+1.5z^{-1}+0.54z^{-2})}$$

Solution Let

$$H_1(z) = \frac{1+0.7z^{-1}}{1+0.8z^{-1}} \quad \text{and} \quad H_2(z) = \frac{1+1.6z^{-1}+0.6375z^{-2}}{1+1.5z^{-1}+0.54z^{-2}}$$

Then, from Eqs. (7.170a, 7.170b)

$$F_1(z) = \frac{1}{1+0.8z^{-1}} \quad \text{and} \quad F_2(z) = \frac{1+0.7z^{-1}}{1+0.8z^{-1}} \frac{1}{1+1.5z^{-1}+0.54z^{-2}}$$

Now, using Eq. (7.172)

$$s_1^2 = \|F_1\|_2^2 = \frac{1}{2\pi j} \oint F_1(z)F_1(z^{-1})z^{-1}dz = 2.7778$$

and

$$s_2^2 = \|F_2\|_2^2 = \frac{1}{2\pi j} \oint F_2(z)F_2(z^{-1})z^{-1}dz = 53.349$$

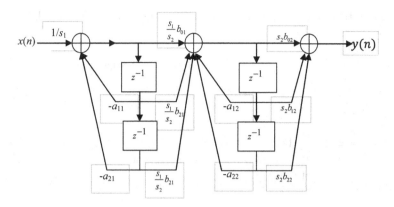

Fig. 7.52 L_2-norm scaled realization of $H(z)$ of Example 7.28

Hence,

$$s_1 = 1.6667 \quad \text{and} \quad s_2 = 7.3041$$

The required scaled cascaded realization is shown in Fig. 7.52, where

$$a_{11} = 0.8, \quad b_{01} = 1, \quad b_{11} = 0.7$$
$$a_{12} = 1.5, \quad a_{22} = 0.54, \quad b_{01} = 1, \quad b_{12} = 1.6, \quad b_{22} = 0.6375$$

The following example illustrates the scaling of cascaded canonic section realization of a digital IIR filter using MATLAB:

Example 7.29 Consider the following transfer function and obtain cascade realization of it with scaling using L_2 norm.

$$H(z) = \frac{\left(1 - 0.8z^{-1} + 0.6z^{-2}\right)\left(1 + 1.9z^{-1} + 2.4z^{-2}\right)\left(1 + 1.75z^{-1} + 2.1z^{-2}\right)}{\left(1 + 0.2z^{-1} - 0.3z^{-2}\right)\left(1 + 0.8z^{-1} + 0.7z^{-2}\right)\left(1 + 1.2z^{-1} + 0.8z^{-2}\right)}$$

Solution The following MATLAB program with flag option 2 can be used to compute the scaling factors using L2 norm approach.

Program 7.11 Scale Factors of a Transfer Function Realized by Cascaded Second-Order Canonic Sections Using $L1$, $L2$, and Loo Norms.

```
clear;clc;
b1 = [1 -0.8 0.6]; %numerator coefficients of the first section
b2 = [1 1.9 2.4]; %numerator coefficients of the second section
b3 = [1 1.75 2.1];%numerator coefficients of the third section
a1 = [1 0.2 -0.3]; %denominator coefficients of the first section
a2 = [1 0.8 0.7]; %denominator coefficients of the second section
a3 = [1 1.2 0.8]; %denominator coefficients of the third section
b = [b1; b2; b3]; a = [a1; a2; a3];
```

flag=input('Enter the value of flag = 0 for L1 norm, 1 for L2 norm and 2 for Loo norm=');

```
    A = 1; B = 1;
    for i=1:size(b,1)                    %loop for each stage
    A = conv(A,a(i,:));
    if i>1
        B = conv(B,b(i-1,:));
    end
    if (flag==0|flag==1)
                    [f,t]=impz(B,A); %impulse response
                    if (flag)
    s(i) = sqrt(sum(f.^2));
    else
        s(i) = sum(abs(f));
    end
    elseif (flag==2)
                    [f,w] = freqz(B,A); % frequency response
                    s(i) = max(abs(f));
    end
    end
```

The scaling factors obtained by using L2 norm for the above program are given below:

$$s_1 = 1.0939, \quad s_2 = 2.9977 \quad \text{and} \quad s_3 = 12.1411.$$

The scaled cascaded realization is shown in Fig. 7.51, where

$$
\begin{array}{lllll}
a_{11} = 0.2, & a_{21} = -0.3, & b_{01} = 1, & b_{11} = -0.8, & b_{21} = 0.6 \\
a_{12} = 0.8, & a_{22} = 0.7, & b_{02} = 1, & b_{12} = 1.9, & b_{22} = 2.4 \\
a_{13} = 1.2, & a_{22} = 0.8, & b_{02} = 1, & b_{12} = 1.75, & b_{22} = 2.1
\end{array}
$$

7.6.4 Pole-Zero Pairing and Ordering of the Cascade Form

For a high-order transfer function, there are a variety of ways in which the poles and zeros can be paired to form the second-order IIR functions, and a number of different ways these IIR filters can be ordered to realize a the given IIR filter in the cascade form. Each of these realizations will have different output noise power due to product quantization. We now derive a closed-form expression for the output noise variance due to product round-off errors in a cascade form. For this purpose, we consider the quantization noise model of the scaled cascade form IIR digital filter, with R second-order sections, as shown in Fig. 7.53.

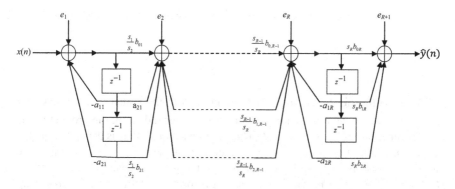

Fig. 7.53 Noise model of a scaled cascade realization of an IIR filter with R second-order sections

The various noise sources $e_k(n)$ arise out of the quantization of the product signals feeding the kth adder. It is assumed that all the individual noise sources are uncorrelated. Let $\hat{H}_i(z)$ be the transfer function of the scaled ith section of the cascade. Then, using the procedure adopted in Sect. 7.5.6 for deriving the expression for the output variance of a cascade structure, we see that the variance of the noise output for the scaled cascade structure of Fig. 7.53 is given by

$$\sigma_\epsilon^2 = \sigma_e^2 \left[3\hat{\eta}_1 + 5 \sum_{k=2}^{R} \hat{\eta}_k + 3 \right] \tag{7.173}$$

where

$$\sigma_e^2 = \frac{q^2}{12} = \text{Variance of the individual noise sources}$$

$$\hat{\eta}_k = \frac{1}{2\pi j} \oint \hat{G}_k(z) \hat{G}_k(z^{-1}) z^{-1} dz$$

and

$$\hat{G}_k(z) = \prod_{i=k}^{R} \left[\hat{H}_i(z) \right]$$

Now,

$$\hat{\eta}_k = \frac{1}{2\pi j} \oint \hat{G}_k(z) \hat{G}_k(z^{-1}) z^{-1} dz = \left\| \hat{G}_k \right\|_2^2$$

Hence, the output noise variance may be written as

$$\sigma_\epsilon^2 = \sigma_e^2 \left[3 \|\hat{G}_1\|_2^2 + 5 \sum_{k=2}^{R} \|\hat{G}_k\|_2^2 + 3 \right] \tag{7.174}$$

The noise transfer function $\hat{G}_k(z)$ is the product of the scaled transfer functions $\hat{H}_i(z), i = k, \ldots, R$, and the scaling factor for $\hat{G}_k$ is dependent on the product of the transfer functions $H_i(z), i = 1, \ldots, k$. Thus, each term in (7.174) is dependent on the transfer function of each of the second-order sections. Hence, in order to minimize the output noise power, we have to minimize the norms of each of the functions $H_i(z)$ by appropriately pairing the poles and zeros. To achieve this, the following rules for pole-zero pairing and ordering are proposed in [2].

(i)   The pole that is closest to the unit circle must be paired with the zero that is closest to it in the z-plane.

(ii) The above rule must be continually applied until all the poles and zeros have been paired.

(iii) The resultant second-order sections must be ordered either in the increasing nearness to the unit circle or in the decreasing nearness to the unit circle.

Pairing a pole that is more close to the unit circle with an adjacent zero reduces the peak gain of the section formed by the pole-zero pair.

To illustrate the above rules, consider an elliptic IIR filter with the following specifications:

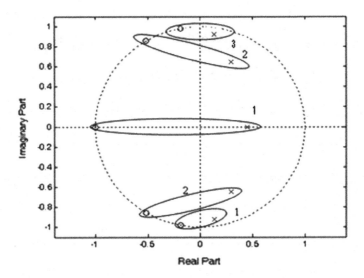

Fig. 7.54 Pole-zero pairing and ordering

passband edge at 0.45π, passband ripple of 0.5 dB, and minimum stopband attenuation of 45 dB.

For the given specifications, the poles, zeros, and the gain constant of the desired filter are obtained using the following MATLAB command.

[z,p,k]=ellip(5,0.5,45,0.45)

The pole-zero pairing and ordering from the least peaked to the most peaked is shown in Fig. 7.54.

7.7 Limit Cycles in IIR Digital Filters

When a stable IIR digital filter is excited by specific input signals, such as zero or constant inputs, it may exhibit an unstable behavior because of the nonlinearities caused by the quantization of the arithmetic operations. However, nonlinearities due to finite precision arithmetic operations often cause periodic oscillations to occur in the output. Such oscillations in recursive systems are called limit cycles, and the system does not return to the normal operation till the input amplitude is adequately large. In IIR filters, limit cycles arise due to the feedback path, whereas there is no scope for limit cycles in FIR structures in view of the absence of any feedback path. There are two different forms of limit cycles, namely limit cycles due to round-off or truncation of products and overflow.

7.7.1 Limit Cycles Due to Round-off and Truncation of Products

The round-off and truncation errors can cause oscillations in the filter output even when the input is zero. These oscillations are referred to as the limit cycles due to product round-off or truncation errors. These limit cycles are now illustrated through the following examples.

Example 7.30 Consider the first-order IIR system described by the difference equation

$$y(n) = ay(n-1) + x(n) \tag{7.175}$$

Investigate for limit cycles when implemented using a signed four-bit fractional arithmetic with a quantization step of 2^{-3}, with $x(0) = 0.875$ and $y(-1) = 0$, when (i) $a = 0.5$ and (ii) $a = -0.5$.

Solution Quantization step size $= 2^{-3}$

Code word length = four bits (including the sign bit)

Table 7.2 Limit cycle behavior of the first-order IIR digital filter

n	$a = 0.100,\ \hat{y}(-1) = 0$		$a = 1.100,\ \hat{y}(-1) = 0$	
	$a\,\hat{y}(n-1)$	$\hat{y}(n) = y(n)$ after rounding	$a\,\hat{y}(n-1)$	$\hat{y}(n) = y(n)$ after rounding
0	0	0.111	0	0.111
1	0.011100	0.100	1.011100	1.100
2	0.010000	0.010	0.010000	0.010
3	0.001000	0.001	1.001000	1.001
4	0.000100	0.001	0.000100	0.001
5	0.000100	0.001	1.000100	1.001
6	0.000100	0.001	0.000100	0.001
7	0.000100	0.001	1.000100	1.001
8	0.000100	0.001	0.000100	0.001

Fig. 7.55 Illustration of limit cycles in a first-order IIR digital filter for the case $a = 0.5$

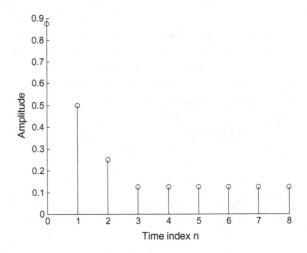

Table 7.2 shows the first nine output samples for the two different pole positions, $z = 0.5$ and $z = -0.5$, for an impulse with $x(0) = 0.875 = (0.111)_2$ and $x(n) = 0$ for $n > 0$ and $y(-1) = 0$, when the quantization used for the product is rounding. It can be observed that the steady-state output in the first case is a nonzero constant with period of 1, whereas in the second case it is with a period of 2. On the other hand, with infinite precision, the ideal output goes to zero as $n \rightarrow \infty$. The first nine output samples of the first-order IIR filter for the two cases are shown in Figs. 7.55 and 7.56, respectively.

The following MATLAB program illustrates the limit cycle process for the above example. In this program, flag=2 is used to develop the decimal equivalent of the binary representation of the filter coefficient magnitude after rounding.

Program 7.12 Limit Cycles in First-Order IIR Filter

```
clear all;close all;clc;
a = input('Enter the value of filter coefficient = ');
y0 = input('Enter the initial condition = ');
x = input('Enter the value of x[0] = ');
flag=input('Enter 1 for truncation, 2 for rounding=');
yi=y0;
for n = 1:9
   y(n) = truncround(a*yi,3,flag) + x;
   yi = y(n); x=0;
end
y(2:9)=y(1:8);
y(1)=y0;
k = 0:8;
stem(k,y)
ylabel('Amplitude'); xlabel('Time index n')
```

Example 7.31 Consider the first-order IIR system described by

$$y(n) = 0.625y(n-1) + x(n) \qquad (7.176)$$

Investigate for limit cycles when implemented for zero input using a signed four-bit fractional arithmetic

(i) when the product is rounded and
(ii) when the product is truncated.

The initial condition is $y(0) = 1/4$.

Fig. 7.56 Illustration of limit cycles in a first-order IIR digital filter for the case $a = -0.5$

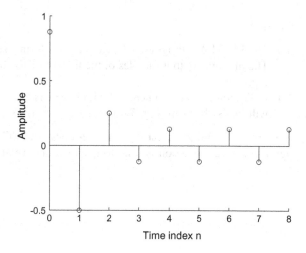

Fig. 7.57 Limit cycle in the first-order IIR digital filter of Example 7.31 with product round-off

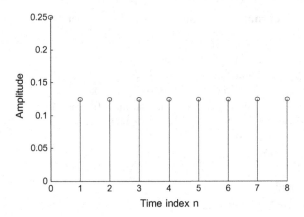

Fig. 7.58 Limit cycles in the first-order IIR digital filter of Example 7.31 with product truncation

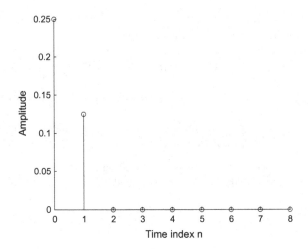

Solution

(i) The MATLAB Program 7.12 can be used to investigate the limit cycles. The first nine output samples of the first-order IIR filter obtained are shown in Fig. 7.57.

(ii) In Program 7.12, if truncround with flag=1 is used for this example, the result will be as shown in Fig. 7.56.

From Fig. 7.58, it is clear that the filter does not exhibit overflow limit cycles, if sign magnitude truncation is used to quantize the product.

7.7.2 Overflow Limit Cycles

An overflow occurs when the sum of two or more binary numbers exceeds the word length available in digital filters implemented using finite precision arithmetic. The overflow may result in limit cycle-like oscillations. Such limit cycles are referred to as overflow limit cycles, which can have a much more severe effect compared to that of the limit cycles due to round-off errors in multiplication. The following example illustrates the generation of overflow limit cycles in a second-order all-pole IIR digital filter.

Example 7.32 Consider an all-pole second-order IIR digital filter described by the following difference equation

$$y(n) = -0.75y(n-1) + 0.75y(n-2) + x(n) \qquad (7.177)$$

Investigate for the overflow limit cycles when implemented using a sign magnitude four-bit arithmetic

(i) with rounding of the sum of the products by a single quantizer and
(ii) with truncation of the sum of the products by a single quantizer

The initial conditions are $y(-1) = -0.5$ and $y(-2) = -0.125$. The input $x(n) = 0$ for $n \geq 0$.

Solution The MATLAB Program 7.13 given below is used to investigate the overflow limit cycles.

Fig. 7.59 Illustration of the overflow limit cycles in the second-order IIR digital filter of Example 7.32 with quantization due to rounding

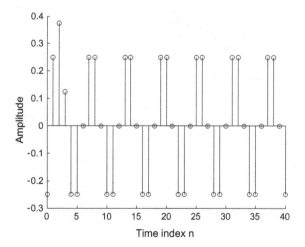

Program 7.13 Overflow Limit Cycles

```
clear all; close all;clc;
a(1)=-0.75;a(2)=0.75;%filter coefficients
yi1 = -0.5; yi2 = -0.125;
flag=input('enter 1 for truncation, 2 for rounding=');
for n = 1:41;
    y(n) = - a(1)*yi1 - a(2)*yi2;
    y(n) = truncround(y(n),3,flag);
    yi2 = yi1; yi1 = y(n);
end
k = 0:40;
stem(k,y)
xlabel('Time index n');ylabel('Amplitude')
```

(i) In the above program, flag=2 is used to perform the rounding operation on the sum of products. Figure 7.59 shows the output generated by the program illustrating the generation of overflow limit cycles with zero input.

(ii) In the above program, flag=1 is used for truncation, and the result is shown in Fig. 7.60.

From Fig. 7.60, we see that the filter described by Eq. (7.177) does not exhibit overflow limit cycles, if sign magnitude truncation is used to quantize the sum of the products.

Fig. 7.60 Illustration of the overflow limit cycles in the second-order IIR digital filter of Example 7.32 with quantization due to truncation

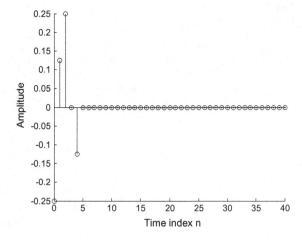

7.8 Quantization Effect in FFT Computation

7.8.1 Direct Computation of the DFT

The DFT of a finite duration sequence $x(n)$, $0 \leq n \leq N - 1$, is given by

$$X(k) = \sum_{n=0}^{N-1} x(n) W_N^{kn} \quad k = 0, 1, \ldots, N - 1 \qquad (7.178)$$

where $W_n = e^{-j2\pi/N}$. Generally, it is assumed that $x(n)$ is a complex-valued sequence. As a result, the product $x(n) W_N^{kn}$ needs four real multiplications. Hence, there are four sources of round-off errors for each complex-valued multiplication. In direct computation of a particular sample in the DFT, N complex-valued multiplications are required. For that reason, to compute a single sample in the DFT, the total number of real multiplications required is $4N$. Thus, it results in $4N$ quantization error sources.

From Eq. (7.102), we know that the variance of each quantization error (for round-off) is given by

$$\sigma_e^2 = \frac{q^2}{12} = \frac{2^{-2b}}{12} \qquad (7.179)$$

Since there are $4N$ multiplications required in computing one DFT sample, the variance of the quantization error in computing one sample is

$$\sigma_q^2 = 4N\sigma_e^2 = \frac{N}{3} 2^{-2b} \qquad (7.180)$$

To overcome the problem of overflow, the input sequence $x(n)$ has to be scaled. In order to analyze the effect of scaling, we assume that the signal sequence $x(n)$ is white and each value of the sequence is uniformly distributed in the range $(-1/N)$ to $(1/N)$ after scaling [10]. Then, the variance of the input signal is given by

$$\sigma_x^2 = \frac{q^2}{12} = \frac{[(1/N) - (-1/N)]^2}{12} = \frac{1}{3N^2} \qquad (7.181)$$

and the variance of the corresponding output signal is

$$\sigma_X^2 = N\sigma_x^2 = \frac{1}{3N} \qquad (7.182)$$

Accordingly, the signal-to-noise ratio is

$$SNR = \frac{\sigma_x^2}{\sigma_q^2} = \frac{2^{2b}}{N^2} \qquad (7.183)$$

It is seen from Eq. (7.183) that as a consequence of scaling and round-off error, there is a reduction in the SNR by a factor of N^2. Thus, for a desired SNR, the above equation can be used to determine the word length required to compute an N-point DFT.

Example 7.33 Find the word length required to compute the DFT of a 512-point sequence with a SNR of 40 dB.

Solution The range of the sequence is $N = 2^9$. Hence, from Eq. (7.183), the SNR is

$$10 \log_{10} 2^{2b-18} = 40$$

$$b = 9 + \frac{2}{\log_{10} 2} = 15.644$$

or

$$b = 16 \quad \text{bits}$$

Thus, a 16-bit word length is required to compute the DFT of a 512-point sequence with a SNR of 40 dB.

7.8.2   FFT Computation

Consider the computation of a single DFT sample as shown in Fig. 7.61 for an eight-point DIT DFT. From this figure, it can be observed that the computation of a single DFT sample requires three stages, and in general, we need $v = \log_2 N$ stages in the case of an N-point DFT. In general, there are $N/2$ butterflies in the first stage of the FFT, $N/4$ in the second stage, $N/8$ in the third stage, and so on, until the last stage, where only one is left. Thus, the number of butterflies involved for each output point is $(1 + 2 + 2^2 + \cdots + 2^{v-1}) = (N - 1)$. For example, in Fig. 7.61, the DFT sample $X(2)$ is connected to $(8 - 1) = 7$ butterflies. Since the input sequence and the twiddle factors are complex, four real multiplications are required for each complex-valued multiplication. Hence, for each complex-valued multiplication, there are four round–off error sources each having the same variance, since the magnitudes of the twiddle factors are unity. Thus, the variance of the round-off error for each complex-valued multiplication is

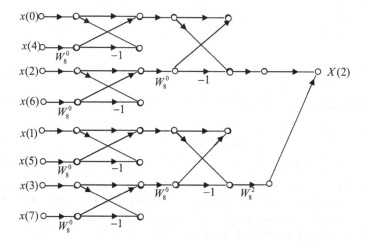

Fig. 7.61 Butterflies that affect the computation of the single DFT sample $X(2)$

$$4\frac{q^2}{12} = \frac{2^{-2b}}{3}.$$

Since the total number of butterflies needed per DFT sample is $(N-1)$, the variance of the total round-off error per DFT sample at the output is given by

$$\sigma_q^2 = \frac{N-1}{3}2^{-2b}$$

For large values of N, the above equation may be written as

$$\sigma_q^2 \approx \frac{N}{3}2^{-2b} \tag{7.184}$$

which is exactly the same expression as (7.180) for the case of direct DFT computation.

As in the case of direct DFT computation, we have to scale the input sequence to prevent overflow. However, instead of scaling the input by $(1/N)$, we can distribute the total scaling of $1/N$ into each of the $v = \log_2 N$ stages of the FFT algorithm. If the input signals at each stage are scaled by $(1/2)$, then we obtain an overall scaling of $(1/2)^v = (1/N)$. Each scaling by a factor of $(1/2)$ reduces the round-off noise variance by a factor of $(1/4)$. It can be shown that the total round-off noise variance at the output is given by [10]

$$\sigma_q^2 = \frac{2}{3}2^{-2b}\left[1 - \left(\frac{1}{2}\right)^v\right]$$

Since the term $(1/2)^v$ is negligible for large N, the above equation can be approximated as

$$\sigma_q^2 \approx \frac{2}{3} 2^{-2b} \tag{7.185}$$

Thus, the SNR is given by

$$\text{SNR} = \frac{\sigma_X^2}{\sigma_q^2} = \frac{2^{2b}}{2N} = \frac{2^{2b-1}}{N} \tag{7.186}$$

Hence, by distributing the scaling into each stage, the SNR is increased by a factor of N compared with that of the direct DFT computation. Equation (7.186) can be used to calculate the word length needed to obtain a particular SNR in the computation of an N-point DFT using FFT.

Example 7.34 Find the word length required to compute a 512-point FFT with an SNR of 40 dB.

Solution The length of the FFT is $N = 2^9$. According to Eq. (7.186), the SNR can be written as

$$10\log_{10} 2^{2b-9-1} = 40$$
$$b = 5 + \frac{2}{\log_{10} 2} = 11.644$$

or

$$b = 12\,\text{bits}$$

Thus, a word length of 12 bits is required to compute a 512-point FFT with a SNR of 40 dB, as compared to the 16 bits required in the direct computation of a 512-point DFT.

7.9 Problems

1. (a) For the SFG shown in Fig. P7.1, find the system function $Y(z)/X(z)$.
   (b) Obtain its transpose and find its system function.
   (c) Obtain block diagram representations for the given SFG and its transpose

2. Draw a signal flow diagram for the following system function. Your diagram should use a minimum number of delay units. Be sure to mark the input $x(n)$ and the output $y(n)$ on your diagram.

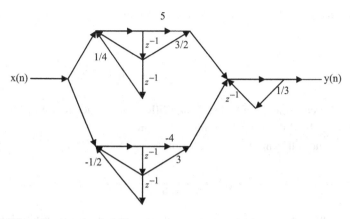

Fig. P7.1 Signal flow graph of problem 2

$$H(z) = \frac{0.2(1+z^{-1})^6}{\left(1 - 2z^{-1} + \frac{7}{8}z^{-2}\right)\left(1 + z^{-1} + \frac{1}{2}z^{-2}\right)\left(1 - \frac{1}{2}z^{-1} + z^{-2}\right)}$$

3. Obtain two different cascade realizations for the system described by the system function

$$H(z) = 10\frac{\left(1 - \frac{2}{3}z^{-1}\right)}{\left(1 - \frac{7}{8}z^{-1} + \frac{3}{32}z^{-2}\right)}\frac{\left(1 + \frac{3}{2}z^{-1} - z^{-2}\right)}{\left(1 - z^{-1} + \frac{1}{2}z^{-2}\right)}$$

4. Obtain a parallel realization for the system function of Problem 2.
5. Obtain cascade and parallel realizations of the system described by the system function

$$H(z) = \frac{0.1432 + 0.4256z^{-1} + 0.4296z^{-2} + 0.1432z^{-3}}{1 - 0.1801z^{-1} + 0.3419z^{-2} - 0.0165z^{-3}}$$

6.

(a) Obtain the lattice realization of the system whose transfer function is

$$H(z) = \frac{1}{1 + 0.75z^{-1} + 0.25z^{-2} + 0.25z^{-3}}$$

and check for its BIBO stability

(b) Using the realization of part (a), obtain a realization for the allpass filter

$$H(z) = \frac{0.25 + 0.25z^{-1} + 0.75z^{-2} + z^{-3}}{1 + 0.75z^{-1} + 0.25z^{-2} + 0.25z^{-3}}$$

7. Obtain the direct form II and lattice realizations of the system whose transfer function is

$$H(z) = \frac{1 + 2z^{-1} + z^{-2}}{(1 - 0.5z^{-1})(1 - 0.25z^{-1})}$$

8. The lattice coefficients of a three-stage FIR lattice structure are $k_1 = \frac{1}{4}, k_2 = \frac{1}{4}$ and $k_3 = \frac{1}{3}$. Find the FIR direct form structure.

9. Consider the following transfer function.

$$H(z) = \frac{1}{1 - 0.9z^{-1} + 0.2z^{-2}}$$

If the coefficients are quantized so that they can be expressed in four-bit binary form (including the sign bit), find the pole positions for the cascade and direct forms with quantized coefficients (a) with truncation and (b) with rounding.

10. Consider the transfer function of Problem 9. Find the sensitivities of the poles with respect to the coefficients for the direct as well as the cascade form.

11. For the IIR digital filter with a transfer function

$$H(z) = \frac{1 - 0.25z^{-1}}{(1 - 0.75z^{-1})(1 - 0.5z^{-1})}$$

obtain all the possible cascade and parallel realizations, using direct form II structure for each of the first-order IIR functions. Determine the variance of the output round-off noise due the product round-off quantization. Determine the structure which has the lowest round-off noise.

12. Design an elliptic IIR lowpass digital filter with the following specifications: passband edge at 0.45π, stopband edge at 0.5π, passband ripple of 1 dB, and minimum stopband attenuation of 45 dB. Obtain the corresponding scaled cascade realization using L_2 norm.

13. Design an elliptic IIR lowpass digital filter with the following specifications: passband edge at 0.35π, stopband edge at 0.45π, passband ripple of 0.5 dB, and minimum stopband attenuation of 40 dB. Obtain the scaled parallel realization using L_2 norm.

7.10 MATLAB Exercises

1. Consider the following transfer function of an IIR digital filter. Modify MATLAB Program 7.7 to study the effect on the gain responses and pole-zero locations when the filter coefficients are truncated or rounded to six

bits and realized in direct form. Plot gain responses and pole-zero locations for quantized and unquantized coefficients of the filter. Comment on the results.

$$H(z) = \frac{(1+0.2189z^{-1}+z^{-2})}{(1-0.0127z^{-1}+0.9443z^{-2})} \frac{(1-0.5291z^{-1}+z^{-2})}{(1-0.1731z^{-1}+0.7252z^{-2})}$$
$$\times \frac{(1+1.5947z^{-1}+z^{-2})}{(1-0.6152z^{-1}+0.2581z^{-2})}$$

2. Consider the transfer function same as given above. Modify MATLAB Program 7.7 to study the effect on the gain responses and pole-zero locations when the filter coefficients are truncated or rounded to six bits and realized in cascade form. Plot gain responses and pole-zero locations for quantized and unquantized coefficients of the filter. Comment on the results.

3. Design an elliptic IIR lowpass digital filter with the following specifications: passband edge at $0.5\,\pi$, stopband edge at $0.55\,\pi$, passband ripple of 1 dB, and minimum stopband attenuation of 35 dB. Determine the suitable coefficient word length to maintain stability and to satisfy the frequency response specifications when realized in cascade form.

4. Determine the optimum pole-zero pairing and their ordering for the following transfer function to minimize its output under L_2 norm scaling.

$$H(z) = \frac{(z^2-1.0166z+1)}{(z^2-1.4461z+0.7957)} \frac{(0.05634z+0.05634)}{(z-0.683)}$$

5. Consider an all-pole first-order IIR digital filter described by the following difference equation

$$y(n) = ay(n-1) + x(n)$$

Modify MATLAB Program 7.13 to investigate the overflow limit cycles when implemented using a sign magnitude four-bit arithmetic

a. with a rounding of the sum of products, and $a = 0.75$.
b. with truncation of the sum of products, and $a = 0.75$.
c. with a rounding of the sum of products, and $a = -0.75$.
d. with truncation of the sum of products, and $a = -0.75$.

The initial conditions are $y(0) = 6$, the input $x(n) = 0$ for $n \geq 0$.

References

1. S.J. Mason, H.J. Zimmerman, *Electronic Circuits, Signals and Systems* (Wiley, New York, 1960), pp. 122–123
2. L.B. Jackson, *Digital Filters and Signal Processing* (Kluwer Academic, Norwell, M.A., 1996)
3. L.B. Lawrence, K.V. Mirna, A new and interesting class of limit cycles in recursive digital filters, in *Proceedings of the IEEE International Symposium on Circuit and Systems*, 1977, pp. 191–194
4. A.V. Oppenheim, R. Schafer, *Discrete-Time Signal Processing* (Prentice Hall, Upper Saddle River, NJ, 1989)
5. A.H. Gray, J.D. Markel, Digital lattice and ladder filter synthesis. IEEE Trans. Acoust. Speech Signal Process. ASSP **21**, 491–500 (1973)
6. A.H. Gray, J.D. Markel, A normalized digital filter structure. IEEE Trans. Acoust. Speech Signal Process. ASSP **23**, 268–277 (1975)
7. B. Liu, T. Kaneko, Error analysis of digital filters realized in floating point arithmetic. Proc. IEEE **57**, 1735–1747 (1969)
8. J. Weinstein, A.V. Openheim, A comparison of roundoff noise in fixed point and floating point digital filter realizations. Proc. IEEE **57**, 1181–1183 (1969)
9. E.P.F. Kan, J.K. Agarwal, Error analysis in digital filters employing floating point arithmetic. IEEE Trans. Circ. Theor. CT **18**, 678–686 (1971)
10. J.G. Proakis, D.G. Manolakis, *Digital Signal Processing Principles, Algorithms and Applications*, 3rd edn. (Prentice-Hall, India, 2004)
11. Antoniou, A. *Digital filters: Analysis and Design* (McGraw Hill Book Co., New York, 1979)

Chapter 8
Basics of Multirate Digital Signal Processing

A continuous-time signal can be represented by a discrete-time signal consisting of a sequence of samples $x(n) = x_a(nT)$. It is often necessary to change the sampling rate of a discrete-time signal, i.e., to obtain a new discrete-time representation of the underlying continuous-time signal of the form $x'(n) = x_a(nT')$. One approach to obtain the sequence $x'(n)$ from $x(n)$ is to reconstruct $x_a(t)$ from $x(n)$ and then resample $x_a(t)$ with period T' to obtain $x'(n)$. However this is not a desirable approach, because of the non-ideal analog reconstruction filter, DAC, and ADC that would be used in a practical implementation. Thus, it is of interest to consider methods that involve only discrete-time operation.

8.1 Advantages of Multirate Signal Processing

There are several advantages of multirate DSP. The following are some of them.

- **Processing at various sampling rates**—It enables the processing of a signal with different sampling rates. For example, broadcasting requires 32 kHz sampling rate, whereas compact disk and digital audiotape require sampling rates of 44.1 and 48 kHz, respectively. The sampling frequency of the digital audio signal can be inherently varied by using multirate signal processing techniques. Another example is a telephone system which requires signal translation between the time-division multiplexing (TDM) and frequency-division multiplexing (FDM) formats. In a TDM–FDM translator, the sampling rate of the TDM speech signal (8 kHz) is increased to that of the FDM, whereas in a FDM–TDM translator, the sampling rate of the FDM is reduced to that of the TDM by using multirate DSP.
- **Simple anti-imaging analog filters**—Simple anti-imaging analog filters can be implemented. For example, to reproduce an analog audio signal of frequency 22 kHz, from a digital audio signal of frequency 44.1 kHz, and to remove the

© Springer Nature Singapore Pte Ltd. 2018
K. D. Rao and M. N. S. Swamy, *Digital Signal Processing*,
https://doi.org/10.1007/978-981-10-8081-4_8

images outside 22 kHz, a complicated anti-imaging analog filter with sharp cutoff frequency is required. Instead, by oversampling the audio signal, a simple anti-imaging analog filter can be used in the place of an expensive anti-imaging analog filter.

- **Highly reduced filter orders**—A narrow band digital FIR filter order will be very large requiring a huge number of coefficients to meet the design specifications. The multirate DSP implements a narrow band filter very effectively using filters of highly reduced orders, i.e., with less number of filter coefficients.
- **Subband Decomposition**—For example, in subband coding of speech signals, the signal is subdivided into different frequency bands. Multirate techniques are used to achieve a reduction in the transmission rate of the digitized speech signal and to reconstruct the original speech signal at a higher rate from the low-rate-encoded speech signal. Another example is in the subband adaptive filtering, resulting in increased convergence speed in applications such as echo cancellation and adaptive channel equalization.

8.2 Multirate Signal Processing Concepts

8.2.1 Down-Sampling: Decimation by an Integer Factor

The block diagram representation of a down-sampler, also known as a sampling rate compressor, is depicted in Fig. 8.1.

The down-sampling operation is implemented by defining a new sequence $x_d(n)$ in which every Mth sample of the input sequence is kept and $(M - 1)$ in-between samples are removed to obtain the output sequence; i.e., $x_d(n)$ is identical to the sequence obtained from $x_a(t)$ with a sampling period $T' = MT$

$$x_d(n) = x(nM) \tag{8.1}$$

For example, if $x(n) = \{2, 6, 3, 0, 1, 2, -5, 2, 4, 7, -1, 1, -2, \ldots\}$ $x_d(n) = \{2, 1, 4, -2, \ldots\}$ for $M = 4$, i.e., $M - 1 = 3$ samples are left in between the samples of x (n) to get $x_d(n)$.

Example 8.1 MATLAB Program 8.1 given below is used to illustrate down-sampling by an integer factor of 4 of a sum of two sinusoidal sequences, each of length 50, with normalized frequencies of 0.2 and 0.35 Hz.

Fig. 8.1 Block diagram representation of a down-sampler

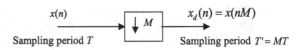

$x(n)$

Sampling period T

$\downarrow M$

$x_d(n) = x(nM)$

Sampling period $T' = MT$

Program 8.1 Illustration of down-sampling of sum of two sinusoidal sequences of length 50 by an integer factor 4

```
len = 50;%output length
M = 4;%down sampling factor
f0 = 0.2;f1 = 0.35;% input signals frequencies
title('input sequence');
n = 1:len;
m = 1:len*M;
x = sin(2*pi*f0*m) + sin(2*pi*f1*m);
y = x([1:M:length(x)]);
subplot(2,1,1)
stem(n,x(1:len));
title('input sequence');
xlabel('Time index n');
ylabel('Amplitude');
subplot(2,1,2)
stem(n,y);
xlabel('Time index n');
ylabel('Amplitude');title(['Output sequence down-sampled by', num2str(M)])
```

Figure 8.2 shows the results of the down-sampling.

The relation between the Fourier transform of the output and the input of a factor-M down-sampler is obtained as follows:

We first define an intermediate sequence $x_1(n)$ whose sample values are the same as that of $x(n)$ at the values of n that are multiples of M and are zeros at other values of n; otherwise,

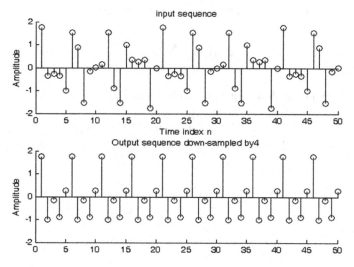

Fig. 8.2 Illustration of down-sampling by factor 4

$$x_1(n) = \begin{cases} x(n), & n = 0, \pm M, \pm 2M, \ldots \\ 0 & \text{otherwise} \end{cases}. \tag{8.2}$$

Then, the Fourier transform

$$X_d(e^{j\omega}) = \sum_{n=-\infty}^{\infty} x(nM)e^{-j\omega n} = \sum_{n=-\infty}^{\infty} x_1(nM)e^{-j\omega n} \tag{8.3}$$

Let $k = nM$, and then Eq. (8.3) can rewritten as

$$X_d(e^{j\omega}) = \sum_{k=-\infty}^{\infty} x_1(k)e^{-j\omega k/M} = X_1\left(e^{j\omega/M}\right) \tag{8.4}$$

Now, $x_1(nM)$ can be written as

$$x_1(nM) = x(n)w(n) \tag{8.5}$$

where

$$\begin{aligned} w(n) &= 1, & n = 0, \pm M, \pm 2M, \ldots \\ w(n) &= 0, & \text{otherwise} \end{aligned} \tag{8.6}$$

Thus, Eq. (8.5) can be written as

$$X_1\left(e^{j\omega}\right) = \sum_{n=-\infty}^{\infty} x(n)w(n)e^{-j\omega n} \tag{8.7}$$

The quantity $w(n)$ can be represented conveniently as

$$w(n) = \frac{1}{M} \sum_{k=0}^{M-1} e^{\frac{-j2\pi kn}{M}}, \tag{8.8}$$

since the RHS in the above equation can be written as

$$\frac{1}{M} \frac{1 - e^{-j2\pi n}}{1 - e^{-j2\pi n/M}} = \begin{cases} 1 & \text{if } n = \pm M, \pm 2M, \ldots \\ 0 & \text{otherwise} \end{cases}$$

Substituting Eq. (8.8) in Eq. (8.7), we obtain

$$
\begin{aligned}
X_1\left(e^{j\omega}\right) &= \frac{1}{M}\sum_{k=0}^{M-1}\sum_{n=-\infty}^{\infty} x(n)e^{-j2\pi kn/M}e^{-j\omega n} \\
&= \frac{1}{M}\sum_{k=0}^{M-1}\sum_{n=-\infty}^{\infty} x(n)e^{-j\omega n-\frac{j2\pi kn}{M}} \\
&= \frac{1}{M}\sum_{k=0}^{M-1} X\left(e^{j\omega}e^{\frac{j2\pi k}{M}}\right)
\end{aligned}
\tag{8.9}
$$

Substitution of Eq. (8.9) in Eq. (8.4) results in

$$
X_d\left(e^{j\omega}\right) = \frac{1}{M}\sum_{k=0}^{M-1} X\left(e^{\frac{j\omega}{M}}e^{\frac{j2\pi k}{M}}\right)
\tag{8.10}
$$

$$
= \frac{1}{M}\left[X\left(e^{\frac{j\omega}{M}}\right) + \sum_{k=1}^{M-1} X\left(e^{\frac{j\omega}{M}}e^{j2\pi}e^{\frac{j2\pi(k-M)}{M}}\right)\right]
\tag{8.11}
$$

$$
= \frac{1}{M}\left[X\left(e^{\frac{j\omega}{M}}\right) + \sum_{q=1}^{M-1}\left(e^{\frac{j\omega}{M}}e^{\frac{-j2\pi q}{M}}\right)\right]
\tag{8.12}
$$

Hence, the relation between the Fourier transform of the output and the input of a factor-M down-sampler given by Eq. (8.10) can be rewritten as

$$
X_d(e^{j\omega}) = \frac{1}{M}\sum_{k=0}^{M-1} X\left(e^{\frac{j(\omega - 2\pi k)}{M}}\right)
\tag{8.13}
$$

which is composed of M copies of the periodic Fourier transform $X(e^{j\omega})$, frequency scaled by M and shifted by integer multiples of 2π. The sampling rate compressor reduces the sampling rate from F_T to F_T/M. In the frequency domain, it is equivalent to multiplying the original signal bandwidth by a factor M. The effect of down-sampling on a signal band-limited to $\pi/2$ for a down-sampling factor $M = 3$ is shown in Fig. 8.3. From Fig. 8.3, it can be seen that there is aliasing. Hence, if the signal is not band-limited to $\frac{\pi}{M}$, down-sampling results in aliasing.

Aliasing due to a factor-of-M down-sampling is absent if and only if the signal $x(n)$ is band-limited to $\pm\pi/M$ by means of a lowpass filter (LPF), called the decimator filter $H(z)$, with unity gain and cutoff frequency of π/M, as shown in Fig. 8.4. The system of Fig. 8.4 is often called a decimator.

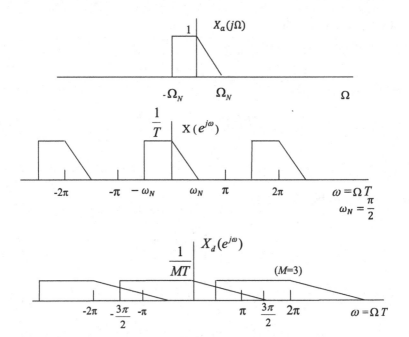

Fig. 8.3 Illustration of down-sampling with aliasing

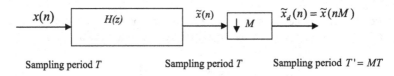

Fig. 8.4 Block diagram representation of a decimator

Due to the decimation filter, there is no aliasing at the decimator output with the output spectrum $\tilde{X}_d(e^{j\omega})$ as shown in Fig. 8.5 with a down-sampling factor of 3.

Example 8.2 Consider an input sequence

$$x(n) = \left(\frac{1}{2}\right)^n u(n)$$

If $x(n)$ is the input to a down-sampler with down-sampling factor of 2, determine the output spectrum $Y(e^{j\omega})$ of the down-sampler and comment on the result.

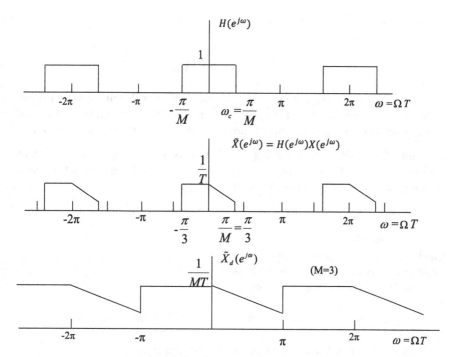

Fig. 8.5 Illustration of down-sampling with no aliasing

Solution From Eq. (8.13), the output spectrum $Y(e^{j\omega})$ for $M = 2$ is given by

$$Y\left(e^{j\omega}\right) = \frac{1}{2}\sum_{k=0}^{1} X\left(e^{j(\omega-2\pi k)/2}\right)$$

$$= \frac{1}{2}\left[X\left(e^{\frac{j\omega}{2}}\right) + X\left(e^{\frac{j(\omega-2\pi)}{2}}\right)\right]$$

$$= \frac{1}{2}\left[X\left(e^{\frac{j\omega}{2}}\right) + X\left(e^{\frac{j\omega}{2}}e^{-j\pi}\right)\right]$$

$$= \frac{1}{2}\left[X\left(e^{\frac{j\omega}{2}}\right) + X\left(-e^{\frac{j\omega}{2}}\right)\right]$$

$$= \frac{1}{2}\left[X\left(e^{\frac{j\omega}{2}}\right) + X\left(-e^{\frac{j\omega}{2}}\right)\right]$$

$$= \frac{1}{2}\left[\frac{1}{1 - 0.5e^{\frac{-j\omega}{2}}} + \frac{1}{1 - 0.5e^{\frac{-j\omega}{2}}}\right]$$

$$= \frac{1}{1 - 0.25e^{-j\omega}}$$

If $y(n) = x(2n) = \left(\frac{1}{2}\right)^{2n}u(n)$, using the definition of DTFT, it can be easily shown that the output spectrum $Y(e^{j\omega})$ is the DTFT of $x(2n)$.

Example 8.3 Consider an input sequence

$$x(n) = \frac{f_0}{2}\left(\text{sinc}\left(\frac{f_0}{2}(n-d)\right)\cos\left(\frac{f_0}{2}(n-d)\right)\right)^2$$

where f_0 is the normalized frequency and d is one-half of the length of the sequence. Down-sample the input sequence by a factor of 5, and plot the frequency response of the input sequence and the down-sampled sequence. Comment on the results.

Solution MATLAB Program 8.2 given below is used to generate the input sequence of length 1024, to down-sample it by a factor of 5 and to plot frequency responses of the input sequence and down-sampled sequence.

Program 8.2 Illustration of the effect of down-sampling on the frequency response

```
clear all; close all;
len=1024;% input sequence length
M=input('Enter the down-sampling factor');
f0=input('Enter the normalized frequency');
n=1:len;
x=(f0/2)*(sinc((f0/2)*(n-512)).*cos((f0/2)*(n-512))).^2;%input sequence
f=-3:1/512:3
hx=freqz(x,1,f*pi); %frequency response of input sequence
xd=x([1:M:length(x)]);% down-sampled input sequence
hxd=freqz(xd,1,f*pi);% frequency response of down-sampled input sequence
subplot(2,1,1)
plot(f,abs(hx))
title('Frequency response of input sequence');
xlabel('\omega/\pi');
ylabel('Magnitude');
subplot(2,1,2)
plot(f,abs(hxd))
title('frequency response of downsampled sequence');
xlabel('\omega/\pi');
ylabel('Magnitude');
```

The frequency responses obtained from the above program for $M = 5$ and $f_0 = 0.25$ are shown in Fig. 8.6. It is observed that there is aliasing in the frequency response of the down-sampled sequence

Example 8.4 Design a decimator considering the input sequence given in Example 8.3. Assume input sequence length, down-sampling factor, and f_0 to be the same as used in Example 8.2. Comment on the frequency response of the decimator output sequence.

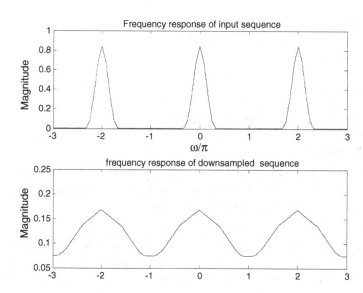

Fig. 8.6 Effect of down-sampling on the frequency response

Solution MATLAB Program 8.3 given below is used to design a decimator and to plot the frequency responses of the anti-aliasing filter input and output sequences, and the decimator output sequence.

Program 8.3 Illustration of the design of a decimator

```
clear all; close all;
len=1024;%input sequence length length
M=input('Enter the down-sampling factor');
f0=input('Enter the normalized frequency');
n=1:len;
x=(f0/2)*(sinc((f0/2)*(n-512)).*cos((f0/2)*(n-512)))).^2;%input sequence
f = -3:1/512:3
fl=fir1(127,1/M);%design of anti-aliasing filter
hf=freqz(fl,1,f*pi);%frequency response of anti-aliasing filter
xf=filter(fl,1,x);% output of anti-aliasing filter
hxf=freqz(xf,1,f*pi);% frequency response of anti-aliasing filter output sequence
j=1:len/M;
xr=xf(j*M); %output of decimator
hxr=freqz(xr,1,f*pi);% frequency response of decimator output sequence
subplot(3,1,1)
plot(f,abs(hf))
title('Frequency response of anti-aliasing filter input sequence');
xlabel('\omega/\pi');
ylabel('Magnitude');
subplot(3,1,2)
```

```
plot(f,abs(hxf))
title('Frequency response of anti-aliasing filter output sequence');
xlabel('\omega/\pi');
ylabel('Magnitude');
subplot(3,1,3)
plot(f,abs(hxr))
title('Frequency response decimator output sequence');
xlabel('\omega/\pi');
ylabel('Magnitude');
```

The frequency responses obtained from the above program for $M = 5$ and $f_0 = 0.25$ are shown in Fig. 8.7. It is observed that the frequency response of the decimator output sequence is free from aliasing.

Example 8.5 A speech signal $x(t)$ is digitized at a sampling rate of 16 kHz. The speech signal was destroyed once the sequence $x(n)$ was stored on a magnetic tape. Later, it was required to obtain the speech signal sampled at the standard 8 kHz used in telephony. Develop a method to do this using discrete-time processing.

Solution The speech signal $x(t)$ is sampled at 16 kHz to obtain $x(n)$. The sampling period T of $x(n)$ is $T = \frac{1}{16} \times 10^{-3}$ s. To obtain the speech signal with a sampling period $T' = \frac{1}{8} \times 10^{-3}$ from $x(n)$, the sampling time period T is to be increased by a factor of 2. Thus, the down-sampling scheme, shown in Fig. 8.8, is used to obtain from $x(n)$, the speech signal, with the sampling period $T' = \frac{1}{8} \times 10^{-3}$.

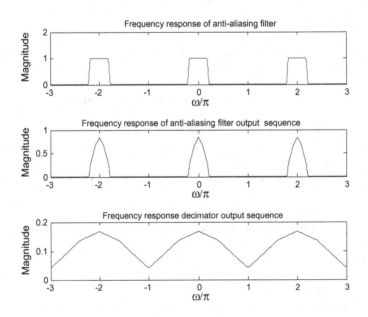

Fig. 8.7 Illustration of a decimator design

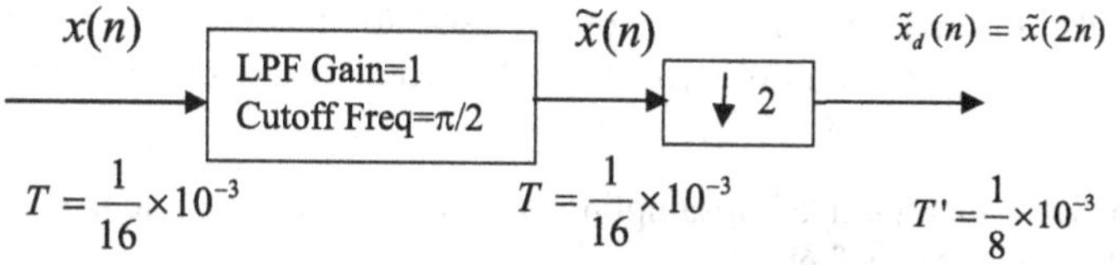

$$x(n)$$

$$\boxed{\begin{array}{c} \text{LPF Gain=1} \\ \text{Cutoff Freq=}\pi/2 \end{array}}$$

$$\tilde{x}(n)$$

$$\boxed{\downarrow 2}$$

$$\tilde{x}_d(n) = \tilde{x}(2n)$$

$$T = \frac{1}{16} \times 10^{-3} \qquad T = \frac{1}{16} \times 10^{-3} \qquad T' = \frac{1}{8} \times 10^{-3}$$

Fig. 8.8 Down-sampling schemes for changing the sampling rate by 2

8.2.2   Up Sampling: Interpolation by an Integer Factor

The block diagram representation of an upsampler, also called a sampling rate expander or simply an interpolator, is shown in Fig. 8.9.

The output of an upsampler is given by

$$x_e(n) = \sum_{k=-\infty}^{\infty} x(k)\delta(n - kL) = x\left(\frac{n}{L}\right) \quad n = 0, \pm L, \pm 2L, \ldots$$

$$= 0 \qquad\qquad\qquad \text{otherwise}$$

(8.14)

Equation (8.14) implies that the output of an upsampler can be obtained by inserting $(L - 1)$ equidistant zero-valued samples between two consecutive samples of the input sequence $x(n)$; i.e., $x_e(n)$ is identical to the sequence obtained from $x_a(t)$ with a sampling period $T' = T/L$. For example, if $x_e(n) = \{2, 1, 4, -2, \ldots\}$, then $x_e(n) = \{2, 0, 0, 0, 1, 0, 0, 0, 4, 0, 0, 0, -2, 0, 0, 0, \ldots\}$ for $L = 4$, i.e., $L - 1 = 3$ zero-valued samples are inserted in between the samples of $x(n)$ to get $x_e(n)$.

Fig. 8.9 Block diagram
representation of an
upsampler

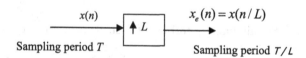

$$x(n)$$

$$\boxed{\uparrow L}$$

$$x_e(n) = x(n/L)$$

Sampling period T

Sampling period T/L

Example 8.6 Program 8.4 given below is used to illustrate upsampling by an integer factor of 4 of a sum of two sinusoidal sequences, each of length 50, with normalized frequencies of 0.2 and 0.35 Hz.

Program 8.4 Illustration of upsampling of sum of two sinusoidal sequences of length 50 by an integer factor 4

```
len=50;%output length
L=4;%up sampling factor
f0=0.2;f1 = 0.35% input signals frequencies
title('input sequence');
n=1:len;
x=sin(2*pi*f0*n) + sin(2*pi*f1*n);
y=zeros(1,L*length(x));
y([1:L:length(y)])=x;
subplot(2,1,1)
stem(n,x);
title('input sequence');
xlabel('Time index n');
ylabel('Amplitude');
subplot(2,1,2)
stem(n,y(1:length(x)));
title(['Output sequence up-sampled by', num2str(L)]);
xlabel('Time index n');ylabel('Amplitude');
```

Figure 8.10 shows the results of the upsampling.
By definition, the Fourier transform of $x_e(n)$ is given by

$$X_e(e^{j\omega}) = \sum_{n=-\infty}^{\infty} x_e(n)e^{-j\omega n} \qquad (8.15)$$

Substituting Eq. (8.14) in the above equation, we obtain

$$X_e(e^{j\omega}) = \sum_{k=-\infty}^{\infty}\sum_{n=-\infty}^{\infty} x(k)\delta(n-kL)e^{-j\omega n} \qquad (8.16)$$

By rearranging the summations, Eq. (8.16) can be rewritten as

$$X_e(e^{j\omega}) = \sum_{k=-\infty}^{\infty} x(k)\sum_{n=-\infty}^{\infty} \delta(n-kL)e^{-j\omega n} \qquad (8.17)$$

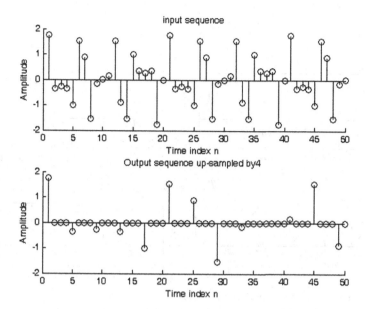

Fig. 8.10 Illustration of upsampling by factor 4

From the definition of the Fourier transform of $\delta(n - kL)$,

$$\sum_{n=-\infty}^{\infty} \delta(n - kL)e^{-j\omega n} = e^{-j\omega Lk} \tag{8.18}$$

Thus, Eq. (8.17) becomes

$$
\begin{aligned}
X_e\left(e^{j\omega}\right) &= \sum_{k=-\infty}^{\infty} x(k)e^{-j\omega kL} \\
&= X\left(e^{j\omega L}\right)
\end{aligned} \tag{8.19}
$$

Hence, the interpolator output spectrum is obtained by replacing ω by ωL in the input spectrum; i.e., ω is now normalized by $\omega = \Omega T'$, as shown in Fig. 8.11. The upsampling leads to a periodic repetition of the basic spectrum, causing $(L - 1)$ additional images of the input spectrum in the baseband, as shown in Fig. 8.11.

The unwanted images in the spectra of the upsampled signal $x_e(n)$ must be removed by using an ideal lowpass filter, called the interpolation filter, $H(z)$, with gain L and cutoff frequency of π/L, as shown in Fig. 8.12.

The use of the interpolation filter removes the images in the output spectrum $X_e(e^{j\omega})$ of the interpolator, as shown in Fig. 8.13.

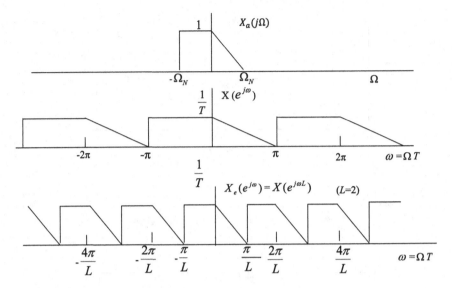

Fig. 8.11 Illustration of upsampling with images

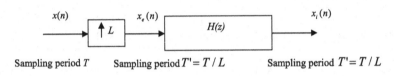

Fig. 8.12 Block diagram representation of an interpolator

Example 8.7 Design an interpolator considering the output sequence of the decimator of Example 8.3 as the input to the interpolator. Assume the upsampling factor to be the same as the down-sampling factor used in Example 8.3. Comment on the frequency response of the interpolator output sequence.

Solution MATLAB Program 8.5 given below is used to design an interpolator and to plot frequency responses of the upsampling sequence, anti-imaging filter, and interpolator output sequence.

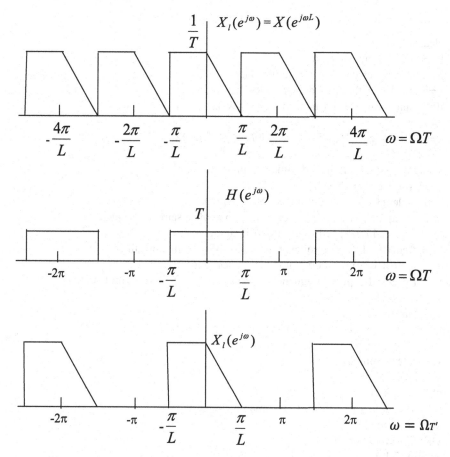

Fig. 8.13 Illustration of upsampling with no images

Program 8.5 Illustration of the design of interpolator

```
clear all; close all;
len=1024;%output length
M=input('Enter the down sampling factor');
L=input('Enter the upsampling factor');
f0=input('Enter the normalized frequency');
n=1:len;
x=(f0/2)*(sinc(f0/2*(n-512)).*cos(f0/2*(n-512))).^2;
f=-3:1/512:3
xr=decimate(x,M,'fir');% decimator output sequence
xe=zeros(1,L*length(xr));
xe([1:L:length(xe)]) = xr;%upsampled sequence
hxe = freqz(xe,1,f*pi);%frequency response of upsampled sequence
fl = L*fir1(117,1/L);%design of anti-imaging filter
hf = freqz(fl,1,f*pi);%frequency response of anti-imaging filter
xi = filter(fl,1,xe);% interpolator output sequence
hi = freqz(xi,1,f*pi);% frequency response of interpolator output sequence
subplot(3,1,1)
plot(f,abs(hxe))
title('Frequency response of upsampled sequence');
xlabel('\omega/\pi');
ylabel('Magnitude');
subplot(3,1,2)
plot(f,abs(hf))
title('Frequency response of anti-imaging filter');
xlabel('\omega/\pi');
ylabel('Magnitude');
subplot(3,1,3)
plot(f,abs(hi))
title('Frequency response interpolator output sequence');
xlabel('\omega/\pi');
ylabel('Magnitude');
```

The frequency responses obtained from the above program for $L = 5$ are shown in Fig. 8.14. It is observed that the frequency response of the interpolator output sequence is the same as the frequency response of the input sequence to the decimator of Example 8.4 as shown in Fig. 8.14.

Example 8.8 Consider the following upsampling scheme shown in Fig. 8.15. In this system, the output $y_1(n)$ is obtained by direct convolution of $x_e(n)$ and $h(n)$. A proposed implementation of the above system with the preceding choice of $h(n)$ is shown in Fig. 8.16. The impulse responses $h_1(n)$ and $h_2(n)$ are restricted to be zero outside the range $0 \leq n \leq 2$. Determine a choice for $h_1(n)$ and $h_2(n)$ so that $y_1(n)$ and $y_2(n)$ are equal.

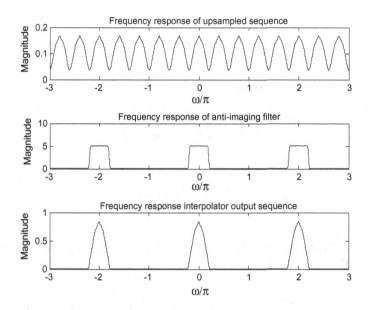

Fig. 8.14 Illustration of an interpolator design

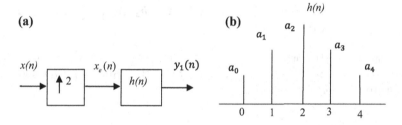

Fig. 8.15 a Upsampling scheme and **b** impulse response of the interpolation filter $h(n)$

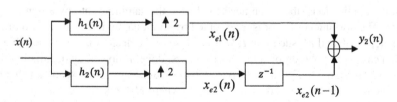

Fig. 8.16 Proposed implementation of the scheme shown in Fig. 8.15

Solution

$$y_1(n) = h(n) * x_e(n) = \sum_{k=-\infty}^{\infty} x_e(n-k)h(k) = a_0 x_e(n) + a_1 x_e(n-1) + a_2 x_e(n-2)$$
$$+ a_3 x_e(n-3) + a_4 x_e(n-4)$$

$$x_e(n) = \begin{cases} x\left(\frac{n}{2}\right) & \text{for } n \text{ even} \\ 0 & \text{for } n \text{ odd} \end{cases}$$

Thus,

$$y_1(n) = \begin{cases} a_0 x\left(\frac{n}{2}\right) + a_2 x\left(\frac{n}{2} - 1\right) + a_4 x\left(\frac{n}{2} - 2\right) & \text{for } n \text{ even} \\ a_1 x\left(\frac{n}{2} - \frac{1}{2}\right) + a_3 x\left(\frac{n}{2} - \frac{3}{2}\right) & \text{for } n \text{ odd} \end{cases}$$

$$x_{e1}(n) = h_1\left(\frac{n}{2}\right) * x\left(\frac{n}{2}\right) = h_1(0)x\left(\frac{n}{2}\right) + h_1(1)x\left(\frac{n}{2} - 1\right) + h_1(2)x\left(\frac{n}{2} - 2\right) \quad \text{for } n \text{ even}$$
$$= 0 \qquad\qquad\qquad\qquad\qquad\qquad\qquad\qquad\qquad\qquad\qquad\qquad\qquad\qquad\qquad\quad \text{for } n \text{ odd}$$

$$x_{e2}(n) = h_2\left(\frac{n}{2}\right) * x\left(\frac{n}{2}\right) = h_2(0)x\left(\frac{n}{2}\right) + h_2(1)x\left(\frac{n}{2} - 1\right) + h_2(2)x\left(\frac{n}{2} - 2\right) \quad \text{for } n \text{ even}$$
$$= 0 \qquad\qquad\qquad\qquad\qquad\qquad\qquad\qquad\qquad\qquad\qquad\qquad\qquad\qquad\qquad\quad \text{for } n \text{ odd}$$

$$x_{e2}(n-1) = \boldsymbol{h_2(0)}x\left(\frac{n}{2} - \frac{1}{2}\right) + \boldsymbol{h_2(1)}x\left(\frac{n}{2} - \frac{3}{2}\right) + \boldsymbol{h_2(2)}x\left(\frac{n}{2} - \frac{5}{2}\right) \quad \text{for } n \text{ odd}$$
$$\boldsymbol{y_2(n) = x_{e1}(n) + x_{e2}(n-1)}$$

Comparing $x_{e1}(n)$ and $x_{e2}(n-1)$ with $y_1(n)$, we see that $x_{e1}(n)$ gives even samples $h_1(0) = a_0; h_1(1) = a_2; h_1(2) = a_4$. Similarly, $x_{e2}(n-1)$ gives odd samples $h_2(0) = a_1; h_2(1) = a_3; h_2(2) = 0$.

8.2.3 Changing the Sampling Rate by a Non-integer Factor

In some applications, there is a need to change the sampling rate by a non-integer factor. The non-integer factor is represented by a rational number, i.e., a ratio of two integers, say L and M, such that L/M is as close to the desired factor as possible. In such a case, interpolation by a factor of L is first done, followed by decimation by a factor of M. Figure 8.17a shows a system that produces an output sequence that has an effective sampling period of $T' = TM/L$, where $H_I(z)$ is an ideal lowpass filter with gain L and cutoff frequency of π/L, and $H_2(z)$ is also an ideal lowpass filter with unity gain and cutoff frequency of π/M. Instead of using two lowpass filters as shown in Fig. 8.17a, a single lowpass filter $H(z)$ with gain L is adequate to serve both as the interpolation filter and the decimation filter, depending on which one of the two stopband frequencies $\frac{\pi}{L}$ or $\frac{\pi}{M}$ is a minimum [1]. Thus, the lowpass filter shown in Fig. 8.17b has a normalized stopband cutoff frequency at $\min(\frac{\pi}{L}, \frac{\pi}{M})$.

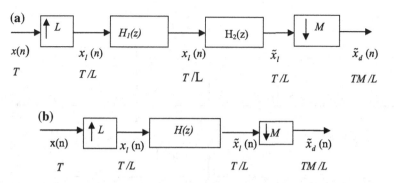

(a)

$x(n)$ $x_1(n)$ $x_1(n)$ $\tilde{x}_1$ $\tilde{x}_d(n)$

T T/L T/L T/L TM/L

(b)

$x(n)$ $x_1(n)$ $\tilde{x}_1(n)$ $\tilde{x}_d(n)$

T T/L T/L TM/L

Fig. 8.17 a Block diagram representation of a system for changing the sampling rate by a non-integer factor and **b** block diagram representation of a system for changing the sampling rate by a non-integer factor with a single lowpass filter

Example 8.9 Consider the system shown in Fig. 8.18, where $H(e^{j\omega})$ is an ideal LTI lowpass filter with cutoff of $(\pi/3)$. If the Fourier transform of $x(n)$ is as shown in Fig. 8.19, sketch the spectrums of $x_e(n)$ and $y_e(n)$.

Solution The spectrums of $x_e(n)$ and $y_e(n)$ are shown in Fig. 8.20.

Example 8.10 The data from a compact disk system is at a rate of 44.1 kHz and is to be transferred to a digital audiotape at 48 kHz. Develop a method to do this using discrete-time signal processing.

Solution This can be achieved by increasing the data rate of the CD by a factor of 48/44.1, a non-integer. To do this, one would, therefore, use the arrangement of Fig. 8.21 with $L = 480$ and $M = 441$. Such large values of L normally imply that $H(z)$ has a very high order. A multistage design is more convenient in such cases.

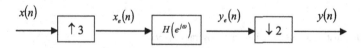

Fig. 8.18 Schematic diagrams for changing the sampling rate by 3/2

Fig. 8.19 Fourier transform of $x(n)$

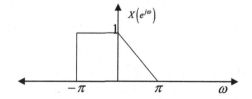

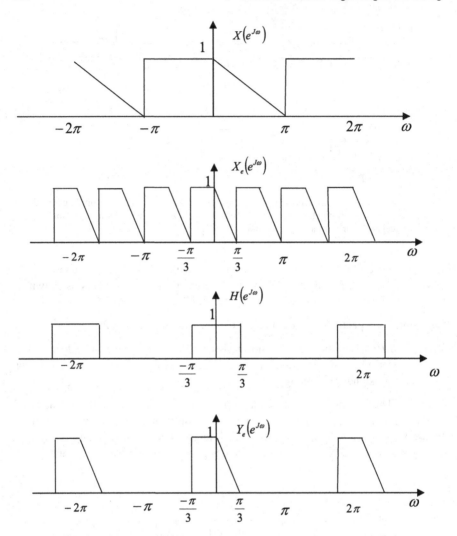

Fig. 8.20 Illustration of changing sampling rate by 3/2

Fig. 8.21 Scheme for changing the sampling rate by a non-integer factor

8.2.4 Sampling Rate Conversion Via Multistage Approach

So far, we have discussed single-stage decimators and interpolators. In the case of a single-stage decimator (Fig. 8.4) or a single-stage interpolator (Fig. 8.12), a large value of the decimation factor M or the interpolation factor L indicates a large change in the sampling rate. In such a case, a multistage implementation is more effective than a single-stage implementation, since the multistage approach results in tremendous computational savings and cheaper anti-aliasing (decimation) or anti-imaging (interpolation) filters at each stage. A multistage decimator and a multistage interpolator with k stages are shown in Fig. 8.22a, b, respectively. In Fig. 8.22, the overall decimation factor M or the interpolation factor L is the product of the decimation or the interpolation factors of the individual stages, i.e., $M = M_1 M_2 M_3 \ldots M_k$ or $L = L_1 L_2 L_3 \ldots L_k$, where M_i and L_j are integers.

8.3 Practical Sampling Rate Converter Design

In the design of a multirate system, either IIR or FIR filters can be used, but FIR is mostly preferred because of its computational efficiency, linear phase response, low sensitivity to finite word-length effects, and simplicity of implementation. The performance of multistage decimators and interpolators depends on various parameters such as the number of stages, the decimation and interpolation factors at each stage, the filter requirements and their actual designs at each stage. Illustrative design examples choosing these parameters are discussed in Sect. 8.3.3. The overall filter requirements as well as the filter requirements at each stage for a decimator and an interpolator are specified in Sects. 8.3.1 and 8.3.2.

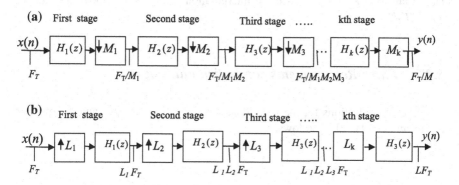

Fig. 8.22 Structures of **a** multistage decimator and **b** multistage interpolator

8.3.1 Overall Filter Specifications

To avoid aliasing after reducing the sampling rate, the overall decimation filter specifications are:

$$\text{Passband} \quad 0 \leq f \leq f_p \tag{8.20a}$$

$$\text{stopband} \quad F_T/2M \leq f \leq F_T/2 \tag{8.20b}$$

$$\text{passband ripple} \quad \delta_p \tag{8.20c}$$

$$\text{stopband ripple} \quad \delta_s \tag{8.20d}$$

In the above specifications, $f_p < F_T/2M$, F_T being the initial input sampling frequency, and the final stopband edge is restricted to be less than or equal to $F_T/2$ to protect the entire baseband, from 0 to $F_T/2$, from aliasing. The typical highest frequency of interest in the original signal is f_p.

In the case of interpolation, the overall filter requirements are

$$\text{passband} \quad 0 \leq f \leq f_p \tag{8.21a}$$

$$\text{stopband} \quad F_T/2 \leq f \leq LF_T/2 \tag{8.21b}$$

$$\text{passband ripple} \quad \delta_p \tag{8.21c}$$

$$\text{stopband ripple} \quad \delta_s \tag{8.21d}$$

We note that in the above specifications $f_p < F_T/2$, a passband gain of L is required as the interpolator reduces the passband gain to $1/L$. After increasing the sampling rate to LF_T, the highest valid frequency is $LF_T/2$. But it is necessary to band-limit to $F_T/2$ or less, since the interpolation filter has to remove the images above $F_T/2$.

8.3.2 Filter Requirements for Individual Stages

The filter specifications for the individual stages of the decimator and interpolator are formulated [1, 2] as follows:

Filter requirements for individual stages in a multistage decimation filter
In order to satisfy the overall filter requirements, the filter specifications at each
stage in a multistage decimator are:

> passband : $f \in [0, f_p]$
>
> stopband : $f \in (F_i - (F_T/2M), F_{i-1}/2)$, $i = 1, 2, \ldots, k$
>
> passband ripple : δ_p/k
>
> stopband ripple : δ_s

where F_i is the ith stage output sampling frequency, F_T is the input sampling
frequency to the first stage of the multistage decimator, k is the number of stages,
and M is the decimation factor.

Filter requirements for individual stages in a multistage interpolator
For a multistage interpolator, the filter requirements for each stage to ensure that the
overall filter requirements are satisfied are:

> passband : $f \in [0, f_p]$
>
> stopband : $f \in (F_i - (F_0/2L), F_{i-1}/2)$ $i = k, k - 1, \ldots, 2, 1$
>
> passband ripple : δ_p/k
>
> stopband ripple : δ_s

where F_i is the ith stage input sampling frequency, F_0 is the output sampling
frequency of the last stage of the multistage interpolator, k is the number of stages,
and L is the interpolation factor.

Design of practical multistage sampling rate converter
The steps involved in designing a multistage sampling rate converter are as follows:

Step1: State the specifications to be satisfied by the overall decimation or
        interpolation filters.
Step2: Decide on the optimum number of stages for efficient implementation.
Step3: Find the decimation or interpolation factors for all the individual stages.
Step4: Devise suitable filters for each of the individual stages.

8.3.3   Illustrative Design Examples

Example 8.11 A signal at a sampling frequency of 2.048 kHz is to be decimated by
a factor of 32. The anti-aliasing filter should satisfy the following specifications:

passband ripple: 0.00115

stopband ripple: 0.0001

stopband edge frequency: 32 Hz

passband edge frequency: 30 Hz

Design a suitable decimator.

Solution From the specifications, the following can be determined.
Δf = transition width normalized to the sampling frequency = $(32 - 30)/2048 = 9.766 \times 10^{-4}$

δ_p = passband ripple = 0.00115, δ_s = stopband ripple = 0.0001

An estimate of the filter order for the single-stage decimator can be computed by using the following formula [3]:

$$N = \{D_\infty(\delta_p, \delta_s)/\Delta f\} - (\delta_p, \delta_s)\Delta f \qquad (8.22a)$$

where

$$D_\infty(\delta_p, \delta_s) = \log(\delta_s)) * [a1(\log(\delta_p))^2 + a2\log(\delta_p) + a3)]$$
$$+ [a4(\log(\delta_p))^2 + a5\log(\delta_p) + a6]$$

$$f(\delta_p, \delta_s) = 11.01217 + 0.51244[\log(\delta_p/\delta_s)];$$
$$a1 = 5.309\ddot{I} \times 10^{-3}, a2 = 7.114\ddot{I} \times 10^{-2}, a3 = -4761\ddot{I} \times 10^{-1}$$
$$a4 = -2.66\ddot{I} \times 10^{-3}, a5 = -5.9418\ddot{I}\,10^{-1}, a6 = -4.278\ddot{I} \times 10^{-1}$$

A simple approximation formula for order N is given by [4].

$$N \equiv \frac{-20\log_{10}\left(\sqrt{\delta_p\delta_s}\right) - 13}{14.6(\omega_s - \omega_p)/2\pi} \qquad (8.22b)$$

For the given specifications, it is found that $N = 3946$. It is obvious that N is too large for all practical purposes and the design of the lowpass filter for a single-stage decimator is not possible. This example makes the need for a multistage decimator design very evident. As such, a three-stage decimator is designed as shown in Fig. 8.23.

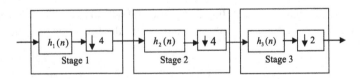

Fig. 8.23 Three-stage decimator for Example 8.8

Table 8.1 Specifications of the individual filters

	First stage	Second stage	Third stage
Passband edge (Hz)	30	30	30
Stopband edge (Hz)	480	96	32
Passband ripple	0.00115/ 3 = 0.000384	0.00115/ 3 = 0.000384	0.00115/ 3 = 0.000384
Stopband ripple	0.0001	0.0001	0.0001

Following the procedure given in Sect. 8.3.2, the specifications of the individual filters for the three stages are as given in Table 8.1.

The following MATLAB programs can be used to design and plot the magnitude responses of the individual decimation filters of the three-stage decimator.

Program 8.6 To design and plot the log magnitude response of the first-stage filter.

```
clear;clf;
st1=2048/4;
ss=2048/(2*32);
fedge=[30 (st1-ss)];
mval=[1 0];
dev=[(0.00115/3) 0.0001];
FT=2048;
[N,fpts,mag,wt]=firpmord(fedge,mval,dev,FT);
b=firpm(20,fpts,mag,wt);
[h,omega]=freqz(b,1,512);
plot(omega/(2*pi),20*log10(abs(h)));grid;
xlabel('Frequency');ylabel('Gain, dB')
```

Program 8.7 To design and plot the log magnitude response of the second-stage filter.

```
clear;clf;
st1=2048/(4*4);
ss=2048/(2*32);
fedge=[30 (st1-ss)];
mval=[1 0];
dev=[(0.00115/3) 0.0001];
FT=2048/4;
[N,fpts,mag,wt] = firpmord(fedge,mval,dev,FT);
b=firpm(N,fpts,mag,wt);
[h,omega]=freqz(b,1,512);
plot(omega/(2*pi),20*log10(abs(h)));grid;
xlabel('Frequency');ylabel('Gain, dB')
```

Program 8.8 To design and plot the log magnitude response of the third-stage filter.

```
clear;clf;
st1=2048/(4*4*2);
ss=2048/(2*32);
fedge=[30 (st1-ss)];
mval=[1 0];
dev=[(0.00115/3) 0.0001];
FT=2048/(4*4);
[N,fpts,mag,wt]=firpmord(fedge,mval,dev,FT);
b=firpm(N,fpts,mag,wt);
[h,omega]=freqz(b,1,512);
plot(omega/(2*pi),20*log10(abs(h)));grid;
xlabel('Frequency');ylabel('Gain, dB')
```

The log magnitude responses of the individual filters and the filter orders satisfying the specifications for the three-stage decimator are shown in Figs. 8.24, 8.25, and 8.26 respectively.

Example 8.12 Design an efficient two-stage decimator to convert a single-bit stream at 3072 kHz into a multi bit stream at 48 kHz for which the passband and stopband ripples for the decimator are 0.001 and 0.0001, respectively. The passband ranges from 0 to 20 kHz.

Solution The block diagram of the two-stage decimator is shown in Fig. 8.27.

An optimum design is one which leads to the least computational effort, for example, as measured by the number of multiplications per second (MPS) or the total storage requirements (TSR) for the coefficients:

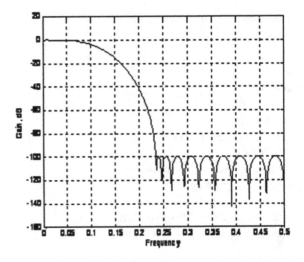

Fig. 8.24 Log magnitude response of the first-stage decimation filter; filter order $N1 = 23$

Fig. 8.25 Log magnitude response of the second-stage decimation filter; filter order $N2 = 32$

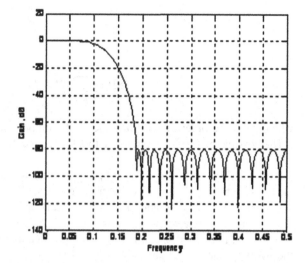

Fig. 8.26 Log magnitude response of the third-stage decimation filter; filter order $N3 = 269$

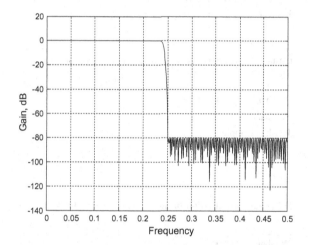

Fig. 8.27 A two-stage decimator

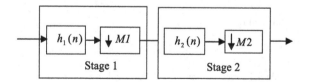

$$\text{MPS} = \sum_{i=1}^{2} (N_i + 1)F_i \tag{8.23a}$$

$$\text{TSR} = \sum_{i=1}^{2} (N_i + 1) \tag{8.23b}$$

where N_i is the order of the filter and F_i the output sampling frequency of stage i. The possible pairs of decimation factors for M_1 and M_2 are (8, 8), (16, 4), and (32, 2).

Case (a): For $M_1 = 8$ and $M_2 = 8$, the specifications of the individual filters for the two-stage decimator are as shown in Table 8.2.

Case (b): For $M_1 = 16$ and $M_2 = 4$, the specifications of the individual filters for the two-stage decimator are as shown in Table 8.3.

Case (c): For $M_1 = 32$ and $M_2 = 2$, the specifications of the individual filters for the two-stage decimator are as shown in Table 8.4.

For Case (a), the order of the filters as well as the output sampling frequencies of the two stages are given in Table 8.5.

Table 8.2 Specification of the individual filters for Case (a)

	First stage	Second stage
Passband edge (kHz)	20	20
Stopband edge (kHz)	360	24
Passband ripple	0.0005	0.0005
Stopband ripple	0.0001	0.0001

Table 8.3 Specification of the individual filters for Case (b)

	First stage	Second stage
Passband edge (kHz)	20	20
Stopband edge (kHz)	168	24
Passband ripple	0.0005	0.0005
Stopband ripple	0.0001	0.0001

Table 8.4 Specification of the individual filters for Case (c)

	First stage	Second stage
Passband edge (kHz)	20	20
Stopband edge (kHz)	72	24
Passband ripple	0.0005	0.0005
Stopband ripple	0.0001	0.0001

Table 8.5 Order of the individual filters and the output sampling frequencies of the stages for Case (a)

Stages	Filter order (N)	Output sampling frequency (F_i) in kHz
First stage	35	384
Second stage	422	48

Table 8.6 Order of the individual filters and the output sampling frequencies of the stages for Case (b)

Stages	Filter order (N)	Output sampling frequency (F_i) in kHz
First stage	92	192
Second stage	212	48

Table 8.7 Order of the individual filters and the output sampling frequencies of the stages for Case (c)

Stages	Filter order (N)	Output sampling frequency (F_i) in kHz
First stage	260	96
Second stage	106	48

Table 8.8 Multiplications per second (MPS) and total storage requirements (TSR)

Decimation factors	$\text{TSR} = \sum_{i=1}^{I}(N_i+1)f_i$	$\text{MPS} = \sum_{i=1}^{I}(N_i+1)f_i$
8×8	459	$34,128 \times 10^3$
16×4	306	$28,080 \times 10^3$
32×2	368	$30,192 \times 10^3$

For Case (b), the order of the filters as well as the output sampling frequencies of the two stages are given in Table 8.6.

For Case (c), the order of the filters as well as the output sampling frequencies of the two stages are given in Table 8.7.

The computational and storage complexities for the two-stage decimator with different decimation factors are shown in Table 8.8.

From Table 8.8, it is observed that $M_1 = 16$ and $M_2 = 4$ are the optimum decimation factors for the two-stage decimator. The log magnitude responses of the individual filters in the two-stage decimator corresponding to the optimal decimation factors are shown in Figs. 8.28, 8.29, and 8.30, respectively.

Fig. 8.28 Log magnitude
response of the first-stage
decimation filter, $N_1 = 92$

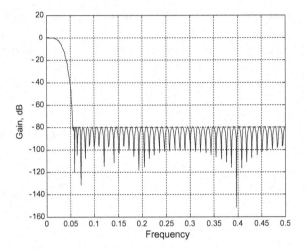

Fig. 8.29 Log magnitude
response of the second-stage
decimation filter, $N_2 = 212$

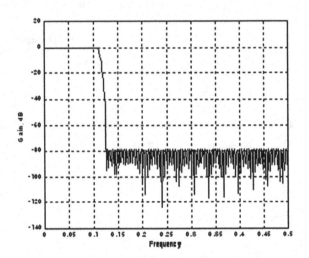

Example 8.13 The overall specifications of an interpolator are as follows:

Base band: $0 - 20$ kHz
Input sampling frequency: 44.1 kHz
Output sampling frequency: 176.4 kHz
Stop band ripple: 40 dB
Pass band ripple: 1 dB
Transition bandwidth: 2 kHz
Stop band edge frequency: 22.05 kHz

Fig. 8.30 Block diagram of
the single-stage interpolator

Compare the filter orders for single-stage and multistage implementation of the interpolator.

Solution The single-stage interpolator is shown in Fig. 8.30.

The following MATLAB programs 8.9, 8.10, and 8.11 can be used to design and plot the log magnitude responses of the single-stage and two-stage interpolation filters.

Program 8.9 To design and plot log magnitude response of the single-stage interpolation filter

```
clear;clf;
fedge=[20.5e+03 22.5e+03];
mval=[1 0];
dev=[0.1087 0.01];
FT=176.4e+03;
[N,fpts,mag,wt]=firpmord(fedge,mval,dev,FT);
b=firpm(N,fpts,mag,wt);
[h,omega]=freqz(b,1,512);
plot(omega/(2*pi),20*log10(abs(h)));grid;
xlabel('Frequency');ylabel('Gain, dB');
```

In the above program, the function firpmord gives the initial order; if initial order does not satisfy the desired specifications, we have to increase the filter order until the specifications are satisfied. In this case, we obtain the initial order of the filter as 108, but the specifications are met when the order is increased to 118. The log magnitude response of the single-stage filter for Example 8.12 is shown in Fig. 8.31.

Fig. 8.31 Log magnitude
response of the single-stage
interpolation filter

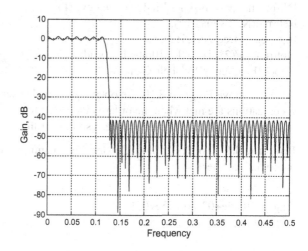

Fig. 8.32 Block diagram of
the two-stage interpolator

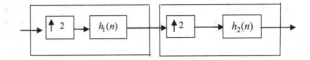

Table 8.9 Specifications of
the individual filters

	$h_1(n)$	$h_2(n)$
Passband edge (kHz)	20.05	20.05
Stopband edge (kHz)	22.05	66.155
Passband ripple	0.05435	0.05435
Stopband ripple	0.01	0.01

The block diagram of two-stage interpolator is shown in Fig. 8.32.

The specifications of the individual filters for the two-stage interpolator, for example, are shown in Table 8.9.

The transition widths, passband and stopband ripples for the individual stages are

$$\Delta f_1 = (22.05 - 20.05)/88.2; \quad \delta p_1 = (0.1087)/2$$
$$\Delta f_2 = (66.15 - 20.05)/176.4; \quad \delta p_2 = (0.1087)/2$$
$$\delta s_1 = 0.01; \delta s_2 = 0.01$$

Program 8.10 To design and plot the log magnitude response of the second-stage interpolation filter

```
clear;clf;
fedge=[20.05e+03 22.05e+03];
mval=[1 0];
dev=[0.5*0.1087 0.01];
FT=0.5*176.4e+03;
[N,fpts,mag,wt]=firpmord(fedge,mval,dev,FT);
b=firpm(N,fpts,mag,wt);
[h,omega]=freqz(b,1,512);
plot(omega/(2*pi),20*log10(abs(h)));grid;
xlabel('Frequency');ylabel('Gain, dB')
```

In the above Program 8.10, we obtain the initial order of the filter as 64, but the specifications are met when the order is increased to 69. We now use Program 8.11 to design and plot the log magnitude response of the first-stage interpolation filter.

Program 8.11 To design and plot the log magnitude response of the first-stage interpolation filter

```
clear;clf;
fedge=[20.05e+03 66.155e+03];
mval=[1 0];
```

```
dev=[0.5*0.1087 0.01];
FT=176.4e+03;
[N,fpts,mag,wt]=firpmord(fedge,mval,dev,FT);
b=firpm(N,fpts,mag,wt);
[h,omega]=freqz(b,1,512);
plot(omega/(2*pi),20*log10(abs(h)));grid;
xlabel('Frequency');ylabel('Gain, dB')
```

In the above program, we obtain the initial order of the filter as 3, but the specifications are met when the order is increased to 6. The log magnitude responses of the individual filters for the two-stage interpolator are shown in Figs. 8.33 and 8.34, respectively (Table 8.10).

Fig. 8.33 Log magnitude response of the first-stage interpolation filter, $N1 = 69$

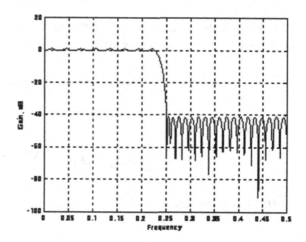

Fig. 8.34 Log magnitude response of the second-stage interpolation filter, $N2 = 6$

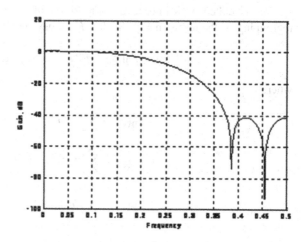

Table 8.10 Comparison of the filter orders for the single-stage and two-stage interpolators	No. of stages	Filters	Filter orders
	1	$h(n)$	118
	2	$h_1 n$	69
		$h_2(n)$	6

8.4   Polyphase Decomposition

The purpose of polyphase decomposition is to implement filter banks with reduced computational complexity. Consider an FIR filter in the z-domain:

$$H(z) = \sum_{n=0}^{\infty} h(n) z^{-n} \tag{8.24a}$$

Separating the even-numbered coefficients of $h(n)$ from the odd-numbered coefficients, $H(z)$ can be represented as

$$H(z) = \sum_{n=0}^{\infty} h(2n) z^{-2n} + z^{-1} \sum_{n=0}^{\infty} h(2n+1) z^{-2n}$$

Defining

$$E_0(z) = \sum_{n=0}^{\infty} h(2n) z^{-n}, E_1(z) = \sum_{n=0}^{\infty} h(2n+1) z^{-n}$$

$H(z)$ can be rewritten as

$$H(z) = E_0(z^2) + z^{-1} E_1(z^2) \tag{8.24b}$$

This is called a two-phase decomposition of $H(z)$. Similarly, by a different regrouping of the terms in Eq. (8.24a), $H(z)$ may be rewritten as

$$H(z) = E_0(z^3) + z^{-1} E_1(z^3) + z^{-2} E_2(z^3) \tag{8.24c}$$

where

$$E_0(z) = \sum_{0}^{\infty} h(3n) z^{-n},$$

$$E_1(z) = \sum_{0}^{\infty} h(3n+1) z^{-n} \tag{8.24d}$$

$$E_2(z) = \sum_{0}^{\infty} h(3n+2) z^{-n}$$

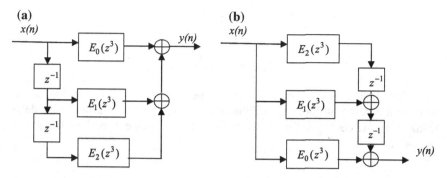

Fig. 8.35 A three-branch polyphase realization of an FIR filter. **a** Type 1 and **b** Type 2, which is the transpose of the structure in (**a**)

The above is a three-phase decomposition of $H(z)$. A direct way of realizing H (z) given by Eq. (8.24c) is shown in Fig. 8.35a. An alternate direct realization of H (z) given by Eq. (8.24c) is shown in Fig. 8.35b. It may be observed that the realizations given in Fig. 8.35a, b are simply transposes of each other. We shall refer to the structure in Fig. 8.35a as the Type 1 polyphase realization of $H(z)$ and that in Fig. 8.35b as the Type 2 polyphase realization.

In the general case, for any given integer M, an M-phase polyphase decomposition of an FIR filter $H(z)$ can be obtained by rearranging $H(z)$ as

$$H(z) = \sum_{0}^{M-1} z^{-k} E_k(z^M) \tag{8.25}$$

One can easily obtain, for $H(z)$ given by Eq. (8.25), both the Type 1 and Type 2 M-branch polyphase realizations, similar to the ones shown in Fig. 8.35a, b for the case $M = 3$. Of course, the two realizations are transposes of each other.

Example 8.14 Develop a three-branch polyphase realization of a length-7 FIR filter using minimum number of delays.

Solution The transfer function $H(z)$ for a length-7 FIR filter is given by

$$H(z) = h(0) + h(1)z^{-1} + h(2)z^{-2} + h(3)z^{-3} + h(4)z^{-4} + h(5)z^{-5} + h(6)z^{-6}$$

This can be expressed as a three-branch polyphase decomposition given by

$$H(z) = E_0(z^3) + z^{-1}E_1(z^3) + z^{-2}E_2(z^3)$$

where the polyphase transfer functions $E_0(z)$, $E_1(z)$ and $E_2(z)$ are given by

$$E_0 = h(0) + h(3)z^{-1} + h(6)z^{-2}$$
$$E_1 = h(1) + h(4)z^{-1}$$
$$E_2 = h(2) + h(5)z^{-1}$$

A three-branch Type 1 polyphase realization of $H(z)$ is shown in Fig. 8.36a, wherein the delays in the three subfilters are shared to obtain a minimum delay realization. The corresponding transpose realization, viz the Type 2 realization, is shown in Fig. 8.36b.

8.4.1 Structures for Decimators and Interpolators

Consider the polyphase Type 1 realization of the decimation filter of Fig. 8.4. If the lowpass filter $H(z)$ is realized using Fig. 8.37, the decimator structure can be represented in the form of Fig. 8.38.

Consider the cascade equivalences shown in Figs. 8.39 and 8.40, which are useful in the development of efficient structures for polyphase realization of decimators and interpolators.

The validity of these equivalences can be verified as follows:

For Equivalence 1, it can be noted that for Fig. 8.39b,

$$v_2\left(e^{j\omega}\right) = H\left(e^{j\omega M}\right)X\left(e^{j\omega}\right) \tag{8.26}$$

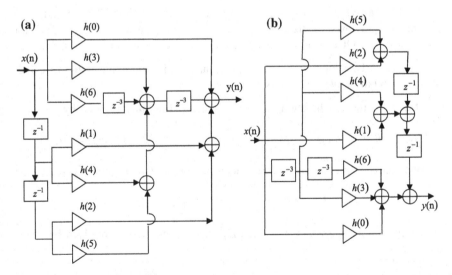

Fig. 8.36 A three-branch realization of a length-7 FIR filter using minimum number of delays. **a** Type 1 realization and **b** Type 2, which is the transpose of the structure in (**a**)

Fig. 8.37 Type 1 realization
of an FIR filter

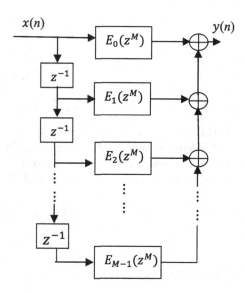

Fig. 8.38 Decimator
implementation based on
Type 1 polyphase realization

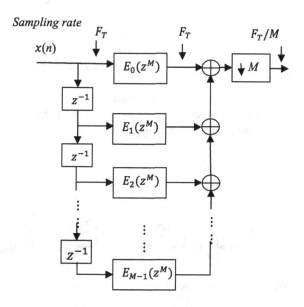

$$Y\left(e^{j\omega}\right) = \frac{1}{M}\sum_{k=0}^{M-1} v_2\left(e^{\frac{j(\omega-2\pi k)}{M}}\right)$$

$$Y\left(e^{j\omega}\right) = \frac{1}{M}\sum_{k=0}^{M-1} X\left(e^{\frac{j(\omega-2\pi k)}{M}}\right) H\left(e^{j(\omega-2\pi k)}\right) \tag{8.27}$$

Since $H\left(e^{j(\omega-2\pi k)}\right) = H(e^{j\omega})$, Eq. (8.27) reduces to

$$Y\left(e^{j\omega}\right) = H\left(e^{j\omega}\right) \frac{1}{M} \sum_{k=0}^{M-1} X\left(e^{j\left(\frac{\omega}{M}-2\pi k/m\right)}\right)$$

$$= H\left(e^{j\omega}\right) V_1\left(e^{j\omega}\right). \tag{8.28}$$

This corresponds to Fig. 8.39a.

A similar identity applies to up-sampling. Specifically, using Eq. (8.19) from Sect. 8.2.2, it is also straightforward to show the equivalence of the two systems in Fig. 8.40. We have from Eq. (8.19) and Fig. 8.40b.

$$Y\left(e^{j\omega}\right) = V_2\left(e^{j\omega L}\right)$$

$$= X\left(e^{j\omega L}\right) H\left(e^{j\omega L}\right). \tag{8.29}$$

Since from Eq. (8.19)

$$V_2\left(e^{j\omega}\right) = X\left(e^{j\omega L}\right) \tag{8.30}$$

it follows that Eq. (8.29) is equivalently

$$Y\left(e^{j\omega}\right) = H\left(e^{j\omega L}\right) V_2(e^{j\omega}) \tag{8.31}$$

This corresponds to Fig. 8.40a.

Using Equivalence 1 shown in Fig. 8.39, the structure shown in Fig. 8.38 can be represented as shown in Fig. 8.41.

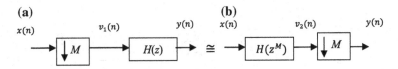

(a)
$x(n) \quad\quad v_1(n) \quad\quad y(n)$
$\boxed{\downarrow M} \quad \boxed{H(z)} \quad \cong$

(b)
$x(n) \quad\quad v_2(n) \quad\quad y(n)$
$\boxed{H(z^M)} \quad \boxed{\downarrow M}$

Fig. 8.39 Cascade Equivalence 1

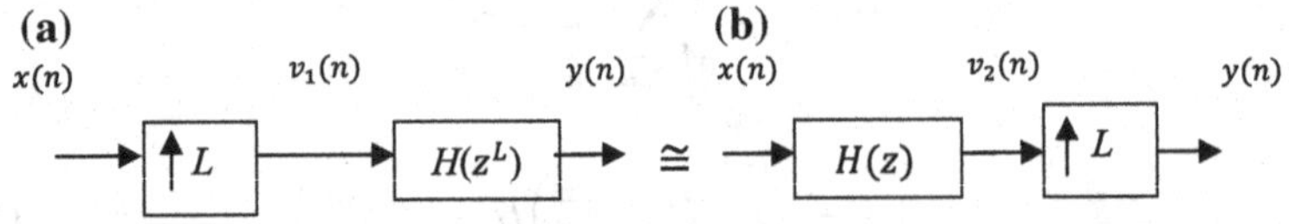

(a)
$x(n) \quad\quad v_1(n) \quad\quad y(n)$
$\boxed{\uparrow L} \quad \boxed{H(z^L)} \quad \cong$

(b)
$x(n) \quad\quad v_2(n) \quad\quad y(n)$
$\boxed{H(z)} \quad \boxed{\uparrow L}$

Fig. 8.40 Cascade Equivalence 2

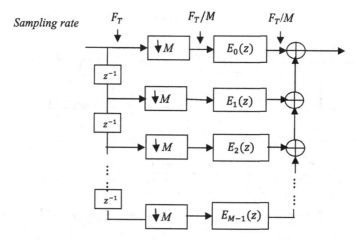

Fig. 8.41 Decimator implementation based on polyphase decomposition using Cascade Equivalence 1

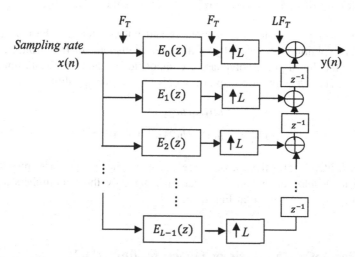

Fig. 8.42 Interpolator implementation based on Type 2 polyphase realization using Cascade Equivalence 2

Similarly, using Equivalence 2 shown in Fig. 8.40, the structure for the polyphase decomposition in the realization of the interpolation filter can be represented as shown in Fig. 8.42.

Example 8.15 Develop an efficient realization of a factor-of-2 decimator by exploiting the linear phase symmetry of a length-6 decimator filter.

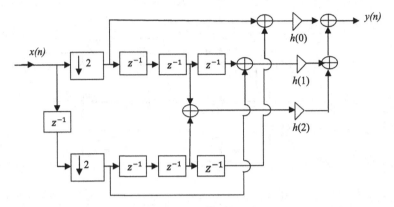

Fig. 8.43 An efficient realization of the factor-of-2 decimator of Example 8.12

Solution The transfer function $H(Z)$ for a length-6 linear phase FIR lowpass filter with symmetric impulse response is given by

$$H(z) = h(0) + h(1)z^{-1} + h(2)z^{-2} + h(2)z^{-3} + h(1)z^{-4} + h(0)z^{-5} \qquad (8.32)$$

For a linear phase filter, the decimators and interpolators can be realized efficiently by exploiting the symmetry of the filter coefficients of $H(z)$ from Sect. 6.2. A two-channel linear phase polyphase decomposition of the FIR transfer function $H(z)$ with a symmetric impulse response yields the following subfilters

$$E_0(z) = h(0) + h(2)z^{-1} + h(1)z^{-2}$$
$$E_1(z) = h(1) + h(2)z^{-1} + h(0)z^{-2}$$

The subfilter $E_1(z)$ is the mirror image of the subfilter $E_0(z)$. These relations can be used to develop an efficient realization using only three multipliers and five two-input adders, as shown in Fig. 8.43.

8.5 Resolution Analysis of Oversampling ADC

8.5.1 Reduction of ADC Quantization Noise by Oversampling

Consider an n-bit ADC sampling analog signal $x(t)$ at sampling frequency of F_T as shown in Fig. 8.44a. The power spectral density of the quantization noise with an assumption of uniform probability distribution is shown in Fig. 8.44b.

(a)   **(b)**

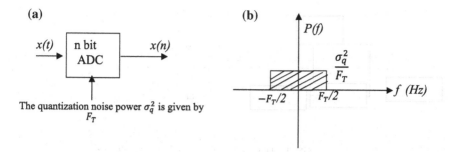

The quantization noise power σ_q^2 is given by F_T

Fig. 8.44   a An n-bit ADC and **b** power spectral density of quantization noise

The quantization noise power σ_q^2 is given by

$$\sigma_q^2 = \int_{-\infty}^{\infty} P(f)\mathrm{d}f = \frac{\sigma_q^2}{F_T} \times F_T \qquad (8.33)$$

The quantization noise power can be expressed by

$$\sigma_q^2 = \frac{\text{quantization step}^2}{12} = \frac{A^2}{12} \times \frac{1}{2^{2n}} = \frac{A^2}{12} 2^{-2n} \qquad (8.34)$$

The ADC quantization noise is reduced by oversampling an analog signal at a sampling frequency F_{Tos} higher than the minimum rate needed to satisfy the Nyquist criterion ($2f_{max}$), where F_{Tos} is the oversampling frequency and f_{max} is the maximum frequency of the analog signal. The block diagram of an oversampling ADC is shown in Fig. 8.45.

The power spectral density of the quantization noise with oversampling ADC is shown in Fig. 8.46. The quantization noise power is spread with decreased level of over a wider frequency range.

After the decimation process with the decimation filter, only a portion of the quantization noise power in the in-band frequency range ($-fmax$ and $fmax$) is kept. Thus, the shaded area in Fig. 8.46 is the quantization noise power with oversampling ADC.

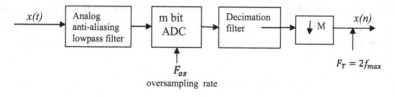

Fig. 8.45 An m-bit oversampling ADC

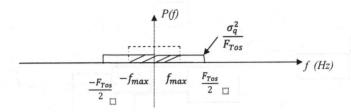

Fig. 8.46 Power spectral density of quantization noise with oversampling ADC

The quantization noise power σ_q^2 with oversampling ADC is given by

$$\text{quantization noise power} = \int\limits_{-\infty}^{\infty} P(f)\mathrm{d}f = \frac{2f_{\max}}{F_{os}} \times \sigma_q^2 \qquad (8.35)$$

Since σ_q^2 for an m-bit ADC is given by $\frac{A^2}{12}2^{-2m}$, Eq. (8.35) is rewritten as

$$\text{quantization noise power} = \int\limits_{-\infty}^{\infty} P(f)\mathrm{d}f = \frac{2f_{\max}}{F_{os}}\frac{A^2}{12}2^{-2m} \qquad (8.36)$$

Assuming that the regular ADC and the oversampling ADC are equivalent and their quantization noise powers are the same, we obtain

$$\frac{A^2}{12}2^{-2n} = \frac{2f_{\max}}{F_{os}}\frac{A^2}{12}2^{-2m} \qquad (8.37)$$

Equation (8.37) leads to the following two useful relations

$$n = m + \frac{1}{2}\log_2\left(\frac{F_{os}}{2f_{\max}}\right) \qquad (8.38)$$

$$F_{os} = 2f_{\max}2^{2(n-m)} \qquad (8.39)$$

Example 8.16 Considering an oversampling ADC system with maximum analog signal frequency of 4 kHz and ADC resolution of eight bits, determine the oversampling rate to improve the ADC resolution to 12-bit resolution.

Solution $m = 8; n = 12$.

Using Eq. (8.39), the oversampling rate is

$$F_{os} = 2f_{max}2^{(n-m)} = 8000 \times 2^{2(12-8)}$$

$$= 8000 \times 2^8 = 2048 \text{ kHz}$$

Example 8.17 Considering an eight-bit ADC with oversampling rate of 256 kHz and analog signal maximum frequency of 2 kHz, determine the equivalent ADC resolution.

Solution

$$f_{max} = 2 \text{ kHz}, \text{m} = 8, F_{os} = 256 \text{ kHz}$$

Using Eq. (8.38), the equivalent resolution becomes

$$n = 8 + \frac{1}{2}\log_2\left(\frac{256,000}{4000}\right) = 11 \text{ bits}$$

8.5.2 Sigma-Delta Modulation ADC

A sigma-delta modulation ADC with oversampling is shown in Fig. 8.47.

The first-order sigma-delta modulator [5] is shown in Fig. 8.48.

Assuming that the DAC is ideal, it can be replaced by unity gain transfer function. Then, the z-domain output $Y(z)$ of the modulator is given by

$$Y(z) = X(z)z^{-1} + E(z)(1 - z^{-1}) \tag{8.40}$$

so that $H_x(z) = z^{-1}$ and $H_e(z) = (1 - z^{-1})$
where

$H_x(z)$ is the signal transfer function
$H_e(z)$ is the noise transfer function

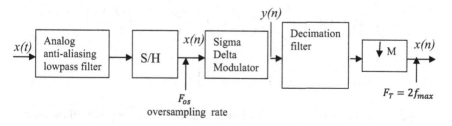

Fig. 8.47 Block diagram of sigma-delta modulation ADC

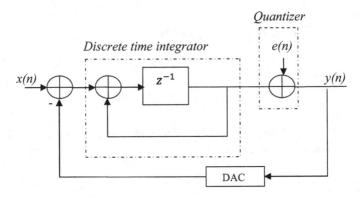

Fig. 8.48 Block diagram of first-order sigma-delta modulator

The corresponding time domain output of the modulator output is

$$y(n) = x(n-1) + e(n) - e(n-1) \tag{8.41}$$

If a stationary random process with power spectral density $P(f)$ is the input to a linear filter with transfer function, $H(f)$, the power spectral density of the output random process is $P(f)|H(f)|^2$. Consequently,

$$P_{xy}(f) = P_x(f)|H_x(f)|^2 \tag{8.42}$$

$$P_{ey}(f) = P_e(f)|H_e(f)|^2 \tag{8.43}$$

$$H_e(f) = 1 - e^{-j2\pi f} \tag{8.44}$$

$$1 - e^{-j2\pi f} = 1 - \left(1 + \frac{(-j2\pi f)}{1!} + \frac{(-j2\pi f)^2}{2!} + \cdots\right) \approx j2\pi f \tag{8.45}$$

Hence, the in-band noise power at the output of a first-order sigma-delta modulator in the frequency range $(-f_{max}, f_{max})$ is given by

$$
\sigma_{ey}^2 = \int_{-\infty}^{\infty} P_{ey}(f) df = \int_{-f_{max}}^{f_{max}} \sigma_q^2 \left(\frac{2\pi f}{F_{os}}\right)^2 df = \left.\frac{\sigma_q^2}{3}(2\pi)^2 \left(\left(\frac{f}{F_{os}}\right)^3\right)\right|_{-f_{max}}^{f_{max}} \tag{8.46}
$$

$$
= \sigma_q^2 \frac{\pi^2}{3} \left(\frac{2f_{max}}{F_{os}}\right)^3 = \frac{\pi^2}{3} \frac{A^2}{12} 2^{-2m} \left(\frac{2f_{max}}{F_{os}}\right)^3
$$

Equating this in-band noise power to the quantization noise power of the regular ADC, we obtain

$$\frac{A^2}{12}2^{-2n} = \frac{\pi^2}{3}\frac{A^2}{12}2^{-2m}\left(\frac{2f_{\max}}{F_{os}}\right)^3 \tag{8.47}$$

Equation (8.37) leads to the following two useful relations

$$n = m + 1.5\log_2\left(\frac{F_{os}}{2f_{\max}}\right) - 0.5\log_2\left(\frac{\pi^2}{3}\right) \tag{8.48}$$

$$\left(\frac{F_{os}}{2f_{\max}}\right)^3 = \frac{\pi^2}{3}2^{2(n-m)} \tag{8.49}$$

The second-order sigma-delta modulation ADC is the most widely used one. The block diagram of a second-order sigma-delta modulator is shown in Fig. 8.49. The second-order sigma-delta modulator realizes

$$H_x(z) = z^{-1} \text{ and } H_e(z) = \left(1 - z^{-1}\right)^2$$

so that

$$Y(z) = X(z)z^{-1} + E(z)\left(1 - z^{-1}\right)^2 \tag{8.50}$$

Hence, the in-band noise power at the output of the second-order sigma-delta modulator in the frequency range $(-f_{\max}, f_{\max})$ is given by

$$\sigma_{ey}^2 = \int_{-\infty}^{\infty} P_{ey}(f)df = \int_{-f_{\max}}^{f_{\max}} \sigma_q^2\left(\frac{2\pi f}{F_{os}}\right)^4 df = \frac{\sigma_q^2}{5}(2\pi)^4\left(\left(\frac{f}{F_{os}}\right)^5\right)\Bigg|_{-f_{\max}}^{f_{\max}}$$

$$= \sigma_q^2\frac{\pi^4}{5}\left(\frac{2f_{\max}}{F_{os}}\right)^5 = \frac{\pi^4}{5}\frac{A^2}{12}2^{-2m}\left(\frac{2f_{\max}}{F_{os}}\right)^5 \tag{8.51}$$

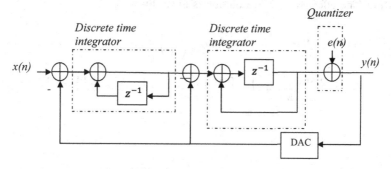

Fig. 8.49 Block diagram of second-order sigma-delta modulator

Equating this in-band noise power to the quantization noise power of the regular ADC, we obtain

$$\frac{A^2}{12}2^{-2n} = \frac{\pi^4}{5}\frac{A^2}{12}2^{-2m}\left(\frac{2f_{max}}{F_{os}}\right)^5 \tag{8.52}$$

Equation (8.37) leads to the following two useful relations

$$n = m + 2.5\log_2\left(\frac{F_{os}}{2f_{max}}\right) - 0.5\log_2\left(\frac{\pi^4}{5}\right) \tag{8.53}$$

$$\left(\frac{F_{os}}{2f_{max}}\right)^5 = \frac{\pi^2}{5}2^{2(n-m)} \tag{8.54}$$

In general, a Kth-order sigma-delta modulator realizes

$$H_x(z) = z^{-1} \text{ and } H_e(z) = (1 - z^{-1})^K$$

so that

$$Y(z) = X(z)z^{-1} + E(z)(1 - z^{-1})^K \tag{8.55}$$

and results in the following useful relations

$$n = m + \frac{1}{2}(2K + 1)\log_2\left(\frac{F_{os}}{2f_{max}}\right) - \frac{1}{2}\log_2\left(\frac{\pi^{2K}}{2K + 1}\right) \tag{8.56}$$

$$\left(\frac{F_{os}}{2f_{max}}\right)^{2K+1} = \frac{\pi^{2K}}{2K + 1}2^{2(n-m)} \tag{8.57}$$

Example 8.18 Considering a second-order SDM oversampling one-bit ADC system with sampling rate of 512 kHz and maximum analog signal frequency of 4 kHz, determine the effective ADC resolution.

Solution The effective ADC resolution is given by

$$n = m + 2.5\log_2\left(\frac{F_{os}}{2f_{max}}\right) - 0.5\log_2\left(\frac{\pi^4}{5}\right)$$

$$= m + 2.5\log_2\left(\frac{512}{8}\right) - 0.5\log_2\left(\frac{\pi^4}{5}\right) = 1 + 15 - 2.14 \approx 14\,\text{bits}.$$

8.6 Design of Multirate Bandpass and Bandstop Filters

A narrow bandpass filter is defined as a bandpass filter with a narrow passband. The narrow bandpass filter is implemented using modulation techniques. The techniques used for designing narrow passband lowpass filters may be readily extended to bandpass, highpass, and bandstop filter designs. A very simple and straight forward approach to the design of a narrowband bandpass filter is the quadrature modulation. In quadrature modulation, the center frequency of the passband is modulated to baseband (zero frequency), filtered by a narrow lowpass filter, and then demodulated to the original center frequency. Quadrature modulation requires that the input signal entering the lowpass filter contains real and imaginary parts. Thus, the narrow bandpass filter shown in Fig. 8.50a can be realized using the quadrature modulation structure shown in Fig. 8.50b.

In order to realize the bandpass filter using Fig. 8.50b, the following constraints [1] must be made on the bandpass response.

1. Symmetry around

$$\omega = \omega_0 = 2\pi f_0 = 2\pi(f_{s1} + f_{s2})/2 \qquad (8.58)$$

where ω_0 is the center frequency of the bandpass filter in radians/sec, f_{s1}, f_{s2} are the stopband frequencies of the bandpass filter, f_{p1}, f_{p2} are the passband frequencies of the bandpass filter, and

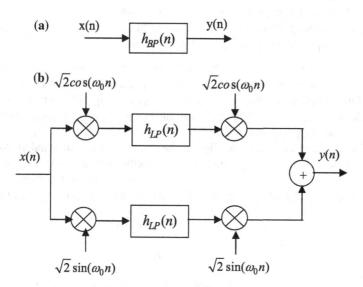

Fig. 8.50 a Block diagram of a narrowband bandpass filter. **b** A quadrature modulation structure for bandpass filtering using lowpass filter

$$f_0 - f_{p1} = f_{p2} - f_0$$
$$f_0 - f_{s1} = f_{s2} - f_0$$

(8.59)

or, equivalently,

$$f_{p1} - f_{s1} = f_{s2} - f_{p2}$$

(8.60)

Thus, the widths of the lower and upper transition bands must be equal.
2. Ripple symmetry:

$$\delta_{s1} = \delta_{s2}$$

(8.61)

If the constraints of Eqs. (8.59) to (8.61) are met, the lowpass filter specifications become

$$\tilde{\delta}_p = \delta_p$$
$$\tilde{\delta}_s = \delta_{s1} = \delta_{s2}$$
$$f_p = f_{p2} - f_0$$
$$f_s = f_{s2} - f_0$$

(8.62)

where

δ_{s1}, δ_{s2} are the stopband ripple values of the bandpass filter.
$\tilde{\delta}_p$ is the passband ripple value of the lowpass filter.
δ_p is the passband ripple value of the bandpass filter.
$\tilde{\delta}_s$ is the stopband ripple value of the lowpass filter.
f_p is the passband frequency of the lowpass filter.
f_s is the stopband frequency of the lowpass filter.

The desired bandpass filter is achieved by the structure of Fig. 8.50b. To achieve efficiency, the lowpass filters used in the quadrature modulation structure are realized in a multirate, multistage structure as shown in Fig. 8.51, where $h_{11}(n), \ldots,$ $h_{1N}(n), h_{21}(n), \ldots, h_{2N}(n)$ are the decimator filters and $g_{1N}(n), \ldots, g_{11}(n), g_{2N}(n), \ldots,$ $g_{21}(n)$ are the interpolation filters.

Multirate technique can be applied to the implementation of standard lowpass and bandpass filters, yielding structures whose efficiency increases as the signal bandwidth (i.e., the width of the passband) decreases. These techniques can be applied to narrow stopband highpass and bandstop filters, by realizing such filters as

$$H_{\mathrm{HP}}(e^{j\omega}) = 1 - H_{\mathrm{LP}}(e^{j\omega})$$

(8.63)

$$H_{\mathrm{BS}}(e^{j\omega}) = 1 - H_{\mathrm{BP}}(e^{j\omega})$$

(8.64)

The structure for the highpass filter, as illustrated in Fig. 8.52, consists of lowpass filtering the signal $x(n)$ and subtracting the filtered signal from the unfiltered input. In practice, the signal $x(n)$ must be delayed by the delay of the filter, as

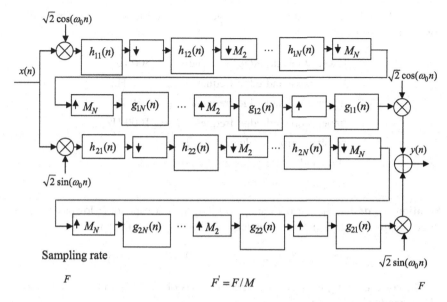

Fig. 8.51 Multirate, multistage quadrature structure of the narrowband bandpass filter

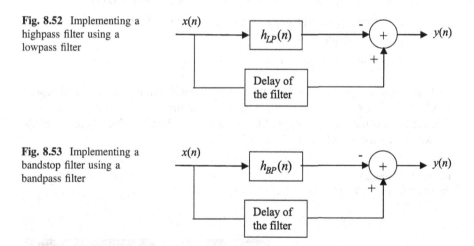

Fig. 8.52 Implementing a highpass filter using a lowpass filter

Fig. 8.53 Implementing a bandstop filter using a bandpass filter

shown in Fig. 8.52, before the difference is taken. Thus, it is important to design the filter to have a flat delay of an integer number of samples for this method to work.

In the bandstop filtering scheme shown in Fig. 8.53, the narrow notch frequencies are removed from the signal. In such a case, the equivalent bandpass bandwidth is quite small and a multirate structure can achieve very high efficiency compared to that of a standard notch filter

Example 8.19 Design a multirate multistage equiripple FIR narrow bandpass filter for the following specifications:

lower stopband edge frequency $f_{s1} = 800$ Hz

lower passband edge frequency $f_{p1} = 840$ Hz

upper passband edge frequency $f_{p2} = 960$ Hz

upper stopband edge frequency $f_{s2} = 1000$ Hz

passband ripple value $\alpha_p = 1$ dB

stopband ripple value $\alpha_s = 40$ dB

sampling frequency $f_s = 22{,}050$ Hz.

Solution To design the narrow bandpass filter, the corresponding lowpass filter is to be designed first. The center frequency of the bandpass filter is

$$f_0 = (f_{s1} + f_{s2})/2 = (800 + 1000)/2 = 900 \text{ Hz.}$$

The corresponding lowpass filter specifications obtained from the specifications of the bandpass filter are

passband edge frequency $f_p = f_{p2} - f_0 = 960 - 900 = 60$ Hz

stopband edge frequency $f_s = f_{s2} - f_0 = 1000 - 900 = 100$ Hz

passband ripple $\alpha_p = 1$ dB

stopband ripple $\alpha_s = 40$ dB

The lowpass filter satisfying the above specifications is designed using MATLAB SPTOOL. The magnitude response of the lowpass filter is shown in Fig. 8.54. The order of the lowpass filter with wt = [1 5.75] is 826, where wt stands for weight vector as defined in Sect. 6.7.

Fig. 8.54 Magnitude response of equiripple FIR lowpass filter

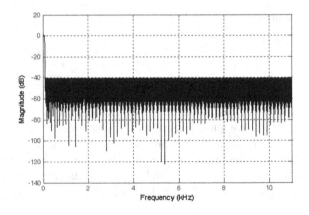

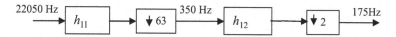

Fig. 8.55 Two-stage decimator for the lowpass filter

Table 8.11 Specifications of the individual filters for the decimator		First stage $h_{11}(n)$	Second stage $h_{12}(n)$
	Passband edge (kHz)	60	60
	Stopband edge (kHz)	262.5	87.5
	Passband ripple	0.10283588	0.10283588
	Stopband ripple	0.01	0.01

For the desired bandpass filter, two lowpass filters of order 826 are required in the quadrature modulation structure. For an efficient realization of the desired narrow bandpass filter, multirate multistage structure shown in Fig. 8.51 with two stages is chosen. The decimation and interpolation factors for the two-stage structure are $M1 = 63$ and $M2 = 2$. The two-stage decimator structure for the lowpass filter in the real channel is shown in Fig. 8.55.

Following the procedure given in Sect. 8.3.2, the specifications for the individual filters of the decimator are determined and these are provided in Table 8.11.

The filters $h_{11}(n)$ and $h_{12}(n)$ are designed using MATLAB SPTOOL for the specifications given in Table 8.11. The magnitude responses of the filters $h_{11}(n)$ and $h_{12}(n)$ are shown in Figs. 8.56 and 8.57, respectively. The order of the lowpass filter $h_{11}(n)$ satisfying the specifications with weighted vector wt = [1 6.75] is 185. The order of the lowpass filter $h_{12}(n)$ satisfying the specifications with weighted vector wt = [1 5.75] is 20.

Fig. 8.56 Magnitude response of the lowpass filter $h_{11}(n)$

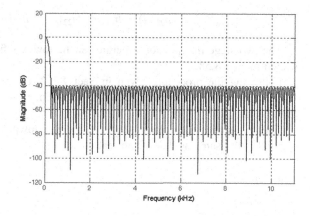

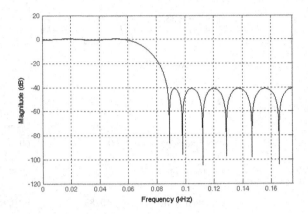

Fig. 8.57 Magnitude response of the lowpass filter $h_{12}(n)$

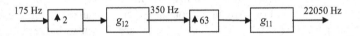

Fig. 8.58 Two-stage interpolator for the lowpass filter

Table 8.12 Specifications of the individual filters for the interpolator

	Second stage $g_{12}(n)$	First stage $g_{11}(n)$
Passband edge	60 Hz	60 Hz
Stopband edge	87.5 Hz	262.5 Hz
Passband ripple	0.10283588	0.10283588
Stopband ripple	0.01	0.01

The two-stage decimator structure for lowpass filter in the imaginary channel is the same as that of the real channel with

$$h_{21}(n) = h_{11}(n); \ h_{22}(n) = h_{12}(n)$$

The two-stage interpolator structure for the lowpass filter in the real channel is shown in Fig. 8.58.

Following the procedure given in Sect. 8.3.2, the specifications for the individual filters of the interpolator are determined and these are provided in Table 8.12.

The two-stage interpolator structure for the lowpass filter in the imaginary channel is the same as that of the real channel with

$$g_{21}(n) = g_{11}(n);$$

$$g_{22}(n) = g_{12}(n)$$

From the specifications given in Tables 8.11 and 8.12, it is observed that

$$g_{11}(n) = h_{11}(n);$$

$$g_{12}(n) = h_{12}(n)$$

As a consequence, the multirate two-stage quadrature modulation structure for the designed narrow bandpass filter is as shown in Fig. 8.59.

Example 8.20 Design the equiripple FIR bandstop filter with the following specifications

> lower passband edge frequency f_{p1} = 800 Hz
> lower stopband edge frequency f_{s1} = 840 Hz
> upper stopband edge frequency f_{s2} = 960 Hz
> upper passband edge frequency f_{p2} = 1000 Hz
> passband ripple value α_p = 1 dB
> stopband ripple value α_s = 40 dB
> sampling frequency f_s = 22050 Hz

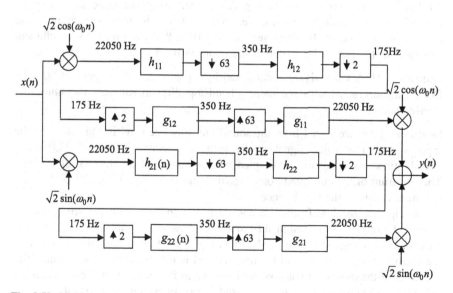

Fig. 8.59 Multirate two-stage structure for narrow bandpass filter of Example 8.19

Solution From the specifications of the desired narrowband bandstop filter, the corresponding bandpass filter specifications can be obtained as

$$\text{lower stopband edge frequency} f_{s1} = 800 \text{ Hz}$$
$$\text{lower passband edge frequency} f_{p1} = 840 \text{ Hz}$$
$$\text{upper passband edge frequency} f_{p2} = 960 \text{ Hz}$$
$$\text{upper stopband edge frequency} f_{s2} = 1000 \text{ Hz}$$
$$\text{passband ripple value } \alpha_p = 1 \text{ dB}$$
$$\text{stopband ripplevalue } \alpha_s = 40 \text{ dB}$$

The center frequency of the bandpass filter is

$$f_0 = (f_{s1} + f_{s2})/2 = (800 + 1000)/2 = 900 \text{ Hz}$$

The corresponding lowpass filter specifications are

$$\text{Passband edge frequency } f_p = f_{p2} - f_0 = 960 - 900 = 60 \text{ Hz}$$
$$\text{Stopband edge frequency} f_s = f_{s2} - f_0 = 1000 - 900 = 100 \text{ Hz}$$
$$\text{passband ripple value } \alpha_p = 1 \text{ dB}$$
$$\text{stopband ripple value } \alpha_s = 40 \text{ dB}$$

An equiripple FIR lowpass filter with its impulse response satisfying the above desired specifications is designed first. The filter order with weighted vector wt = [1 5.75] is 826. Incorporating this lowpass filter in the modulation structure for bandpass filtering shown in Fig. 8.50b and using the structure shown in Fig. 8.53, the bandstop filter can be realized. However, by incorporating multirate and two-stage structure for bandpass filtering (Fig. 8.59) in Fig. 8.53, an efficient structure for the realization of the desired narrow bandstop filter can be obtained.

Example 8.21 A voice signal is corrupted by a sinusoidal interference of 900 Hz. Design a multirate, multistage narrow bandstop filter to suppress the sinusoidal interference.

Solution The voice signal from the sound file 'theforce.wav' is considered as the input signal. The voice signal is digitized at a sampling rate of 22,050 Hz. The digitized voice signal is corrupted with a sinusoidal interference of 900 Hz. The spectrum of the corrupted voice signal is shown in Fig. 8.60. The peak in the spectrum is due to the interference.

To suppress the interference, a narrow bandstop filter with the specifications given in Example 8.20 is designed from the solutions of Examples 8.19 and 8.20 as a multirate two-stage structure, as shown in Fig. 8.61. In Fig. 8.61, the input $x(n)$ is the corrupted voice signal, and the output $x_r(n)$ is the recovered voice signal. The spectrum of the recovered voice signal is shown in Fig. 8.62. It can be seen that the spike due to the interference is suppressed in the recovered voice signal.

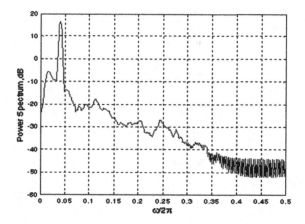

Fig. 8.60 Spectrum of the voice signal from the sound file 'theforce.wav', corrupted by a sinusoidal interference of 900 Hz

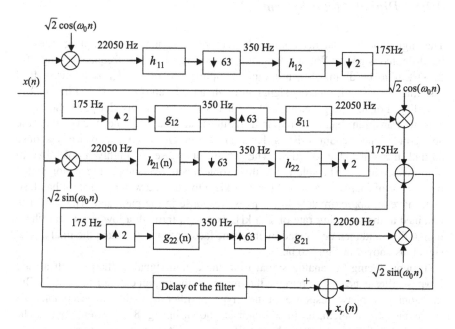

Fig. 8.61 Multirate two-stage structure for the narrow bandstop filter of Example 8.20

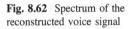

Fig. 8.62 Spectrum of the
reconstructed voice signal

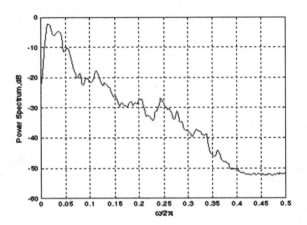

8.7 Application Examples

8.7.1 Digital Audio System

The digital audio systems commonly require to change the sampling rates of
band-limited signals. When an analog music signal in the frequency range 0–
22 kHz is to be digitized, a minimum sampling rate of 44 kHz is required. It is
essential to pass the analog signal through an anti-aliasing analog filter before
sampling. For this purpose, an analog filter with a reasonably flat passband and a
narrow transition band is required. Almost all the possible filters with the
above-mentioned characteristics have an extremely nonlinear phase response
around the bandedge 22 kHz. The nonlinear phase is highly intolerable in
high-quality music. The solution to this problem is to oversample the analog signal
by a factor of four (i.e., at $4 \times 44 = 176$ kHz) to obtain a wider transition band, so
as to have an approximately linear phase response in the passband. The sequence
obtained at the sampling rate of 176 kHz is passed through a lowpass linear phase
digital filter and then decimated by the same factor of four, to obtain the final digital
signal, as shown in Fig. 8.63a.

In reproducing the analog signal from the digital signal, a sharp cutoff analog
lowpass filter is needed to remove the images in the region exterior to 22 kHz. But
the nonlinear phase response of this type of filter is highly unacceptable. To
overcome this problem, an interpolator as shown in Fig. 8.63b is employed. The
interpolation filter is a linear phase FIR lowpass filter following which D/A conversion is performed. This is followed by an analog filter such as a simple Bessel
filter [6] with an approximately linear phase in the pass band.

For better quality digital audio, delta-sigma modulation techniques are used to
design single-bit ADCs with high resolution and high speed. The single-bit ADC
eliminates the analog circuitry such as the analog anti-aliasing filters, and sample
and hold circuits at the front end of a digital audio system. First, the delta-sigma

(a)

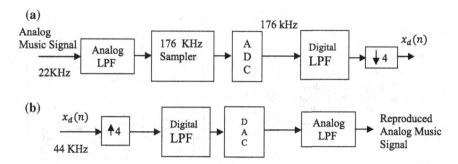

(b)

Fig. **8.63** **a** Scheme for ADC stage of a digital audio system and **b** scheme for D/A stage of a digital audio system

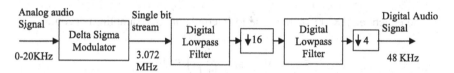

Fig. **8.64** Down-sampling of single-bit stream

modulator converts the analog audio signal into a single-bit stream, at a rate of 3.072 MHz. Then, the single-bit stream is decimated to 48 kHz with a decimation factor $M = 64$, using an efficient two-stage decimator, as shown in Fig. 8.64.

8.7.2 Compact Disk Player

For most studio work, the sampling rate is 48 kHz, whereas for CD mastering, the rate is 44.1 kHz. To convert from studio frequency to CD mastering standards, one would use an interpolator with $L = 441$ and a decimator with $M = 480$. Such large values of L and M normally imply very high-order filters. A multistage design is preferable in such cases. The CD contains 16-bit words' information of the digital signal at a sampling rate of 44.1 kHz. Image frequency bands centered at multiples of the sampling frequency of 44.1 kHz will result if the 16-bit words are directly transformed into analog form. As the resulting image frequencies are beyond the baseband of 0–20 kHz, they are not audible, but may cause overloading to the amplifier and loudspeaker of the player.

A common approach to solve this problem is to oversample the digital signal by a factor of 4 (4 × 44.1 kHz = 176 kHz) before it is applied to the 14-bit DAC. The image frequencies can be easily filtered, since they are now raised to higher frequencies. A simple third-order Bessel filter is used after the DAC conversion as shown in Fig. 8.65.

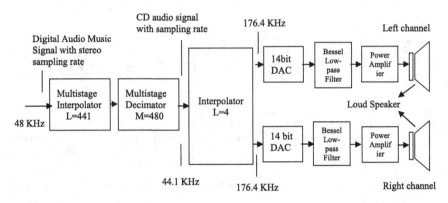

Fig. 8.65 Reproduction of an audio signal from compact disk

The effective ADC resolution due to interpolator (oversampling) becomes

$$n = 14 + 0.5\log_2\left(\frac{176.4}{44.1}\right) = 15 \text{ bits}$$

If a first-order sigma-delta modulator is added to the 14-bit DAC as shown below in the system shown in Fig. 8.59

> 14 bit DAC
> first order SDM

The first-order SDM pushes the quantization noise to the higher frequency range, and hence, the effective ADC resolution now becomes

$$n = 14 + 0.5(2K+1)\log_2\left(\frac{176.4}{44.1}\right) - 0.5\log_2\left(\frac{\pi^{2K}}{2K+1}\right)$$

since $K = 1$ for the first-order SDM, $n \approx 16$ bits.

8.8 Problems

1. Verify the down-sampler and upsampler for linearity and time invariance.
2. Consider the system shown in Fig. P8.1a,
   If $M = 8$, $\omega = \frac{\pi}{4}$ and the spectrum of $x(n)$ is as given in Fig. P8.1b. Determine the spectrum of $y(n)$.
3. A speech signal $s(t)$ is digitized at a sampling rate of 10 kHz. The speech signal was destroyed once the sequence $s(n)$ was stored on a magnetic tape. Later, it is

(a)

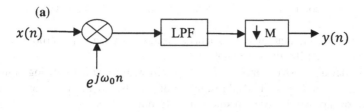

(b)

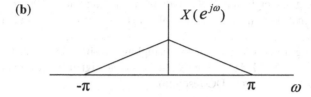

Fig. P8.1

required to obtain the speech signal sampled at the standard 8 kHz used in telephony. Develop a method to do this using discrete-time processing.

4. The sampling frequency 64 Hz of a signal is to be decimated by a factor of 64 to bring it down to 1 Hz. The anti-aliasing filter should satisfy the following specifications:

passband edge frequency = 0.45 Hz
stopband edge frequency = 0.5 Hz
passband ripple value = 0.01
minimum stopband ripple value = 60 dB
Design a two-stage decimator with decimation factors 16 and 4 for the first and second stages, respectively. Compare its computational complexity with that of a single-stage decimator. Use Eq. (8.22a) to compute the orders of the filters.

5. Design a two-stage interpolator to increase the sampling rate from 600 Hz to 9 kHz, and compare its complexity with a single-stage interpolator. The interpolator is to be designed as an FIR filter with a passband edge at 200 Hz and a stopband ripple of 0.001. Use the Kaiser window method to estimate the order of the FIR filter.

6. It is desired to reduce the sampling rate of a digital signal by a factor of 20.

i. If $\omega_p = \frac{0.9\pi}{20}$ and $\omega_s = \frac{\pi}{20}$ are chosen for a single-stage decimator scheme, what are the appropriate values for $\omega_{p1}, \omega_{s1}, \omega_{p2}, \omega_{s2}$ for the implementation of a two-stage decimator scheme with decimation factors 5 and 4 for the first and second stages, respectively?

ii. With the stopband attenuation of 60 dB and with the Kaiser window method being used to estimate the impulse response of the filters, determine as to which scheme has a lower computational complexity.

7. Develop a computationally efficient realization of a factor 3 decimator using a linear phase FIR filter of order 11.

8. Develop a computationally efficient realization of a factor 4 interpolator using a linear phase FIR filter of order 15.

9. Considering an oversampling ADC system with maximum analog signal frequency of 20 kHz and ADC resolution of 14 bits, determine the oversampling rate to improve the ADC resolution to 16-bit resolution.

10. Considering an eight-bit ADC with oversampling rate of 80,000 kHz and analog signal maximum frequency of 4 kHz, determine the equivalent ADC resolution.

11. Considering a first-order SDM oversampling two-bit ADC system with sampling rate of 512 kHz and maximum analog signal frequency of 4 kHz, determine the effective ADC resolution.

8.9   MATLAB Exercises

1. Write a MATLAB program to study the operation of a factor-of-5 down-sampler on a square-wave input sequence. Choose the input length to be 50. Plot the input and output sequences.

2. Write a MATLAB program to study the operation of a factor-of-5 interpolator on a length 50 sinusoidal sequence of normalized frequency 0.95. Plot the input and output sequences.

3. The overall specifications for a decimator are as follows:

   Passband edge frequency = 800 Hz
   Stopband edge frequency = 1250 Hz
   Maximum passband ripple value α_p = 0.1 dB
   Minimum stopband ripple α_s = 60.0 dB
   Input sampling frequency = 50 kHz
   Output sampling frequency = 2500 Hz
   Design a two-stage decimator, and compare its complexity with the single-stage decimator. Plot the magnitude responses of the filters of the single-stage and two-stage decimators.

4. The overall specifications for an interpolator are as follows:

   Passband edge frequency = 53 Hz
   Stopband edge frequency = 55 Hz
   Passband ripple α_p = 0.05 dB
   Stopband ripple α_s = 30 dB
   Input sampling frequency = 125 Hz
   Output sampling frequency = 500 Hz

Design a suitable two-stage interpolator for the specifications. Compare its computational complexity with a single-stage interpolator. Plot the magnitude responses of the filters of the single-stage and two-stage interpolators.

5. Design a multirate two-stage equiripple FIR narrow bandpass filter for the following specifications:

lower stopband edge frequency $f_{s1} = 1845$ Hz
lower passband edge frequency $f_{p1} = 1865$ Hz
upper passband edge frequency $f_{p2} = 1885$ Hz
upper stopband edge frequency $f_{s2} = 1905$ Hz
passband ripple value $\alpha_p = 0.5$ dB
stopband ripple value $\alpha_s = 60$ dB
sampling frequency $f_T = 7500$ Hz.

6. Design a multirate two-stage equiripple FIR narrow bandstop filter with the following specifications:

lower passband edge frequency $f_{p1} = 1845$ Hz
lower stopband edge frequency $f_{s1} = 1865$ Hz
upper stopband edge frequency $f_{s2} = 1885$ Hz
upper passband edge frequency $f_{p2} = 1905$ Hz
passband ripple value $\alpha_p = 0.5$ dB
stopband ripple value $\alpha_s = 60$ dB
sampling frequency $f_T = 7500$ Hz.

References

1. R.E. Crochiere, L.R. Rabiner, *Multirate Digital Signal Processing* (Prentice-Hall, Englewood Cliffs, NJ, 1983)
2. E.C. Ifeachor, B.W. Jervis, *Digital Signal Processing—A Practical Approach* (Prentice Hall, 2002)
3. O. Herrmann, L.R. Rabiner, D.S.K. Chan, Practical design rules for optimum finite impulse response lowpass digital filters. Bell. Syst. Tech. J. 769–799 (1973)
4. F. Kaiser, Nonrecursive digital filter design using the $-I_o$ n window function. In *Proc. 1974 IEEE International Symposium on Circuits and Systems*, San Francisco CA, 20–23 April 1974
5. P.M. Aziz, H.V. Sorensen, J. Van der Spiegel, An overview of Sigma-Delta converters, IEEE Signal Processing Magazine, January 1996
6. R. Raut, M.N.S. Swamy, Modern analog filter analysis and design, Wiley-VCH, 2010

Chapter 9
Multirate Filter Banks

A set of digital lowpass, bandpass, and highpass filters with a common input or a common output signal, as shown in Fig. 9.1, is called a digital filter bank. The structure of an M-band analysis filter bank is shown in Fig. 9.1a. Each subfilter $H_k(z)$ is called an analysis filter. The analysis filters $H_k(z)$ for $k = 0, 1, \ldots, M - 1$ decompose the input signal $x(n)$ into a set of M subband signals $v_k(z)$. Each subband signal occupies a portion of the original frequency band. Figure 9.1b shows an M-band synthesis filter bank which is used for the dual operation of that of the analysis filter bank. The synthesis bank combines a set of subband signals $\hat{v}_k(n)$ into a reconstructed signal $y(n)$. In Fig. 9.1b, each subfilter $G_k(z)$ is called a synthesis filter.

9.1 Uniform DFT Filter Banks

Let $H_0(z)$ be the transfer function of a causal lowpass digital filter given by

$$H_0(z) = \sum_{n=0}^{\infty} h_0(n)z^{-n} \tag{9.1}$$

where $h_0(n)$ is the real impulse response of the filter.

Now, consider the causal impulse response $h_k(n)$, $0 \le k \le M - 1$, with its frequency response as

$$H_k(e^{j\omega}) = H_0(e^{j(\omega - 2\pi k/M)}) \quad 0 \le k \le M - 1 \tag{9.2}$$

where M is an integer constant. From Eq. (9.2), it is seen that the frequency response of $H_k(z)$ is obtained by shifting the response of the lowpass filter $H_0(z)$ to the right, by an amount $2\pi k/M$. Thus, the responses of the $M - 1$ filters

© Springer Nature Singapore Pte Ltd. 2018
K. D. Rao and M. N. S. Swamy, *Digital Signal Processing*,
https://doi.org/10.1007/978-981-10-8081-4_9

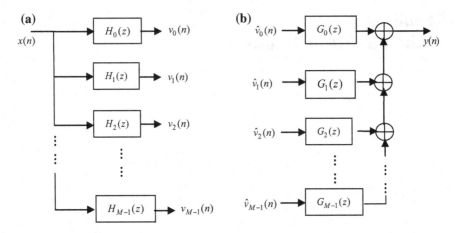

Fig. 9.1 **a** Analysis filter bank. **b** Synthesis filter bank

$H_1(z), H_2(z), \ldots, H_{M-1}(z)$ are uniformly shifted versions of the response of the basic prototype filter $H_0(z)$. Equation (9.2) can be represented in the z-domain as

$$H_k(z) = H_0\big(ze^{-j2\pi k/M}\big) \quad 0 \le k \le M-1. \tag{9.3}$$

The M filters $H_k(z)$ defined by Eq. (9.3) can be used as the analysis filters in the analysis filter bank of Fig. 9.1a or as the synthesis filters $G_k(z)$ in the synthesis filter bank of Fig. 9.1b. Since the set of magnitude responses $|H_k(e^{j\omega})|$, $0 \le k \le M-1$, are uniformly shifted versions of the basic prototype $|H_0(e^{j\omega})|$, that is,

$$\left|H_k(e^{j\omega})\right| = \left|H_0\left(e^{j\left(\omega - \frac{2\pi k}{M}\right)}\right)\right| \tag{9.4}$$

the filter bank obtained is called a uniform filter bank.

The inverse z-transform of $H_0(z)$ yields the impulse response $h_0(n)$. From the z-transform properties and the inverse z-transform of Eq. (9.3), the impulse response $h_k(n)$ is defined as

$$h_k(n) = h_0(n)e^{\frac{j2\pi k}{M}} \quad 0 \le k \le M-1. \tag{9.5}$$

9.1.1 Lth-Band Filter

Consider the interpolator shown in Fig. 8.10 with an interpolation factor of L. The input–output relation of the filter is given by

$$Y(z) = H(z)X(z^L) \qquad (9.6)$$

where $H(z)$ is the transfer function of the lowpass filter. The L-band polyphase realization of the interpolation filter $H(z)$ is given by

$$H(z) = \sum_{l=o}^{L-1} z^{-l} E_l(z^L).$$

If it is assumed that the kth polyphase component of $H(z)$ is a constant, i.e.,

$E_k(z^l) = \alpha$, then

$$H(z) = \sum_{l=o}^{k-1} z^{-l} E_l(z^L) + \alpha z^{-k} + \sum_{l=k+1}^{L-1} z^{-l} E_l(z^L). \qquad (9.7)$$

Then, $Y(z)$ can be written as

$$Y(z) = \alpha z^{-k} X(z^L) + \sum_{\substack{l=o \\ l \neq k}}^{L-1} z^{-l} E_l(z^L) X(z^L). \qquad (9.8)$$

Hence, $y(Ln + k) = \alpha x(n)$ implying that for all values of n, there is no distortion in the input samples that appear at the output, with the values of the in-between $(L - 1)$ samples being determined by interpolation. A filter which satisfies the above property is called an Lth-band filter or Nyquist filter. Since its impulse response has many zero-valued samples, it is computationally more efficient than other lowpass filters of the same order and is often used for single-rate and multirate signal processing applications. Consider the example of the Lth-band filter for $k = 0$. The impulse response of such a filter satisfies the condition [1].

$$h(Ln) = \begin{cases} \alpha & n = 0 \\ 0 & \text{otherwise} \end{cases} \qquad (9.9)$$

9.1.2 Design of Linear Phase Lth-Band FIR Filter

The windowing method detailed in Sect. 6.3 can be readily used to design a lowpass linear phase Lth-band FIR filter with cutoff frequency at $\omega_c = \pi/L$. In this method, the impulse response coefficients of the lowpass filter are given by

$$h(n) = h_{LP}(n)w(n) \tag{9.10}$$

where $h_{LP}(n)$ is the impulse response of an ideal lowpass filter with cutoff frequency of π/L, and $w(n)$ is a suitable window function. Equation (9.9) is satisfied if

$$h_{LP}(n) = 0 \quad \text{for } n = \pm L, \pm 2L, \ldots \tag{9.11}$$

Now by substituting $\omega_c = \pi/L$ in Eq. (6.2), the impulse response $h_{LP}(n)$ of an ideal Lth-band filter can be expressed as

$$h_{LP}(n) = \frac{\sin\left(\frac{\pi n}{L}\right)}{\pi n} \quad -\infty \leq n \leq \infty. \tag{9.12}$$

Example 9.1 Design a four-channel uniform DFT analysis filter bank using a linear phase FIR filters of length 21. Use the Hamming window for designing the FIR filter.

Solution Program 9.1 given below is used to design a four-channel uniform DFT filter bank. The input data requested by the program is the desired filter length and the value of L. The program determines the impulse response coefficients of the Lth-band filter using the expression given by Eq. (9.12) and computes and plots the gain response of the designed filter as shown in Fig. 9.2.

Program 9.1: Design a four-channel uniform DFT filter bank

```
% design of a 4-channel uniform DFT filter bank
len= input('Type in the filter length=');
L=input('Type in the value of L=');
k=(len-1)/2;
n=-k:k;
b=sinc(n/L)/L; %Generate the truncated impulse response of the ideal
%lowpass filter
win=hamming(len); % Generate the window sequence
h0=b.*win'; % Generate the coefficients of the windowed filter
j=sqrt(-1);
for i=1:21

h1(i)=j^(i)*h0(i);
h2(i)=(-1)^i*h0(i);
h3(i)=(-j)^(i)*h0(i);

end
[H0z,w]=freqz(h0,1, 512,'whole');
H0=abs(H0z);
M0=20*log10(H0);
```

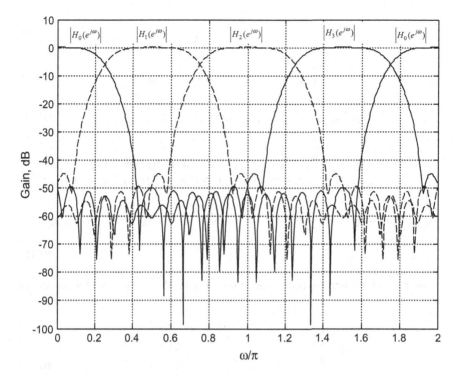

Fig. 9.2 Gain responses of the four-channel analysis filters for a uniform DFT filter bank using Hamming window

```
[H1z,w]=freqz(h1, 1, 512,'whole');
H1=abs(H1z);
M1=20*log10(H1);
[H2z,w]=freqz(h2, 1, 512,'whole');
H2=abs(H2z);
M2=20*log10(H2);
[H3z, w]=freqz(h3, 1, 512,'whole');
H3=abs(H3z);
M3=20*log10(H3);
plot(w/pi, M0, '-k', w/pi, M1, '–k', w/pi, M2, '–k', w/pi, M3, '-k');grid
xlabel('\omega/\pi');
ylabel('Gain, dB')
```

9.2 Polyphase Implementations of Uniform Filter Banks

An M-band polyphase representation of the lowpass prototype transfer function $H_0(z)$ is given by

$$H_0(z) = \sum_{l=0}^{M-1} z^{-l} E_l(z^M) \tag{9.13}$$

where $E_l(z)$, the polyphase component of $H_0(z)$, is given by

$$E_l(z) = \sum_{n=0}^{\infty} e_l(n) z^{-n} = \sum_{n=0}^{\infty} h_0[l+nM] z^{-n}, \quad 0 \le l \le M-1. \tag{9.14}$$

Replacing z by $z e^{-j2\pi k/M}$ in Eq. (9.13), and using the identity $e^{j2\pi k} = 1$, $0 \le k \le M-1$, the M-band polyphase decomposition of $H_k(z)$ can be obtained as

$$H_k(z) = \sum_{i=0}^{M-1} z^{-l} e^{j2\pi kl/M} E_l(z^M e^{-j2\pi kM/M}) = \sum_{l=0}^{M-1} z^{-l} e^{j2\pi kl/M} E_l(z^M) \quad 0 \le k \le M-1. \tag{9.15}$$

After some mathematical manipulations, Eq. (9.15) can be written in matrix form as

$$\begin{bmatrix} H_0(z) \\ H_1(z) \\ H_2(z) \\ \vdots \\ H_{M-1}(z) \end{bmatrix} = MD^{-1} \begin{bmatrix} E_0(z^M) \\ z^{-1} E_1(z^M) \\ z^{-2} E_2(z^M) \\ \vdots \\ z^{-(M-1)} E_{M-1}(z^M) \end{bmatrix}. \tag{9.16}$$

where D denotes the DFT matrix:

$$D = \begin{bmatrix} 1 & 1 & 1 & \cdots & 1 \\ 1 & e^{-j2\pi/M} & e^{-j4\pi/M} & \cdots & e^{-j2\pi(M-1)/M} \\ 1 & e^{-j4\pi/M} & e^{-j8\pi/M} & \cdots & e^{-j4\pi(M-1)/M} \\ \vdots & \vdots & \vdots & \cdots & \vdots \\ 1 & e^{-j2\pi(M-1)/M} & e^{-j4\pi(M-1)/M} & \cdots & e^{-j2\pi(M-1)^2/M} \end{bmatrix} \tag{9.17}$$

Equation (9.16) represents an M-band analysis uniform DFT filter bank. An implementation of Eq. (9.16) is shown in Fig. 9.3, where $H_k(z) = V_k(z)/X(z)$, and is known as the polyphase decomposition of a uniform DFT analysis filter bank.

Example 9.2 The four-channel analysis filter bank of Fig. 9.3 is characterized by the set of four transfer functions $H_k(z) = \frac{Y_k(z)}{X_{k(z)}}$, $k = 0, 1, 2, 3$. The transfer functions of the four subfilters are given by

Fig. 9.3 Polyphase implementation of a uniform DFT analysis filter bank

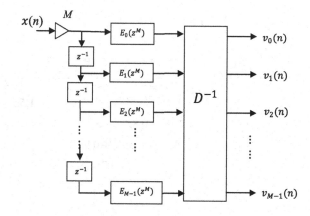

$$E_0(z) = 1 + 0.3z^{-1} - 0.8z^{-2} \quad E_1(z) = 2 - 1.5z^{-1} + 3.1z^{-2}$$
$$E_2(z) = 4 - 0.9z^{-1} + 2.3z^{-2} \quad E_3(z) = 1 + 3.7z^{-1} + 1.7z^{-2}.$$

Find the four transfer functions $H_0(z), H_1(z), H_2(z), H_3(z)$.

Solution

$$D = \begin{bmatrix} 1 & 1 & 1 & \cdots & 1 \\ 1 & e^{-j2\pi/M} & e^{-j4\pi/M} & \cdots & e^{-j2\pi(M-1)/M} \\ 1 & e^{-j4\pi/M} & e^{-j8\pi/M} & \cdots & e^{-j4\pi(M-1)/M} \\ \vdots & \vdots & \vdots & \cdots & \vdots \\ 1 & e^{-j2\pi(M-1)/M} & e^{-j4\pi(M-1)/M} & \cdots & e^{-j2\pi(M-1)^2/M} \end{bmatrix}$$

For $M = 4$, D becomes

$$D = \begin{bmatrix} 1 & 1 & 1 & 1 \\ 1 & e^{-j\pi/2} & e^{-j\pi} & e^{-j3\pi/2} \\ 1 & e^{-j\pi} & e^{-j2\pi} & e^{-j3\pi} \\ 1 & e^{-j3\pi/2} & e^{-j3\pi} & e^{-j9\pi/2} \end{bmatrix} = \begin{bmatrix} 1 & 1 & 1 & 1 \\ 1 & -j & -1 & j \\ 1 & -1 & 1 & -1 \\ 1 & j & -1 & -j \end{bmatrix}$$

$$D^{-1} = \begin{bmatrix} 0.25 & 0.25 & 0.25 & 0.25 \\ 0.25 & 0.25i & -0.25 & -0.25i \\ 0.25 & -0.25 & 0.25 & -0.25 \\ 0.25 & -0.25i & -0.25 & +0.25i \end{bmatrix}$$

$$E_0(z^4) = 1 + 0.3z^{-4} - 0.8z^{-8} \quad E_1(z^4) = 2 - 1.5z^{-4} + 3.1z^{-8}$$
$$E_2(z^4) = 4 - 0.9z^{-4} + 2.3z^{-8} \quad E_3(z^4) = 1 + 3.7z^{-4} + 1.7z^{-8}$$

Using the above results in Eq. (9.16), the analysis filters $H_0(z), H_1(z), H_2(z), H_3(z)$ are given by

$$H_0(z) = (1 + 2z^{-1} + 4z^{-2} + z^{-3} + 0.3z^{-4} - 1.5z^{-5} - 0.9z^{-6}$$
$$+ 3.7z^{-7} - 0.8z^{-8} + 3.1z^{-9} + 2.3z^{-10} + 1.7z^{-11})$$
$$H_1(z) = [(1 - 4z^{-2} + 0.3z^{-4} + 0.9z^{-6} - 0.8z^{-8} - 2.3z^{-10}) + i(2z^{-1}$$
$$- z^{-3} - 1.5z^{-5} - 3.7z^{-7} + 3.1z^{-9} - 1.7z^{-11})]$$
$$H_2(z) = (1 - 2z^{-1} + 4z^{-2} - z^{-3} + 0.3z^{-4} + 1.5z^{-5} - 0.9z^{-6}$$
$$- 3.7z^{-7} - 0.8z^{-8} - 3.1z^{-9} + 2.3z^{-10} - 1.7z^{-11})$$
$$H_3(z) = [(1 - 4z^{-2} + 0.3z^{-4} + 0.9z^{-6} - .8z^{-8} - 2.3z^{-10}) - i(2z^{-1}$$
$$- z^{-3} - 1.5z^{-5} - 3.7z^{-7} + 3.1z^{-9} - 1.7z^{-11})]$$

9.3 Two-Channel Quadrature Mirror Filter (QMF) Bank

9.3.1 The Filter Bank Structure

Multirate analysis–synthesis systems based on filter banks are now widely used for time–frequency decomposition and reconstruction in many applications, especially speech and image processing and communications. FIR quadrature mirror filter (QMF) bank is an important class of filter banks widely used in multirate analysis–synthesis systems. It requires signal decomposition into subbands and reconstruction of the signal from coded subbands. The structure of a two-channel FIR QMF banks is shown in Fig. 9.4.

$H_0(z)$ and $H_1(z)$ are called analysis filters, while $G_0(z)$ and $G_1(z)$ are called synthesis filters. The analysis bank channelizes the input signal into two subbands using the analysis filters. The synthesis bank reconstructs the subband signals using synthesis filters. The combined structure is called a two-channel quadrature mirror filter (QMF) bank. Theoretically, $H_0(Z)$ and $H_1(Z)$ should be ideal lowpass and ideal highpass filters with cutoff at $\pi/2$. In other words, $H_0(z)$ should be a mirror image of the filter $H_1(z)$ with respect to the quadrature frequency $2\pi/4$, justifying the name quadrature mirror filters (see Fig. 9.5).

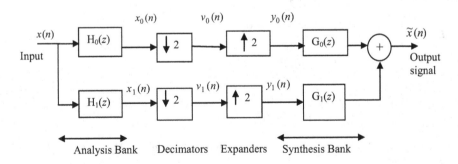

Fig. 9.4 Two-channel QMF bank

Fig. 9.5 Amplitude
responses of H_0 and H_1

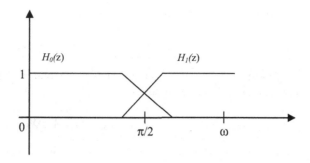

   In practice, the analysis filters have nonzero transition bandwidth and stopband
gain. The subband signals are, therefore, not band-limited, and aliasing will occur
after their decimation. However, with a careful choice of synthesis filters, it is
possible to eliminate the effect of aliasing. Apart from aliasing, the output signal
might suffer from phase or amplitude distortion. Perfect reconstruction (PR) filter
banks are systems where there is no error at the output; i.e., the output will be a
delayed copy of the input. Systems with small amount of aliasing and distortion are
known as near-perfect reconstruction (NPR) filter banks.

9.3.2   Analysis of Two-Channel QMF Bank

The input–output relation of the two-channel filter bank shown in Fig. 9.4 can be
derived quite easily. From Fig. 9.4, the following equation can be written.

$$X_k(z) = H_k(z)X(z) \quad k = 0, 1 \qquad (9.18)$$

The z-transform of the decimated signals is then found as

$$V_k(z) = \frac{1}{2}\left[X_K\left(z^{\frac{1}{2}}\right) + X_k\left(-z^{\frac{1}{2}}\right)\right], \quad k = 0, 1. \qquad (9.19)$$

The z-transform of $y_k(n)$ is

$$Y_k(z) = V_k(z^2) = \frac{1}{2}[X_k(z) + X_k(-z)]$$
$$= \frac{1}{2}[H_k(z)X(z) + H_k(-z)X(-z)], \quad k = 0, 1 \qquad (9.20)$$

The reconstructed signal is given by

$$\hat{X}(z) = G_0(z)Y_0(z) + G_1(z)Y_1(z).$$

Substituting Eq. (9.20) in the above, we obtain the final expression of the reconstructed signal, after some manipulations, to be

$$\hat{X}(z) = T(z)X(z) + A(z)X(-z) \qquad (9.21)$$

where

$$T(z) = \frac{1}{2}[G_0(z)H_0(z) + G_1(z)H_1(z)] \qquad (9.22)$$

$$A(z) = \frac{1}{2}[G_0(z)H_0(-z) + G_1(z)H_1(-z)]. \qquad (9.23)$$

$T(z)$ and $A(z)$ are called distortion transfer function (amplitude and phase distortions) and aliasing transfer function, respectively. It is clear that aliasing can be completely removed by making $A(z) = 0$. This is usually done by proper choice of synthesis filters. For a perfect reconstruction filter bank, the output of the filter bank should be a delayed copy of the input. In order to do so, the synthesis filters, $G_0(z)$ and $G_1(z)$, should satisfy the following conditions:

$$G_0(z)H_0(z) + G_1(z)H_1(z) = z^{-k_d} \qquad (9.24a)$$

$$G_0(z)H_0(-z) + G_1(z)H_1(-z) = 0 \qquad (9.24b)$$

k_d is the reconstruction delay, which is an integer. Equation (9.24b) ensures zero aliasing. One way of realizing zero aliasing is to choose

$$G_0(z) = H_1(-z) \text{ and } G_1(z) = -H_0(-z).$$

A simple way of achieving alias-free two-channel filter bank is to choose $H_1(z) = H_0(-z)$. Substituting this in the above equation, we get

$$G_0(z) = H_0(z) \text{ and } G_1(z) = -H_1(z) = -H_0(-z). \qquad (9.25)$$

Moreover, since $H_0(z)$ is a lowpass filter, $G_0(z)$ is also a lowpass filter and $G_1(z)$ a highpass filter. The distortion transfer function $T(z)$ becomes:

$$T(z) = \frac{1}{2}\left[H_0^2(z) - H_1^2(z)\right] = \frac{1}{2}\left[H_0^2(z) - H_0^2(-z)\right]. \qquad (9.26)$$

The corresponding input–output relation is given as follows:

$$\hat{X}(z) = \frac{1}{2}\left[H_0^2(z) - H_0^2(-z)\right]X(z) \tag{9.27}$$

and the perfect reconstruction condition is given by

$$H_0^2(z) - H_0^2(-z) = z^{-k_d}. \tag{9.28}$$

Example 9.3 Show that the two-channel filter bank of Fig. 9.4 is a perfect reconstruction system for the following analysis and synthesis filters:

$$H_0(z) = 2 - z^{-1},\ H_1(z) = 2 + 3z^{-1},\ G_0(z) = -1 + 1.5z^{-1},\ G_1(z) = 1 + 0.5z^{-1}.$$

Solution For a perfect reconstruction filter bank, the output of the filter bank should be a delayed copy of the input. For this, the analysis and synthesis filters should satisfy the following conditions:

$$\begin{aligned} G_0(z)H_0(z) + G_1(z)H_1(z) &= z^{-k_d} \\ G_0(z)H_0(-z) + G_1(z)H_1(-z) &= 0 \end{aligned}.$$

For perfect reconstruction condition, we should have

$$\begin{aligned} G_0(z)H_0(z) + G_1(z)H_1(z) &= (2 - z^{-1})(-1 + 1.5z^{-1}) + (2 + 3z^{-1})(1 + 0.5z^{-1}) \\ &= -2 + z^{-1} + 3z^{-1} - 1.5z^{-2} + 2 + 3z^{-1} + z^{-1} + 1.5z^{-2} \\ &= 8z^{-1} \end{aligned}$$

and for alias-free condition

$$\begin{aligned} G_0(z)H_0(-z) + G_1(z)H_1(-z) &= (2 + z^{-1})(-1 + 1.5z^{-1}) + (2 - 3z^{-1})(1 + 0.5z^{-1}) \\ &= -2 - z^{-1} + 3z^{-1} + 1.5z^{-2} + 2 - 3z^{-1} + z^{-1} - 1.5z^{-2}. \\ &= 0 \end{aligned}$$

Thus, the two-channel filter bank is an alias-free perfect reconstruction system. It should be noted that in this example, conditions (9.25) are not satisfied, but the two-channel filter bank is still a PR system.

Example 9.4 Consider a two-channel QMF bank with the analysis and synthesis filters given by

$$H_0(z) = 2 + 6z^{-1} + z^{-2} + 5z^{-3} + z^{-5}, H_1(z) = H_0(-z);$$
$$G_0(z) = H_0(z); G_1(z) = -H_1(z).$$

(i) Is the QMF filter bank alias-free?
(ii) Is the QMF filter bank a perfect reconstruction system?

Solution

$$H_0(z) = 2 + 6z^{-1} + z^{-2} + 5z^{-3} + z^{-5}$$
$$H_1(z) = H_0(-z) = 2 - 6z^{-1} + z^{-2} - 5z^{-3} - z^{-5}$$
$$G_0(z) = H_0(z) = 2 + 6z^{-1} + z^{-2} + 5z^{-3} + z^{-5}$$
$$G_1(z) = -H_1(z) = -2 + 6z^{-1} - z^{-2} + 5z^{-3} + z^{-5}$$

(i) Alias function $A(z)$ is

$$A(z) = G_0(z)H_0(-z) + G_1(z)H_1(-z)$$
$$= [(-4 + 32z^{-2} + 59z^{-4} + 37z^{-6} + 10z^{-8} + z^{-10})$$
$$+ (4 - 32z^{-2} - 59z^{-4} - 37z^{-6} - 10z^{-8} - z^{-10})] = 0$$

Thus, the system is alias-free.

(ii) Distortion function $T(z)$ is

$$T(z) = \frac{1}{2}[G_0(z)H_0(z) + G_1(z)H_1(z)] = \frac{1}{2}[(4 + 24z^{-1} + 40z^{-2} + 32z^{-3}$$
$$+ 61z^{-4} + 14z^{-5} + 37z^{-6} \quad + 2z^{-7} + 10z^{-8} + z^{-10})$$
$$+ (-4 + 24z^{-1} - 40z^{-2} + 32z^{-3} - 61z^{-4} + 14z^{-5} - 37z^{-6} + 2z^{-7}$$
$$- 10z^{-8} - z^{-10})] = 24z^{-1} + 32z^{-3} + 14z^{-5} + 2z^{-7}$$

The output of the filter bank is not a delayed copy of the input. Hence, it is not a perfect reconstruction system.

9.3.3   Alias-Free and Perfect Reconstruction for M-Channel QMF Bank

The basic M-channel QMF bank is shown in Fig. 9.6.

For an M-channel QMF bank, the reconstructed signal $Y(z)$ can be shown to be [1, 2].

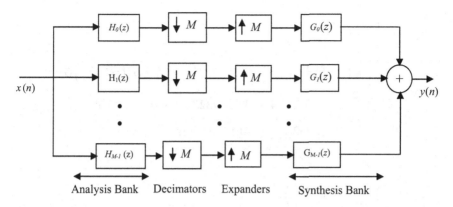

Fig. 9.6 Basic M-channel QMF bank structure

$$Y(z) = a_0(z)X(z) + \sum_{l=1}^{M-1} a_l(z)X(ze^{-j2\pi l/M}) \tag{9.29}$$

where $X(ze^{-j2\pi l/M})$, $l > 0$ are the alias terms, and $a_l(z)$ is given by

$$a_l(z) = \frac{1}{M}\sum_{k=0}^{M-1} H_k(ze^{-j2\pi l/M})G_k(z), \quad 0 \le l \le M-1 \tag{9.30}$$

The aliasing effect can be completely eliminated at the output if and only if $a_l(z) = 0$, $1 \le l \le M-1$. Under such a condition, the QMF bank becomes a linear time-invariant (LTI) system $Y(z) = T(z)X(z)$, where the distortion transfer function $T(z)$ is given by

$$T(z) = a_0(z) = \frac{1}{M}\sum_{k=0}^{M-1} H_k(z)G_k(z). \tag{9.31}$$

Example 9.5 Show that the three-channel QMF bank is an alias-free and perfect reconstruction system for the following analysis and synthesis filters: $H_0(z) = 1$, $H_1(z) = 2+z^{-1}$, $H_2(z) = 3+2z^{-1}+z^{-2}$, $G_0(z) = 1-2z^{-1}+z^{-2}$, $G_1(z) = -2+z^{-1}$, and $G_2(z) = 1$.

Solution Aliasing function:

$$a_l(z) = \frac{1}{M}\sum_{k=0}^{M-1} H_k(ze^{-j2\pi l/M})G_k(z), \quad 0 \le l \le M-1.$$

The aliasing effect can be completely eliminated at the output $Y(z)$ if and only if $a_l(z) = 0$, $1 \le l \le M-1$

$$a_1(z) = \frac{1}{3} \sum_{k=0}^{2} H_k\left(ze^{-j2\pi/3}\right) G_k(z),$$

$$= \frac{1}{3}\left[H_0\left(ze^{-j2\pi/3}\right)G_0(z) + H_1\left(ze^{-j2\pi/3}\right)G_1(z) + H_2\left(ze^{-j2\pi/3}\right)G_2(z)\right]$$

$$= \frac{1}{3}[z^{-2} - 2z^{-1} + 1 - 4 + (3 - 1.732i)z^{-1} - (0.5 - 0.866i)z^{-2}$$

$$+ 3 - (1 - 1.732i)z^{-1} - (0.5 + 0.866i)z^{-2}]$$

$$= 0$$

$$a_2(z) = \frac{1}{3} \sum_{k=0}^{2} H_k\left(ze^{-j4\pi/3}\right) G_k(z),$$

$$= \frac{1}{3}\left[H_0\left(ze^{-j4\pi/3}\right)G_0(z) + H_1\left(ze^{-j4\pi/3}\right)G_1(z) + H_2\left(ze^{-j4\pi/3}\right)G_2(z)\right]$$

$$= \frac{1}{3}[z^{-2} - 2z^{-1} + 1 - 4 + (3 + 1.732i)z^{-1} - (0.5 + 0.866i)z^{-2}$$

$$+ 3 - (1 + 1.732i)z^{-1} - (0.5 - 0.866i)z^{-2}]$$

$$= 0$$

Thus, the system is alias-free.

Distortion function:

$$T(z) = \frac{1}{M} \sum_{k=0}^{M-1} H_k(z)G_k(z)$$

$$= \frac{1}{3}[z^{-2} - 2z^{-1} + 1 - 4 + z^{-2} + 3 + 2z^{-1} + z^{-2}]$$

$$= \frac{1}{3} \times 3z^{-2} = z^{-2}$$

Hence, the three-channel QMF bank is alias-free and a perfect reconstruction system.

9.3.4   Polyphase Representation of M-Channel QMF Banks

It is known from Sect. 8.4 that the kth analysis filter $H_k(z)$ can be represented in the polyphase form as [see Eq. (8.25)]

$$H_k(z) = \sum_{l=0}^{M-1} z^{-l} E_{kl}(z^M) \qquad 0 \le k \le M - 1. \tag{9.32a}$$

We can write the above equation as

$$
\begin{bmatrix} H_0(z) \\ H_1(z) \\ \vdots \\ H_{M-1}(z) \end{bmatrix} = \begin{bmatrix} E_{00}(Z^M) & E_{01}(Z^M) & E_{02}(Z^M) & \cdots & E_{0,M-1}(Z^M) \\ E_{10}(Z^M) & E_{11}(Z^M) & E_{12}(Z^M) & \cdots & E_{1,M-1}(Z^M) \\ E_{20}(Z^M) & E_{21}(Z^M) & E_{22}(Z^M) & \cdots & E_{2,M-1}(Z^M) \\ \vdots & \vdots & \vdots & \cdots & \vdots \\ E_{M-1,0}(Z^M) & E_{M-1,1}(Z^M) & E_{M-1,2}(Z^M) & \cdots & E_{M-1,M-1}(Z^M) \end{bmatrix}
$$

$$
\times \begin{bmatrix} 1 \\ z^{-1} \\ \vdots \\ z^{-(M-1)} \end{bmatrix}
$$

$$(9.32b)$$

Equation (9.32b) can be rewritten as

$$h(z) = E(z^M)e(z) \qquad (9.33)$$

where

$$h(z) = [H_0(z)\, H_1(z) \ldots H_{M-1}(z)]^T \qquad (9.34)$$

$$e(z) = \left[1 z^{-1} z^{-2} \ldots z^{-(M-1)} \right]^T \qquad (9.35)$$

and

$$
E(z) \begin{bmatrix} E_{00}(z) & E_{01}(z) & E_{02}(z) & \cdots & E_{0,M-1}(z) \\ E_{10}(z) & E_{11}(z) & E_{12}(z) & \cdots & E_{1,M-1}(z) \\ E_{20}(z) & E_{21}(z) & E_{22}(z) & \cdots & E_{2,M-1}(z) \\ \vdots & \vdots & \vdots & \cdots & \vdots \\ E_{M-1,0} & E_{M-1,1}(z) & E_{M-1,2}(z) & \cdots & E_{M-1,M-1}(z) \end{bmatrix}. \qquad (9.36)
$$

The $M \times M$ matrix $E(z)$ is called a Type 1 polyphase component matrix. The corresponding Type 1 polyphase representation of the analysis filter bank is shown in Fig. 9.7.

In a similar manner, the M synthesis filters can be represented in the following polyphase form:

$$G_k(z) = \sum_{l=0}^{M-1} z^{-(M-1-l)} R_{lk}(z^M) \quad 0 \le k \le M - 1 \qquad (9.37)$$

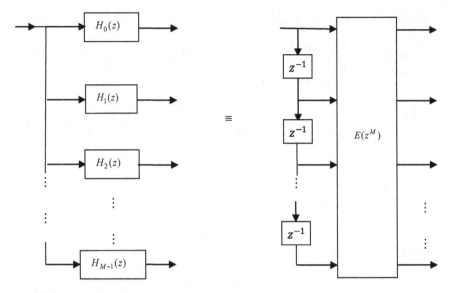

Fig. 9.7 Type 1 polyphase representation of an analysis filter bank

The above set of M equations can be rewritten in a matrix form as

$$[G_0(z)G_1(z)\ldots G_{M-1}(z)] = \left[z^{-(M-1)}z^{-(M-2)}\ldots 1\right]$$

$$\begin{bmatrix} R_{00}(z^M) & R_{01}(z^M) & R_{02}(z^M) & \ldots & R_{0,M-1}(z^M) \\ R_{10}(z^M) & R_{11}(z^M) & R_{12}(z^M) & \ldots & R_{1,M-1}(z^M) \\ R_{20}(z^M) & R_{21}(z^M) & R_{22}(z^M) & \ldots & R_{2,M-1}(z^M) \\ \vdots & \vdots & \vdots & \ldots & \vdots \\ R_{M-1,0}(z^M) & R_{M-1,1}(z^M) & R_{M-1,2}(z^M) & \ldots & R_{M-1,M-1}(z^M) \end{bmatrix}.$$

$$(9.38a)$$

The above equation can be written as

$$g^T(z) = \tilde{e}^T(z)R(z^M) \tag{9.38b}$$

where

$$g(z) = [G_0(z)\, G_1(z)\, \ldots\, G_{M-1}(z)]^T \tag{9.39}$$

$$\tilde{e}(z) = \left[z^{-(M-1)}\, z^{-(M-2)}\, \ldots\, z^{-1}\, 1\right]^T = z^{-(M-1)}e(z) \tag{9.40}$$

$$\boldsymbol{R}(z) = \begin{bmatrix} R_{00}(z) & R_{01}(z) & R_{02}(z) & \cdots & R_{0,M-1}(z) \\ R_{10}(z) & R_{11}(z) & R_{12}(z) & \cdots & R_{1,M-1}(z) \\ R_{20}(z) & R_{21}(z) & R_{22}(z) & \cdots & R_{2,M-1}(z) \\ \vdots & \vdots & \vdots & \cdots & \vdots \\ R_{M-1,0} & R_{M-1,1}(z) & R_{M-1,2}(z) & \cdots & R_{M-1,M-1}(z) \end{bmatrix}. \tag{9.41}$$

The matrix $\boldsymbol{R}(z)$ is called the Type 2 polyphase component matrix for the synthesis bank. The corresponding Type 2 polyphase representation of the synthesis filter bank is shown in Fig. 9.8.

Using these two representations in the M-channel QMF bank of Fig. 9.6, an equivalent representation of Fig. 9.6 may be obtained and this is shown in Fig. 9.9.

The transfer matrix $\boldsymbol{E}(z^M)$ can be moved past the decimators by replacing z^M with z using the cascade equivalence of Fig. 8.39. Similarly, $\boldsymbol{R}(z^M)$ can be moved past the interpolators using the equivalence of Fig. 8.40. This results in the equivalent representation shown in Fig. 9.10.

If the two matrices $\boldsymbol{E}(z)$ and $\boldsymbol{R}(z)$ satisfy the condition

$$\boldsymbol{R}(z)\boldsymbol{E}(z) = \boldsymbol{I} \tag{9.42}$$

where $\boldsymbol{I}$ is an $M \times M$ identity matrix, then the structure of Fig. 9.10 reduces to that shown in Fig. 9.11.

Comparing Figs. 9.11 and 9.6, we see that the QMF bank of Fig. 9.11 can be considered as a special case of an M-channel QMF bank if we set

$$H_k(z) = z^{-k}, \quad G_k(z) = z^{-(M-1-k)}, \quad 0 \le k \le M - 1. \tag{9.43}$$

Substituting the above in Eq. (9.30), we get

$$a_l(z) = \frac{1}{M} \sum_{k=0}^{M-1} z^{-k} e^{j2\pi lk/M} z^{-(M-1-k)} = z^{-(M-1)} \left(\frac{1}{M} \sum_{k=0}^{M-1} e^{j2\pi lk/M} \right). \tag{9.44}$$

Since

$$\sum_{k=0}^{m-1} e^{j2\pi kl/M} = \frac{1 - e^{j\pi kl}}{1 - e^{j\pi kl/M}} = \begin{cases} \frac{1}{M} & l = 0 \\ 0 & l > 0 \end{cases}$$

it follows that $a_0(z) = z^{-(M-1)}$ and $a_l(z) = 0$ for $l \ne 0$. Hence, from Eq. (9.31), we see that $T(z) = z^{-(M-1)}$. Hence, the structure of Fig. 9.10 is a perfect reconstruction M-channel QMF bank, if the condition given by Eq. (9.42) is satisfied, and may be realized by the structure of Fig. 9.11.

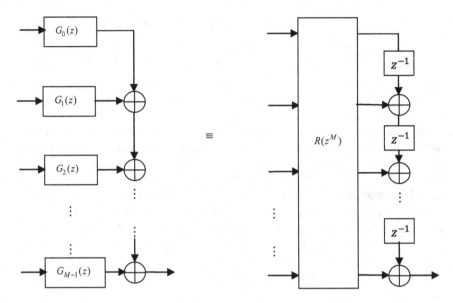

Fig. 9.8 Type 2 polyphase representation of a synthesis filter bank

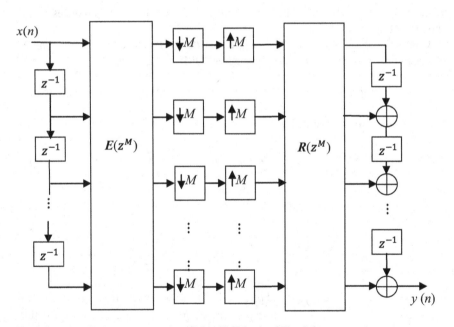

Fig. 9.9 An equivalent representation of the QMF bank of Fig. 9.6

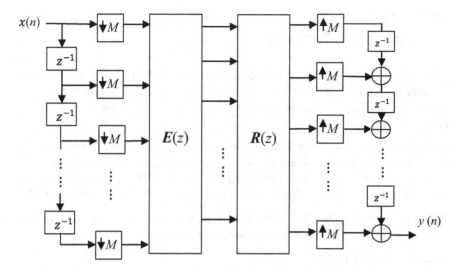

Fig. 9.10 Equivalent polyphase representation of analysis and synthesis filters for M-channel QMF bank

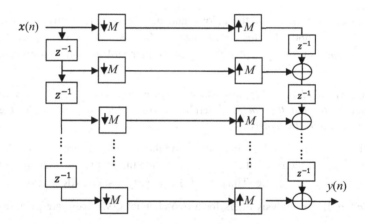

Fig. 9.11 A perfect reconstruction M-channel QMF bank

The most general condition for perfect reconstruction is given [1, 2] by

$$\boldsymbol{R}(z)\boldsymbol{E}(z) = cz^{-m_0}\boldsymbol{I} \tag{9.45}$$

or

$$\boldsymbol{R}(z) = cz^{-m_0}\boldsymbol{E}^{-1}(z). \tag{9.46}$$

9.3.5   Conditions for Existence of FIR Analysis/Synthesis Filters for Perfect Reconstruction

Definition 1 An $M \times M$ polyphase component matrix $E(z)$ is non-singular if its determinant is not equal to zero; i.e., the determinant can be a scalar (constant) or a polynomial in z [3]. It is well known that if $E(z)$ is non-singular, then the inverse of $E(z)$ exists.

Definition 2 A non-singular $M \times M$ polyphase component matrix $E(z)$ is said to be unimodular if its determinant is a scalar (constant), i.e., not a function of z [3].

With these definitions, it can be stated that 'an $M \times M$ polyphase component matrix $E(z)$ has an $M \times M$ FIR inverse if $E(z)$ is non-singular and unimodular.'

Lemma 1 *If an $M \times M$ FIR polyphase component matrix $E(z)$ is unimodular, then its inverse is also an FIR polyphase component matrix of the same order.*

Proof The inverse of $E(z)$ is given by

$$E^{-1}(z) = \frac{\mathrm{adj}(E(z))}{\det(E(z))} \tag{9.47}$$

where adj and det stand for the adjoint and determinant, respectively. Since the matrix $E(z)$ is unimodular, the determinant of $E(z)$ is a scalar (constant). Hence, $E^{-1}(z)$ is clearly an FIR polyphase component matrix of the same order as that of $E(z)$.

Lemma 2 *If an $M \times M$ FIR polyphase component matrix $E(z)$ is non-singular, non-unimodular, and the roots of $\det(E(z))$ are inside the unit circle, then its inverse is a stable IIR matrix.*

Proof If $E(z)$ is non-singular and non-unimodular, then the determinant of $E(z)$ is a polynomial in z. If the roots of $\det(E(z))$ are inside the unit circle, then the poles of $E^{-1}(z)$ are inside unit circle. Thus, $E^{-1}(z)$ is clearly a stable IIR matrix.

Example 9.6 If the analysis filters for a two-channel QMF bank are given by

$$H_0(z) = 1 + z^{-1} + z^{-2} - z^{-3}, \; H_1(z) = 1 + z^{-1} - z^{-2} + z^{-3}$$

find the corresponding synthesis filters for a perfect reconstruction system.

Solution

$$H_0(z) = 1 + z^{-1} + z^{-2} - z^{-3} = 1 + z^{-2} + z^{-1}(1 - z^{-2}) = E_{00}(z^2) + z^{-1}E_{01}(z^2)$$
$$H_1(z) = 1 + z^{-1} - z^{-2} + z^{-3} = 1 - z^{-2} + z^{-1}(1 + z^{-2}) = E_{10}(z^2) + z^{-1}E_{11}(z^2)$$
$$\boldsymbol{h}(z) = \boldsymbol{E}(z^2)\boldsymbol{e}(z)$$

where

$$h(z) = [H_0(z)\, H_1(z)]^T, \quad e(z) = \begin{bmatrix} 1 & z^{-1} \end{bmatrix}^T$$

$$E(z) = \begin{bmatrix} 1+z^{-1} & 1-z^{-1} \\ 1-z^{-1} & 1+z^{-1} \end{bmatrix}$$

For perfect reconstruction, we should have from Eq. (9.46)

$$R(z) = c z^{-m_0} E^{-1}(z)$$

Now,

$$E^{-1}(z) = \frac{1}{4z^{-1}} \begin{bmatrix} 1+z^{-1} & -1+z^{-1} \\ -1+z^{-1} & 1+z^{-1} \end{bmatrix}$$

Hence,

$$R(z) = c z^{-m_0} E^{-1}(z) = \frac{c z^{-m_0}}{4z^{-1}} \begin{bmatrix} 1+z^{-1} & -1+z^{-1} \\ -1+z^{-1} & 1+z^{-1} \end{bmatrix}$$

so that the perfect reconstruction condition holds. Choosing $c = 4$ and $m_0 = 1$, this becomes

$$R(z) = \begin{bmatrix} 1+z^{-1} & -1+z^{-1} \\ -1+z^{-1} & 1+z^{-1} \end{bmatrix}$$

Now, the synthesis filters can be determined by using Eqs. (9.38a) and (9.38b)

$$g^T(z) = \tilde{e}^T(z) R(z^M)$$

where

$$g(z) = [G_0(z)\, G_1(z)]^T$$

$$\tilde{e}(z) = \begin{bmatrix} z^{-1} & 1 \end{bmatrix}^T$$

Thus,

$$[G_0(z)\, G_1(z)] = \begin{bmatrix} z^{-1} & 1 \end{bmatrix} \begin{bmatrix} 1+z^{-2} & -1+z^{-2} \\ -1+z^{-2} & 1+z^{-2} \end{bmatrix}$$

Thus, the synthesis filters are

$$G_0(z) = -1 + z^{-1} + z^{-2} + z^{-3}; \quad G_1(z) = 1 - z^{-1} + z^{-2} + z^{-3}$$

Example 9.7 The analysis filters of a four-channel perfect reconstruction QMF bank are given by

$$\begin{bmatrix} H_0(z) \\ H_1(z) \\ H_2(z) \\ H_3(z) \end{bmatrix} = \begin{bmatrix} 1 & 2 & 3 & 4 \\ 3 & 2 & 1 & 5 \\ 2 & 1 & 4 & 3 \\ 4 & 2 & 3 & 1 \end{bmatrix} \begin{bmatrix} 1 \\ z^{-1} \\ z^{-2} \\ z^{-3} \end{bmatrix}.$$

Determine the synthesis filters of the perfect reconstruction system with an input–output relation $y(n) = 4x(n - 3)$.

Solution Taking the z-transform of the input–output relation, we have $Y(z) = 4z^{-3} X(z)$

$$E(z^4) = \begin{bmatrix} 1 & 2 & 3 & 4 \\ 3 & 2 & 1 & 5 \\ 2 & 1 & 4 & 3 \\ 4 & 2 & 3 & 1 \end{bmatrix}$$

Hence,

$$R(z^4) = 4 E^{-1}(z^4)$$

Thus,

$$R(z^4) = \begin{bmatrix} -1.6 & 0.8 & 0.64 & 0.48 \\ 2.93 & -0.8 & -3.04 & 1.39 \\ 0.267 & -0.8 & 0.96 & 0.05 \\ -0.267 & 0.8 & 0.64 & -0.85 \end{bmatrix}$$

Using Eqs. (9.38a), (9.38b), (9.39), and (9.40), the synthesis filters are obtained as

$$[G_0(z) \ G_1(z) \ G_2(z) \ G_3(z)] = [z^{-3} \ z^{-2} \ z^{-1} \ 1] \begin{bmatrix} -1.6 & 0.8 & 0.64 & 0.48 \\ 2.93 & -0.8 & -3.04 & 1.39 \\ 0.267 & -0.8 & 0.96 & 0.05 \\ -0.267 & 0.8 & 0.64 & -0.85 \end{bmatrix}$$

Hence, the synthesis filters are

$$G_0(z) = -0.267 + 0.267z^{-1} + 2.93z^{-2} - 1.6z^{-3}$$
$$G_1(z) = 0.8 - 0.8z^{-1} - 0.8z^{-2} + 0.8z^{-3}$$
$$G_2(z) = 0.64 - 0.96z^{-1} - 3.04z^{-2} + 0.64z^{-3}$$
$$G_3(z) = -0.85 - 0.05z^{-1} - 1.39z^{-2} + 0.48z^{-3}$$

Example 9.8 Consider a two-channel QMF bank with the analysis filters given by

$$H_0(z) = 2 + 6z^{-1} + z^{-2} + 5z^{-3} + z^{-5}, \; H_1(z) = H_0(-z);$$

(i) Is it possible to construct FIR synthesis filters for perfect reconstruction? If so find them.
(ii) If not, find stable IIR synthesis filters for perfect reconstruction.

Solution

(i) For perfect reconstruction, from Eq. (9.46) we have

$$R(z) = cz^{-m_0} E^{-1}(z)$$

$$E(z) = \begin{bmatrix} 2 + z^{-1} & (6 + 5z^{-1} + z^{-2}) \\ 2 + z^{-1} & -(6 + 5z^{-1} + z^{-2}) \end{bmatrix}$$

The determinant of $E(z) = -2(2 + z^{-1})(6 + 5z^{-1} + z^{-2})$. Since the determinant is non-unimodular, it is not possible to find FIR synthesis filters for perfect reconstruction.

(ii)

$$E^{-1}(z) = \frac{1}{-(2 + z^{-1})(6 + 5z^{-1} + z^{-2})} \begin{bmatrix} -(6 + 5z^{-1} + z^{-2}) & -(6 + 5z^{-1} + z^{-2}) \\ -(2 + z^{-1}) & (2 + z^{-1}) \end{bmatrix}$$

$$R(z) = E^{-1}(z) = \frac{1}{2} \begin{bmatrix} \dfrac{1}{2 + z^{-1}} & \dfrac{1}{2 + z^{-1}} \\ \dfrac{1}{6 + 5z^{-1} + z^{-2}} & \dfrac{-1}{6 + 5z^{-1} + z^{-2}} \end{bmatrix}$$

Now, the synthesis filters can be determined by using Eqs. (9.38a) and (9.38b):

$$[G_0(z) \; G_1(z)] = [z^{-1} \; 1] \frac{1}{2} \begin{bmatrix} \dfrac{1}{2 + z^{-1}} & \dfrac{1}{2 + z^{-1}} \\ \dfrac{1}{6 + 5z^{-1} + z^{-2}} & \dfrac{-1}{6 + 5z^{-1} + z^{-2}} \end{bmatrix}.$$

The stable IIR synthesis filters for perfect reconstruction are

$$G_0(z) = \frac{1}{2} \left[\frac{2 + 7z^{-1} + 5z^{-2} + z^{-3}}{(2 + z^{-1})(6 + 5z^{-1} + z^{-2})} \right]$$

$$G_1(z) = \frac{1}{2} \left[\frac{-2 + 5z^{-1} + 5z^{-2} + z^{-3}}{(2 + z^{-1})(6 + 5z^{-1} + z^{-2})} \right].$$

Example 9.9 Consider the four branch QMF bank structure with Type 1 polyphase component matrix given by

$$E(z) = \begin{bmatrix} 1 & 2 & 3 & 2 \\ 2 & 13 & 9 & 7 \\ 3 & 9 & 11 & 10 \\ 2 & 7 & 10 & 15 \end{bmatrix}.$$

Determine the Typ 2 polyphase component matrix $R(z)$ such that the four-channel QMF structure is a perfect reconstruction system with an input–output relation $y(n) = 3x(n - 3)$.

Solution Applying z-transform to both sides of the relation $y(n) = 3x(n - 3)$, we have

$$Y(z) = 3z^{-3}X(z).$$

From the above equation, $c = 3$ and $m_0 = 3$. Hence, the Type 2 polyphase component matrix becomes

$$R(z) = 3z^{-3}E^{-1}(z)$$

$$E^{-1}(z) = \begin{bmatrix} 38.9999 & 4.3333 & -19.3333 & 5.6666 \\ 4.3333 & 0.6666 & -2.3333 & 0.6666 \\ -19.3333 & -2.3333 & 9.9999 & -3.0000 \\ 5.6666 & 0.6666 & -2.9999 & 0.9999 \end{bmatrix}$$

Hence,

$$R(z) = 3z^{-3} \begin{bmatrix} 38.9999 & 4.3333 & -19.3333 & 5.6666 \\ 4.3333 & 0.6666 & -2.3333 & 0.6666 \\ -19.3333 & -2.3333 & 9.9999 & -3.0000 \\ 5.6666 & 0.6666 & -2.9999 & 0.9999 \end{bmatrix}$$

9.4 Methods for Designing Linear Phase FIR PR QMF Banks

9.4.1 Johnston Method

Let $H_0(z)$ be a real coefficient transfer function of a linear phase FIR filter of order N given by

$$H_0(z) = \sum_{n=0}^{N} h_0(n) z^{-n}. \qquad (9.48)$$

Since $H_0(z)$ has to be a lowpass filter, its impulse response coefficients must satisfy the symmetry condition $h_0(n) = h_0(N-n)$. Hence, $H_0(e^{j\omega}) = e^{-j\omega N/2} |H_0(\omega)|$, where $|H_0(\omega)|$ is a real function of ω. By making use of Eq. (9.26) and the fact that $|H_0(e^{j\omega})|$ is an even function of ω, we get

$$T(e^{j\omega}) = \frac{e^{-j\omega N/2}}{2} \left(|H_0(e^{j\omega})|^2 - (-1)^N |H_0(e^{j(\pi-\omega)})|^2 \right). \qquad (9.49)$$

If N is even, then $T(e^{j\omega}) = 0$ at $\omega = \pi/2$ resulting in severe amplitude distortion at the output of the bank. So N must be chosen to be odd. Since $|H_1(e^{j\omega})|^2 = |H_0(e^{j(\pi-\omega)})|^2$, for odd N, Eq. (9.49) becomes

$$T(e^{j\omega}) = \frac{e^{-jN\omega}}{2} \left(|H_0(e^{j\omega})^2| + |H_1(e^{j\omega})^2| \right). \qquad (9.50)$$

For odd N, if $|H_0(e^{j\omega})^2| + |H_1(e^{j\omega})^2| = 1$, the above equation satisfies the perfect reconstruction condition as given in Eq. (9.28). Therefore, to minimize the amplitude distortion, an optimization method is required to iteratively adjust the filter coefficients $h_0(n)$ of $H_0(z)$ such that the constraint

$$|H_0(e^{j\omega})|^2 + |H_1(e^{j\omega})|^2 \cong 1 \qquad (9.51)$$

is satisfied for all values of ω [4]. Toward this end, Johnston has minimized the following objective function, designed a large class of linear phase FIR lowpass filters $H_0(z)$ with a wide range, and tabulated the impulse response coefficients in [4]. These tables can also be found in [5].

$$\emptyset = \alpha \int_{\omega_s}^{\pi} |H_0(e^{j\omega})|^2 d\omega + (1-\alpha) \int_0^{\pi} (1 - |H_0(e^{j\omega})|^2 - |H_1(e^{j\omega})|^2)^2 d\omega \qquad (9.52)$$

where $0 < \alpha < 1$ and $\omega_s = \left(\frac{\pi}{2}\right) + \varepsilon$ for small $\varepsilon > 0$.

Example 9.10 Verify the performance of the analysis filters for Johnston's 16A filter.

Solution The MATLAB Program 9.2 given below is used to verify the performance of the analysis filters. The input to the program is first half of the filter coefficients as tabulated in [5]. The program determines the second half by using the MATLAB function fliplr. The gain responses of the analysis filters for Johnston's 16A filter are shown in Fig. 9.12. The amplitude distortion function

$$\left|H_0\left(e^{j\omega}\right)\right|^2 + \left|H_1\left(e^{j\omega}\right)\right|^2$$

in dB is shown in Fig. 9.13. From Figs. 9.12 and 9.13, it can be seen that the stop band edge frequency of the 16A filter is 0.78π corresponding to a transition bandwidth $(\omega_s - \frac{\pi}{2})/\pi = 0.14$. The minimum stop band attenuation is 60 dB. The amplitude distortion is very close to zero for all ω, with a peak value of 0.005 dB.

Program 9.2: Frequency Response of Johnston's 16A QMF

```
clear;clf;
H0=[.0010501670  -.0050545260  -.0025897560  .027641400  -.0096663760  -
.090392230 .097798170 0.48102840 ];% Johnston's 16A filter impulse response
coefficients;
H0=[H0 fliplr(H0)];
%Generate the complementary highpass filter
L=length(H0);
H1= [ ];
for k=1:L

H1(k)=((-1)^k)*H0(k);
```

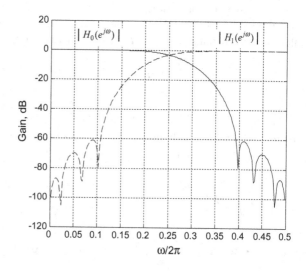

Fig. 9.12 Magnitude response of Johnston's 16A filter

Fig. 9.13 Reconstruction error in dB for Johnston's 16A filter

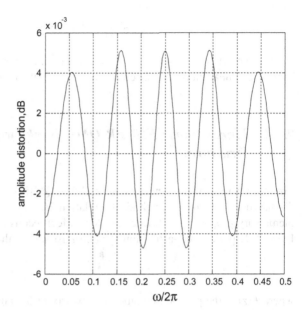

end
%Compute the gain responses of the two filters
[H0z, w]=freqz(H0, 1, 256);
h0=abs(H0z);
g0=20*log10(h0);
[H1z, w]=freqz(H1, 1, 256);
h1=abs(H1z);
g1=20*log10(h1);
figure(1),plot(w/(2*pi), g0, '-', w/(2*pi), g1, '–');
grid
xlabel('\omega/2\pi'); ylabel('Gain, dB')
%compute the sum of the squared-magnitude responses
for i=1:256

sum(i)=(h0(i)*h0(i))+(h1(i)*h1(i));

end
d=10*log10(sum);
plot the ampitude distortion
figure(2),plot(w/(2*pi),d);grid;
xlabel('\omega/2\pi');ylabel('amplitude distortion,dB');

In the Johnston method of design, the analysis filters are power complementary, since the objective function is minimized with the power complementary condition given below as the constraint.

$$\left|H_0(e^{j\omega})\right|^2 + \left|H_1(e^{j\omega})\right|^2 = 1 \qquad (9.53)$$

However, it is not possible to realize a perfect reconstruction two-channel QMF bank with linear phase power complementary analysis filters [6].

9.4.2 Linear Phase FIR PR QMF Banks with Lattice Structures

The design method for linear phase FIR PR QMF banks with lattice structure is discussed in [2]. In this method, the relation $H_1(z) = H_0(-z)$ or the power complementary property is not necessary for perfect reconstruction in FIR QMF banks. Every FIR perfect reconstruction system must satisfy the following condition

$$\det E(z) = \alpha z^{-k}, \quad \alpha \neq 0, k = \text{integer} \qquad (9.54)$$

where $E(z)$ is the polyphase component matrix of the analysis filters. The above is really a necessary and sufficient condition.

Apart from perfect reconstruction, the impulse response coefficients satisfy the conditions

$$h_0(n) = h_0(N - n), h_1(n) = -h_1(N - n). \qquad (9.55)$$

so that the filter has linear phase.

An objective function of the following form that reflects the passbands and stopbands of both the filters is defined

$$\phi = \int_0^{\omega_p} \left[1 - \left|H_0(e^{j\omega})\right|\right]^2 d\omega + \int_{\omega_s}^{\pi} \left|H_0(e^{j\omega})\right|^2 d\omega + \int_{\omega_s}^{\pi} \left[1 - \left|H_1(e^{j\omega})\right|\right]^2 d\omega$$

$$+ \int_0^{\omega_p} \left|H_1(e^{j\omega})\right|^2 d\omega. \qquad (9.56)$$

An optimization method to minimize the above objective function and the filter coefficients is given in [7].

Example 9.11 Verify the performance of the lattice PR QMF banks.

Solution The coefficients $h_0(n)$ and $h_1(n)$ of the lattice filter [7] as well as $h_0(n)$ of Johnston's 64D filter [7] given in Table 9.1 are used in the MATLAB Program 9.2 to verify the performance of the two filters. The gain responses of the analysis filters

Table 9.1 Lattice filter and Johnston's 64D filter coefficients

PR lattice [7]			Johnston's 64D [5]
n	Filter coefficients $h_0(n)$	Filter coefficients $h_1(n)$	Filter coefficients $h_0(n)$
0	−2.8047649e−008	2.7701557e−008	3.5961890e−005
1	4.6974271e−009	−4.6394635e−009	−1.1235150e−004
2	−9.0320484e−006	8.8828616e−006	−1.1045870e−004
3	1.5395771e−006	−1.5142595e−006	2.7902770e−004
4	−3.8484265e−004	3.6797322e−004	2.2984380e−004
5	7.3072834e−005	−7.0105005e−005	−5.9535630e−004
6	−5.4844134e−004	3.3979746e−005	−3.8236310e−004
7	4.4822759e−004	−3.4610132e−004	1.1382600e−003
8	−3.1420371e−004	−2.9314606e−006	5.3085390e−004
9	6.0057016e−005	4.6300165e−004	−1.9861770e−003
10	9.0093327e−004	−9.6019270e−004	−6.2437240e−004
11	1.6949632e−004	−5.3867082e−004	3.2358770e−003
12	−1.7113556e−003	2.6665204e−003	5.7431590e−004
13	−8.8586615e−004	1.3763566e−003	−4.9891470e−003
14	2.1821453e−003	−3.9203747e−003	−2.5847670e−004
15	1.8133626e−003	−3.2204698e−003	7.3671710e−003
16	−2.5292917e−003	5.2062396e−003	−4.8579350e−004
17	−4.1596795e−003	6.1298979e−003	−1.0506890e−002
18	3.5887675e−003	−6.2739943e−004	1.8947140e−003
19	7.6713480e−003	−1.0972496e−002	1.4593960e−002
20	−5.7266130e−003	6.2975035e−003	−4.3136740e−003
21	−1.2745794e−002	1.8004225e−002	−1.9943650e−002
22	9.5185739e−003	−4.2379771e−003	8.2875600e−003
23	2.0342217e−002	−2.7168914e−002	2.7166550e−002
24	−1.6072096e−002	−1.0169911e−003	−1.4853970e−002
25	−3.1588257e−002	3.9969784e−002	−3.7649730e−002
26	2.7807655e−002	1.2025400e−002	2.6447000e−002
27	5.0150999e−002	−6.0718905e−002	5.5432450e−002
28	−5.2720604e−002	−3.6911412e−002	−5.0954870e−002
29	−9.3506916e−002	1.0590613e−001	−9.7790960e−002
30	1.4064635e−001	1.2541815e−001	1.3823630e−001
31	4.5677058e−001	−4.7085952e−001	4.6009810e−001

of the PR lattice pair and Johnston's 64D pair are shown in Figs. 9.14 and 9.15, respectively. The amplitude distortion function in dB is shown in Fig. 9.16. In this design, the transition bandwidth is $\frac{2\omega_s - \pi}{4\pi} = 0.043$. The Johnston's 64D filter also has an order 63 and the same transition bandwidth. For comparison, from Figs. 9.14 and 9.15, we see that Johnston's 64D filter offers a minimum stopband attenuation of 65 dB, whereas the PR lattice offers stopband attenuation of 42.5 dB.

The plots of $|H_0(e^{j\omega})|^2 + |H_1(e^{j\omega})|^2$ for the lattice pair and 64D Johnston's pair are shown in Fig. 9.16. From Fig. 9.16, it is observed that $|H_0(e^{j\omega})|^2 + |H_1(e^{j\omega})|^2$ is very flat for Johnston's design but not for the linear phase PR pair. In spite of this, the linear phase lattice structure enjoys perfect reconstruction because the quantity

Fig. 9.14 Gain response of
the lattice PR QMF filter bank

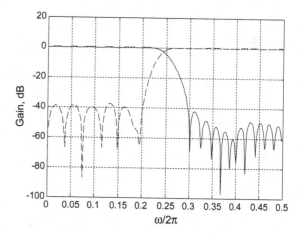

Fig. 9.15 Gain response of
Johnston's 64D QMF filter
bank

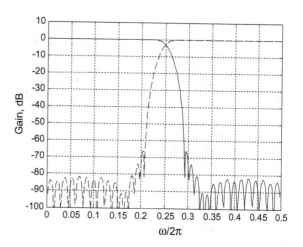

Fig. 9.16 Amplitude
distortion function of lattice
PR and Johnson's 64D QMF
filter banks in dB

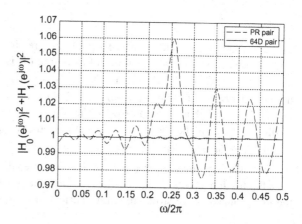

$|H_0(e^{j\omega})|^2 + |H_1(e^{j\omega})|^2$ is not proportional to the amplitude distortion unlike in Johnston's design.

9.4.3 Design of Perfect Reconstruction Two-Channel FIR Filter Bank Using MATLAB

The MATLAB function **firpr2chfb** can be used to design four FIR filters for the analysis (*H0* and *H1*) and synthesis (*G0* and *G1*) sections of a two-channel perfect reconstruction filter bank. The design corresponds to power-symmetric filter banks. The various forms of the function firpr2chfb are:

$$[H0, H1, G0, G1] = \text{firpr2chfb}(N, \text{fp})$$
$$[H0, H1, G0, G1] = \text{firpr2chfb}(N, \text{dev}, `\text{dev}')$$
$$[H0, H1, G0, G1] = \text{firpr2chfb}(`\text{minorder}', \text{fp}, \text{dev})$$

The basic form [*H0*, *H1*, *G0*, *G1*] = firpr2chfb(N,fp) is used to design $H_0(z), H_1(z), G_0(z)$, and $G_1(z)$, N is the order of all four filters, and it must be an odd integer. fp is the passband edge for the lowpass filters $H_0(z)$ and $G_0(z)$; it must be less than 0.5. $H_1(z)$ and $G_1(z)$ are highpass filters with passband edge given by 1-fp.

The option [*H0*, *H1*, *G0*, *G1*] = firpr2chfb(N,dev,'dev') designs the four filters such that the maximum stopband ripple of $H_0(z)$ is given by the scalar Dev. The stopband ripple of $H_1(z)$ will also be given by dev, while the maximum stopband ripple for both $G_0(z)$ and $G_1(z)$ will be 2*dev.

The other option [*H0*, *H1*, *G0*, *G1*] = firpr2chfb('minorder',fp,dev) designs the four filters such that $H_0(z)$ meets the passband edge fp and the stopband ripple dev with minimum order.

The squared magnitude responses of $H_0(z)$ and $H_1(z)$ are found using the MATLAB command fvtool(H0, 1, H1, 1, G0, 1, G1, 1).

The power complementary condition given in Eq. (9.53) for perfect reconstruction can be verified by the following MATLAB stem function.

```
stem(1/2*conv(G0,H0)+1/2*conv(G1,H1))

n=0:N;
stem(1/2*conv((-1).^n.*H0,G0)+1/2*conv((-1).^n.*H1,G1))
stem(1/2*conv((-1).^n.*G0,H0)+1/2*conv((-1).^n.*G1,H1))
stem(1/2*conv((-1).^n.*G0,(-1).^n.*H0)+1/2*conv((-1).^n.*G1,(-1).^n.*H1))
stem(conv((-1).^n.*H1,H0)-conv((-1).^n.*H0,H1))
```

The following example illustrates the design of perfect reconstruction two-channel QMF bank using MATLAB.

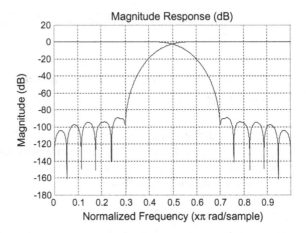

Fig. 9.17 Magnitude-squared responses of the analysis filters $H_0(z)$ and $H_1(z)$ in the perfect reconstruction QMF bank

Example 9.12 Design a linear phase two-channel QMF bank with the filters order $N = 31$, normalized passband edge frequency $f_p = 0.3$.

Solution The following MATLAB commands are used for design

$$[H0, H1, G0, G1] = \text{firpr2chfb}(31, 0.3);$$
$$\text{fvtool}(H0, 1, H1, 1);$$

The magnitude-squared responses of the analysis filters $H_0(z)$ and $H_1(z)$ in the perfect reconstruction filter bank are shown in Fig. 9.17.

9.5   Tree-Structured Filter Banks

9.5.1   Maximally Decimated Filter Banks

Consider a general M-channel non-uniform filter bank as shown in Fig. 9.18. If the integers M_i are such that

$$\sum_{i=0}^{M-1} \frac{1}{M_i} = 1 \tag{9.57}$$

then the system is said to be maximally decimated [2]. A maximally decimated filter bank with equal passband widths is shown in Fig. 9.19.

9.5.2 Tree-Structured Filter Banks with Equal Passband Width

A multiband analysis–synthesis filter bank can be generated by iterating a two-channel QMF bank. If the two-channel QMF bank satisfies perfect reconstruction condition, then the generated multiband structure also has the perfect reconstruction property [1]. A four-channel QMF bank as shown in Fig. 9.20 can be generated by inserting a two-channel maximally decimated QMF bank (Fig. 9.4) in each channel of a two-channel maximally decimated QMF bank between the down-sampler and the upsampler.

Figure 9.20 exhibits a tree structure in which a signal is split into two subbands and decimated. After decimation, each subband is again split into two subbands and decimated. This is achieved by use of two-channel analysis banks. By use of two-channel synthesis banks, the subbands are then recombined, two at a time. The overall system is often referred to as maximally decimated tree-structured filter bank. The upper two-channel QMF bank and the lower two-channel QMF bank at the second level in Fig. 9.20 may not be identical. In such a case, to compensate for

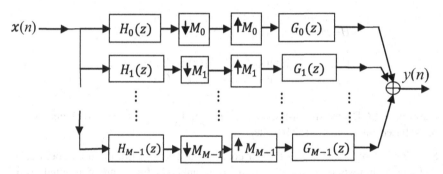

Fig. 9.18 M-channel structure for QMF bank

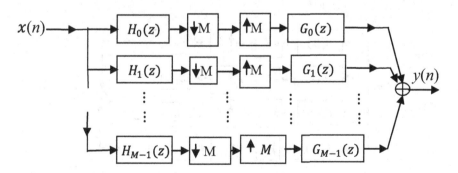

Fig. 9.19 Maximally decimated filter bank

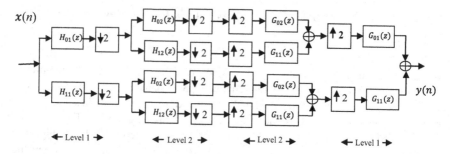

Fig. 9.20 A four-channel tree-structured QMF bank with equal passband widths

the unequal gains and unequal delays, it is necessary to insert appropriate scale factors and delays at proper places to ensure perfect reconstruction of the overall system.

Figure 9.21 shows an equivalent non-tree structure for the four-channel QMF system of Fig. 9.20. The analysis and synthesis filters in non-tree equivalent structure of Fig. 9.21 are related to the analysis and synthesis filters of tree-structured filter bank of Fig. 9.20 as follows:

$$
\begin{array}{ll}
H_0(z) = H_{01}(z)H_{02}(z^2) & H_1(z) = H_{01}(z)H_{12}(z^2) \\
H_2(z) = H_{11}(z)H_{02}(z^2) & H_3(z) = H_{11}(z)H_{12}(z^2)
\end{array}
\tag{9.58}
$$

$$
\begin{array}{ll}
G_0(z) = G_{01}(z)G_{02}(z^2) & G_1(z) = G_{01}(z)G_{12}(z^2) \\
G_2(z) = G_{11}(z)G_{02}(z^2) & G_3(z) = G_{11}(z)G_{12}(z^2)
\end{array}
\tag{9.59}
$$

Example 9.13 Design a four-channel QMF bank by iterating the two-channel QMF bank based on Johnston's 24B filter.

Solution From Johnston's 24B filter coefficients given in [5], the filter coefficients and the gain response of each of the four analysis filters are computed using Eq. (9.58) and Program 9.3 listed below.

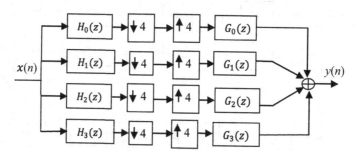

Fig. 9.21 Equivalent four-channel tree-structured QMF bank of Fig. 9.20

Program 9.3 Frequency responses of a tree-structured Johnston's 24B QMF filters with equal passband widths

```
clear;clf;
%Type in prototype lowpass filter coefficients
HL=[.00038330960 -.0013929110 -.0013738610 .0064858790 .0014464610 -
.019019930 .0038915220 .044239760 -.025615330 -.098297830 .11603550
.47312890];
HL=[HL fliplr(HL)];
%Generate the complementary highpass filter
L=length(HL);
for k=1:L

HH(k)=((-1)^k)*HL(k);

end
%Determine the coefficients of the four filters
H10=zeros(1, 2*length(HL));
H10([1:2:length(H10)])=HL;
H11=zeros(1,2*length(HH));
H11([1:2:length(H11)])=HH;
C0=conv(HL, H10);
C1=conv(HL, H11);
C2=conv(HH, H10);
C3=conv(HH, H11);
%Determine the frequency responses
[H0z, w]=freqz(C0, 1, 256);
h0=abs(H0z);
M0=20*log10(h0);
[H1z, w]=freqz(C1, 1, 256);
h1=abs(H1z);
M1=20*log10(h1);
[H2z, w]=freqz(C2, 1, 256);
h2=abs(H2z);
M2=20*log10(h2);
[H3z, w]=freqz(C3, 1, 256);
h3=abs(H3z);
M3=20*log10(h3);
plot(w/pi, M0, '-', w/pi, M1, '--', w/pi, M2, '-', w/pi, M3, '-');grid
xlabel('\omega/\pi'); ylabel('Gain, dB')
axis([0 1 -100 10]);
```

Figure 9.22 shows the gain response of each of the four analysis filters for Johnston's 24B QMF bank.

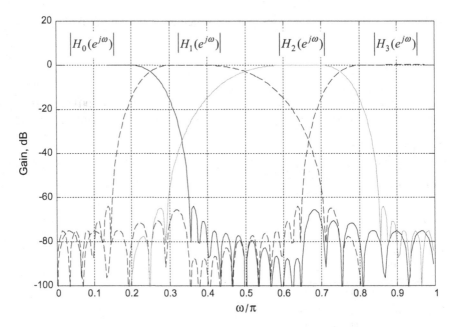

Fig. 9.22 Gain responses of the four analysis filters in tree-structured Johnston's 24B QMF bank with equal passband widths

9.5.3 Tree-Structured Filter Banks with Unequal Passband Widths

Consider a four-channel tree-structured QMF bank with equal passband widths as shown in Fig. 9.20. A five-channel maximally decimated tree-structured QMF bank can be generated by inserting another two-channel maximally decimated QMF bank in the top subband between the down-sampler and the upsampler. The resulting analysis filter bank of a five-channel filter bank is shown in Fig. 9.23.

An equivalent representation of the five-channel analysis filter bank of the QMF system of Fig. 9.23 is shown in Fig. 9.24. The analysis filters in the equivalent representation of Fig. 9.24 are related to the analysis filters of Fig. 9.23 as follows:

$$
\begin{aligned}
&H_0(z) = H_{01}(z)H_{02}(z^2)H_{03}(z^4) \quad H_1(z) = H_{01}(z)H_{02}(z^2)H_{13}(z^4) \\
&H_2(z) = H_{01}(z)H_{12}(z^2) \qquad\qquad\; H_3(z) = H_{11}(z)H_{02}(z^2) \\
&H_4(z) = H_{11}(z)H_{12}(z^2)
\end{aligned}
\tag{9.60}
$$

These structures belong to the non-uniform class of QMF banks due to unequal passband widths. The non-uniform filter banks are mostly used in speech and image coding.

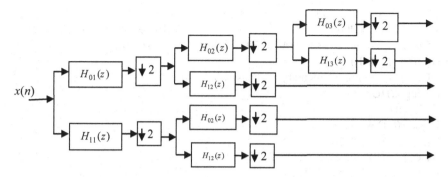

Fig. 9.23 Three-stage cascade realization of a five-channel analysis filter bank from the four-channel one of Fig. 9.20

Fig. 9.24 Equivalent representation of the five-channel analysis filter bank of Fig. 9.23

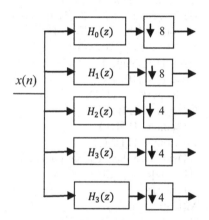

Example 9.14 Design a five-channel QMF bank by iterating the two-channel QMF bank based on Johnston's 32D filter.

Solution From Johnston's 32D filter coefficients given in [5], the filter coefficients and the gain response of each of the five analysis filters are computed using Eq. (9.60) and Program 9.4 listed below.

Program 9.4: Frequency response of non-uniform tree-structured Johnston's 32D QMF filters

clear;
clf;
%Type in prototype lowpass filter coeffients
%B1=input('Filter coefficients= ');

```
HL=[.0022451390  -.0039711520  -.0019696720  .0081819410  .00084268330  -
.014228990 .0020694700 .022704150 -.0079617310 -.034964400 .019472180
.054812130 -.044524230 -.099338590 .13297250 .46367410];
HL=[HL fliplr(HL)];
%Generate the complementary highpass filter
L=length(HL);
for k=1:L
HH(k)=((-1)^k)*HL(k);
end
%Determine the coefficients of the four filters
H10=zeros(1, 2*length(HL));
H10([1:2:length(H10)])=HL;
H11=zeros(1,2*length(HH));
H11([1:2:length(H11)])=HH;
H100=zeros(1,2*length(H10));
H100([1:2:length(H100)])=H10;
H101=zeros(1,2*length(H11));
H101([1:2:length(H101)])=H11;
C0=conv(H10,H100);
C1=conv(H10,H101);
C2=conv(HL,H11);
C3=conv(HH,H10);
C4=conv(HH,H11);
%Determine the frequency responses
[H0z, w]=freqz(C0, 1, 256);
h0=abs(H0z);
M0=20*log10(h0);
[H1z, w]=freqz(C1, 1, 256);
h1=abs(H1z);
M1=20*log10(h1);
[H2z, w]=freqz(C2, 1, 256);
h2=abs(H2z);
M2=20*log10(h2);
[H3z, w]=freqz(C3, 1, 256);
h3=abs(H3z);
M3=20*log10(h3);
[H4z, w]=freqz(C4, 1, 256);
h4=abs(H4z);
M4=20*log10(h4);
plot(w(1:128)/(2*pi), M0(1:128), '-k', w(1:128)/(2*pi), M1(1:128), '-k', w/(2*pi),
M2, '-k',w/(2*pi),M3,'-k',w/(2*pi),M4,'-k');
%plot(w/(pi), M0,'-');
grid
xlabel('\omega/2\pi'); ylabel('Gain, dB');
```

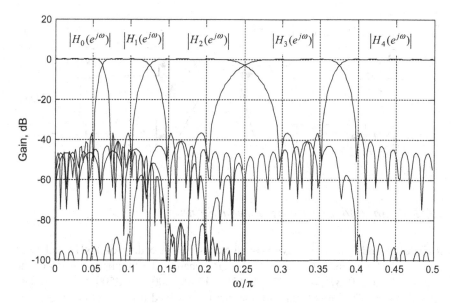

Fig. 9.25 Gain responses of a tree-structured Johnston's 32D QMF bank with unequal passband widths

Figure 9.25 shows the gain response of each of the five analysis filters for Johnston's 32D QMF bank.

9.6 Application Examples

9.6.1 Transmultiplexers

The time-division multiplex (TDM) and the frequency-division multiplexing (FDM) are two different telephone systems usually preferred for short-haul and long-haul communication, respectively. In digital telephone networks, it is necessary to translate signals between the TDM and FDM formats. This is achieved by the transmultiplexer shown in Fig. 9.26.

It consists of an N-channel synthesis filter bank at the input end followed by an N-channel analysis filter bank at the output end. In a typical TDM-to-FDM format translation, the digitized speech signals are interpolated by a factor of M, modulated by single-sideband modulation, digitally summed, and then converted into an FDM analog signal by D/A conversion. At the receiving end, the analog signal is converted into a digital signal by A/D conversion and passed through a bank of M single-sideband demodulators, whose outputs are then decimated, resulting in the low-frequency speech signals.

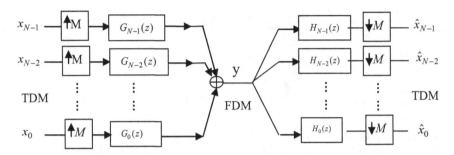

Fig. 9.26 Transmultiplexer system

9.6.2 Subband Coding of Speech Signals

Richardson and Jayant [8] have investigated subband coding of 7-kHz wideband audio at 56 kbits/s based on a five-band QMF bank. The frequency ranges of the five bands are 0–875 Hz, 875–1750 Hz, 1750–3500 Hz, 3500–5250 Hz, and 5250–7000 Hz. The five-band partition can be obtained using the three-stage cascade realization of a five-channel analysis filter bank shown in Fig. 9.23. The analysis and synthesis filter banks used in the subband encoding and decoding of speech signals can be represented as shown in Fig. 9.27.

9.6.3 Analog Voice Privacy System

Analog voice privacy systems are intended to communicate speech over standard analog telephone links while at the same time ensuring the voice privacy. Although the channel signal is analog, all of the signal processing is done digitally. Figure 9.28a, b illustrates a full-duplex voice privacy system's transmitter and receiver, respectively. The main idea here is to split the signal $s(n)$ into M subband

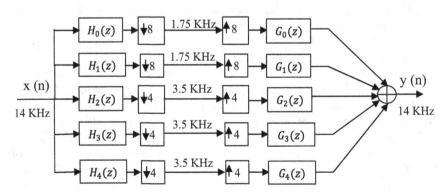

Fig. 9.27 Maximally decimated five-channel QMF filter bank with unequal passband widths

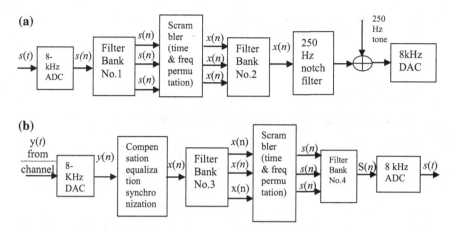

Fig. 9.28   a A full-duplex voice privacy system transmitter. **b** A full-duplex voice privacy system receiver

signals $s_i(n)$ and then divide each subband signal into segments in time domain. These segments are then permuted and recombined into a single encrypted signal $x(n)$, which can then be transmitted after D/A conversion. If there are three subbands and 18 time segments in each subband, then there are 54! possible permutations, which make it very difficult for someone who does not have the key for decryption to eavesdrop. At the receiver end, $x(n)$ is again split into subbands, and the time segments of the subbands are de-permuted to get $s_i(n)$, which can be interpolated and recombined through the synthesis filter banks. More details of the system can be found in [9].

9.7   Problems

1. Design a four-channel uniform DFT filter bank using the following polyphase components.

$$E_0(z) = 0.00163694 - 0.01121888z^{-1} + 0.06311487z^{-2} + 0.22088513z^{-3}$$
$$- 0.02725549z^{-4} + 0.00382693z^{-5}$$
$$E_1(z) = 0.00313959 - 0.025174873z^{-1} + 0.147532912z^{-2} + 0.147532912z^{-3}$$
$$- 0.025174873z^{-4} + 0.00313959z^{-5}$$
$$E_2(z) = 0.00382693 - 0.02725549z^{-1} + 0.22088513z^{-2} + 0.06311487z^{-3}$$
$$- 0.01121888z^{-4} + 0.00163694z^{-5}$$
$$E_3(z) = 0.25z^{-2}$$

2. Show that the two-channel QMF bank of Fig. 9.4 is a perfect reconstruction system for the following analysis and synthesis filters:

$$H_0(z) = 5 + 6z^{-1}, \ H_1(z) = -3 + 4z^{-1}, \ G_0(z) = 1.5 + 2z^{-1},$$

$$G_1(z) = 2.5 - 3z^{-1}.$$

3. If the analysis filters for a three-channel QMF bank are given by

$$H_0(z) = 5 + 2z^{-1} + z^{-3} + 2z^{-4} + z^{-5}$$
$$H_1(z) = 2 + z^{-1} + 2z^{-3} + 4z^{-4} + 2z^{-5}$$
$$H_2(z) = z^{-3} + 2z^{-4} + z^{-5}.$$

find the corresponding synthesis filters for the perfect reconstruction QMF system.

4. The analysis filters of a three-channel QMF filter bank are

$$H_0(z) = 1, \ H_1(z) = 2 + z^{-1} + z^{-5}, \ H_2(z) = 3 + z^{-1} + 2z^{-2}$$

   (i) Can you determine the FIR synthesis filters $G_0(z)$ and $G_1(z)$ so that the two-channel QMF bank is an alias-free and perfect reconstruction system. If so find them.
   (ii) If not, find the set of stable IIR filters for an alias-free and perfect reconstruction system.

5. The synthesis filters for a two-channel perfect reconstruction QMF bank are given by

$$\begin{bmatrix} G_0(z) \\ G_1(z) \end{bmatrix} = \begin{bmatrix} -1 + z^{-2} & 1 + z^{-2} \\ 1 + z^{-2} & -1 + z^{-2} \end{bmatrix} \begin{bmatrix} 1 \\ z^{-1} \end{bmatrix}.$$

Find the corresponding analysis filters $H_0(z)$ and $H_1(z)$.

6. The analysis filters of a two-channel QMF filter bank are

$$H_0(z) = 12 + 4z^{-1} + 10z^{-2} + 2z^{-3} + 2z^{-4}$$
$$H_1(z) = H_0(-z)$$

   (i) Can you determine the FIR synthesis filters $G_0(z)$ and $G_1(z)$ so that the two-channel QMF bank is an alias-free and perfect reconstruction system. If so find them.
   (ii) If not, find the set of stable IIR filters for an alias-free and perfect reconstruction system.

9.8 MATLAB Exercises

1. Verify the performance of the analysis filters for Johnston's 48D filter.
2. Design a two-channel uniform DFT filter bank using a linear phase FIR filter of length 23. Design the filter using the function firpm of MATLAB.
3. Design a three-channel QMF bank by iterating the two-channel QMF bank based on Johnston's 48D filter [4, 5]. Plot the gain responses of three analysis filters, $H_0(z)$, $H_1(z)$, and $H_2(z)$ on the same diagram. Comment on the results.
4. Design a four-channel QMF bank by iterating the two-channel QMF bank based on Johnston's 16B filter [4, 5]. Plot the gain responses of the four analysis filters $H_0(z)$, $H_1(z)$, $H_2(z)$, and $H_{23}(z)$ on the same diagram.
5. Compare the performance of a linear phase PR QMF lattice bank having a transition width of 0.172π and of length 64 with that of Johnston's 32D filter. Plot the gain responses of the two analysis filters $H_0(z)$ and $H_1(z)$.

References

1. S.K. Mitra, *Digital Signal Processing—A Computer Based Approach* (MC Graw-Hill Inc, NY, 2006)
2. P.P. Vaidyanathan, *Multirate Systems and Filter Banks* (Pearson Education, London, 2004)
3. T. Kailath, *Linear Systems* (Prentice-Hall, Englewood Cliffs, NJ, 1980)
4. J.D. Johnson, A filter family designed for use in quadrature mirror filter banks. In Proceedings of IEEE International Conference on Acoustics, Speech and Signal Processing, pp. 291–294, April 1980
5. R.E. Crochiere, L.R. Rabiner, *Multirate Digital Signal Processing* (Prentice-Hall, Englewood Cliffs, NJ, 1983)
6. P.P. Crochiere, On power-complementary FIR filter. IEEE Trans. Circuits Systems, CAS **32**, 1308–1310 (1985)
7. T.Q. Ngugen, P.P. Vaidyanathan, Two-channel perfect reconstruction FIR QMF structures which yield linear phase FIR analysis and synthesis filter. IEEE Trans. Acoust. Speech Signal Process. **ASSP-37**, 676–690 (1989)
8. E.B. Richardson, N.S. Jayant, Subband coding with adaptive prediction for 56 Kbits/s audio. IEEE Trans. Acoust. Speech Signal Process. **ASSP-34**, 691–696 (1986)
9. R.V. Cox, J.M. Tribolet, Analog voice privacy system using TFSP scrambling: full-duplex and half-duplex. Bell Syst. Tech. J. **62**, 47–61 (1983)

Chapter 10
Discrete Wavelet Transforms

Fourier transform has been extensively used in signal processing to analyze stationary signals. A serious drawback of the Fourier transform is that it cannot reflect the time evolution of the frequency. Further, the Fourier basis functions are localized in frequency but not in time. Small frequency changes in the Fourier transform will produce changes everywhere in the time domain. This gives rise to the need for a time and frequency localization method. This can be achieved by short-time Fourier transform (STFT) in which the Fourier transform is applied to a windowed portion of the signal and then slid the time window $w(t)$ across the original signal $x(t)$. If the Gaussian window is selected, the STFT becomes the Gabor transform. However, with the STFT, the resolution in time and resolution in frequency cannot be made arbitrarily small at the same time because of the Heisenberg uncertainty principle. In contrast to the STFT, the wavelet transform uses short windows at high frequencies to give time resolution and long windows at low frequencies to give good frequency resolution.

The wavelet transform was first introduced by Grossman and Morlet [1] and used for seismic data evaluation. Since then, various types of wavelet transforms and applications have emerged [2–9]. The wavelets have advantages over traditional Fourier methods in analyzing signals with discontinuities and sharp spikes. Most of the data analysis applications use the continuous-time wavelet transform (CWT), which yields an affine invariant time–frequency representation. However, the discrete wavelet transform (DWT) is the most famous version, since it has excellent signal compaction properties for many classes of real-world signals and it is computationally very efficient. Further, the implementation of DWT is simple because it depends on the perfect reconstruction filter banks, upsampling and down-sampling. Hence, it has been applied to several different fields such as signal processing, image processing, communications and pattern recognition.

This chapter emphasizes the time–frequency representation, short-time Fourier transform (STFT), inverse STFT (ISTFT), scaling functions and wavelets, discrete wavelet transform (DWT), multiresolution analysis (MRA), generation of orthogonal and biorthogonal scaling functions and wavelets, computation of

© Springer Nature Singapore Pte Ltd. 2018
K. D. Rao and M. N. S. Swamy, *Digital Signal Processing*,
https://doi.org/10.1007/978-981-10-8081-4_10

one-dimensional DWT and two-dimensional DWT, wavelet packets, and some application examples of DWT.

10.1 Time–Frequency Representation of Signals

The Fourier transform (FT) is acceptable for stationary signals, i.e., signals whose components do not change in time, but unacceptable for non-stationary signals wherein information on different frequency components occurs in time.

Time–frequency representations (TFRs) map a one-dimensional signal in time and frequency into a two-dimensional function in time and frequency.

$$x(t) \leftrightarrow T_x(t,f) \qquad (10.1)$$

The domain of time–frequency representation is often called the time–frequency (TF) plane.

To demonstrate the importance of TFRs, consider (i) a sinusoidal stationary signal consisting of two frequencies 0.22 and 0.34 Hz, which exist for all time and (ii) a non-stationary signal with a frequency 0.22 Hz existing for half the time and with a frequency 0.34 Hz exiting for the other half. To generate the stationary signal and its spectra, the following MATLAB Program 10.1 is used.

Program 10.1 Generation of a stationary signal and its spectra

```
clear all;close all;clc;
N=128;R=128;n=0:N-1;fr1=0.22;fr2=0.34;
x=sin(2*pi*n*fr1)+sin(2*pi*n*fr2);
Fx=fft(x,R);omega=2*(127/128);
k=linspace(0,omega/2,128);
figure,plot((x));xlabel('Time');ylabel('Amplitude');
title('Stationary Signal');
figure,plot(k,abs(Fx));xlabel('Frequency');ylabel('Magnitude');
title('Fourier Transform of Stationary Signal');
```

To generate the non-stationary signal and its spectra, the following MATLAB Program 10.2 is used.

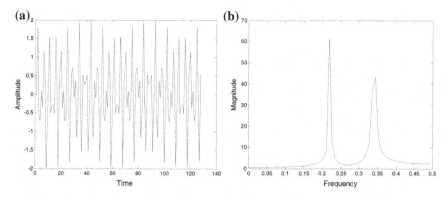

Fig. 10.1 **a** Stationary signal consisting of two frequencies, 0.22 and 0.34 Hz, which exist for all time. **b** Fourier transform of the stationary signal

Program 10.2 Generation of a non-stationary signal and its spectra

```
clear all;close all;clc;
N=128;R=128;n=0:N-1;fr1=0.22;fr2=0.34;
x1=sin(2*pi*n*fr1);x2=sin(2*pi*n*fr2);x3=x1(1:N/2);
x3((N/2)+1:N)=x2(1:N/2);
Fx=fft(x3,R);
omega=2*pi*(127/128);k=linspace(0,omega/(2*pi),128);
figure,plot((x3));xlabel('Time');ylabel('Amplitude');title('Non-stationary Signal');
figure,plot(k,abs(Fx));xlabel('Frequency');ylabel('Magnitude');
title('Fourier Transform of Non-stationary Signal');
```

The stationary signal and its spectra generated from Program 10.1 are shown in Fig. 10.1a, b. The non-stationary signal and its spectra generated from Program 10.2 are shown in Fig. 10.2a, b. Due to the fact that the time information is lost, the spectra for the stationary and non-stationary signals appear identical. Ideally, we would like the TFR to display data as shown in Fig. 10.3a, b, respectively.

10.2 Short-Time Fourier Transform (STFT)

The STFT of a sequence $x(n)$ can be expressed mathematically as

$$X_{STFT}\left(n, e^{j\omega}\right) = \sum_{m=-\infty}^{\infty} x(m)w(n-m)e^{-j\omega m} \qquad (10.2)$$

where $w(n)$ is window sequence.

Let the window length be M in the range $0 \leq m \leq M - 1$. Then, $X_{STFT}\left(n, e^{j\omega}\right)$ sampled at N equally spaced frequencies $\omega_k = \frac{2\pi k}{N}$, with $N \geq M$ can be defined as

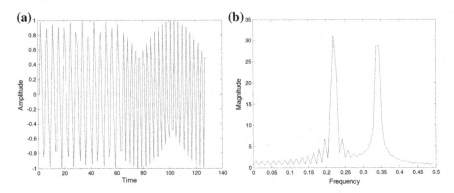

Fig. 10.2 **a** Non-stationary signal with a frequency 0.22 Hz existing for half the time and with a frequency 0.34 Hz existing for the other half. **b** Fourier transform of the non-stationary signal

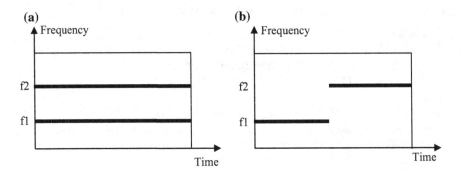

Fig. 10.3 **a** Ideal TFR for the stationary signal. **b** Ideal TFR for the non-stationary signal

$$
\begin{aligned}
X_{\text{STFT}}(n,k) &= X_{\text{STFT}}\left(n,e^{j\omega}\right)\Big|_{\omega=2\pi k/N} = X_{\text{STFT}}\left(n,e^{j2\pi k/N}\right) \\
&= \sum_{m=0}^{M-1} x(m)w(n-m)e^{-j2\pi mk/N},\ 0 \le k \le N-1.
\end{aligned}
\tag{10.3}
$$

The *signal processing toolbox* of MATLAB includes the function *specgram* for the computation of the STFT of a signal. The following MATLAB Program 10.2a is used to compute the STFT of non-stationary signal shown in Fig. 10.2.

Fig. 10.4 STFT of
Fig. 10.2a

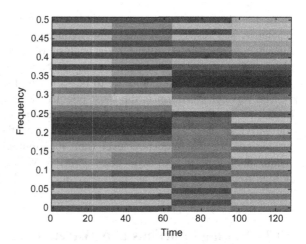

Program 10.2a

```
N=128;n=0:N-1;fr1=0.22;fr2=0.34;
x1=sin(2*pi*n*fr1);x2=sin(2*pi*n*fr2);
x3=x1(1:N/2); x3((N/2)+1:N)=x2(1:N/2);
specgram(x3,64,1,hamming(32),0);
```

The STFT obtained is shown in Fig. 10.4. From the spectrogram, the time–
frequency information of the non-stationary signal, with a frequency of 0.22 Hz
existing for half the time and a frequency of 0.34 Hz existing for the other half, is
very clear.

10.2.1 Inverse STFT

Equation (10.3) indicates that $X_{\text{STFT}}(n,k)$ is the DFT of $x(m)w(n-m)$. The
$X_{STFT}(n,k)$ is periodic in k with a period N. Applying the IDFT to Eq. (10.3), we
obtain

$$x(m)w(n-m) = \frac{1}{N}\sum_{k=0}^{N-1} X_{\text{STFT}}(n,k)e^{j2\pi mk/N}, \quad 0 \le m \le M-1. \quad (10.4)$$

Thus, the inverse STFT (ISTFT) can be defined as

$$x(m) = \frac{1}{Nw(n-m)}\sum_{k=0}^{N-1} X_{\text{STFT}}(n,k)e^{j2\pi mk/N}, \quad 0 \le m \le M-1. \quad (10.5)$$

In the STFT, the fixed duration window is accompanied by a fixed frequency resolution and results in a fixed time–frequency resolution. The fixed time–frequency resolution is a bottleneck for analysis of signals with discontinuities and sharp spikes, since the discontinuities are to be resolved sharply in time at high frequencies and slow variations are to be seen sharply at low frequencies.

The wavelet transform is the most popular alternative to the STFT. A linear expansion of a signal is obtained using scales and shifts of a prototype wavelet. In the wavelets, the scales used are powers of 2 and the frequency localization is proportional to the frequency level (i.e., logarithmic). Consequently finer time localization results at high frequencies. Thus, the wavelet transform is sharp in time at high frequencies as well as sharp in frequency at low frequencies.

10.3   Scaling Functions and Wavelets

Before we introduce the mathematical descriptions of scaling functions and wavelets, we start with the following preliminaries.

10.3.1   Expansion of a Signal in Series Form

Let a real function $x(t)$ be expressed as linear combination of expansion functions $\{\varphi_i(t)\}$ in the form

$$x(t) = \sum_i \alpha_i \varphi_i(t) \tag{10.6}$$

where i is an integer, α_i's are real coefficients, and the sum may be finite or infinite. If the expansion is unique, then the functions $\varphi_i(t)$ are called the basis functions and the set $\{\varphi_i(t)\}$ as the basis for the class of functions $x(t)$ that can be expressed in the form (10.6). The functions $x(t)$ that can be expressed in that form constitute a function space, which is referred to as the closed span of the expansion set and is denoted by

$$V = \operatorname*{span}_i\{\varphi_i(t)\}.$$

For any function space V and the expansion function set $\{\varphi_i(t)\}$, there exist a set of 'dual functions' $\{\tilde{\varphi}_i(t)\}$, which can be used to determine the expansion coefficients α_i for any $x(t) \in V$:

$$\alpha_i = \langle x(t), \; \varphi_i(t) \rangle = \int x(t)\tilde{\varphi}_i(t)\, dt \qquad (10.7)$$

If $\{\varphi_i(t)\}$ is an orthonormal set, then

$$\langle \varphi_i(t), \; \varphi_j(t) \rangle = \delta_{ij} = \begin{cases} 0 & i \neq j \\ 1 & i = j \end{cases} \qquad (10.8)$$

Hence, $\varphi_i(t) = \tilde{\varphi}_i(t)$, that is, the basis and its dual are equivalent. However, if $\{\varphi_i(t)\}$ is not orthonormal, but is an orthogonal basis for V, then

$$\langle \varphi_i(t), \; \tilde{\varphi}_j(t) \rangle = 0, \; i \neq j \qquad (10.9)$$

and the basis functions and their duals are called biorthogonal. In such a case,

$$\langle \varphi_i(t), \; \tilde{\varphi}_j(t) \rangle = \delta_{ij} = \begin{cases} 0 & i \neq j \\ 1 & i = j \end{cases} \qquad (10.10)$$

10.3.2 Scaling Functions

Consider the set $\varphi_{j,k}$ given by

$$\varphi_{j,k}(t) = 2^{j/2}\varphi(2^j t - k) \qquad (10.11)$$

j and k being integers, and $\varphi(t)$ is the set of all measurable, square-integrable functions. The integer k determines the shift of $\varphi(t)$ along the t-axis, j determines the width of $\varphi(t)$ along the t-axis, and $2^{j/2}$ the height or the amplitude. The function $\varphi(t)$ is called the *scaling function,* since the shape of $\varphi_{j,k}(t)$ can be controlled by changing j. We denote the subspace spanned over k for a given j as $V_j = \underset{k}{\mathrm{span}}\{\varphi_{j,k}(t)\}$. If $x(t) \in V_j$, then it can be expressed as

$$x(t) = \sum_k \alpha_k \varphi_{j,k}(t) \qquad (10.12)$$

Example 10.1 Consider the Haar scaling function given by

$$\varphi(t) = \begin{cases} 1 & 0 \leq t < 1 \\ 0 & \text{otherwise} \end{cases} \qquad (10.13)$$

Sketch $\varphi_{0,0}(t), \varphi_{0,1}(t), \varphi_{1,0}(t)$ and $\varphi_{1,1}(t)$. Show that

$$\varphi_{0,k}(t) = \frac{1}{\sqrt{2}} \varphi_{1,2k}(t) + \frac{1}{\sqrt{2}} \varphi_{1,2k+1}(t)$$

Solution From Eq. (10.11), we have

$$\varphi_{0,0}(t) = \varphi(t), \varphi_{0,1}(t) = \varphi(t-1), \varphi_{1,0}(t) = \sqrt{2}\varphi(2t), \varphi_{1,1}(t) = \sqrt{2}\varphi(2t-1)$$
$$\varphi_{0,k}(t) = \varphi(t-k), \varphi_{1,2k}(t) = \sqrt{2}\varphi(2t-2k), \varphi_{1,2k+1}(t) = \sqrt{2}\varphi(2t-2k-1)$$

The various functions are shown in Fig. 10.5.
It is seen from Fig. 10.10e, f that

$$\varphi_{0,k}(t) = \frac{1}{\sqrt{2}} \varphi_{1,2k}(t) + \frac{1}{\sqrt{2}} \varphi_{1,2k+1}(t). \tag{10.14}$$

Hence, any function $x(t)$ that is an element of V_0 is also an element of V_1, that is, $V_0 \subset V_1$. Similarly, it can be shown that $V_1 \subset V_2$, etc. Hence,

$$V_0 \subset V_1 \subset V_2 \subset V_3 \ldots \tag{10.15}$$

From Eq. (10.14), we also have

$$\varphi_{0,0}(t) = \frac{1}{\sqrt{2}} \varphi_{1,0}(t) + \frac{1}{\sqrt{2}} \varphi_{1,1}(t) \tag{10.16}$$

Hence,

$$\varphi(t) = \varphi(2t) + \varphi(2t-1) \tag{10.17}$$

It is also observed that the Haar scaling functions are orthogonal within each scale, that is,

$$\langle \varphi_{j,k}(t), \varphi_{j,l}(t) \rangle = \delta_{kl} \tag{10.18}$$

10.3.3   Wavelet Functions

Given the scaling functions $\{\varphi_{j,k}(t)\}$, as defined in the previous section, we now define a set of wavelet functions $\{\psi_{j,k}(t)\}$ by

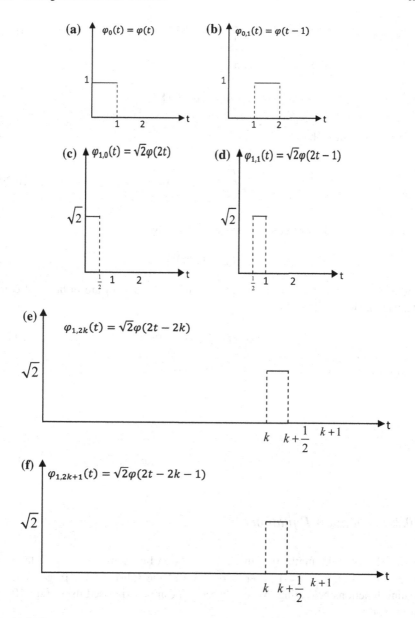

Fig. 10.5 Haar scaling functions. **a** $\varphi_{0,0}(t)$, **b** $\varphi_{0,1}(t)$, **c** $\varphi_{1,0}(t)$, **d** $\varphi_{1,1}(t)$, **e** $\varphi_{1,2k}(t)$, and **f** $\varphi_{1,2k+1}(t)$

$$\{\psi_{j,k}(t)\} = 2^{j/2}\psi(2^jt - k) \tag{10.19}$$

so that the space W_j between V_j and V_{j+1}, i.e.,

$$W_j = \operatorname*{span}_{k}\{\psi_{j,k}(t)\}.$$

We also assume that

$$\int\limits_{-\infty}^{\infty} \psi(t)\mathrm{d}t = 0 \tag{10.20}$$

The scaling and wavelet subspaces are related by

$$V_{j+1} = V_j \oplus W_j \tag{10.21}$$

where $\oplus$ represents the direct sum of V_j and W_j; V_j and W_j are orthogonal complements in V_{j+1}. Thus,

$$\langle \varphi_{j,k}(t), \psi_{j,l}(t) \rangle = 0 \tag{10.22}$$

Since j is arbitrary, Eq. (10.21) can be rewritten as

$$V_{j+1} = V_{j-1} \oplus W_{j-1} \oplus W_j$$

or

$$V_{j+1} = V_k \oplus W_k \oplus \cdots \cdots \oplus W_j \quad \text{for any } k \leq j. \tag{10.23}$$

10.3.4 Dilation Equations

It was shown in Example 10.1 that the expansion functions $\{\varphi_{j,k}(t)\}$ of the subspace V_0 can be expressed in terms of those of the subspace V_1. In general, the scaling functions belonging to the subspace V_j can be expressed using Eq. (10.12) in the form

$$\varphi_{j,k}(t) = \sum_{m} \alpha_m \varphi_{j+1,m}(t)$$

Substituting for $\varphi_{j+1,m}(t)$ using Eq. (10.11) in the above equation, we get

$$\varphi_{j,k}(t) = \sum_m \alpha_m 2^{(j+1)/2} \varphi(2^{j+1}t - m)$$

Since $\varphi_{0,0}(t) = \varphi(t)$, we may rewrite the above equation, after changing m to k and α_m to $h_0(k)$, as

$$\varphi(t) = \sqrt{2} \sum_k h_0(k)\varphi(2t - k) \tag{10.24}$$

The above equation is called the *dilation equation or refinement equation*, and the coefficients $h_0(k)$ as the *scaling function coefficients*. Equation (10.22) shows that the expansion functions of any subspace can be expressed in terms of those of the next higher resolution space. This is important in terms of multiresolution analysis.

From Eq. (10.21), we know that the wavelet space W_j resides within the next higher scaling function space V_{j+1}. Hence, we can represent the wavelet basis as a linear combination of the scaling functions as

$$\psi(t) = \sqrt{2} \sum_k h_1(k)\varphi(2t - k) \tag{10.25}$$

where the coefficients $h_1(k)$ are called the *wavelet coefficients*. For each set of scaling functions coefficients $h_0(k)$, there is a corresponding set of coefficients $h_1(k)$ satisfying (10.25).

Example 10.2 It will be shown later that the Haar wavelet function $\psi(t)$ is given by

$$\psi(t) = \varphi(2t) - \varphi(2t - 1) \tag{10.26a}$$

where $\varphi(t)$ is given by (10.13); that is, $h_1(0) = \frac{1}{\sqrt{2}}$ and $h_1(1) = -\frac{1}{\sqrt{2}}$. Sketch $\psi_{0,0}(t)$, $\psi_{0,1}(t)$, $\psi_{1,0}(t)$, and $\psi_{1,1}(t)$.

Solution From (10.26a) and (10.13), we see that the Haar wavelet function $\psi(t)$ is given by

$$\psi(t) = \begin{cases} 1 & 0 \le t < 0.5 \\ -1 & 0.5 \le t < 1 \\ 0 & \text{otherwise} \end{cases} \tag{10.26b}$$

It is clear from (10.19) that

$$\psi_{0,0}(t) = \psi(t), \; \psi_{0,1}(t) = \psi(t - 1), \; \psi_{1,0}(t) = \sqrt{2}\psi(2t) \text{ and } \psi_{1,1}(t) \\ = \sqrt{2}\psi(2t - 1).$$

These wavelet functions are shown in Fig. 10.6.

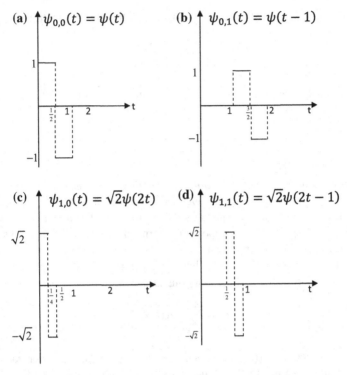

Fig. 10.6 Haar wavelet functions **a** $\psi_{0,0}(t)$, **b** $\psi_{0,1}(t)$, **c** $\psi_{1,0}(t)$, and **d** $\psi_{1,1}(t)$

10.4   The Discrete Wavelet Transform (DWT)

Discrete wavelet transform (DWT) maps continuous-time functions into a set of numbers. The forward (analysis) equations are given by the following inner products of $x(t)$, $\varphi_{j,k}(t)$ and $\psi_{j,k}(t)$.

$$c_{j,k} = \langle x(t),\ \varphi_{j,k}(t) \rangle = \int x(t)\varphi_{j,k}(t)\,\mathrm{d}t \qquad (10.27\text{a})$$

$$d_{j,k} = \langle x(t),\ \psi_{j,k}(t) \rangle = \int x(t)\psi_{j,k}(t)\,\mathrm{d}t \qquad (10.27\text{b})$$

The coefficients $c_{j,k}$ in the DWT are called the *smooth* or *approximation coefficients* and the coefficients $d_{j,k}$ as the *detail* or *wavelet coefficients*. The inverse DWT (IDWT) or synthesis equation is given by

$$x(t) = \sum_k c_{j,k}\varphi_{j,k}(t) + \sum_{j=J}^{\infty}\sum_k d_{j,k}\psi_{j,k}(t) \qquad (10.28)$$

where J is the starting index, usually equal to zero. The DWT decomposes a given signal $x(t)$ into its constituent components $c_{j,k}$ and $d_{j,k}$. The inverse DWT reconstructs the signal $x(t)$ from its constituent components $c_{j,k}$ and $d_{j,k}$.

Example 10.3 Using Haar wavelets, find the approximation and detailed coefficients for the function $x(t)$ given by

$$x(t) = \begin{cases} t & 0 \le t < 1 \\ 0 & \text{otherwise} \end{cases}$$

Solution From Eqs. (10.27a) and (10.27b), we get

$$c_{0,0} = \langle x(t), \varphi_{0,0}(t) \rangle = \int_0^1 t\varphi_{0,0}(t)\mathrm{d}t = \int_0^1 t\mathrm{d}t = \frac{1}{2}$$

$$d_{0,0} = \langle x(t), \psi_{0,0}(t) \rangle = \int_0^1 t\psi_{0,0}(t) = \int_0^{1/2} t\mathrm{d}t - \int_{1/2}^1 t\mathrm{d}t = -\frac{1}{4}$$

$$d_{1,0} = \langle x(t), \psi_{1,0}(t) \rangle = \int_0^1 t\psi_{1,0}(t)\mathrm{d}t = \sqrt{2}\int_0^{1/4} t\mathrm{d}t - \sqrt{2}\int_{1/4}^{1/2} t\mathrm{d}t = -\frac{\sqrt{2}}{8}$$

$$d_{1,1} = \langle x(t), 1mu\psi_{1,1}(t) \rangle = \int_0^1 t\psi_{1,1}(t)\mathrm{d}t = \sqrt{2}\int_{1/2}^{3/4} t\mathrm{d}t - \sqrt{2}\int_{3/4}^1 t\mathrm{d}t = -\frac{\sqrt{2}}{16}$$

Hence,

$$x(t) = \left[\frac{1}{2}\varphi_{0,0}(t)\right] + \left[-\frac{1}{4}\psi_{0,0}(t)\right] + \left[-\frac{\sqrt{2}}{8}\psi_{1,0}(t) - \frac{\sqrt{2}}{16}\psi_{1,1}(t)\right] + \ldots\ldots$$

The first term in square brackets corresponds to V_0, the second to W_0, the third to W_1, and so on. The wavelet expansion is shown in Fig. 10.7.

Example 10.4 Using Haar wavelets, find the approximation and detailed coefficients for the function $x(t)$ shown in Fig. 10.8.

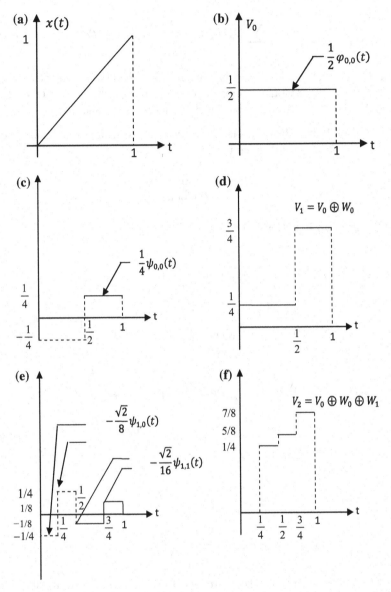

Fig. 10.7 Wavelet expansion of $x(t)$ of Example 10.3. **a** $x(t)$, **b** $\frac{1}{2}\varphi_{0,0}(t)$, **c** $-\frac{1}{4}\psi_{0,0}(t)$, **d** $V_1 = V_0 \oplus W_0$, **e** W_1, and **f** $V_2 = V_1 \oplus W_1$

Fig. 10.8 $x(t)$ of Example
10.4

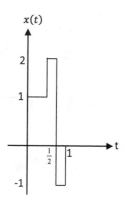

Solution Using Eq. (10.27a) and (10.27b), we have

$$c_{0,0} = \int_0^{1/2} \varphi_{0,0}(t)\mathrm{d}t + \int_{1/2}^{3/4} 2\varphi_{0,0}(t)t\mathrm{d}t - \int_{3/4}^{1} \varphi_{0,0}(t)\mathrm{d}t = \frac{3}{4}$$

$$d_{0,0} = \int_0^{1/2} \psi_{0,0}(t)\mathrm{d}t + \int_{1/2}^{3/4} 2\psi_{0,0}(t)t\mathrm{d}t - \int_{3/4}^{1} \psi_{0,0}(t)\mathrm{d}t = \frac{1}{4}$$

$$d_{1,0} = \int_0^{1/2} \psi_{1,0}(t)\mathrm{d}t + \int_{1/2}^{3/4} 2\psi_{1,0}(t)t\mathrm{d}t - \int_{3/4}^{1} \psi_{1,0}(t)\mathrm{d}t = 0$$

$$d_{1,1} = \int_0^{1/2} \psi_{1,1}(t)\mathrm{d}t + \int_{1/2}^{3/4} 2\psi_{1,1}(t)t\mathrm{d}t - \int_{3/4}^{1} \psi_{1,1}(t)\mathrm{d}t = \frac{3\sqrt{2}}{4}$$

Therefore,

$$x(t) = \left[\frac{3}{4}\varphi_{0,0}(t)\right] + \left[\frac{1}{4}\psi_{0,0}(t)\right] + \left[\frac{3\sqrt{2}}{4}\psi_{1,1}(t)\right] = V_0 \oplus W_0 \oplus W_1$$

Figure 10.9 shows the wavelet representation of $x(t)$.

From the previous two examples, we see that if we have a function $x(t) \subset V_{j+1}$, then we can decompose it as a sum of functions starting with a lower resolution approximation followed by a series of wavelet functions representing the remaining details. It should be noted that at the higher resolution, the approximation of the signal is closer to the actual signal $x(t)$ with more details of the signal.

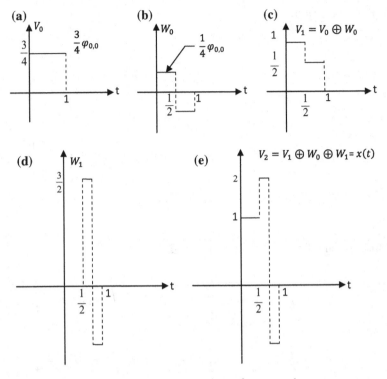

Fig. 10.9 Wavelet expansion of $x(t)$ of Example 10.4. **a** $\frac{3}{4}\varphi_{0,0}(t)$, **b** $\frac{1}{4}\psi_{0,0}(t)$, **c** $V_1 = V_0 \oplus W_0$, **d** W_1, and **e** $V_2 = V_1 \oplus W_1 = V_0 \oplus W_0 \oplus W_1$

10.4.1   Computation of Wavelet Coefficients

We know from the dilation Eq. (10.24) that

$$\varphi(t) = \sqrt{2}\sum_n h_0(n)\varphi(2t - n)$$

Scaling t by 2^j and translating it by k, we have

$$\varphi(2^j t - k) = \sqrt{2}\sum_n h_0(n)\varphi\big(2(2^j t - k) - n\big)$$

Letting $n + 2k = m$, we get

$$\varphi(2^j t - k) = \sqrt{2}\sum_m h_0(m - 2k)\varphi(2^{j+1} t - m)$$

or

$$\varphi_{j,k}(t) = 2^{j/2}\varphi\left(2^j t - k\right) = \sum_m h_0(m - 2k)2^{(j+1)/2}\varphi\left(2^{j+1}t - m\right)$$

$$= \sum_m h_0(m - 2k)\varphi_{j+1,k}(t)$$

Substituting the above relation for $\varphi_{j,k}(t)$ in Eq. (10.27a), we have the approximating coefficients $c_{j,k}$ as

$$c_{j,k} = \int x(t)\varphi_{j,k}(t)\,dt = \int x(t)\sum_m h_0(m - 2k)\varphi_{j+1,k}(t)\,dt$$

$$= \sum_m h_0(m - 2k)\int x(t)\varphi_{j+1,k}(t)\,dt$$

Hence,

$$c_{j,k} = \sum_m h_0(m - 2k)c_{j+1,k} \tag{10.29}$$

Similarly, starting with Eq. (10.25) for $\psi(t)$, namely

$$\psi(t) = \sqrt{2}\sum_n h_1(n)\varphi(2t - n)$$

we can derive the relation

$$\psi_{j,k}(t) = \sum_m h_1(m - 2k)\varphi_{j+1,k}(t)$$

Using the above in Eq. (10.27b), we can show that the detailed coefficients $d_{j,k}$ can be expressed as

$$d_{j,k} = \sum_m h_1(m - 2k)c_{j+1,k} \tag{10.30}$$

From Eqs. (10.29) and (10.30), we see that $c_{j+1,k}$ provides enough information to find all the lower-scale coefficients. From the approximation coefficients $c_{j+1,k}$ at level $(j+1)$, we can determine both the approximation and detailed coefficients $c_{j,k}$ and $d_{j,k}$ at level j using (10.29) and (10.30), Thus, all the lower coefficients can be computed by iteration.

10.5 Multiresolution Analysis

Multiresolution analysis (MRA) is the key idea behind DWT and many other algorithms for a fast computation of the DWT. The basic idea of the MRA is to decompose the signal successively along its approximation coefficients. Let us recall that $x(t)$ can be accurately represented at the resolution $(j + 1)$. Hence, we can replace $c_{j+1,k}$ by the samples of $x(t)$; let us say $x(m)$. Then, using (10.29), we have

$$c_{j,k} = \sum_m h_0(m - 2k)x(m)$$

and

$$d_{j,k} = \sum_m h_1(m - 2k)x(m)$$

Hence, in a DWT, the approximation coefficients $c_{j,k}$ are obtained by convolving the signal with a LP filter and down-sampling the result by 2. Similarly, the detailed coefficients $d_{j,k}$ are obtained by convolving the signal with a HP filter and down-sampling the result by 2. This process is repeated along the approximation coefficients resulting in $\{c_{1,k}, d_{1,k}, \ldots, d_{j,k}\}$ as the final representation of the signal.

The number of approximation and detailed coefficients at each stage of decomposition is equal to half the number of approximation coefficients at the previous stage (corresponding to the next higher resolution) due to down-sampling by a factor of 2 at each stage. Suppose a discrete signal has L samples, then its DWT decomposition generally will have $(\frac{L}{2} + \frac{L}{4} + \frac{L}{8} + \cdots + \frac{L}{2^j})$ detailed coefficients and $\frac{L}{2^j}$ approximation coefficients.

However, due to the restrictions on the maximum value of j by the number of signal samples L, the total number of DWT coefficients (including the approximation and detailed coefficients) is equal to L.

The entire iterative process of the decomposition is equivalent to passing $x(n)$ through a multirate filter bank, as shown in Fig. 10.10. This process is called *decomposition* or *analysis*.

10.6 Wavelet Reconstruction

The process of assembling the decomposition components back into the original signal without loss of information is called *reconstruction* or *synthesis*. The inverse discrete wavelet transform (IDWT) is the mathematical manipulation that effects the reconstruction, and is given by

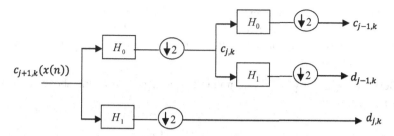

Fig. 10.10 Two stages in a DWT decomposition

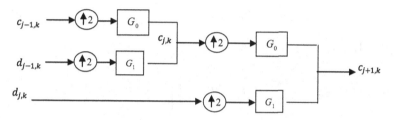

Fig. 10.11 Two stages in IDWT reconstruction

$$c_{j+1,k} = \sum_m g_0(k - 2m)c_{j,m} + \sum_m g_1(k - 2m)d_{j,m} \qquad (10.31)$$

The above relation leads to an iterative procedure for the reconstruction of the original signal from the DWT coefficients. In the reconstruction procedure, at each stage the approximation and detailed coefficients at a particular resolution are upsampled by a factor of 2 and convolved with a LP filter and a HP filter, respectively. Then, the approximation coefficients at the next higher resolution are obtained by adding the outputs of both the filters. Thus, the reconstruction procedure involves in passing the DWT coefficients through a multirate filter bank as shown in Fig. 10.11.

10.7 Required Properties of Wavelets

10.7.1 Orthogonality

Orthogonality is a desirable property for the basis functions to ensure unique and non-redundant representation, as well as perfect reconstruction of the signal from its DWT coefficients. Since the multiresolution decomposition of a signal requires basis functions having more than one scale, the following orthogonality conditions should hold.

1. Orthogonality of the scaling functions within each scale
2. Orthogonality of the wavelets within each scale and across scales
3. Mutual orthogonality of the scaling functions and the wavelets, within each scale and across scales.

Denoting $\varphi_{0,m}(t)$ by $\varphi_m(t)$ and noting that $\varphi_{0,0}(t) = \varphi(t)$, the above conditions may be written as

$$\langle \varphi(t), \varphi_m(t) \rangle = \delta_m \tag{10.32a}$$

$$\langle \psi_{j,k}(t), \psi_{l,m}(t) \rangle = \delta_{jl} \tag{10.32b}$$

$$\langle \varphi_{j,k}(t), \psi_{l,m}(t) \rangle = 0 \tag{10.32c}$$

Using Eqs. (10.24) in (10.32a), we get

$$\langle \varphi(t), \varphi_m(t) \rangle = \int \left[\sqrt{2} \sum_k h_0(k) \varphi(2t - k) \right] \left[\sqrt{2} \sum_l h_0(l) \varphi(2t - 2m - l) \right] = \delta_m$$

or

$$\langle \varphi(t), \varphi_m(t) \rangle = \sum_k \sum_l 2h_0(k) h_0(l) \int \varphi(2t - k) \varphi(2t - 2m - l) = \delta_m$$

Letting $t' = 2t$ in the above equation, we get

$$\langle \varphi(t), \varphi_m(t) \rangle = \sum_k \sum_l 2h_0(k) h_0(l) \int \varphi(t' - k) \varphi(t' - 2m - l) = \delta_m$$

or

$$\langle \varphi(t), \varphi_m(t) \rangle = \sum_k \sum_l 2h_0(k) h_0(l) \delta_{k-(2m+l)} = \delta_m$$

That is,

$$\sum_k h_0(k) h_0(k - 2m) = \delta_m$$

Therefore, we have

$$\sum_k h_0^2(k) = 1 \tag{10.33a}$$

And

$$\sum_k h_0(k)h_0(k-2m) = 0, \quad m \neq 0 \qquad (10.33b)$$

Similarly, using (10.25), we can establish from (10.32b) that

$$\sum_k h_1^2(k) = 1 \qquad (10.34)$$

Also, from (10.32c) we can deduce that

$$\sum_k h_0(k)h_1(k) = 0 \qquad (10.35)$$

In terms of $H_0(\omega)$ and $H_1(\omega)$, the filters' power complementary conditions become

$$|H_0(\omega)|^2 + |H_0(\omega+\pi)|^2 = 1 \qquad (10.36a)$$

$$|H_1(\omega)|^2 + |H_1(\omega+\pi)|^2 = 1 \qquad (10.36b)$$

$$H_0(\omega)H_1^*(\omega) + H_0(\omega+\pi)H_1^*(\omega+\pi) = 0 \qquad (10.36c)$$

where the * denotes complex conjugation. We also have from Eq. (10.24):

$$\int_{-\infty}^{\infty} \varphi(t)dt = \int_{-\infty}^{\infty} \sqrt{2}\sum_k h_0(k)\varphi(2t-k)dt$$

Letting $t' = 2t$ in the above equation, we get

$$\int_{-\infty}^{\infty} \varphi(t)dt = \frac{1}{\sqrt{2}}\sum_k h_0(k) \int_{-\infty}^{\infty} \varphi(t)dt \qquad (10.37)$$

Since the average value of the scaling function is not zero and in general is unity, that is,

$$\int_{-\infty}^{\infty} \varphi(t)dt = 1 \qquad (10.38)$$

we get from (10.37) that

$$\sum_k h_0(k) = \sqrt{2} \tag{10.39}$$

Similarly, using Eqs. (10.24) and (10.25), we can deduce that

$$\sum_k h_1(k) = 0 \tag{10.40}$$

10.7.2   Regularity Condition

The scaling filter $H_0(z)$, which is a lowpass filter, is said to be *p-regular* if it has p zeros at $z = e^{j\pi}$ (i.e., at $\omega = \pi$). Thus, $H_0(z)$ is of the form

$$H_0(z) = \frac{(1+z^{-1})^p}{2} Q(z) \tag{10.41}$$

where $Q(z)$ has no zeros at $z = -1$. If the length of the filter $H_0(z)$ is N, then it is a polynomial of degree $N - 1$; since $z = -1$ is a zero of order p for $H_0(z)$, $Q(z)$ is a polynomial of degree $(N-1-p)$. The degree of regularity p is limited by the condition $(1 \le p \le N/2)$, since $(N/2)$ conditions are required to satisfy the orthogonality conditions given by (10.33a, 10.33b). Daubechies has used these degrees of freedom to obtain maximum regularity for a given N or to get a minimum length N for a given regularity p.

Taking the Fourier transform of the dilation Eq. (10.24).

$$\varphi(t) = \sqrt{2} \sum_k h_0(k)\varphi(2t - k)$$

we get

$$\Phi(\omega) = \sqrt{2} \sum_k h_0(k)e^{-j\omega k/2}\frac{1}{2}\Phi\left(\frac{\omega}{2}\right)$$

$$= \frac{1}{\sqrt{2}}\left[\sum_k h_0(k)e^{-\frac{j\omega k}{2}}\right]\Phi\left(\frac{\omega}{2}\right)$$

hence,

$$\Phi(\omega) = \frac{1}{\sqrt{2}}H_0\left(\frac{\omega}{2}\right)\Phi\left(\frac{\omega}{2}\right) \tag{10.42}$$

It is seen from Eq. (10.42) that the Fourier transform of $\varphi(t)$ is related to the frequency response of the lowpass filter $H_0(z)$. If $H_0(z)$ has a zero of high order at $z = -1$ or $\omega = \pi$, then the Fourier transform of $\varphi(t)$ should fall off very rapidly, and hence $\varphi(t)$ should be smooth. Thus, higher the regularity, smoother the scaling function $\varphi(t)$. It may be mentioned in passing that $H_1(z)$, which is a highpass filter, would contain multiple zeros at $z = 1$.

We now define the qth moments of $\varphi(t)$ and $\psi(t)$ as

$$m_0(q) = \int t^q \varphi(t)\, dt \qquad (10.43a)$$

$$m_1(q) = \int t^q \psi(t)\, dt \qquad (10.43b)$$

Then, the following theorem can be established [10, 11].

Theorem *The scaling filter*
$H_0(z)$ is p-regular if and only if all the moments of the wavelets are zero, that is,

$$m_1(q) = 0, \text{ for } q = 0, 1, 2, .., p - 1$$

According to Strang's accuracy condition [12], $\Phi(\omega)$ must have zeros of the highest order when $\omega = 2v\pi$, $v = 1, 2, \ldots$ From (10.42), we can deduce that

$$\Phi(\omega) = (\frac{1}{\sqrt{2}})^i H_0\left(\frac{\omega}{2}\right) H_0\left(\frac{\omega}{2^2}\right) \ldots H_0\left(\frac{\omega}{2^i}\right) \Phi\left(\frac{\omega}{2^i}\right)$$

thus,

$$\Phi(2\pi) = (\frac{1}{\sqrt{2}})^i H_0(\pi) H_0\left(\frac{\pi}{2}\right) \ldots H_0\left(\frac{\pi}{2^{i-1}}\right) \Phi\left(\frac{\pi}{2^{i-1}}\right)$$

and $H_0(\omega)$ will have a zero of order p at $\omega = \pi$; that is, $H_0(z)$ is p-regular if

$$\frac{d^m}{d\omega^m} H_0(\omega) = 0 \quad \text{for } m = 0, 1, 2, \ldots p - 1 \qquad (10.44)$$

Since

$$H_0(\omega) = \sum_k h_0(k) e^{-j\omega k}$$

We obtain from Eq. (10.44) that

$$\sum_k h_0(k)(-jk)^m e^{-j\pi k} = 0 \quad \text{for } m = 0, 1, 2, \ldots p - 1$$

or

$$\sum_k h_0(k)(-1)^k k^m = 0 \quad \text{for } m = 0, 1, 2, \ldots p - 1 \qquad (10.45)$$

The above relation is called the regularity condition for $H_0(z)$ to be p-regular, that is, condition for the first p moments of the wavelets to be zero. For the case when $m = 0$, we have

$$\sum_k (-1)^k h_0(k) = 0 \qquad (10.46)$$

From (10.39) and (10.45), it follows that

$$\sum_k h_0(2k) = \sum_k h_0(2k + 1) = \frac{1}{\sqrt{2}} \qquad (10.47)$$

10.8 Generation of Daubechies Orthonormal Scaling Functions and Wavelets

Actually, we look for the filters rather than the wavelets as the wavelets can be termed as perfect reconstruction filters. Daubechies has derived perfect reconstruction filters with desirable maximally flat property.

10.8.1 Relation Between Decomposition Filters (H_0, H_1) and Reconstruction Filters (G_0, G_1) for Perfect Reconstruction

For perfect reconstruction, the distortion term $T(z)$ and the aliasing term $A(z)$ should satisfy the following (see Chap. 9)

$$T(z) = H_0(z)G_0(z) + H_1(z)G_1(z) = 2z^{-d} \qquad (10.48)$$

$$A(z) = H_0(-z)G_0(z) + H_1(-z)G_1(z) = 0 \qquad (10.49)$$

The alias term becomes zero if we choose the following relationship between the four filters $H_0(z)$, $H_1(z)$, $G_0(z)$, and $G_1(z)$.

$$G_0(z) = H_1(-z), \ G_1(z) = -H_0(-z) \tag{10.50}$$

Substituting Eq. (10.50) in Eq. (10.48), we have

$$H_0(z)G_0(z) - H_0(-z)G_0(-z) = 2z^{-d} \tag{10.51}$$

Setting $P_0(z) = H_0(z)G_0(z)$, we get

$$P_0(z) - P_0(-z) = 2z^{-d} \tag{10.52}$$

Now, the two-step design procedure for obtaining a perfect reconstruction filter bank is as follows:

Step 1: Design a lowpass filter P_0 satisfying Eq. (10.52).
Step 2: Factorize P_0 into $H_0 G_0$. Then, find H_1 and G_1 using the conditions in Eq. (10.50).

For the factorization of P_0, one of the possibilities is that H_0 is minimum phase and G_0 maximum phase or vice versa. Thus, the coefficients of G_0 are in the reverse order of those for H_0. That is, if

$$\{h_0(n)\} = \{h_0(0), h_0(1), \ldots \ldots h_0(n-1), h_0(n)\} \tag{10.53a}$$

then the coefficients of the filter are G_0 given by

$$\{g_0(n)\} = \{h_0(n), h_0(n-1), \ldots \ldots h_0(1), h_0(0)\} \tag{10.53b}$$

After getting H_0 and G_0, we can find the H_1 and G_1 by using the following relations

$$H_1(z) = G_0(-z) \tag{10.54}$$

$$G_1(z) = -H_0(-z) \tag{10.55}$$

The above frequency-domain relations can be expressed in time domain as follows.

$$h_1(n) = (-1)^n g_0(n) \tag{10.56}$$

$$g_1(n) = (-1)^{n+1} h_0(n) \tag{10.57}$$

10.8.2   *Daubechies Wavelet Filter Coefficients*

Coefficient Domain solution

The stepwise procedure for coefficient domain is as follows

Step 1: Select an even number of coefficients such that $2N = 2, 4, 6, 8$, etc.
Step 2: Form N nonlinear equations using the orthonormality conditions given by (10.33a, 10.33b).
Step 3: Form N linear equations such that all the remaining degrees of freedom are used to impose as much regularity as possible for $H_0(z)$; that is, introduce as many vanishing moments (N) as possible starting from the zeroth using the relation (10.45).
Step 4: Solve the $2N$ equations simultaneously for the $2N$ unknown coefficients to obtain $\{h_0(n)\}$.
Step 5: Obtain the coefficients $\{g_0(n)\}$ from $\{h_0(n)\}$ using Eq. (10.54).
Step 6: Now, obtain the coefficients $\{h_1(n)\}$ and $\{g_1(n)\}$ from $\{g_0(n)\}$ and $\{h_0(n)\}$ using Eqs. (10.56) and (10.57), respectively.

Case (a) N = 1 The above procedure yields the trivial case of Haar wavelets. From (10.33a, 10.33b) and (10.45), we have

$$h_0^2(0) + h_0^2(1) = 1$$
$$h_0(0) - h_0(1) = 1$$

$\vdots$

Thus,

$$h_0(0) = h_0(1) = \frac{1}{\sqrt{2}} \tag{10.58}$$

Hence, from Eq. (10.54),

$$\{g_0(n)\} = \left\{ \frac{1}{\sqrt{2}}, \frac{1}{\sqrt{2}} \right\}$$

Thus, from (10.56)

$$h_1(0) = \frac{1}{\sqrt{2}}, \; h_1(1) = \frac{-1}{\sqrt{2}} \tag{10.59}$$

One can verify that $\{h_0(n)\}$ and $\{h_1(n)\}$ satisfy the relations (10.39) and (10.40), respectively.

Case (b) N = 2 From the orthogonality conditions (10.33a, 10.33b), we have

$$h_0^2(0) + h_0^2(1) + h_0^2(2) + h_0^2(3) = 1 \tag{10.60a}$$

and

$$h_0(2)h_0(0) + h_0(3)h_0(1) = 0 \tag{10.60b}$$

Further, the regularity condition (10.45) gives

$$h_0(0) - h_0(1) + h_0(2) - h_0(3) = 0 \tag{10.61a}$$

and

$$h_0(0) - 1h_0(1) + 2h_0(2) - 3h_0(3) = 0 \tag{10.61b}$$

Solving the above four equations, we get

$$\{h_0(n)\} = \left\{ \frac{1+\sqrt{3}}{4\sqrt{2}}, \frac{3+\sqrt{3}}{4\sqrt{2}}, \frac{3-\sqrt{3}}{4\sqrt{2}}, \frac{1-\sqrt{3}}{4\sqrt{2}} \right\} \tag{10.62}$$

Hence, from Eq. (10.53b),

$$\{g_0(n)\} = \left\{ \frac{1-\sqrt{3}}{4\sqrt{2}}, \frac{3-\sqrt{3}}{4\sqrt{2}}, \frac{3+\sqrt{3}}{4\sqrt{2}}, \frac{1+\sqrt{3}}{4\sqrt{2}} \right\} \tag{10.63}$$

Thus, from (10.56)

$$\{h_1(n)\} = \left\{ \frac{1-\sqrt{3}}{4\sqrt{2}}, -\frac{3-\sqrt{3}}{4\sqrt{2}}, \frac{3+\sqrt{3}}{4\sqrt{2}}, -\frac{1+\sqrt{3}}{4\sqrt{2}} \right\} \tag{10.64}$$

Again, one can verify that $\{h_0(n)\}$ and $\{h_1(n)\}$ satisfy the relations (10.39) and (10.40), respectively.

For large values of N, it is more convenient to determine the sets of the coefficients $\{h_0(n)\}$ and $\{h_1(n)\}$ indirectly through spectral factorization in the frequency domain.

Frequency-domain solutions Daubechies approach [3, 6] provides an elegant way for the design of FIR filters based on spectral factorization. The idea is to have $P_0(z)$ such that it is of the form

$$P_0(z) = z^{-(2N-1)} P(z) \tag{10.65}$$

where

$$P(z) = 2\left(\frac{1+z^{-1}}{2}\right)^N \left(\frac{1+z}{2}\right)^N R(z) \tag{10.66}$$

The function $R(z)$ is given by [3, 6]

$$R(z) = \sum_{k=0}^{N-1}\left(\binom{N-1+k}{k}\left(\frac{1}{2}-\frac{1}{4}(z+z^{-1})\right)^k\right) \tag{10.67}$$

It is seen from Eq. (10.67) that the zeros of $R(z)$ occur in reciprocal pairs, and hence can be written as

$$R(z) = Q(z)Q(z^{-1}) \tag{10.68}$$

We can obtain $Q(z)$ using spectral factorization, where there is a degree of freedom of choice in selecting the factors for $Q(z)$. We can use a minimal phase factorization (by choosing the zeros within the unit circle), or a maximal phase factorization (by choosing the zeros outside of the unit circle), or a combination of the two. Hence, Eq. (10.66) can be rewritten as

$$P(z) = 2\left(\frac{1+z^{-1}}{2}\right)^N \left(\frac{1+z}{2}\right)^N Q(z)Q(z^{-1}) \tag{10.69}$$

In view of Eq. (10.65), the perfect reconstruction condition given by (10.52) reduces to

$$[P(z) + P(-z)]z^{-(2N-1)} = 2z^{-d} \tag{10.70}$$

which can be satisfied with $d = (2\,N - 1)$ and

$$P(z) + P(-z) = 2 \tag{10.71}$$

That is, $P(z)$ is to be designed as a half-band filter. Further, from Eqs. (10.66) and (10.69), we can
write

$$P_0(z) = \left[\sqrt{2}\left(\frac{1+z^{-1}}{2}\right)^N Q(z)\right]\left[z^{-(2N-1)}.\sqrt{2}\left(\frac{1+z}{2}\right)^N Q(z^{-1})\right] \tag{10.72}$$

Since $P_0(z) = H_0(z)G_0(z)$, we may identify $H_0(z)$ and $G_0(z)$ to be

$$H_0(z) = \sqrt{2}\left(\frac{1+z^{-1}}{2}\right)^N Q(z) \tag{10.73}$$

and

$$G_0(z) = z^{-(2N-1)}\cdot\sqrt{2}\left(\frac{1+z}{2}\right)^N Q(z^{-1}) = z^{-(2N-1)}H_0(z^{-1}) \tag{10.74}$$

It can be seen that if $Q(z)$ is chosen as a minimum phase function, then $H_0(z)$ corresponds to a minimum phase filter and $G_0(z)$ to a maximum phase one. Thus, we see from Eq. (10.73) that once we find the coefficients for the analysis LP filter $H_0(z)$, the coefficients of the synthesis LP filter $G_0(z)$ are given by

$$g_0(n) = h_0(2N - 1 - n) \tag{10.75}$$

Of course, the coefficients of the analysis and synthesis HP filters are given by (10.56) and (10.57), respectively. The design of Daubechies-4 (D4) and Daubechies-8 (D8) filters is considered below.

Daubechies-4 (D4) filter
For the D4 filter, $N = 2$. Then from Eq. (10.67),

$$R(z) = 2 - \frac{z}{2} - \frac{z^{-1}}{2}$$

The zeros of $R(z)$ are given by $\alpha = 2 - \sqrt{3}$ and $\left(\frac{1}{\alpha}\right) = 2 + \sqrt{3}$. Hence, $R(z)$ can be written as

$$R(z) = \frac{1}{2}\left(1 - \alpha z^{-1}\right)\left(\frac{1}{\alpha} - z\right) = \frac{(1 - \alpha z^{-1})}{\sqrt{2\alpha}} \cdot \frac{(1 - \alpha z)}{\sqrt{2\alpha}}$$

Choosing the minimal phase factorization for Q(z), we have

$$Q(z) = \frac{(1 - \alpha z^{-1})}{\sqrt{2\alpha}}$$

Hence, from Eq. (10.73), the analysis LP filter is given by

$$H_0(z) = \sqrt{2}\left(\frac{1+z^{-1}}{2}\right)^2\frac{(1 - \alpha z^{-1})}{\sqrt{2\alpha}} = \frac{1}{4\sqrt{\alpha}}\left[1 + (2 - \alpha)z^{-1} + (1 - 2\alpha)z^{-2} - \alpha z^{-3}\right]$$

Substituting $\alpha = 2 - \sqrt{3}$ in the above, we get the coefficients of $H_0(z)$ to be

$$h_0(n) = \left\{ \frac{1+\sqrt{3}}{4\sqrt{2}}, \frac{3+\sqrt{3}}{4\sqrt{2}}, \frac{3-\sqrt{3}}{4\sqrt{2}}, \frac{1-\sqrt{3}}{4\sqrt{2}} \right\}$$

the same as that given by (10.62). Also, from (10.75), the coefficients of the synthesis LP filter $G_0(z)$ are given by

$$g_0(n) = \left\{ \frac{1-\sqrt{3}}{4\sqrt{2}}, \frac{3-\sqrt{3}}{4\sqrt{2}}, \frac{3+\sqrt{3}}{4\sqrt{2}}, \frac{1+\sqrt{3}}{4\sqrt{2}} \right\}$$

Finally, the coefficients of the HP analysis and synthesis filters may be found using (10.56) and (10.57), respectively, as

$$h_1(n) = \left\{ \frac{1-\sqrt{3}}{4\sqrt{2}}, -\frac{3-\sqrt{3}}{4\sqrt{2}}, \frac{3+\sqrt{3}}{4\sqrt{2}}, -\frac{1+\sqrt{3}}{4\sqrt{2}} \right\}$$

and

$$g_1(n) = \left\{ -\frac{1+\sqrt{3}}{4\sqrt{2}}, \frac{3+\sqrt{3}}{4\sqrt{2}}, -\frac{3-\sqrt{3}}{4\sqrt{2}}, \frac{1-\sqrt{3}}{4\sqrt{2}} \right\}$$

Further,

$$P(z) = \frac{1}{16} \left[16 + 9(z+z^{-1}) - (z^3 + z^{-3}) \right]$$

and hence, satisfies the perfect reconstruction condition given by Eq. (10.71).

Daubechies-8 (D8) filter

For D8 filter $N = 4$. Then from Eq. (10.67),

$$R(z) = -\frac{5}{16}z^3 + \frac{5}{2}z^2 - \frac{131}{16}z + 13 - \frac{131}{16}z^{-1} + \frac{5}{2}z^{-2} - \frac{5}{16}z^{-3}$$

The zeros of $R(z)$ are given by $\left(\alpha, \beta, \gamma, \frac{1}{\alpha}, \frac{1}{\beta}, \frac{1}{\gamma}\right)$, where $\alpha = 0.3289, \beta = 0.2841 + j0.2432$ and $\gamma = 0.2841 - j0.2432$.

The zeros of the polynomial $R(z)$ are shown in Fig. 10.12.

Hence, $R(z)$ can be written as

$$R(z) = -\frac{5}{16}(1-\alpha z^{-1})(1-\beta z^{-1})(1-\gamma z^{-1})\left(\frac{1}{\alpha}-z\right)\left(\frac{1}{\beta}-z\right)\left(\frac{1}{\gamma}-z\right)$$

$$= \frac{5}{16\alpha\beta\gamma}\left[(1-\alpha z^{-1})(1-\beta z^{-1})(1-\gamma z^{-1})(1-\alpha z)(1-\beta z)(1-\gamma z)\right]$$

Fig. 10.12 Zeros of R(z)for
the D8 filter

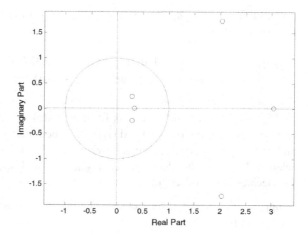

The zeros corresponding to minimal and maximal phase factorizations are shown
in Fig. 10.13.

Choosing the minimal phase factorization

$$Q(z) = \frac{\sqrt{5}}{4\sqrt{\alpha\beta\gamma}}\left(1 - \alpha z^{-1}\right)\left(1 - \beta z^{-1}\right)\left(1 - \gamma z^{-1}\right)$$

Hence, from Eq. (10.73), the analysis LP filter is given by

$$H_0(z) = \sqrt{2}\left(\frac{1 + z^{-1}}{2}\right)^4 \frac{\sqrt{5}}{4\sqrt{\alpha\beta\gamma}}\left(1 - \alpha z^{-1}\right)\left(1 - \beta z^{-1}\right)\left(1 - \gamma z^{-1}\right)$$

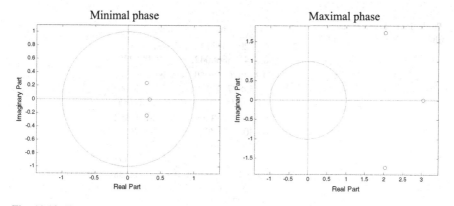

Fig. 10.13 Zeros corresponding to minimal and maximal phase factorizations

Substituting for $\alpha, \beta,$ and γ in the above, we get the coefficients of the analysis filter of $H_0(z)$ as

$\{h_0(n)\} = [0.230377813309, 0.714846570553, 0.630880767930, -0.027983769417,$
$\quad\quad -0.187034811719, 0.030841381836, 0.032883011667, -0.010597401785].$

The coefficients of the synthesis LP filter $G_0(z)$ as well as those of the analysis and synthesis HP filters $H_1(z)$ and $G_1(z)$ may be found using Eqs. (10.75), (10.56), and (10.57), respectively. The following MATLAB Program 10.3 illustrates the relation between the decomposition (h_0, h_1) and the reconstruction (g_0, g_1) Daubechies filters of length 8.

Program 10.3 Relationship between the decomposition and reconstruction filters for perfect reconstruction

```
% clear;clc;
h0=sqrt(2)*dbaux(4);g0=fliplr(h0);p0=conv(h0,g0);
i=0;
for i1=1:8
    h1(i1)=(-1)^(i)*g0(i1);g1(i1)=(-1)^(i+1)*h0(i1);
    i=i+1;
end
p1=conv(h1,g1);distortion=p0+p1;display(distortion)
```

Table 10.1 gives the Daubechies filter coefficients for $N = 2, 3, 4,$ and 5. To generate the scaling function, the stepwise procedure is as follows:

1. Select the lowpass filter h_0 of order N.
2. Set initial scaling function $\varphi = 1$.
3. Upsample $\varphi(n)$ by 2.
4. Convolve h_0 with φ; i.e., let $\varphi = \langle h_0, \varphi \rangle$.
5. Repeat steps 3 and 4 several times.

Table 10.1 Daubechies filters for $N = 2, 3, 4,$ and 5

N	Lowpass filter coefficients
2	0.482962913145, 0.836516303738 0.224143868042, −0.129409522551
3	0.332670552950, 0.806891509311 0.459877502118, −0.135011020010 −0.08544127388, 0.035226291882
4	0.230377813309, 0.714846570553 0.630880767930, −0.027983769417 −0.187034811719, 0.030841381836 0.032883011667, −0.010597401785
5	0.16010, 0. 60383, 0.72431, 0.13843, −0.24229, −0.032245, 0.077571, −0.0062415, −0.012581, 0.0033357

10.8.3   Generation of Daubechies Scaling Functions and Wavelets Using MATLAB

The Daubechies filters can be generated using the MATLAB command dbaux(N) in the wavelet toolbox where N is the order. For $N = 1$, the filter is the Haar filter with length 2, which is the first Daubechies filter. Also, we see that the Nth-order filter has a length of $2N$. The Daubechies scaling functions and wavelets of orders 2, 3, 4, and 5 are shown in Figs. 10.14, 10.15, 10.16, and 10.17, respectively. The following MATLAB Program 10.4 generates the scaling function and wavelet given the order N and the number of iterations.

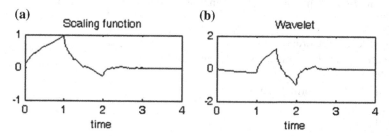

Fig. 10.14 **a** Scaling function corresponding to Daubechies filter for $N = 2$. **b** Wavelet corresponding to Daubechies filter for $N = 2$

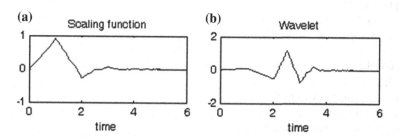

Fig. 10.15 **a** Scaling function corresponding to Daubechies filter for $N = 3$. **b** Wavelet corresponding to Daubechies filter for $N = 3$

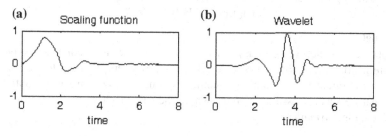

Fig. 10.16 **a** Scaling function corresponding to Daubechies filter for $N = 4$. **b** Wavelet corresponding to Daubechies filter for $N = 4$

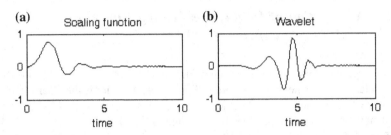

Fig. 10.17 a Scaling function corresponding to Daubechies filter for $N = 5$. **b** Wavelet corresponding to Daubechies filter for $N = 5$

Program 10.4 Orthogonal scaling functions and wavelet generation

```
clear;clc;
N=input('enter the order N=');
iter=input('enter the number of iterations iter=');
lpf=dbaux(N);
lpf=sqrt(2)*lpf; %normalization
lpfr=fliplr(lpf);
for i=0:2*N-1
    hpf(i+1)=(-1)^i*lpfr(i+1);
    end
l=length(lpf);
% Scaling function generation
y= [1];s=y;
for i=1:iter
    s=conv(upsample(s,2),lpf);
    end
%wavelet function generation
x=[1];
w=conv(upsample(x,2),hpf);
K=iter-1;for i=1:K
    w=conv(upsample(w,2),lpf);
    end
time=(1/2)^iter*(1:length(w));
phi=2^((iter-1)/2)*s;w=2^((iter-1)/2)*w;
subplot(321);plot(time,phi);xlabel('time');title('Scaling function');
subplot(322);plot(time,w);xlabel('time');title('Wavelet');
```

Symmlets

Daubechies wavelets are very asymmetric. Symmlets are also Daubechies wavelets, but are nearly symmetric. To construct Daubechies filters, P_0 is factorized so that H_0 is a minimum phase filter. As another option, factorization can be optimized such that H_0 has almost linear phase. This produces much more symmetric filters that are called *Symmlets*.

The wavelet, the scaling function, and the four associated filters for symlet4 are shown in Fig. 10.18. The following MATLAB commands are used to generate Fig. 10.18.

```
[phi,psi]=wavefun('sym4',10);
[lpfd,hpfd]=wfilters('sym4','d');
[lpfr,hpfr]=wfilters('sym4','r');
```

In the above commands, ten indicate the number of iterations, sym4 indicates the wavelet name, d stands for decomposition, and r for reconstruction.

10.9 Biorthogonal Scaling Functions and Wavelets Generation

The decomposition and reconstruction filter banks for the biorthogonal case are shown in Fig. 10.19.

The biorthogonal wavelet transform uses distinct analysis and synthesis scaling function/wavelet pairs. H_0 and H_1 are the filters used for decomposition corresponding to the scaling function $\varphi(t)$ and wavelet function $\psi(t)$. The dual $\widetilde{H}_0$ and $\widetilde{H}_1$, and $\tilde{\varphi}(t)$ and $\tilde{\psi}(t)$ are used for reconstruction. The functions $\varphi(t)$, $\psi(t)$, $\tilde{\varphi}(t)$ and $\tilde{\psi}(t)$ are given by

$$\varphi(t) = \sum_k h_0(k)\varphi(2t - k) \tag{10.76}$$

$$\psi(t) = \sum_k h_1(k)\varphi(2t - k) \tag{10.77}$$

$$\tilde{\varphi}(t) = \sum_k \tilde{h}_0(k)\tilde{\varphi}(2t - k) \tag{10.78}$$

$$\tilde{\psi}(t) = \sum_k \tilde{h}_1(k)\tilde{\varphi}(2t - k) \tag{10.79}$$

In the biorthogonal wavelet decomposition and reconstruction, the scaling functions and wavelets are not orthogonal to each other. But the filters satisfy the reconstruction condition, since $\varphi(t)$ is orthogonal to $\tilde{\psi}(t)$ and $\tilde{\varphi}(t)$ is orthogonal to $\psi(t)$.

For biorthogonal wavelets, the following relations hold good.

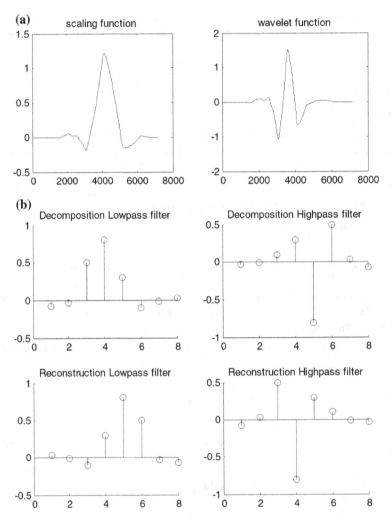

Fig. 10.18 a Scaling and wavelet function for Symlet4. **b** Decomposition and reconstruction filters for Symlet4

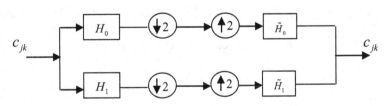

Fig. 10.19 Decomposition and reconstruction filter bank for biorthogonal wavelet

$$\left\langle \tilde{\phi}(t), \phi_m(t) \right\rangle = \delta_m$$

$$\left\langle \tilde{\psi}_{j,k}(t), \psi_{l,m}(t) \right\rangle = \delta_{j-l,k-m} \tag{10.80}$$

$$\left\langle \tilde{\phi}_{j,k}(t), \psi_{l,m}(t) \right\rangle = \left\langle \tilde{\psi}_{j,k}(t), \phi_{l,m}(t) \right\rangle = 0$$

In the frequency domain, the relations are

$$\sum_k \tilde{\phi}(\omega + 2\pi k)\phi^*(\omega + 2\pi k) = 1 \tag{10.81}$$

$$\sum_k \tilde{\psi}(\omega + 2\pi k)\psi^*(\omega + 2\pi k) = 1 \tag{10.82}$$

$$\sum_k \tilde{\psi}(\omega + 2\pi k)\phi^*(\omega + 2\pi k) = 0 \tag{10.83}$$

$$\sum_k \tilde{\phi}(\omega + 2\pi k)\psi^*(\omega + 2\pi k) = 0 \tag{10.84}$$

which reduce to

$$\tilde{H}_0(\omega)H_0^*(\omega) + \tilde{H}_0(\omega + \pi)H_0^*(\omega + \pi) = 1 \tag{10.85}$$

$$\tilde{H}_1(\omega)H_1^*(\omega) + \tilde{H}_1(\omega + \pi)H_1^*(\omega + \pi) = 1 \tag{10.86}$$

$$\tilde{H}_1(\omega)H_0^*(\omega) + \tilde{H}_1(\omega + \pi)H_0^*(\omega + \pi) = 0 \tag{10.87}$$

$$\tilde{H}_0(\omega)H_1^*(\omega) + \tilde{H}_0(\omega + \pi)H_1^*(\omega + \pi) = 0 \tag{10.88}$$

This can be written in matrix form as

$$\begin{bmatrix} \tilde{H}_0(\omega) & \tilde{H}_0(\omega + \pi) \\ \tilde{H}_1(\omega) & \tilde{H}_1(\omega + \pi) \end{bmatrix} \begin{bmatrix} H_0^*(\omega) & H_1^*(\omega) \\ H_0^*(\omega + \pi) & H_1^*(\omega + \pi) \end{bmatrix} = \begin{bmatrix} 1 & 0 \\ 0 & 1 \end{bmatrix} \tag{10.89}$$

It can be seen from [13] that the following relationships satisfy the conditions given in Eq. (10.89)

$$\tilde{H}_1(\omega) = e^{-j\omega}H_0^*(\omega + \pi) \tag{10.90}$$

$$H_1(\omega) = e^{-j\omega}\tilde{H}_0^*(\omega + \pi) \tag{10.91}$$

10.9.1   *Biorthogonal Wavelet Filter Coefficients*

The polynomial $P(z)$ for biorthogonal filters is the same as that for the Daubechies orthogonal filters, which is recalled and written as

$$P(z) = 2\left(\frac{1+z^{-1}}{2}\right)^N \left(\frac{1+z}{2}\right)^N R(z) \tag{10.92}$$

It is not possible to design perfect reconstruction filter banks with linear phase analysis and synthesis filters by factorization as was done in Sect. 10.8.2 for the Daubechies orthonormal filters. However, it is possible to maintain the perfect reconstruction condition with linear phase filters by choosing a different factorization scheme. To this end, we factorize $z^{(-2N-1)}P(z) = H_0(z)\widetilde{H}_0(z)$, where $H_0(z)$ and $\check{H}_0(z)$ are linear phase filters.

The relations between the decomposition and reconstruction filters in the time domain are given [14] by

$$\widetilde{H}_1(z) = -H_0(-z) \tag{10.93}$$

$$H_1(z) = \widetilde{H}_0(-z) \tag{10.94}$$

The relations between the decomposition and reconstruction filters in time domain are given by

$$\tilde{h}_1(n) = (-1)^{n+1} h_0(n) \tag{10.95}$$

$$h_0(n) = (-1)^n \tilde{h}_0(n) \tag{10.96}$$

The condition for perfect reconstruction remains the same as before, namely

$$P(z) + P(-z) = 2$$

For $\underline{N = 2}$

$$P(z) = \frac{1}{16}\left[-z^{-3} + 9z^{-1} + 16 + 9z - z^3\right] \tag{10.97a}$$

$$P_0(z) = z^{-3}P(z) = \frac{1}{16}\left[-1 + 9z^{-2} + 16z^{-3} + 9z^{-4} - z^{-6}\right] \tag{10.97b}$$

The above polynomial has four zeros located at $z = -1$, a zero at $z = 0.26795$ and a zero at $z = 3.7321$.

Equation (10.97b) can be factored in several different ways to determine the linear phase analysis filters $H_0(z)$ and $\widetilde{H}_0(z)$. One way of factorizing the above equation is as follows.

$$H_0(z) = \frac{1}{8}\left(-1 + 2z^{-1} + 6z^{-2} + 2z^{-3} - z^{-4}\right)$$
$$\widetilde{H}_0(z) = \frac{1}{2}\left(1 + 2z^{-1} + z^{-2}\right)$$

Since the length of $H_0(z)$ is 5 and the length of $\check{H}_0(z)$ is 3, the above choice is for the decomposition lowpass and reconstruction lowpass filters of the 5/3 Daubechies filters [3].

Another way of factorizing Eq. (10.97b) is as follows.

$$H_0(z) = \frac{1}{4}\left(-1 + 3z^{-1} + 3z^{-2} - z^{-3}\right)$$
$$\widetilde{H}_0(z) = \frac{1}{4}\left(1 + 3z^{-1} + 3z^{-2} + z^{-3}\right)$$

Since the length of $H_0(z)$ is 4 and the length of $\check{H}_0(z)$ is 4, the above choice is for decomposition lowpass and reconstruction lowpass filters of the 4/4 Daubechies filters [3]. Similarly for N = 4, one possible way of factorizing $z^{-7}P(z)$ will lead to CDF 9/7 biorthogonal filters. They are very frequently used filters for lossy compression applications. The coefficients of the decomposition filters for CDF9/7 biorthogonal wavelet filter are given in Table 10.2.

The following MATLAB Program 10.5 is used to generate the scaling function and wavelet for CDF 9/7 biorthogonal filter.

Table 10.2 Coefficients of decomposition filters for CDF 9/7 biorthogonal wavelet filter	Decomposition lowpass filter	Decomposition highpass filter
	$h_0(0) = 0.85270$	$h_0(0) = 0.78849$
	$h_0(\pm 1) = 0.37740$	$h_0(\pm 1) = -0.41809$
	$h_0(\pm 2) = -0.11062$	$h_0(\pm 2) = -0.040689$
	$h_0(\pm 3) = -0.023849$	$h_0(\pm 3) = 0.064539$
	$h_0(\pm 4) = 0.037828$	

Program 10.5 Biorthogonal scaling function and wavelet generation

```
clear;clc;
[RF,DF]=biorwavf('bior4.4');
% bior4.4 represents biorthogonal filter (CDF 9/7) for
%decomposition and reconstruction
iter=input('enter the number of iterations iter=');
 flag=input('enter 1 for decomposition, and 2 for reconstruction');
 lpfd=sqrt(2)*DF;% Decomposition lowpass filter
lpfr=sqrt(2)*RF;% Reconstruction lowpass filter
 if flag==1
lpf=lpfd;%Decomposition lowpass filter
 i=0;
for i1=1:7
  hpf(i1)=(-1)^(i)*lpfr(i1); % Decomposition highpass filter
  i=i+1;
end
end
if flag==2
    lpf=lpfr;%Reconstruction lowpass filter
 i=0;
for i1=1:9
  hpf(i1)=(-1)^(i+1)*lpfd(i1);% Reconstruction highpass filter
  i=i+1;
end
end
 ls = length(lpf)-1; %support of the scaling function
 lw = (ls +length(hpf)-1)/2; %support of the wavelet function
   elf = lpf*lpf';ehf = hpf*hpf';lpf = lpf/sqrt(elf); hpf = hpf/sqrt(ehf);
   l1 = 2*length(lpf)-1;
%upsample
   s1 = zeros([1 l1]);s1(1:2:l1) = lpf;
 s = sqrt(2)*conv(s1,lpf); %first iteration of the scaling function
   l1 = 2*length(hpf)-1;
%upsample
   w1 = zeros([1 l1]);w1(1:2:l1) = hpf;
   w = sqrt(2)*conv(w1,lpf); %first iteration for the wavelet
 %begin rest of the iterations
  for i=1:(iter-1)
   l1 = 2*length(s)-1;
  %upsample
   s1 = zeros([1 l1]); s1(1:2:l1) = s;
   s = sqrt(2)*conv(s1,lpf); %scaling function
   l1 = 2*length(w)-1;
  %upsample
```

```
w1 = zeros([1 11]);w1(1:2:11) = w;
w = sqrt(2)*conv(w1,lpf); %wavelet function
end
l1 = length(s);  l2 = length(w);s = sqrt(elf)*[s 0];w = sqrt(ehf)*[w 0];
ts = 0:ls/l1:ls; %support for scaling function
tw = 0:lw/l2:lw; %support for wavelet function
%Plot the scaling function and wavelet
subplot(321);plot(ts,s);xlabel('time');title('Scaling function');
subplot(322);  plot(tw,-w);xlabel('time');title('Wavelet');
subplot(322);  plot(tw,w);xlabel('time');title('Wavelet');
```

The scaling functions and wavelets generated from Program 10.5 are shown in Figs. 10.20 and 10.21 for decomposition and reconstruction, respectively.

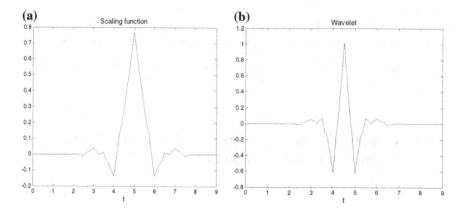

Fig. 10.20 **a** Decomposition of scaling function. **b** Decomposition of wavelet

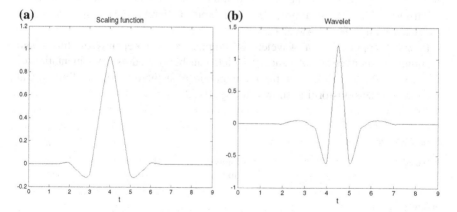

Fig. 10.21 **a** Reconstruction of the scaling function. **b** Reconstruction of the wavelet

10.10   The Impact of Wavelet Properties

Orthogonality: The orthogonality property provides unique and non-redundant representation, and preserves energy in the transform domain. Hence, any quantization scheme optimal in the transform domain is also optimal in the spatial domain.

Linear Phase: Linear phase property handles efficiently the image boundaries, preserves centers of mass, and reduces the blurring of the fine-scale features of the image. Thus, it is an important property for image compression at higher compression ratios.

Compact Support: The ringing effects in the image caused by subsequent quantization can be reduced by the compact support property, and also the compact support lowers the computational complexity due to the finite number of filter coefficients.

Regularity: Higher regularity provides better compression due to greater energy compaction, but with a proportionally increased computational complexity.

The impact of wavelet properties for orthogonal and biorthogonal wavelets is given in Table 10.3.

10.11   Computation of One-Dimensional Discrete Wavelet Transform (1-D DWT) and Inverse DWT (1-D IDWT)

Wavelet transform coefficients can be computed in a way similar to that of the Fourier transform coefficients, by using the wavelets as the basis functions. However, wavelets allow the use of a filter down-sample operation scheme to which samples of the waveform are fed as input. Different filters can be designed associated with different wavelets.

Practical application of wavelets is simple, since it depends on filters and sampling rate conversion. As such wavelets can be viewed as implementation of practical filters. A scheme for the computation of a three-level one-dimensional discrete wavelet transform is shown in Fig. 10.22.

Table 10.3 Impact of wavelet properties

Wavelet family	Orthogonality	Compact support	Linear phase	Real coefficients
Daubechies orthogonal	Yes	Yes	No	Yes
Daubechies biorthogonal	No	Yes	Yes	Yes

This scheme is illustrated using the following MATLAB Program 10.6 for the computation of a three-level discrete wavelet transform of the speech signal mtlb.mat from the MATLAB signal processing toolbox sampled at 7418 Hz.

Program 10.6

```
clear;clc;
load mtlb % mtlb is speech signal consisting of
%4001 samples from signal processing toolbox of MATLAB
L=1024*4;
mtlb(4002:4096)=0; % appended zeros to make the speech signal length is a power
of 2
s=mtlb+1*rand (1, L)';
%define filters h0 and h1
N=2; % order of the filter is chosen as 2
h0=dbaux(N); %[0.3415 0.5915 0.1585 -0.0915];
for i=0:2*N-1
h1(i+1)=(-1)^i*h0(2*N-i);%[-0.0915 -0.1585 0.5915 -0.3415];
end
s0=conv (h0,s);s1=conv (h1,s);% First iteration begins
s0=s0'; s1=s1';
s0=s0(1,2:L+1); s1=s1(1,2:L+1);
s0=reshape (s0, 2, L/2); s1=reshape (s1, 2, L/2);
c21=s0 (1, :); d21=s1 (1, :);
figure(1) subplot(311);plot(s); xlabel('time'); ylabel('s');
subplot(312);plot(c21);xlabel('time');ylabel('c_2_1');
subplot(313);plot(d21);xlabel('time');ylabel('d_2_1'); % First iteration ends
s=c21;L=L/2;%Second iteration begins
s0=conv (h0,s);s1=conv (h1,s);
s0=s0(1,2:L+1);s1=s1(1,2:L+1);
s0=reshape (s0, 2, L/2);
s1=reshape (s1, 2, L/2);
c11=s0 (1, :);d11=s1 (1, :);
figure(2) subplot(311);plot(s);xlabel('time');ylabel('s');
subplot(312);plot(c11);xlabel('time');ylabel('c_1_1');
subplot(313);plot(d11);xlabel('time');ylabel('d_1_1');% Second iteration ends
s=c11;L=L/2;%Third iteration begins
s0=conv (h0,s);s1=conv (h1,s);
s0=s0(1,2:L+1);s1=s1(1,2:L+1);
s0=reshape (s0, 2, L/2);s1=reshape (s1, 2, L/2);
c00=s0 (1, :);d00=s1 (1, :);
figure(3) subplot(311);plot(s);xlabel('time');ylabel('s');
subplot(312);plot(c00);xlabel('time');ylabel('c_0_0');
subplot(313);plot(d00);xlabel('time');ylabel('d_0_0');% Third iteration ends
```

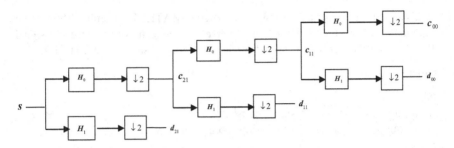

Fig. 10.22 Scheme for the computation of three-level discrete wavelets transform (DWT)

The signals produced after the first-level decomposition are shown in Fig. 10.23a. The signal s is the input to the first stage; the input signal s is filtered and down-sampled by a factor of 2 to produce the output signals c_{21} and d_{21} where c_{21} and d_{21} are the scaling function and the wavelet coefficients, respectively, at this level. The input signal s contains 4096 samples, and the two outputs c_{21} and d_{21} contain 2048 samples each due to the down-sampling by a factor of 2. The second-level decomposition shown in Fig. 10.23b is produced by continuing the process shown in Fig. 10.22 on the scaling function coefficients c_{21} to produce the second-level outputs c_{11} and d_{11}. The outputs c_{11} and d_{11} now contain 1024 samples each. Figure 10.23c shows the third-level decomposition. Here, the third-level output signals c_{00} and d_{00} each contain 512 samples. In the three-level decomposition, from the signals shown in Fig. 10.23a–c, a general trend can be noticed that the c signal resembles more like the original signal at the higher level, and each higher level extracts more and more noise from the signal. This illustrates the signal denoising application of the wavelets.

Inverse Discrete Wavelet Transform

A scheme for a three-level reconstruction of one-dimensional inverse discrete wavelet transform is shown in Fig. 10.24, using which the signal s can be reconstructed.

In the first level of reconstruction, both c_{00} and d_{00} are upsampled, filtered, and then added to produce c_{11}. In the second-level reconstruction, the process is continued with c_{11} and d_{11} to produce c_{21}. Finally, both c_{21} and d_{21} are upsampled, filtered, and then added to produce s.

The following MATLAB Program 10.7 is used to reconstruct the signal s using the inverse discrete wavelet transform scheme shown in Fig. 10.24. The reconstructed signal is shown in Fig. 10.25.

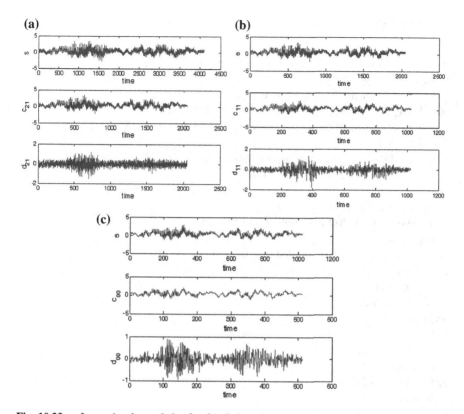

Fig. 10.23 a Input signal s and the first-level decomposition signals c_{21}, d_{21}. **b** Signal s and second-level decomposition signals c_{11}, d_{11}. **c** Signal s and third-level decomposition signals c_{00}, d_{00}

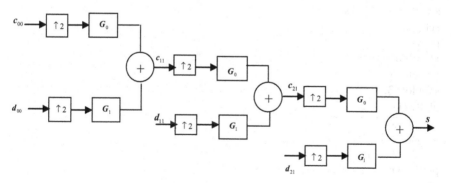

Fig. 10.24 Inverse wavelet transform

Program 10.7

```
clear;clc;
load canddcoeff % c and d coefficients from program 10.6
%define filters g0 and g1
N=2; % order of the filter is chosen as 2
g0=dbaux(N); %[0.3415 0.5915 0.1585 -0.0915];
g0=fliplr(g0);%[-9.1506e-002 1.5849e-001 5.9151e-001 3.4151e-001];
for i=0:2*N-1
g1(i+1)=(-1)^(i+1)*g0(2*N-i);%[-3.4151e-001
5.9151e-001-1.5849e-001-9.1506e-002];
end
db=upsample(d00,2);
dc=conv(db,g1);% d00 upsampling and filtered
cb=upsample(c00,2);
cc=conv(cb,g0);%c00 upsampled and filtered
c11=2*(cc+dc);%
n=length(c11);
c11=c11(1,2:n-2);
db1=upsample(d11,2);
dc1=conv(db1,g1);%d11 upsampled and filtered
cb1=upsample(c11,2);
cc1=conv(cb1,g0);% c11 upsampled and filtered
c21=2*(cc1+dc1);
n=length(c21);
c21=c21(1,2:n-2);
db2=upsample(d21,2);
dc2=conv(db2,g1);% d21 upsampled and filtered
cb2=upsample(c21,2);
cc2=conv(cb2,g0);%c21 upsampled and filtered
s=2*(cc2+dc2);%s
subplot (3, 1, 1);
plot(s);
xlabel('time');
ylabel('s')
subplot (3, 1,2);
plot(c21);
xlabel('time');
ylabel ('c_2_1')
subplot(3,1,3);
plot(d21);
xlabel('time');
ylabel('d_2_1');
```

It is seen that the reconstructed signal in Fig. 10.25 is almost identical to that of Fig. 10.20a, since perfect reconstruction filters were used.

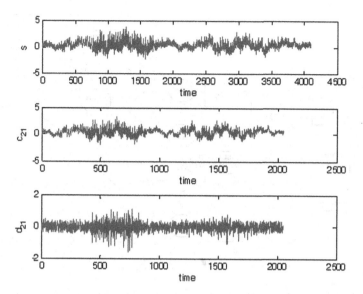

Fig. 10.25 Reconstructed signal

10.12   2-D Discrete Time Wavelet Transform

It is possible to extend the 1-D DWT to two dimensions, wherein the 2-D DWT is separable with two separate scaling functions $\varphi(x)$ and $\varphi(y)$ so that

$$\varphi(x,y) = \varphi(x)\varphi(y) \tag{10.98}$$

The above equation represents image smoothing, and three different kinds of wavelets representing the horizontal, vertical, and diagonal details of the image are expressed as

$$\psi^h(x,y) = \varphi(x)\psi(y) \tag{10.99}$$

$$\psi^v(x,y) = \psi(x)\varphi(y) \tag{10.100}$$

$$\psi^d(x,y) = \psi(x)\psi(y) \tag{10.101}$$

Since the 2-D basis functions have been expressed as products of the corresponding 1-D basis functions, the wavelet decomposition and reconstruction of a 2-D image can be performed by filtering the rows and columns of the image separately.

A scheme for single-level decomposition of images is shown in Fig. 10.26a. In this decomposition scheme, the 2-D DWT at each stage is computed by applying a single-stage 1-D DWT along the rows of the image, followed by a 1-D DWT along

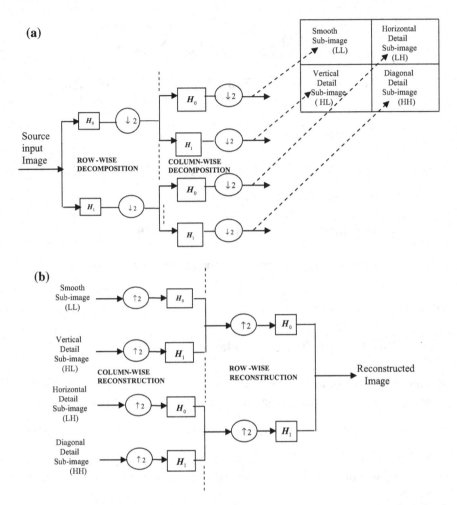

Fig. 10.26 a Scheme for single-level decomposition of images. **b** Scheme for single-level reconstruction of images

the columns of the resulting image. The outputs of the decomposition scheme are termed as low-low (LL), low-high (LH), high-low (HL), and high-high (HH). LL indicates the subband of the image obtained after lowpass filtering and down-sampling both the rows and the columns. HL indicates that the image corresponds to the subband obtained by highpass filtering and down-sampling the rows, and lowpass filtering and down-sampling the columns. This gives the horizontal edge details. LH indicates that the image corresponds to the subband obtained by lowpass filtering and down-sampling the rows, and highpass filtering and down-sampling the columns. It yields the vertical edge details. HH indicates that the image with diagonal details corresponds to the subband obtained by

highpass filtering and down-sampling both the rows and the columns. A scheme for
a single-level reconstruction of images is shown in Fig. 10.26b.

10.12.1   Computation of 2-D DWT and IDWT Using MATLAB

The MATLAB Program 10.8 illustrates the single-level decomposition and
reconstruction of an image.

Program 10.8

```
% 2D DWT and IDWT
load woman2 % loads image file
close all;clf
figure(1);image(X);colormap(map)
axis image; set(gca,'XTick',[],'YTick',[]); title('Original')
% Compute a 1-level decomposition of the image using the 9/7 filters.
wname='bior4.4'
[wc,s] = wavedec2(X,2,'bior4.4');%the CDF 9/7 filters used
a1 = appcoef2(wc,s,wname,1);% extracts approximate coefficients at level 1
[h1,v1,d1]=detcoef2('all',wc,s,1);% extracts horizontal, vertical
%diagonal detail coefficients at level 1
% Display the decomposition up to level 1.
ncolors = size(map,1);              % Number of colors.
sz = size(X);
ca1 = wcodemat(a1,ncolors); cod_a1 = wkeep(ca1, sz/2);
ch1 = wcodemat(h1,ncolors); cod_h1 = wkeep(ch1, sz/2);
cv1 = wcodemat(v1,ncolors); cod_v1 = wkeep(cv1, sz/2);
cd1 = wcodemat(d1,ncolors); cod_d1 = wkeep(cd1, sz/2);
figure(2)
image([cod_a1,cod_h1;cod_v1,cod_d1]);
axis image; set(gca,'XTick',[],'YTick',[]); title('Single stage decomposition');colormap(map)
% The reconstructed branches
ra1 = wrcoef2('a',wc,s,'bior4.4',1);rh1 = wrcoef2('h',wc,s,'bior4.4',1);rv1 =wrcoef2('v',wc,s,'bior4.4',1);
rd1 = wrcoef2('d',wc,s,'bior4.4',1);cod_ra1 = wcodemat(ra1,ncolors);
cod_rh1=wcodemat(rh1,ncolors);
cod_rv1 = wcodemat(rv1,ncolors);cod_rd1 = wcodemat(rd1,ncolors);
figure(3)
image([cod_ra1,cod_rh1;cod_rv1,cod_rd1]);
axis image;
set(gca,'XTick',[],'YTick',[]);
```

title('Single stage reconstruction');
colormap(map)
%Adding together the reconstructed average and
% the reconstructed details give the full reconstructed image
Xhat = ra1 + rh1 + rv1 + rd1;%reconstructed image
sprintf('Reconstruction error (using wrcoef1) = %g', max(max(abs(X-Xhat))))
figure(4); image(Xhat);
axis image;
set(gca,'XTick',[],'YTick',[]);
title('reconstructed')
colormap(map)

The single-level decomposition and reconstruction of the image 'Barbara' are shown in Figs. 10.27 and 10.28, respectively. The reconstruction error is 4.64922e-010.

10.13   Wavelet Packets

The approximation space V can be decomposed into a direct sum of the two orthogonal subspaces defined by their basis functions given by Eqs. (10.24) and (10.25). A disadvantage of the wavelet transformation is the rigid frequency resolution. The 'splitting trick' used in Eqs. (10.24) and (10.25) can be used to decompose the detailed spaces W as well. For example, if we analogously define

(a)   **(b)**

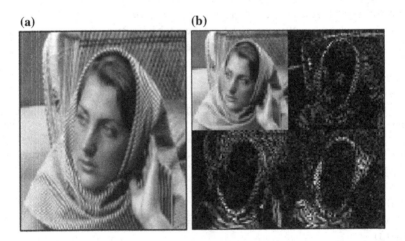

Fig. 10.27 **a** Original image. **b** Single-level decomposition

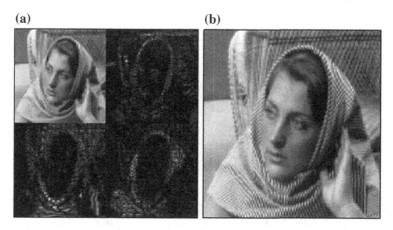

Fig. 10.28 a Single-level reconstruction. **b** Reconstructed image

$$w_2(t) = \sqrt{2} \sum_k h_0(k)\psi(2t-k) \qquad (10.102a)$$

and

$$w_3(t) = \sqrt{2} \sum_k h_1(k)\psi(2t-k) \qquad (10.102b)$$

then $\{w_2(t-k)\}$ and $\{w_3(t-k)\}$ are orthogonal basis functions for the two subspaces whose direct sum is w_1. In general, for $n = 0, 1, 2...$ we define a sequence of functions as follows.

$$w_{2n}(t) = \sqrt{2} \sum_k h_0(k)w_n(2t-k) \qquad (10.103a)$$

and

$$w_{2n+1}(t) = \sqrt{2} \sum_k h_1(k)w_n(2t-k) \qquad (10.103b)$$

Thus, it is possible to obtain an arbitrary frequency resolution within the framework of multiresolution analysis. Such an extension of multiresolution analysis is called as wavelet packet analysis and was first introduced in [7, 15].

Setting $n = 0$ in Eqs. (10.103a, 10.103b), we get $w_0(t) = \varphi(t)$, the scaling function, and $w_1(t) = \psi(t)$, the mother wavelet. So far, the combination of $\varphi(2^j t - k)$ and $\psi(2^j t - k)$ is used to form a basis for V_j. Now, a whole sequence of functions $w_n(t)$ are available at our disposal. Various bases for the function space emerge from the various combinations of $w_n(t)$ and their dilations and translations. A whole collection of orthonormal bases generated from $\{w_n(t)\}$ is called a 'library

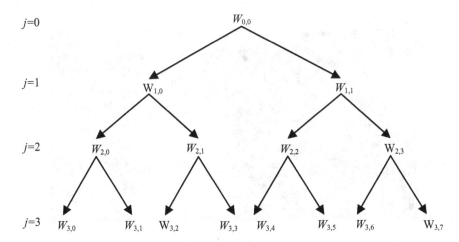

Fig. 10.29 Wavelet packets organized in a tree

of wavelet packet bases,' and the function $w_{n,j,k}(t) = 2^{j/2}w_n(2^j t - k)$ is called a wavelet packet.

The recursive splitting of detailed vector spaces W_j can be represented by a binary tree. The node of the binary tree is labeled as (j, p), where the scale j is the depth of the node in the tree and p is the number of nodes that are on its left at the same depth j. To each node (j, p), a space $W_{j,p}$ is associated. At the root, $W_{0,0} = V_0$ and $\psi_{0,0} = \varphi_0(t)$. Such a binary tree is shown in Fig. 10.29. The tree in Fig. 10.29 is shown for a maximum level decomposition equal to 3. For each scale j, the possible values of parameter p are 0, 1, ..., $2j - 1$.

The two wavelet packet orthogonal bases at the children nodes are defined as:

$$\psi_{j+1,2p}(t) = \sum_n h_0(n)\psi_{j,p}(t - 2^j n) \qquad (10.104)$$

$$\psi_{j+1,2p+1}(t) = \sum_n h_1(n)\psi_{j,p}(t - 2^j n) \qquad (10.105)$$

The following MATLAB command can be used to obtain wavelet packets of CDF9/7 wavelet filter:

[wp,x]=wpfun('bior4.4',7);

The wavelet packets for CDF9/7 wavelet at depth 3 are shown in Fig. 10.30.
Admissible Tree:
Any binary tree where each node can have 0 or 2 children is called an admissible tree, as shown in Fig. 10.31.

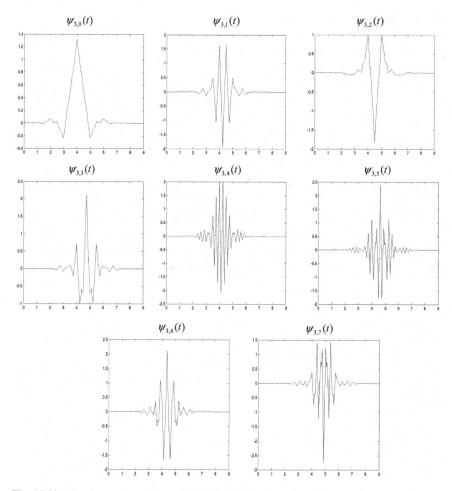

Fig. 10.30 Wavelet packets computed with CDF9/7 biorthogonal filter at a depth of 3

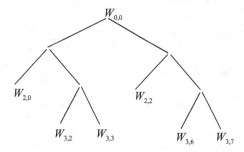

Fig. 10.31 Example of an admissible wavelet packet binary tree

Wavelet packet Decomposition

Wavelet packet coefficients can be computed by the following proposition that generalizes the fast wavelet transform [4].

At the Decomposition:

$$d_{j+1,2p}(k) = d_{j,p} * h_0(2k) \qquad (10.106a)$$

$$d_{j+1,2p+1}(k) = d_{j,p} * h_1(2k) \qquad (10.106b)$$

At the Reconstruction:

$$d_{j,p}(k) = d_{j+1,2p} * g_0(k) + d_{j+1,2p+1} * g_1(k) \qquad (10.107)$$

All of the wavelet packet coefficients can be computed by iterating the above equations along the branches of a wavelet packet tree as shown in Fig. 10.32a. The wavelet packet coefficients $d_{j,p}$ at the top of the tree are recovered as shown in Fig. 10.32b by using Eq. (10.105) for each node inside the tree.

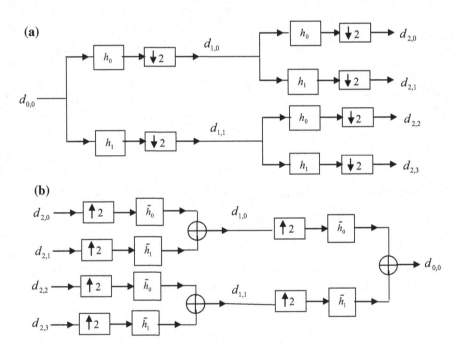

Fig. 10.32 a Wavelet packet filter bank decomposition with successive filtering and subsampling. **b** Reconstruction by inserting zeros and filtering the outputs

10.14 Image Wavelet Packets

In the wavelet-based image compression, only the approximation subband is successively decomposed. With the wavelet packet decomposition, the detailed subbands are also further decomposed and, in turn, each subband creates four more new subbands called approximation subband, horizontal details subband, vertical details subband, and diagonal details subband. Furthermore, each of these four subbands can be recursively decomposed at will. As a result, the decomposition can be represented by a quad-tree.

In image wavelet packets, the elements of wavelet packet bases are product of two separate wavelet packets with the same scale along x_1 and x_2:

$$\psi_{j,p}(x_1 - 2^j n_1) \quad \text{and} \quad \psi_{j,q}(x_2 - 2^j n_2).$$

Each node of this quad-tree corresponds to a separable space labeled by depth j, and two integers p and q.

$$W_{j,p,q} = W_{j,p} \otimes W_{j,q} \tag{10.108}$$

and the separable wavelet packet for $x = (x_1, x_2)$ is

$$\psi_{j,p,q}(x) = \psi_{j,p}(x_1)\psi_{j,q}(x_2) \tag{10.109}$$

The following one-dimensional wavelet packets satisfy [4]

$$\begin{aligned} W_{j,p} &= W_{j+1,2p} \oplus W_{j+1,2p+1} \\ W_{j,q} &= W_{j+1,2q} \oplus W_{j+1,2q+1} \end{aligned} \tag{10.110}$$

Substitution of Eq. (10.110) in Eq. (10.108) yields $W_{j,p,q}$ as the direct sum of the four subspaces

$$W_{j+1,2p,2q}, W_{j+1,2p+1,2q}, W_{j+1,2p,2q+1} \text{ and } W_{j+1,2p+1,2q+1}.$$

These four subspaces are represented by the four children nodes in the quad-tree as shown in Fig. 10.33.

Admissible Quad-Tree

A quad-tree with either zero or four children nodes is called admissible quad-tree [4].

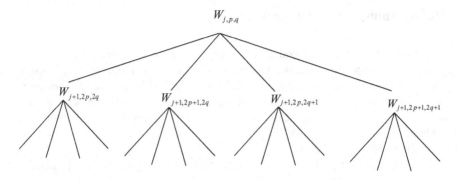

Fig. 10.33 A wavelet packet quad-tree for images

Program 10.9 Wavelet packet image decomposition and restoration at two levels

```
clear;clc;
load lenimage
Xd=im2double(Xd);
imshow(Xd)
i=1;%initial level=1
if(i==1)
T=wpdec2(Xd,i,'bior4.4','shannon')%wavelet packet image decomposition at level
1
for j=1:4
x=wpcoef(T,j);% computes wavelet packet coefficients at terminal nodes of level 1
figure(1)
subplot(5,4,j);%plots wavelet packet decomposition of image at level 1
imshow(mat2gray(x));
end
xr1=wprec2(T);% reconstructed image at level 1
figure (2),imshow(mat2gray(xr1));%plots wavelet packet reconstruction of image at
level 1
i=i+1;
end
T=wpdec2(Xd,2,'bior4.4','shannon'); %wavelet packet image decomposition at
level 2
for j=5:20
x=wpcoef(T,j);%computes wavelet packet coefficients at terminal nodes of level 2
figure(3)
subplot(5,4,j);%plots wavelet packet decomposition of image at level 2
imshow(mat2gray(x));
end
xr2=wprec2(T);% reconstructed image at level 2
figure(4),
imshow(mat2gray(xr2));%plots wavelet packet reconstruction of image at level 1
```

(a)

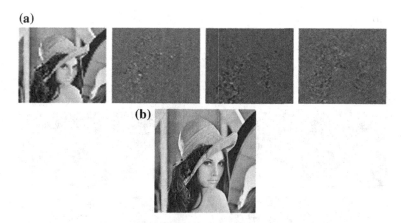

(b)

Fig. 10.34 a Wavelet decomposition of Lenna image at the terminal node at level 1.
b Reconstructed image

The wavelet decomposition and reconstruction of the image 'Lenna' at the terminal node at level 1 and 2 are shown in Figs. 10.34 and 10.35, respectively.

10.15   Some Application Examples of Wavelet Transforms

Many applications of wavelet transforms are available in the literature [16–26]. Denoising, watermarking, and compression are some of the widely used applications of wavelets. These are discussed in the following subsections.

10.15.1   Signal Denoising

The model of a noisy signal $r(n)$ may be written as

$$r(n) = s(n) + \eta(n) \qquad (10.111)$$

where $s(n)$ represents the original signal and $\eta(n)$ the noise. A wavelet transform of the above equation yields the relation

$$W_r = W_a + W_\eta \qquad (10.112)$$

where the vector W_r corresponds to the wavelet coefficients of the noisy signal, W_a contains the wavelet coefficients of the original signal, and W_η are the noise wavelet coefficients. After applying a threshold δ, the modified wavelet coefficients $W_{r\delta}$ of the degraded signal can be obtained. The inverse wavelet transforms of $W_{r\delta}$ yield

(a)

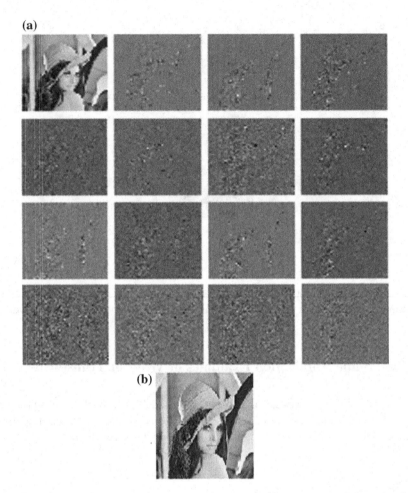

(b)

Fig. 10.35 a Wavelet decomposition of Lenna image at the terminal node at level 2.
b Reconstructed image

the restored image. Choosing the value of the threshold is a very fundamental problem in order to avoid oversmoothing or undersmoothing. The most known threshold estimation methods are: SURE threshold method; universal threshold method; and the method based on SURE and universal threshold methods. The SURE estimator as a threshold is given by [27]

$$\text{SURE}(\delta) = T(\delta) + \frac{N - 2N_0}{N} \sigma^2 \tag{10.113}$$

where $T(\delta) = \frac{1}{N} \| W_{r\delta} - W_r \|$, N is the total number of wavelet coefficients, N_0 is the number of coefficients that were replaced by zeros, and σ^2 is the noise variance.

However, the optimal choice of the threshold depends on the noise energy present. The universal threshold given by [28]

$$\delta = \sqrt{2\log(N)}\sigma \qquad (10.114)$$

uses explicitly the noise energy dependency and chooses threshold proportional to σ. If the amount of noise is different for the various coefficients, then it is difficult to remove it decently by only one threshold. The threshold estimators based on the SURE and universal methods are as follows.

The noise variance σ^2 at each level j is estimated using the following robust estimator [29]:

$$\sigma_j^{-2} = \frac{MAD(W_{rjk},\ k = 1...2^k)}{0.6745} \qquad (10.115)$$

where MAD denotes the median absolute deviation from 0 and the factor 0.6745 is chosen for calibration with the Gaussian distribution. The soft thresholding and hard thresholding are shown in Fig. 10.36a, b, respectively.

In the case of hard thresholding, the estimated coefficients will be

$$
\begin{aligned}
W_{r\delta} &= W_r \quad \text{if } |W_r| \geq \delta \\
&= 0 \qquad \text{otherwise}
\end{aligned} \qquad (10.116)
$$

In the case of soft thresholding, the estimated coefficients will be

$$
\begin{aligned}
W_{r\delta} &= W_r - \delta \quad \text{if } W_r \geq \delta \\
&= 0 \qquad\quad \text{if } |W_r| \leq \delta \\
&= W_r + \delta \quad \text{if } W_r \leq -\delta
\end{aligned} \qquad (10.117)
$$

The following MATLAB Program 10.10 illustrates the denoising of a voice signal from the sound file 'theforce.wav'. The original voice signal, the noisy voice signal, and the reconstructed voice signal using soft and hard thresholdings are shown in Fig. 10.37a–d, respectively.

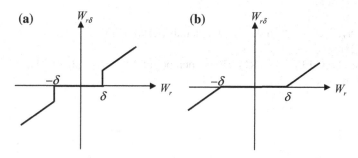

Fig. 10.36 a Hard thresholding and **b** soft thresholding

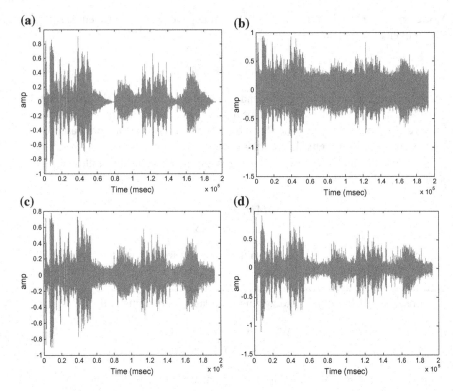

Fig. 10.37 a Original speech signal 'the force.wav'. **b** The noisy speech signal, SNR = 1.2 dB. **c** Denoising speech signal using symlet4 with soft thresholding, SNR = 6.5 dB. **d** Denoising speech signal using symlet4 with hard thresholding, SNR = 6.7 dB

Program 10.10 Voice signal denoising using wavelet transform

[x, fs, bps]=wavread('theforce.wav');% reads the wav file of a voice sigal

N=size(x);

figure(1);plot(x);
xn=x'+0.1*randn(1,N);
figure(2);plot(xn);
[C,S]=wavedec(xn,3,'sym4');sig=median(abs(C))/0.6745;
th=sqrt(2*log10(length(C)))*sig;
[XC,CXC,LXC,PERF0,PERFL2]=wdencmp('gbl',C,S,'sym4',3,th,'s',1);
figure(3);plot(XC)

10.15.2 Image Denoising

The model of a degraded image may be written as

$$r(m, n) = s(m, n) + \eta(m, n) \qquad (10.118)$$

where $s(m, n)$ represents the image pixel in the mth row and nth column, and $\eta(m, n)$ the observation noise. The peak signal-to-noise ratio (PSNR) in dB is defined as

$$\text{PSNR} = 10 \ \log_{10} \frac{P_c^2}{\text{MSE}_c} \qquad (10.119)$$

where P_c and MSE_c are, respectively, the peak pixel value of the noise-free image and the mean square error of the corresponding channel.

The following MATLAB Program 10.11 illustrates denoising of Lenna image using soft thresholding. The original image, the noisy image, and the reconstructed image are shown in Fig. 10.38a–c, respectively.

Program 10.11 Image denoising using wavelet transform

```
load lenimage % loads lenna image
figure(1);imshow(mat2gray(Xd))
rn=10*randn(512);

    Xdn=Xd+rn;

figure(2);
imshow(mat2gray(Xdn))
[C,S]=wavedec2(Xdn(:,:,1),3,'bior4.4');
sig=median(abs(C))/0.6745;
th=sqrt(2*log10(length(C)))*sig;
[XC,CXC,LXC,PERF0,PERFL2]=wdencmp('gbl',C,S,'bior4.4',3,th,'s',1);
figure(3);
imshow(mat2gray(XC))
```

10.15.3 Digital Image Water Marking

Digital watermarking has emerged as a tool for protecting the multimedia data from copyright infringement. In digital watermarking, an imperceptible signal 'mark' is embedded into the host image, which uniquely identifies the ownership. After embedding the watermark, there should be no perceptual degradation. These watermarks should not be removable by an unauthorized person and should be robust against intentional and unintentional attacks. The discrete wavelet transform

(a)   (b)

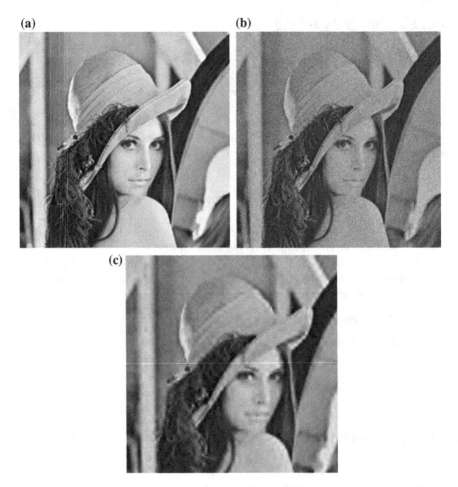

(c)

Fig. 10.38 a Original image. **b** Noisy image (PSNR = 1 dB). **c** Denoised image with soft thresholding (PSNR = 10.5 dB)

(DWT)-based watermarking techniques are gaining popularity, since the DWT has a number of advantages, such as progressive and low bit-rate transmission, quality scalability and region-of-interest (ROI) coding, over the other transforms.

The Watermark Embedding Process The general image watermarking process is shown in Fig. 10.39.

The watermark embedding process [30] can be described in the following steps.

Step 1: Obtian the first-level horizontal detail information (HL1) by decomposing the image using bioorthogonal wavelet transform.

Step 2: Get the second-level detail coefficients (HL2) from HL1 obtained in Step 1 using biorthogonal wavelet transform.

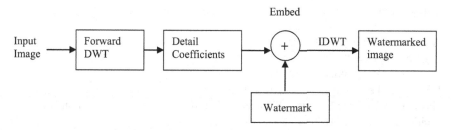

Fig. 10.39 Watermarking scheme

Step 3: Generate a PN sequence whenever a watermark bit is zero, and add it to HL2 resulting from Step 2.

Step 4: Perform the inverse of the DWT (IDWT) twice to obtain the watermarked image.

The Watermark Extraction Process

The watermark extraction process can be described by the following steps:

Step 1: Perform the DWT to obtain the horizontal detail information (HL1).

Step 2: Again, decompose HL1 using DWT to obtain HL2.

Step 3: Compute the correlation between the detail coefficients resulting from Step 2 (HL2) and the PN sequence.

Step 4: Set the watermark pixel value to zero (black) whenever the correlation is greater than a certain threshold.

The following MATLAB Program 10.12 illustrates the watermarking of the Lenna image using biorthogonal wavelet transform and the extraction of the watermark. The original image (input image), the watermark, the watermarked image, the extracted watermark, and the extracted original image are shown in Fig. 10.40a–e, respectively.

Program 10.12

```
clc;
close all;
clear all;
h=imread('lenna.tif'); % reading image from a jpg image
hg=rgb2gray(h); % gray conversion from rgb
figure(1),imshow(hg)
[A1,H1,V1,D1]=dwt2(double(hg),'bior6.8'); % first level decomposition using
biorthogonal wavelet
[A2,H2,V2,D2]=dwt2(H1,'bior6.8'); % second level decomposition using
%biorthogonal wavelet(horizontal coefficients)
w=imread('nertu.jpg');
wg=double(im2bw(w)); % image to binar
```

```
% key required to both embedding & extracting watermark
key=rgb2gray(imread('key.jpg'));
key=imresize(key,[35 1],'nearest');
key=double(key)/(256);
[rw1 cw1 rc cc]=blocksize(wg,H2);
H2=embed(key,wg,H2,rw1,cw1,rc,cc);
i1=idwt2(A2,H2,V2,D2,'bior6.8',[size(A1,1) size(A1,2)]); % inverse transform
i2=idwt2(A1,i1,V1,D1,'bior6.8'); % inverse transform
figure(2),imshow(mat2gray(double(i2)));
%watermark extraction code
[A1,H1,V1,D1]=dwt2(i2,'bior6.8'); % first level decomposition using biorthogonal
wavelet
[A2,H2,V2,D2]=dwt2(H1,'bior6.8'); % second level decomposition using
[rw1 cw1 rc cc]=blocksize(wg,H2);
extract=extract(key,wg,H2,rw1,cw1,rc,cc);
host1=idwt2(A2,H2,V2,D2,'bior6.8',[size(A1,1) size(A1,2)]); % inverse transform
hostextract=idwt2(A1,host1,V1,D1,'bior6.8'); % inverse transform
figure(3);
imshow(extract);
figure(4);
imshow(uint8(hostextract));
```

The functions used in the above program are listed in Appendix at the end of the chapter.

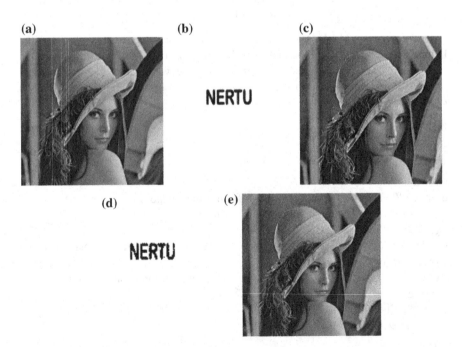

Fig. 10.40 **a** Original image, **b** watermark, **c** watermarked image, **d** extracted watermark, **e** extracted original image

10.15.4   Image Compression

A popular lossy image compression standard is the discrete cosines transform (DCT)-based Joint Photographic Experts Group (JPEG) [31, 32]. The procedure consists of partitioning image into non-overlapping blocks of *8 × 8 pixels* in size and to obtain the DCT of each block. Then, the transformed coefficients of each block are quantized and entropy coded to form the output code stream. It is possible to attain localization of information both in space and frequency by using the DCT of blocks rather than localization of information in frequency alone by using a single DCT of the whole image. In the reconstruction of the image, the output code stream is decoded, dequantized, and undergoes the inverse discrete cosine transform. The disadvantage of the JPEG is the presence of the blocking artifacts in the compressed images at higher compression ratios. This may be due to the non-overlapping blocks.

The JPEG2000 is the new standard for still image compression. It is a discrete wavelet transform (DWT)-based standard. A block diagram of the JPEG2000 encoder is shown in Fig. 10.41a. It is similar to any other transform-based coding scheme. First, the DWT is applied on the input source image and then the transformed coefficients are quantized and entropy decoded to the output code bit stream. The decoder is depicted in Fig. 10.41b. The decoder simply performs inverse operations of the encoder. The output bit stream is first entropy decoded and dequantized. Inverse discrete wavelet transform is performed on the dequantized coefficients and level shifted to produce the reconstructed image. A worthy feature of JPEG2000 is that it can be both lossy and lossless unlike other coding schemes. This depends on the wavelet transform and the quantization applied. In this standard, markers are added in the bit stream to allow error resilience and it also allows tiling of the image. The advantages of tiling are that memory requirements are reduced and they can be used for decoding specific parts of the image instead of the entire image, since the tiles can be reconstructed independently. All samples of the image tile component are DC level shifted by subtracting the same quantity before applying the forward DWT on each tile.

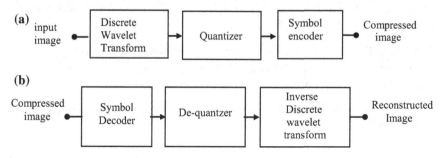

Fig. 10.41 Block diagram of image compression using the JPEG2000 standard: **a** encoder and **b** decoder

(a)   (b)

Fig. 10.42 **a** JPEG compression ratio 100:1. **b** JPEG2000 compression ratio 100:1

For lossless compression, the reversible DWT is applied using a five-tap/ three-tap biorthogonal wavelet filter. In lossy applications, irreversible DWT is employed using a nine-tap/seven-tap biorthogonal wavelet filter. Performance evaluation of the DCT-based JPEG and the DWT-based JPEG2000 for lossy compression of Elaine 512×512 image is shown in Fig. 10.41. From Fig. 10.42, the superiority of the JPEG2000 over JPEG can be clearly observed for image compression at high compression ratios.

10.16   Problems

1. Verify Daubechies wavelet filter coefficients for $N = 5$ using frequency-domain solution.
2. Consider a two-channel perfect reconstruction filter bank with the analysis filters $h_0(n) = \{-1, 2, 6, 2, -1\}/4\sqrt{2}$ and $h_1(n) = \{1, 2, 1\}/2\sqrt{2}$. Consider also the signal $x(n) = \{0, 1, -1, 2, 5, 1, 7, 0\}$.

    (a) Find the corresponding dual (synthesis) filters $\tilde{h}_0(n)$ and $\tilde{h}_1(n)$
    (b) Plot the zeros of the filters $h_0(n), h_1(n), \tilde{h}_0(n)$ and $\tilde{h}_1(n)$
    (c) Plot single-level decomposition and reconstruction signals of $x(n)$.

3. Consider the factorization of $P_0(z)$ in order to obtain orthogonal or biorthogonal filter banks.

(a) Take

$$P_0(z) = -\frac{1}{4}z^3 + \frac{1}{2}z + 1 + \frac{1}{2}z^{-1} - \frac{1}{4}z^{-3}$$

Build an orthogonal filter bank based on this $P_0(z)$. If the function is not positive on the unit circle, apply an adequate correction (see Smith–Barnwell method in Sect. 3.2.3 [9]).

(b) Alternatively, compute a linear phase factorization of $P_0(z)$. In particular, choose

$$H_0(z) = z + 1 + z^{-1}.$$

Give the other filters in this biorthogonal filter bank.

10.17 MATLAB Exercises

1. Modify Program 10.2 to compute STFT of a speech signal.
2. Modify Program 10.8 for a three-level decomposition and reconstruction of an image, and show the images at each level of the decomposition and reconstruction.
3. Modify Program 10.9 for a three-level decomposition of an image using wavelet packets.
4. Modify Program 10.10 for the denoising of a speech signal corrupted by colored noise.
5. Modify Program 10.11 for the denoising of an image corrupted by colored noise.

Appendix

blocksize.m

```
%Here watermark image is referred with 8x8 block size
function [rw1 cw1 rc cc]=blocksize(wg,H2)
w1=mod(size(wg,1),8);
w2=mod(size(wg,2),8);
if(w1==0)
    w1=w1+8;
end;
if(w2==0)
    w2=w2+8;
end;
rw=(size(wg,1)-w1)/8;
rw1=rw;
cw=(size(wg,2)-w2)/8;
cw1=cw;
c1=mod(size(H2,1),rw);
c2=mod(size(H2,2),cw);
if(c1==0)
    c1=mod(size(H2,1),rw+1);
    rw1=rw+1;
end;
if(c2==0)
    c2=mod(size(H2,2),cw+1);
    cw1=cw+1;
end;
rc=(size(H2,1)-c1)/rw1;
cc=(size(H2,2)-c2)/cw1;
embed.m
```

```
function H2=embed(key,wg,H2,rw1,cw1,rc,cc)
g=2;
rand('state',key);
cr1=1;
wr1=1;
wmd1=[];
for i=1:rw1
    wmd2=[];
    cr2=i*rc;
    wr2=i*8;
    if(i==rw1)
```

```
        cr2=size(H2,1);
        wr2=size(wg,1);
    end;
    cc3=1;
    wc3=1;
    for j=1:cw1
        cc4=j*cc;
        wc4=j*8;
        if(j==cw1)
            cc4=size(H2,2);
            wc4=size(wg,2);
        end;
        h=H2(cr1:cr2,cc3:cc4);
        msg=wg(wr1:wr2,wc3:wc4);
        msg=reshape(msg,size(msg,1)*size(msg,2),1);
      for k=1:length(msg)
            pn=3*round(2*(rand(size(h,1),size(h,2))-
0.5)); % generation of PN sequence
            if msg(k)==0
                h=h+g*pn;
            end;
        end;
        wmd2=[wmd2 h];
        cc3=cc4+1;
        wc3=wc4+1;
    end;
    wmd1=[wmd1;wmd2];
    cr1=cr2+1;
    wr1=wr2+1;
end;
H2=wmd1;
extract.m
```

```
function extract=extract(key,wg,H3,rw1,cw1,rc,cc)
g=2;rand('state',key);cr1=1;wr1=1;p=1;correlation=ones(size(wg,1)*size(wg,2),1);
for i=1:rw1
    cr2=i*rc;  wr2=i*8;
    if(i==rw1)
        cr2=size(H3,1); wr2=size(wg,1);
    end;
    cc3=1; wc3=1;
    for j=1:cw1
        cc4=j*cc;wc4=j*8;
        if(j==cw1)
```

```
        cc4=size(H3,2);wc4=size(wg,2);
      end;
    h=H3(cr1:cr2,cc3:cc4);msg=wg(wr1:wr2,wc3:wc4);msg=reshape(msg,size
(msg,1)*size(msg,2),1);
      for k=1:length(msg)
        pn=3*round(2*(rand(size(h,1),size(h,2))-
0.5)); % generation of PN sequence
        correlation(p)=corr2(h,g*pn);p=p+1;
      end;
      cc3=cc4+1;wc3=wc4+1;
    end;
    cr1=cr2+1;wr1=wr2+1;
end;
threshold=mean(abs(correlation));
p=1;wr1=1;
wmd1=[];
for i=1:rw1
    wmd2=[];wr2=i*8;
    if(i==rw1)
        wr2=size(wg,1);
    end;
    wc3=1;
    for j=1:cw1
        wc4=j*8;
        if(j==cw1)
            wc4=size(wg,2);
        end;
        msg=wg(wr1:wr2,wc3:wc4);
        we=ones(size(msg,1),size(msg,2));msg=reshape(msg,size(msg,1)*size
(msg,2),1);
        for k=1:length(msg)
            if(correlation(p)>threshold)
                we(k)=0;
            end;
            p=p+1;
        end;
        wmd2=[wmd2 we];wc3=wc4+1;
    end;
    wmd1=[wmd1;wmd2]; wr1=wr2+1;end;extract=wmd1;
extracthost.m
```

```
function H2=extracthost(key,wg,H2,rw1,cw1,rc,cc)
g=1;
rand('state',key);
```

```
cr1=1;
wr1=1;
wmd1=[];
for i=1:rw1
   wmd2=[];
   cr2=i*rc;
   wr2=i*8;
   if(i==rw1)
      cr2=size(H2,1);
      wr2=size(wg,1);
   end;
   cc3=1;
   wc3=1;
   for j=1:cw1
      cc4=j*cc;
      wc4=j*8;
      if(j==cw1)
         cc4=size(H2,2);
         wc4=size(wg,2);
      end;
      h=H2(cr1:cr2,cc3:cc4);
      msg=wg(wr1:wr2,wc3:wc4);
      msg=reshape(msg,size(msg,1)*size(msg,2),1);
      for k=1:length(msg)
         pn=3*round(2*(rand(size(h,1),size(h,2))-
0.5)); % generation of PN sequence
         if msg(k)==0
            h=h-g*pn;
         end;
      end;
      wmd2=[wmd2 h];
      cc3=cc4+1;
      wc3=wc4+1;
   end;
   wmd1=[wmd1;wmd2];
   cr1=cr2+1;
   wr1=wr2+1;
end;
H2=wmd1;
```

References

1. A. Grossman, J. Morlet, Decompositions of hardy functions into square integrable wavelets of constant shape. SIAM J Math Anal **15**(4), 75–79 (1984)
2. Y. Meyer, *Ondelettes et fonctions splines Seminaire EDP* (Ecole Polytechnique, Paris, 1986)
3. I. Daubechies, Orthonormal bases of compactly supported wavelets. Commun. Pure Appl. Math. **41**(7), 909–996 (1988)
4. S. Mallat, *A Wavelet Tour of Signal Processing*, 2nd edn. (Academic Press, 1999)
5. G.S. Strang, T.Q. Nguyen, *Wavelets and Filter Banks* (Wellesley Cambridge Press, MA, 1996)
6. I. Daubechies, *Ten Lectures on Wavelets* (SIAM, CBMS Series, April 1992)
7. R. Coifman, Y. Meyer, S. Quaker, V. Wickerhauser, *Signal Processing and Compression with Wavelet Packets* (Numerical Algorithms Research Group, New Haven, CT: Yale University, 1990)
8. C.K. Chui, An introduction to wavelets, Academic Press,1992
9. M. Vetterli, J. Kovacevic, *Wavelets and Subband Coding* (Prentice-Hall, 1995)
10. P. Steffen et al., Theory of regular M-band wavelet bases. IEEE Trans. Signal Process. **41**, 3487–3511 (1993)
11. Peter N. Heller, Rank-m wavelet matrices with n vanishing moments. SIAM J. Matrix Anal. **16**, 502–518 (1995)
12. Gilbert Strang, Wavelets and dilation equations. SIAM Rev. **31**, 614–627 (1989)
13. A. Cohen, I. Daubechies, J.C. Feauveau, Biorthogonal bases of compactly supported wavelets. Commun. Pure Appl. Math. **XLV**, 485–560 (1992)
14. S.K. Mitra, *Digital Signal Processing—A Computer Based Approach* (McGraw-Hill Inc, NY, 2006)
15. M.V. Wickerhauser, Acoustic Signal Compression with Wavelet Packets, in *Wavelets: A Tutorial in Theory and Applications*, ed. by C.K. Chui (New York Academic, 1992), pp. 679–700
16. M. Misiti, Y. Misiti, G. Oppenheim, J-M. poggi, *Wavelets and Their Applications* (ISTEUSA, 2007)
17. M.Rao Raghuveer, Ajit S. Bopardikar, *Wavelet Transforms: Introduction to Theory and Applications* (Pearson Education, Asia, 2000)
18. Y. Zhao, M.N.S. Swamy, Technique for designing biorthogonal wavelet filters with an application to image compression. *Electron. Lett.* 35 (18) (September 1999)
19. K.D. Rao, E.I. Plotkin, M.N.S. Swamy, An Hybrid Filter for Restoration of Color Images in the Mixed Noise Environment, in *Proceedings of ICASSP*, vol. 4 (2002), pp. 3680–3683
20. N. Gupta, E.I. Plotkin, M.N.S. Swamy, Wavelet domain-based video noise reduction using temporal dct and hierarchically-adapted thresholding. IEE Proceedings of—Vision, Image and Signal Processing **1**(1), 2–12 (2007)
21. M.I.H. Bhuiyan, M.O. Ahmad, M.N.S. Swamy, Wavelet-based image denoisingwith the normal inverse gaussian prior and linear minimum mean squared error estimator. IET Image Proc. **2**(4), 203–217 (2008)
22. M.I.H. Bhuiyan, M.O. Ahmad, M.N.S. Swamy, Spatially-adaptive thresholding in wavelet domain for despeckling of ultrasound images. IET Image Proc. **3**, 147–162 (2009)
23. S.M.M. Rahman, M.O. Ahmad, M.N.S. Swamy, Video denoising based on inter-frame statistical modeling of wavelet coefficients. IEEE Trans. Circuits Syst. Video Technol. **17**, 187–198 (2007)
24. S.M.M. Rahman, M.O. Ahmad, M.N.S. Swamy, Bayesian wavelet-based imagedenoising using the gauss-hermite expansion. IEEE Trans. Image Process. **17**, 1755–1771 (2008)
25. S.M.M. Rahman, M.O. Ahmad, M.N.S. Swamy, A new statistical detector for dwt-based additive image watermarking using the gauss-hermite expansion. IEEE Trans. Image Process. **18**, 1782–1796 (2009)

26. S.M.M. Rahman, M.O. Ahmad, M.N.S. Swamy, Contrast-based fusion of noisy images using discrete wavelet transform. IET Image Proc. **4**, 374–384 (2010)
27. D.L. Donoho, L.M. Johnstone, Adapting to unknown smoothness via wavelet shrinkage. J Am. Stat. Soc. **90**, 1200–1224 (1995)
28. D.L. Donoho, L.M. Johnstone, Ideal spatial adaptation via wavelet shrinkage. Biometrika **81**, 425–455 (1994)
29. L.M. Johnstone, B.W. Silverman, Wavelet threshold estimators for data with correlated noise. J Roy. Stat. Soc. B **59**, 319–351 (1997)
30. K.D. Rao, New Approach for Digital Image Watermarking and transmission over Bursty WirelessChannels, in *Proceedings of IEEE International Conference on Signal Processing* (Beijing, 2010)
31. G.K. Wallace, The JPEG still picture compression standard. IEEE Trans. Consumer Electron. **38**, xviii–xxxiv (1992)
32. W.B. Pennebaker, J.L. Mitcell, *JPEG: Still Image Data Compression Standard* (Van Nostrand Reinhold, New York, 1993)

Chapter 11
Adaptive Digital Filters

Digital signal processing plays a major role in the current technical advancements. However, standard DSP techniques are not enough to solve the signal processing problems fast enough with acceptable results in the presence of changing environments and changing system requirements. Adaptive filtering techniques must be utilized for accurate solution and a timely convergence in the changing environments and changing system requirements. This chapter discusses principle of adaptive digital filters, cost function, adaptation algorithms, convergence of the adaptation algorithms, and illustration of application of adaptive filters to real-world problems.

11.1 Adaptive Digital Filter Principle

The block diagram of a basic adaptive filter is shown in Fig. 11.1. The input signal $x(n)$ is fed to adaptive filter that produces the output signal $y(n)$. The output signal $y(n)$ is compared to a desired signal $d(n)$ (usually includes noise component also). The error signal $e(n)$ is defined as the difference between the desired signal $d(n)$ and the output signal $y(n)$; i.e.,

$$e(n) = d(n) - y(n) \tag{11.1}$$

The adaptive algorithm uses the error signal to adapt the filter coefficients from time n to time $(n + 1)$ in accordance with a performance criterion. As time n is incremented, the output signal $y(n)$ approaches a better match to the desired signal $d(n)$ through the adaptation process that minimizes the cost function of the error signal $e(n)$.

© Springer Nature Singapore Pte Ltd. 2018
K. D. Rao and M. N. S. Swamy, *Digital Signal Processing*,
https://doi.org/10.1007/978-981-10-8081-4_11

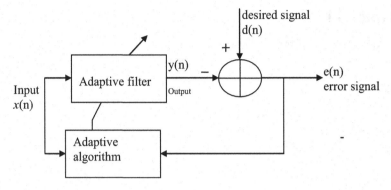

Fig. 11.1 Block diagram of basic adaptive filter

11.2 Adaptive Filter Structures

The input–output relation of an adaptive filter can be characterized by

$$
\begin{aligned}
y(n) = f(W, \ y(n-1), \ y(n-2), \ \ldots, \ y(n-N), \\
x(n), \ x(n-1), \ x(n-2), \ \ldots, \ x(n-M))
\end{aligned}
\tag{11.2}
$$

where $f(.)$ represents a linear or nonlinear function, W is the adaptive filter coefficients (weights) vector.

11.2.1 Adaptive FIR Filter Structure

For an FIR adaptive filter, the vector w is the impulse response of the filter at time index n. The output signal $y(n)$ can be expressed as

$$
\begin{aligned}
y(n) &= \sum_{i=0}^{N} w_i x(n-i) \\
&= W^{\mathrm{T}} X(n)
\end{aligned}
\tag{11.3}
$$

where N is the filter order

$$
X(n) = [x(n) \ x(n-1) \ \ldots \ x(n-N)]^{\mathrm{T}} = \text{input signal vector}
$$
$$
W = [w_0 \ w_1 \ \ldots \ w_N] = \text{filter coefficients vector.}
$$

The filter assumes the same structure as that of an FIR filter structure described in Chap. 7.

11.2.2 Adaptive IIR Filter Structure

The output $y(n)$ of an IIR adaptive filter can be expressed by

$$y(n) = \sum_{i=1}^{N} a_i(n)y(n-i) + \sum_{j=0}^{M} b_i(n)x(n-j)$$
$$= W^T U(n)$$

(11.4)

where

$$U(n) = [y(n-1)\ y(n-2)\dots y(n-N)\quad x(n)\ x(n-1)\dots x(n-M)]^T$$
$$W = [a_1\ a_2\dots a_N\ b_0\ b_1\ b_2\dots b_M]$$

The filter structure is the same as that of IIR filter structures described in Chap. 7.

11.3 Cost Functions

The filter coefficients are adapted minimizing a cost function. The cost function is defined as a norm of the error signal $e(n)$. The most commonly used cost functions are: the mean square cost function, and the exponentially weighted least squares error cost function.

11.3.1 The Mean Square Error (MSE) Cost Function

The mean square error (MSE) cost function is defined as

$$J_{MSE}(n) = E[e^2(n)] = E\left[(d(n) - y(n))^2\right]$$

(11.5)

where E stands for expectation operation.

11.3.2   The Exponentially Weighted Least Square (WLS) Error Function

The exponentially weighted least square error function is defined as

$$J_{WLS}(n) = \sum_{n=0}^{N-1} \lambda^{N-1-n} e^2(n) \qquad (11.6)$$

where N is the total number of samples and λ is an exponential weighting factor having positive value close to 1.

11.4   Algorithms for Adaptive Filters

11.4.1   The LMS Algorithm

It is known that the output $y(n)$ of an adaptive filter with FIR filter structure is given by

$$y(n) = W^T X(n) \qquad (11.7)$$

The error $e(n)$ between the desired signal and the adaptive filter output signal is

$$e(n) = d(n) - W^T X(n) \qquad (11.8)$$

Then, the square of the error is expressed as

$$e^2(n) = d^2(n) - 2d(n)X^T(n)W + W^T X(n)X^T(n)W \qquad (11.9)$$

Now, $J_{MSE}(n)$ becomes

$$\begin{aligned} J_{MSE}(n) &= E[d^2(n)] - 2E[d(n)X^T(n)W] + E[W^T X(n)X^T(n)W] \\ &= \sigma_d^2 - 2R_{dx}^T W + W^T R_{xx} W \end{aligned} \qquad (11.10)$$

where

$\sigma_d^2 = E[d^2(n)]$ is the variance of the desired signal $d(n)$
$R_{dx} = E[d(n)X(n)]$ cross-correlation vector between $d(n)$ and $X(n)$
$R_{xx} = E[X(n)X^T(n)]$ autocorrelation matrix of $X(n)$

The optimal weight vector W_{opt} is obtained by minimizing J_{MSE}. Thus, taking the derivative of $J_{MSE}(n)$ with respect to the weight vector W and setting it to zero leads to the following Wiener solution in optimal filtering

$$\frac{\partial J_{\text{MSE}}(n)}{\partial W} = -2R_{dx} + 2WR_{xx} = 0 \tag{11.11}$$

Solving Eq. (11.11), we obtain W_{opt} as

$$W_{\text{opt}} = R_{xx}^{-1} R_{dx} \tag{11.12}$$

In adaptive filtering, the Wiener solution is obtained using steepest descent method through an iterative procedure as

$$W(n+1) = W(n) - \mu \frac{\partial J_{\text{MSE}}(n)}{\partial W(n)} \tag{11.13}$$

where μ is a step size parameter. It is a positive number that controls the convergence rate and stability of the algorithm.

Equation (11.13) can be rewritten as

$$W(n+1) = W(n) - \mu \frac{\partial E[e^2(n)]}{\partial W(n)} \tag{11.14}$$

where $W(n)$ is the adaptive filter weights vector at time index n.

Replacing $\frac{\partial E[e^2(n)]}{\partial W(n)}$ by $\frac{\partial e^2(n)}{\partial W(n)}$ in the above equation, we arrive at Widrow's LMS algorithm [1] as follows

$$
\begin{aligned}
W(n+1) &= W(n) - \mu \frac{\partial e^2(n)}{\partial W(n)} \\
&= W(n) - \mu \frac{\partial e^2(n)}{\partial e(n)} \frac{\partial e(n)}{\partial W(n)} \\
&= W(n) - 2\mu e(n) \frac{\partial [d(n) - W^{\mathrm{T}} X(n)]}{\partial W(n)} \\
&= W(n) + 2\mu e(n) X(n) \tag{11.15}
\end{aligned}
$$

The Widrow's LMS algorithm has the advantages of low computational complexity, simplicity of implementation, real-time implementation, and does not require statistics of signals.

11.4.2   The NLMS Algorithm

The normalized LMS (NLMS) algorithm is an LMS algorithm [2, 3] with time-varying step size as

$$W(n+1) = W(n) + 2\mu(n)e(n)X(n) \qquad (11.16)$$

The time-varying step size $\mu(n)$ is given by

$$\mu(n) = \frac{\mu}{c + X^{\mathrm{T}}(n)X(n)} \qquad (11.17)$$

where c is a small positive constant to avoid division by zero.

When $\mu = 0$, the updating halts and if $\mu = 1$, the convergence is faster at the cost of high misadjustment.

Making $c = 0$, it can be easily shown that the NLMS converges if

$$0 < \mu < 0.5 \qquad (11.18)$$

11.4.3   The RLS Algorithm

The optimization criterion for the RLS algorithm is to minimize the weighted sum of squares

$$J_{\mathrm{WLS}}(n) = \sum_{i=0}^{n} \lambda^{n-i} e^2(i) \qquad (11.19)$$

for each time n where λ is a weighting factor such that $0 < \lambda \leq 1$.

Differentiating $J_{\mathrm{WLS}}(n)$ with respect to the filter weight vector $W(n)$ at time n, and equating it to zero, we obtain

$$\begin{aligned} 0 &= -2G(n) + 2R(n)W(n) \\ R(n)W(n) &= G(n) \end{aligned} \qquad (11.20)$$

where

$$R(n) = \sum_{i=0}^{n} \lambda^{n-i} X(i) X^{\mathrm{T}}(i) \qquad (11.21)$$

$$G(n) = \sum_{i=0}^{n} \lambda^{n-i} d(i) X(i) \qquad (11.22)$$

The values of $R(n)$ and $G(n)$ can be computed recursively as

$$R(n) = \sum_{i=0}^{n-1} \lambda^{n-i} X(i) X^{\mathrm{T}}(i) + X(n) X^{\mathrm{T}}(n) = \lambda R(n-1) + X(n) X^{\mathrm{T}}(n) \qquad (11.23)$$

$$G(n) = \sum_{i=0}^{n-1} \lambda^{n-i} d(i)X(i) + d(n)X(n) = \lambda G(n-1) + d(n)X(n) \qquad (11.24)$$

If $A = B + CC^T$ where A and B are $N \times N$ matrices, and C is a vector of length N, the well-known matrix inversion lemma [4] yields.

$$A^{-1} = B^{-1} - B^{-1}C(1 + C^T B^{-1} C)^{-1} C^T B^{-1} \qquad (11.25)$$

Hence, $R^{-1}(n)$ is written as

$$R^{-1}(n) = \frac{1}{\lambda}\left[R^{-1}(n-1) - \frac{R^{-1}(n-1)X(n)X^T(n)R^{-1}(n-1)}{\lambda + X^T(n)R^{-1}(n-1)X(n)} \right] \qquad (11.26)$$

Let $P(n) = R^{-1}(n)$, then

$$
\begin{aligned}
W(n) &= P(n)G(n) \\
&= \frac{1}{\lambda}\left[P(n-1) - \frac{P(n-1)X(n)X^T(n)P(n-1)}{\lambda + X^T(n)P(n-1)X(n)} \right][\lambda G(n-1) + d(n)X(n)] \\
&= P(n-1)G(n-1) + \frac{1}{\lambda}P(n-1)d(n)X(n) \\
&\quad - \left[\frac{P(n-1)X(n)X^T(n)P(n-1)}{\lambda + X^T(n)P(n-1)X(n)} \right][\lambda G(n-1) + d(n)X(n)] \\
&= W(n-1) + \frac{1}{\lambda}P(n-1)d(n)X(n) \\
&\quad - \left[\frac{P(n-1)X(n)X^T(n)P(n-1)}{\lambda + X^T(n)P(n-1)X(n)} \right][\lambda G(n-1) + d(n)X(n)]
\end{aligned}
$$

$$(11.27)$$

After some mathematical simplifications, we obtain

$$W(n) = W(n-1) + \frac{(d(n) - X^T(n)W(n-1))P(n-1)X(n)}{\lambda + X^T(n)P(n-1)X(n)} \qquad (11.28)$$

The RLS algorithm is summarized as follows:

Step 1: Initialize $W(0)$ and $P(0)$
Step 2: For $n = 1, 2, \ldots$, compute

$$e(n) = d(n) - X^T(n)W(n-1) \qquad (11.29)$$

$$\alpha(n) = \frac{1}{\lambda + X^T(n)P(n-1)X(n)} \qquad (11.30)$$

$$W(n) = W(n - 1) + \alpha(n)e(n)P(n - 1)X(n) \tag{11.31}$$

$$P(n) = \frac{1}{\lambda}\left[P(n - 1) - \alpha(n)P(n - 1)X(n)X^{\mathrm{T}}(n)P(n - 1)\right] \tag{11.32}$$

11.5   Comparison of the LMS and RLS Algorithms

11.5.1   Computational Complexity

If M filter taps are assumed, LMS needs $(4M + 1)$ additions and $(4M + 3)$ multiplications per update, whereas the exponentially weighted RLS requires $(3M^2 + M - 1)$ additions/subtractions and $(4M^2 + 4M)$ multiplications/divisions per update. Hence, RLS is computationally more expensive than the LMS algorithm.

11.5.2   Rate of Convergence

Convergence of the LMS algorithm

The conditions that must be satisfied for convergence are:

1. The autocorrelation matrix R_{xx} must be positive definite.
2. $0 < \mu < \frac{1}{\lambda_{\max}}$

where $\lambda_{\max}$ is the maximum eigenvalue of the autocorrelation matrix R_{xx}.

In addition, the rate of convergence is related to the eigenvalue spread defined using the condition number of R_{xx} given by

$$\kappa = \frac{\lambda_{\max}}{\lambda_{\min}}, \tag{11.33}$$

where $\lambda_{\min}$ is the minimum eigenvalue of R_{xx}.

The fastest convergence occurs when $\kappa = 1$ corresponding to white noise input.

Convergence of the RLS algorithm

The RLS algorithm does not depend on the eigenvalue spread of the input correlation matrix R_{xx} and has faster convergence. The RLS algorithm is identical to the least squaring filtering for $\lambda = 1$. The forgetting factor $\lambda = 1$ is not to be used in changing conditions as the current values and previous values will have the weighting factor of 1. The forgetting factor is to be chosen as $0.95 < \lambda < 0.9995$ for non-stationary data to smooth out the effect of the old samples.

Comparison of the Convergence of the LMS and RLS Algorithms

The convergence of the LMS and RLS algorithms is demonstrated through the following numerical example. Consider a system with impulse response $h = [1 \; 2]$ and an input signal to the system given by

$$x(n) = u(n) + u(n-1) \tag{11.34}$$

where $u(n)$ is a random signal with unity power. The desired signal $d(n)$ is generated convolving $x(n)$ with h. Then, an adaptive FIR filter with 12 taps and LMS algorithm with 0.05 convergence factor, and RLS algorithm with 0.98 forgetting factor are used. The error convergence of the LMS and RLS algorithms is shown in Figs. 11.2 and 11.3, respectively. The MATLAB program for convergence comparison example is listed in Program 11.1.

Program 11.1

```
% MATLAB program for comparison of convergence of LMS and RLS
%algorithms
clear all
N=1000;
np = 0.01; sp = 1;
h=[1 2];
u = sqrt(sp/2).*randn(1,N+1);
x = u(1:N) + u(2:N+1);   % x(n) is input signal with power 1
d = conv(x,h);
d = d(1:N) + sqrt(np).*randn(1,N);
mu =0.05;
P0 = 10*eye(12); % Initial value of P, i.e., P(0)
        lam =0.98;              % RLS forgetting factor
hrls = adaptfilt.rls(12,lam,P0);
        [yrls,erls] = filter(hrls,x,d);
    hlms = adaptfilt.lms(12,mu);
        [ylms,elms] = filter(hlms,x,d);
        figure(1),plot(elms)
        figure(2), plot(erls)
xlabel('Number of samples');
ylabel('Error')
```

Fig. 11.2 Error convergence
of the LMS algorithm

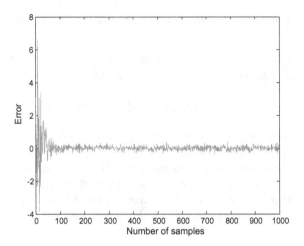

Fig. 11.3 Error convergence
of the RLS algorithm

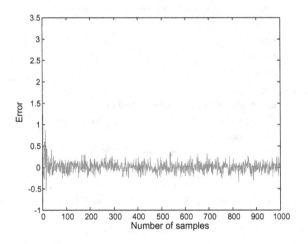

11.6   Applications of Adaptive Filters

11.6.1   Unknown System Identification

The schematic diagram for the application of an adaptive filter for unknown system
identification is shown in Fig. 11.4. After the adaptive filter converges, the output
of the adaptive filter approaches the unknown system output as the same input is fed
to both the unknown system and the adaptive filter and also the transfer function of
the adaptive filter is an approximation of the transfer function of the unknown
system. In this application, $e(n)$ converges to zero.

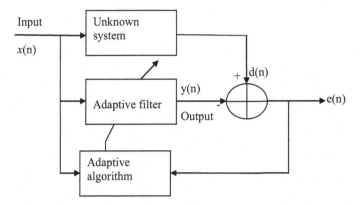

Fig. 11.4 Block diagram of adaptive filter for unknown system identification

The unknown system identification is illustrated through the following example. The unknown system is assumed to be characterized by the following transfer function

$$H(z) = \frac{N(z)}{D(z)}, \tag{11.35}$$

where

$$N(z) = 0.0004z^{10} - 0.002z^8 + 0.004z^6 - 0.004z^4 + 0.002z^2 - 0.0004$$
$$D(z) = z^{10} - 5.4554z^9 + 15.8902z^8 - 30.7264z^7 + 43.3268z^6 - 46.0586z^5$$
$$+ 37.4049z^4 - 22.8909z^3 + 10.2106z^2 - 3.022z + 0.4797$$

An input signal with three tones of 200, 600, and 1000 Hz is assumed. The amplitude spectrum of the input signal is shown in Fig. 11.5. The amplitude spectrum of the output of the unknown system is shown in Fig. 11.6. It may be observed that the unknown system has rejected 600 and 1000 Hz tones of the input signal. An adaptive FIR filter with 31 taps and LMS algorithm with 0.01 convergence factor is used. The error signal $e(n)$ is shown in Fig. 11.7 to demonstrate the convergence of LMS algorithm. The amplitude spectrum of the adaptive filter output is shown in Fig. 11.8. It is observed that the amplitude spectrum of the filter output is almost identical to the amplitude spectrum of the unknown system output. The impulse response of the identified unknown system is given by the coefficients of the adaptive filter. The impulse response of the identified system and its frequency response are shown in Figs. 11.9 and 11.10, respectively. The MATLAB program for the adaptive unknown system estimation is listed in Program 11.2.

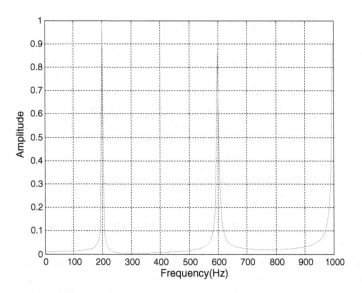

Fig. 11.5 Amplitude spectrum of the input signal

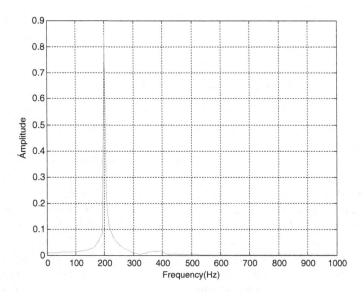

Fig. 11.6 Amplitude spectrum of the output of the unknown system

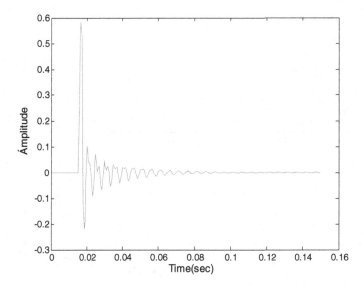

Fig. 11.7 Error signal

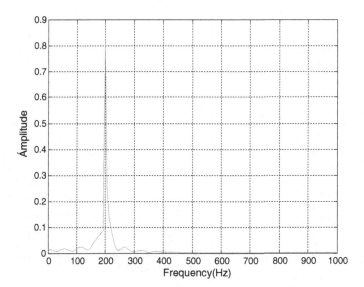

Fig. 11.8 Amplitude spectrum of the adaptive filter output

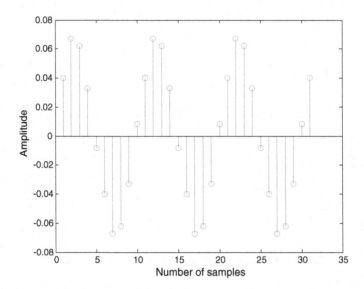

Fig. 11.9 Impulse response of the identified system

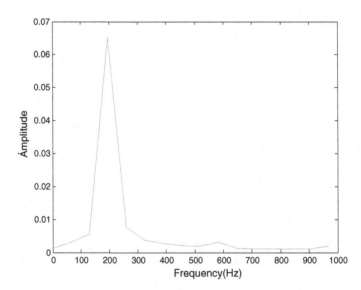

Fig. 11.10 Frequency response of the identified system

Program 11.2

```
% MATLAB program for adaptive filter for unknown system identification
clear all;clc;
num=[ 0.0004  0  -0.0020  0  0.0040  0  -0.0040  0  0.0020  0  -0.0004 ];
den=[1.0  -5.4554  15.8902  -30.7264  43.3268  -46.0586  37.4049 ...
    -22.8909  10.2106  -3.0220  0.4797 ];
Fs=2000;T=1/Fs;
t=0:T:0.15;%timevector
x=cos(2*pi*200*t)+sin(2*pi*600*t)+cos(2*pi*1000*t+pi/4);
d=filter(num,den,x);%unknownsystemoutput
mu=0.01;%Convergencefactor
w=zeros(1,31);y=zeros(1,length(t));%Initializethecoefficientsandoutpute Y
e=y;
for j=32:1:length(t)-1
    sum=0;
    for i=1:1:31
sum=sum+w(i)*x(j-i);
end
y(j)=sum;
e(j)=d(j)-y(j);
for i=1:1:31
    w(i)=w(i)+2*mu*e(j)*x(j-i);
end
end
f=[0:1:length(x)/2]*Fs/length(x);
X=2*abs(fft(x))/length(x);X(1)=X(1)/2;
D=2*abs(fft(d))/length(d);D(1)=D(1)/2;
Y=2*abs(fft(y))/length(y);Y(1)=Y(1)/2;
figure,plot(f,X(1:length(f))); xlabel('Frequency(Hz)');ylabel('Amplitude');
grid;
figure,plot(f,D(1:length(f))); xlabel('Frequency(Hz)');ylabel('Ámplitude');
grid;
figure,plot(f,Y(1:length(f))); xlabel('Frequency(Hz)');ylabel('Ámplitude');
grid;
figure, plot(t,e);xlabel('Time(sec)');ylabel('Ámplitude')
```

11.6.2   Adaptive Interference Canceller

The block diagram of an adaptive interference canceller is shown in Fig. 11.11. The task is to extract the source signal $x(n)$ from the desired signal $d(n)$, which consists of the source signal $x(n)$ and interference $\eta(n)$. A reference interference signal $\eta'(n)$ that is correlated to $\eta(n)$ is to be fed to the adaptive filter to remove $\eta(n)$ from $d(n)$. So long as the reference interference η' remains correlated to the undesired interference $\eta(n)$, the adaptive filter adjusts its coefficients to reduce the value of the difference between $y(n)$ and $d(n)$, removing the interference and resulting in $e(n)$, which converges to the source signal $x(n)$ rather than converging to zero.

The well-known applications are 50 Hz power line interference in the recording of the electrocardiogram (ECG), fetal heartbeat monitoring for removing maternal ECG from Fetal ECG, and echo cancellation.

A.   Power line interference suppression from ECG signal

Electrocardiogram (ECG) is an important tool for physicians in investigating the various activities of the heart. The ECG recording is mostly corrupted by unwanted 50/60 Hz power line interference. Removal of the power line interference from ECG recording is an important task to enhance the ECG recording for diagnosis by the physicians.

An electrocardiogram signal contaminated with 50 Hz power line interference and sampled at 4000 Hz is shown in Fig. 11.12 (over a 6 s interval). The contaminated ECG signal is fed as $d(n)$ to the cancellation system shown in Fig. 11.11, while 50 Hz sinusoidal signal sampled at 4000 Hz is taken as the reference interference η'. The least mean square (LMS) adaptive filter with 31 coefficients and a step size of 0.001 is used to remove the 50 Hz interference. The 50 Hz reference interference signal is used by the adaptive filter to produce an estimate of the 50 Hz interference to be removed from the contaminated ECG. After the system has converged, an estimate of the original ECG signal given by $e(n)$ is shown in Fig. 11.13 over a 6 s interval. The peaks of the ECG are to be counted over a 6 s interval, and the count is to be multiplied by 10 to determine the heart rate. A dynamic threshold is set on the estimated ECG signal so that any signal crossing the threshold is considered as a peak. The detected peaks of the ECG over a 6 s interval are shown in Fig. 11.14. Program 11.3 lists the MATLAB program used for this application.

Fig. 11.11 Block diagram of an adaptive interference canceller

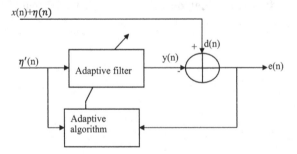

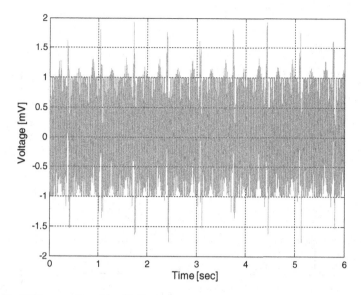

Fig. 11.12 ECG contaminated by 50 Hz power line interference

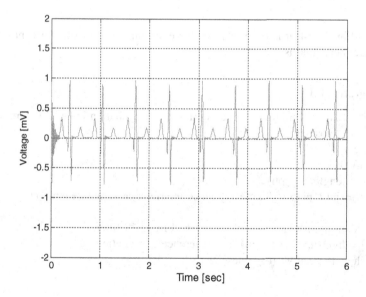

Fig. 11.13 ECG after removal of 50 Hz power line interference

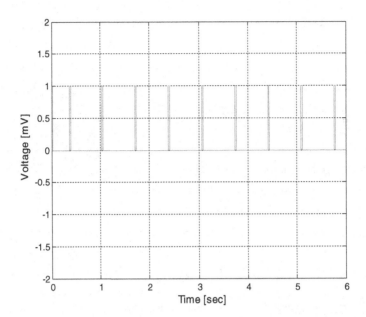

Fig. 11.14 Detection of peaks of ECG over a 6 s interval

From Fig. 11.14, it is observed that there are nine peaks over a 6 s interval. Hence, heart rate is 90.

Program 11.3

```
%MATLAB Program Adaptive interference cancellation in ECG recording
clear all; close all;
load ecg
Fs=4000;
for i=1:length(ecg_pli)
xsin(i)=sin(2*pi*50*i/Fs);% reference interference signal generation
end
h = adaptfilt.lms(31, 0.001);% adaptive FIR LMS filter
[y,e] = filter(h,xsin,ecg_pli);% interference cancellation
thresh = 4*mean(abs(e))*ones(size(e));
peak_e = (e >= thresh);
Time=6;
t = 1/Fs:1/Fs:Time';
figure(1), plot(t,ecg_pli(1:length(t)));% ECG Signal corrupted by power line
interferenc
axis([0 2 -2 2]);
grid;
```

```
figure(2), plot(t,e(1:length(t))); % ECG signal after power line interference %
cancellation
axis([0 2 -2 2]);
grid
figure(3),plot(t,peak_e(1:length(t)));% peaks detection
axis([0 6 -2 2]);
grid;
xlabel('Time [sec]');
ylabel('Voltage [mV]');
```

B. Fetal ECG Monitoring

The block diagram of fetal ECG monitoring is shown in Fig. 11.15a. The fetal heart health is assessed by monitoring the fetal electrocardiogram (fECG). However, the fetal ECG recorded from the maternal abdomen is affected by the maternal electrocardiogram (mECG) being the dominant interference. The ECG signal recorded from chest leads is almost pure maternal ECG, while the ECG recording from abdomen lead is a mixture of maternal and fetal ECGs. Four or more leads (electrodes) are placed on the mother's chest to acquire the maternal ECGs as the reference interference $\eta'(n)$ as shown in Fig. 11.15b. One lead (electrode) is used on the mother's abdomen to record the fetal ECG $d(n)$ contaminated by the maternal ECG being the interference as shown in Fig. 11.15c. An adaptive filter uses the maternal ECGs recorded from the chest leads as the reference interference and predicts the maternal ECG to be subtracted from the contaminated Fetal ECG. After convergence, the fetal ECG with the reduced maternal ECG is given by $e(n)$ as shown in Fig. 11.15d.

C. Adaptive Echo Cancellation

A delayed and distorted version of an original sound or electrical signal is reflected back to the source is referred as echo. Echoes often occur among real-life conversations. The echoes of speech waves can be heard as they are reflected from the floor, wall, and other neighboring objects. These echoes unexpectedly interrupt a conversation. Thus, it is desired to eliminate these echoes. The network and acoustic echoes are the two types of echoes that occur in telephone communication. The network echo results from the impedance mismatch at points along the transmission line, for example, at the hybrids of a public-switched telephony network (PSTN) exchange, where the subscriber two-wire lines are connected to four-wire lines. Acoustic echo is due to the acoustic coupling between the loudspeaker and microphone in hands-free telephone systems; for example, if a communication is between one or more hands-free telephones (or speaker phones), then acoustic feedback paths are set up between the telephone's loudspeaker and

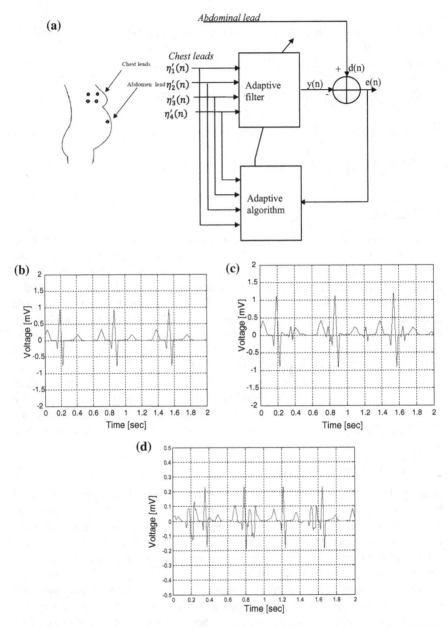

Fig. 11.15 a Adaptive cancellation of maternal ECG in fetal ECG monitoring **b** maternal ECG **c** fetal ECG contaminated by maternal ECG **d** fetal ECG after cancellation of maternal ECG

microphone at each end. Acoustic echo is more hostile than network echo mainly because of the different nature of the echo paths. The solution to these echo problems is to eliminate the echo with an echo suppressor or echo canceller.

   The block diagram of an adaptive echo canceller is shown in Fig. 11.16. Acoustic echo cancellation is important for audio teleconferencing when simultaneous communication (or full-duplex transmission) of speech is necessary The speaker B speech signal and the speaker A speech signal sampled at 16000 Hz as shown in Figs. 11.17 and 11.18 are considered to illustrate adaptive echo cancellation. Here, the speaker B speech signal (x_B) as near-end signal and the speaker A speech signal (x_A) as the far-end signal are considered. The impulse response of echo path A is shown in Fig. 11.19. The speaker B speech signal is convolved with

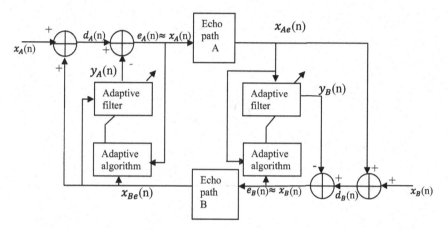

Fig. 11.16 Block diagram of adaptive echo canceller

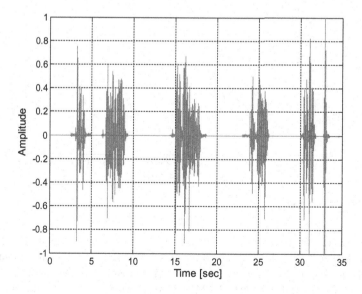

Fig. 11.17 Speaker B speech signal (x_B)

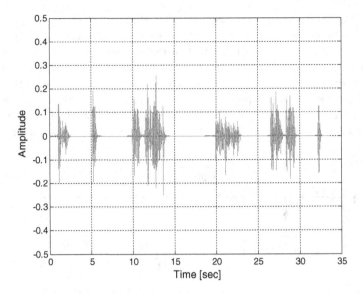

Fig. 11.18 Speaker A speech signal (x_A)

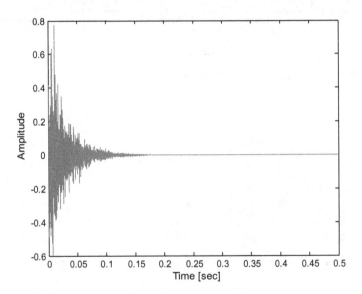

Fig. 11.19 Impulse response

the impulse response to obtain the echoed speaker A speech signal. The echoed speaker A speech signal x_{Ae} is added to the speaker B speech signal to obtain the microphone signal d_B as shown in Fig. 11.20. Both the least mean square (LMS) and the normalized LMS (NLMS) adaptive filters with 31 coefficients and a

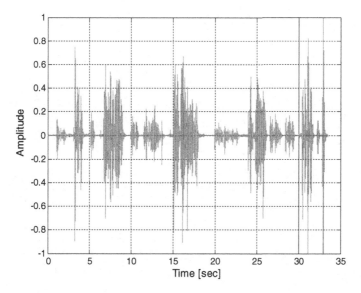

Fig. 11.20   Microphone signal (x_B + echo edx_A)

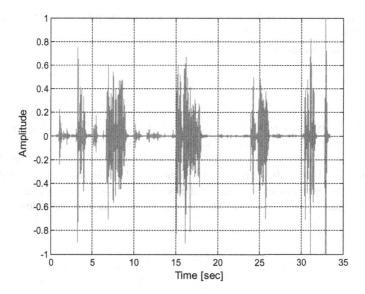

Fig. 11.21   Output of adaptive LMS echo canceller, $\mu = 0.04$

step size of 0.04 are used to estimate the echo signal. The estimated echo signal y_B is subtracted from the microphone signal to obtain echo-free speaker A speech signal. The estimated speaker A speech signal using LMS and NLMS is shown in Figs. 11.21 and 11.22, respectively.

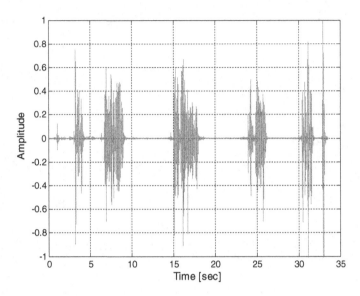

Fig. 11.22 Output of adaptive NLMS echo canceller, $\mu = 1$

Echo return loss enhancement (ERLE) is used as the measure to evaluate echo cancellation quality.

The ELRE measured in dB is defined as

$$\text{ERLE} = 10\log_{10}\frac{P_d(n)}{P_e(n)}$$

where $P_d(n)$ is the instantaneous power of the signal, $d(n)$, and $P_e(n)$ is the instantaneous power of the residual error signal, $e(n)$.

An ERLE in the range of 30–40 dB is considered to be ideal for a good echo canceller. The ERLE obtained using LMS and NLMS is shown in Figs. 11.23 and 11.24, respectively. It can be seen that the performance of NLMS is better than the LMS. Program 11.4 lists the MATLAB program used for this application.

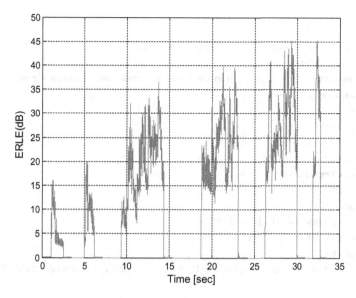

Fig. 11.23 Echo return loss enhancement using LMS, $\mu = 0.04$

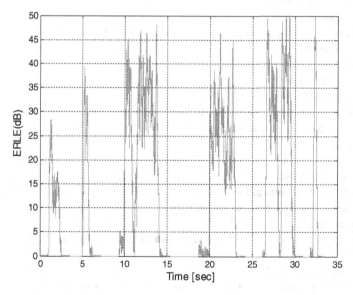

Fig. 11.24 Echo return loss enhancement using NLMS, $\mu = 1$

Program 11.4

```
%MATLAB Program for Adaptive echo cancellation
clear all; clc;
load nearspeech % Speaker B speech signal(near-end speech signal)
n=1:length(v);
t=n/F_T
figure,plot(t,v)
axis([0 35 -1 1])
xlabel('Time [sec]');
ylabel('Amplitude');grid;
F_T = 16000;
M = F_T/2 + 1;
[B,A] = cheby2(4,20,[0.1 0.7]);
IIR = dfilt.df2t( [zeros(1,6) B], A);
h = filter(IIR,...
        (log(0.99*rand(1,M)+0.01).*sign(randn(1,M)).*exp(-0.002*(1:
M)))');
h = h/norm(h)*4;    % Room Impulse Response
figure,plot(0:1/F_T:0.5, h);
xlabel('Time [sec]');
ylabel('Amplitude');
load farspeech% Speaker A speech signal(far-end speech signal)
figure,plot(t,x)
axis([0 35 -0.5 0.5])
xlabel('Time [sec]');
ylabel('Amplitude');grid;
fecho=filter(h,1, x);%echoed speaker A speech signal
ms=v+fecho+0.001*randn(length(v),1);
figure,plot(t,ms)
axis([0 35 -1 1])
xlabel('Time [sec]');
ylabel('Amplitude');grid;
ind=input('enter 1 for LMS,2 for NLMS=')
if (ind==1)
h = adaptfilt.lms(31, 0.04);% adaptive FIR LMS algorithm
end
if(ind==2)
h = adaptfilt.nlms(31,1,1,1);% adaptive FIR NLMS algorithm
end
  [y,e] = filter(h,fecho,ms);% interference cancellation
figure, plot(t,e);
axis([0 35 -1 1])
xlabel('Time [sec]');
ylabel('Amplitude');grid;
```

```
% Compute ERLE
Pd=filter(1,[1, -0.98],ms.^2);
Pe=filter(1,[1, -0.98],e.^2);
erledB=10*log10(Pd./Pe);
figure, plot(t,erledB);
axis([0 35 0 50])
xlabel('Time [sec]');
ylabel('ERLE(dB)');grid;
```

11.7 Problems

1. Show that the LMS algorithm achieves faster convergence regardless of initial values of the adaptive filter coefficients for the white noise input.
2. Show that the convergence of the LMS algorithm depends highly on the initial values of the adaptive filter coefficients for the colored input.
3. Derive the LMS algorithm with the use of orthogonal transform.
4. Compute the eigenvalue spread for the following input

   (i) $x(n)$ is white process with unity power
   (ii) $x(n) = u(n) + u(n-1)$ where $x(n)$ having unity power and $u(n)$ is normal distributed random process.

References

1. B. Widrow et al., Adaptive noise cancelling, principles and applications. Proc. IEEE **63**, 1692–1716 (1975)
2. D.T. Slock, On the convergence behavior of the LMS and the normalized LMS algorithms. IEEE Trans. Signal Process. **41**(9), 2811–2825 (1993)
3. F.F. Yassa, Optimality in the choice of the convergence factor for gradient based adaptive algorithms. IEEE Trans. Acoust. Speech Signal Process. (ASSP). **35**(1), 48–59 (1987)
4. P.S.R. Diniz, *Adaptive Filtering: Algorithms and Practical Implementation*, 3rd edn. (Springer, New York, 2008)

Chapter 12
Spectral Analysis of Signals

The process of determining the frequency contents of a continuous-time signal in the discrete-time domain is known as spectral analysis. Most of the phenomena that occur in nature can be characterized statistically by random processes. Hence, the main objective of spectral analysis is the determination of the power spectrum density (PSD) of a random process. The power is the Fourier transform of the autocorrelation sequence of a stationary random process. The PSD is a function that plays a fundamental role in the analysis of stationary random processes in that it quantifies the distribution of the total power as a function of frequency. The power spectrum also plays an important role in detection, tracking, and classification of periodic or narrowband processes buried in noise. Other applications of spectrum estimation include harmonic analysis and prediction, time series extrapolation and interpolation, spectral smoothing, bandwidth compression, beam forming, and direction finding. The estimation of the PSD is based on a set of observed data samples from the process. Estimating the power spectrum is equivalent to esti- mating the autocorrelation. This chapter deals with the nonparametric methods, parametric methods, and subspace methods for power spectrum estimation. Further, the spectrogram computation of non-stationary signals using STFT is also briefly discussed in this chapter.

12.1 Nonparametric Methods for Power Spectrum Estimation

Classical spectrum estimators do not assume any specific parametric model for the PSD. They are based solely on the estimate of the autocorrelation sequence of the random process from the observed data and hence work in all possible situations, although they do not provide high resolution. In practice, one cannot obtain

© Springer Nature Singapore Pte Ltd. 2018
K. D. Rao and M. N. S. Swamy, *Digital Signal Processing*,
https://doi.org/10.1007/978-981-10-8081-4_12

unlimited data record due to constraints on the data collection process or due to the necessity that the data must be WSS over that particular duration.

When the method for PSD estimation is not based on any assumptions about the generation of the observed samples other than wide-sense stationary, then it is termed a nonparametric estimator.

12.1.1   Periodogram

The periodogram was introduced in [1] searching for hidden periodicities while studying sunspot data. There are two distinct methods to compute the periodogram. One approach is the indirect method. In this approach, first we determine the autocorrelation sequence $r(k)$ from the data sequence $x(n)$ for $-(N-1) \leq k \leq (N-1)$ and then take the DTFT, i.e.,

$$\hat{P}_{\text{PER}}(f) = \sum_{k=-N+1}^{N-1} \hat{r}[k] e^{-j2\pi fk}. \tag{12.1}$$

It is more convenient to write the periodogram directly in terms of the observed samples $x[n]$. It is then defined as

$$\hat{P}_{\text{PER}}(f) = \frac{1}{N} \left| \sum_{n=0}^{N-1} x[n] e^{-j2\pi fn} \right|^2 = \frac{1}{N} |X(f)|^2 \tag{12.2}$$

where $X(f)$ is the Fourier transform of the sequence $x(n)$. Thus, the periodogram is proportional to the squared magnitude of the DTFT of the observed data. In practice, the periodogram is calculated by applying the FFT, which computes it at a discrete set of frequencies.

$$D_f = \left\{ f_k : f_k = \frac{k}{N}, \quad k = 0, 1, 2, \ldots, (N-1) \right\}$$

The periodogram is then expressed by

$$\hat{P}_{\text{PER}}(f_k) = \frac{1}{N} \left| \sum_{n=0}^{N-1} x[n] e^{-j2\pi kn/N} \right|^2 \quad f_k \in D_f. \tag{12.3}$$

To allow for finer frequency spacing in the computed periodogram, we define a zero-padded sequence according to

$$x'[n] = \begin{cases} x[n], & n = 0, 1, \ldots, N-1 \\ 0, & n = N, N+1, \ldots, N' \end{cases}.$$ (12.4)

Then we specify the new set of frequencies $D'_f = \{f_k : f_k = k/N, k \in \{0, 1, 2, \ldots, (N-1)\}\}$, and obtain

$$\hat{P}_{\text{PER}}(f_k) = \frac{1}{N} \left| \sum_{n=0}^{N-1} x[n] e^{-j2\pi kn/N'} \right|^2 \quad f_k \in D'_f.$$ (12.5)

A general property of good estimators is that they yield better estimates when the number of observed data samples increases. Theoretically, if the number of data samples tends to infinity, the estimates should converge to the true values of the estimated parameters. So, in the case of a PSD estimator, as we get more and more data samples, it is desirable that the estimated PSD tends to the true value of the PSD. In other words, if for finite number of data samples the estimator is biased, the bias should tend to zero as $N \to \infty$ as should the variance of the estimate. If this is indeed the case, the estimator is called consistent. Although the periodogram is asymptotically unbiased, it can be shown that it is not a consistent estimator. For example, if $\{\tilde{X}[n]\}$ is real zero mean white Gaussian noise, which is a process whose random variables are independent, Gaussian, and identically distributed with variance σ^2, the variance of $\hat{P}_{\text{PER}}(f)$ is equal to σ^4 regardless of the length N of the observed data sequence. The performance of the periodogram does not improve as N gets larger because as N increases, so does the number of parameters that are estimated, $P(f_0), P(f_1), \ldots, P(f_{N-1})$. In general, the variance of the periodogram at any given frequency is

$$\text{Var}\left[\hat{P}_{\text{PER}}(f)\right] = \text{Cov}\left[\hat{P}_{\text{PER}}(f), \hat{P}_{\text{PER}}(f)\right]$$
$$= P^2(f)\left[1 + \left(\frac{\sin 2\pi Nf}{N \sin 2\pi f}\right)^2\right]$$ (12.6a)

For frequencies not near 0 or 1/2, the above equation reduces to

$$\text{Var}(\hat{P}_{\text{PER}}(f)) \cong P^2(f)$$ (12.6b)

where $P^2(f)$ is the periodogram spectral estimation based on the definition of PSD.

Example 12.1 Consider a random signal composed of two sinusoidal components of frequencies 120 and 280 Hz corrupted with Gaussian distributed random noise. Evaluate its power spectrum using periodogram. Assume sampling frequency $F_s = 1024$ Hz.

Solution The following MATLAB program can be used to evaluate the power spectrum of the considered signal using Bartlett's method.

%Program 12.1

Power spectrum estimation using the periodogram
clear;clc;
N = 512;%total number of samples
k = 0:N-1;
f1 = 120;
f2 = 280;
F_T = 1024;%sampling frequency in Hz
T = 1/F_T;
x = sin(2*pi*f1*k*T) + sin(2*pi*f2*k*T)+2*randn(size(k));%vector of length N
%containing input samples
[pxx,f] = psd(x,length(x),F_T);
plot(f,pxx);grid;
xlabel('Frequency(Hz)');ylabel('Power spectrum');

The power spectrum obtained from the above MATLAB program is shown in Fig. 12.1.

12.1.2 Bartlett Method

In the Bartlett method [2], the observed data is segmented into K non-overlapping segments and the periodogram of each segment is computed. Finally, the average of periodogram of all the segments is evaluated. The Bartlett estimator has a variance that is smaller than the variance of the periodogram.

Fig. 12.1 Power spectrum estimation using the periodogram

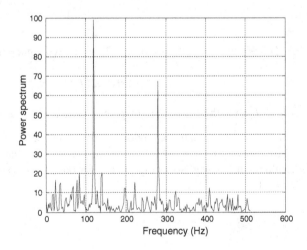

Consider a length N sequence $x(n)$. Then, $x(n)$ can be segmented into K subsequence, each subsequence having a length L. If the ith subsequence is denoted by $x_i(n)$, $0 \leq i < K$, then the ith subsequence can be obtained from the sequence $x(n)$ as

$$x_i(n) = x(iL + n), \quad 0 \leq n \leq L - 1$$

and its periodogram is given by

$$\widehat{P}_i(f) = \frac{1}{L} \left| \sum_{n=0}^{L-1} x_i(n) e^{-j2\pi fn} \right|^2 \quad i = 0, 1, \ldots, K - 1 \tag{12.7}$$

Then the Bartlett spectrum estimator is

$$\widehat{P}_B(f) = \frac{1}{K} \sum_{i=1}^{K} \widehat{P}_i(f) \tag{12.8}$$

$$\mathrm{Var}\left(\widehat{P}_B(f)\right) = \frac{1}{K} P^2(f). \tag{12.9}$$

The variance of the Barlett estimator can be related to the variance of the periodogram as follows.

The Bartlett estimator variance is reduced by a factor of K compared to the variance of the periodogram. However, the reduction in the variance is achieved at the cost of decrease in resolution. Thus, this estimator allows for a trade-off between resolution and variance.

The following example illustrates the computation of the power spectrum of a random signal using the Bartlett method.

Example 12.2 Consider the random signal of Example 12.1 and evaluate its power spectrum using Bartlett's method.

Solution The following MATLAB program can be used to evaluate the power spectrum of the considered signal using Bartlett's method.

Fig. 12.2 Power spectrum estimation using Bartlett's method

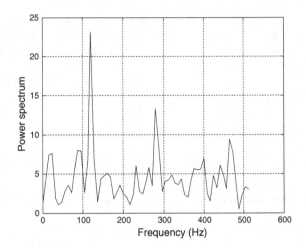

%Program 12.2

Power spectrum estimation using Bartlett's method
clear;clc;
N = 512;%total number of samples
k = 0 : N-1;
f1 = 120;
f2 = 280;
F_T = 1024;%sampling frequency in Hz
T = 1/F_T;
x = sin(2*pi*f1*k*T) + sin(2*pi*f2*k*T)+2*randn(size(k));%vector of length N
%containing input samples
L = 128;%length of subsequence
[pxx,f] = psd(x,L,F_T);
plot(f,pxx);grid;
xlabel('Frequency(Hz)');ylabel('Power spectrum');

The power spectrum obtained from the above MATLAB program is shown in Fig. 12.2.

12.1.2.1 Welch Method

The Welch method [3] is another estimator that exploits the periodogram. It is based on the same idea as the Bartlett's approach of splitting the data into segments and finding the average of their periodogram. The difference is that the segments are overlapped, and the data within a segment is windowed. If a sequence $x(n)$ of length N is segmented into K subsequences, each subsequence having a length L with an overlapping of D samples between the adjacent subsequences, then

$$N = L + D(K-1) \tag{12.10}$$

where N is the total number of observed samples and K the total number of subsequences. Note that if there is no overlap, $K = N/L$, and if there is 50% overlap, $K = 2\,N/L - 1$.

The ith subsequence is defined by

$$x_i(n) = x(n+iD), \quad 0 \le n \le L-1; 0 \le i \le K-1, \tag{12.11}$$

and its periodogram is given by

$$\hat{P}_i(f) = \frac{1}{L} \left| \sum_{n=0}^{L-1} w(n) x_i(n) e^{-j2\pi fn} \right|^2 . \tag{12.12}$$

Here $\hat{P}_i(f)$ is the modified periodogram of the data because the samples $x(n)$ are weighted by a non-rectangular window $w(n)$; the Welch spectrum estimate is then given by

$$\hat{P}_{\text{Wel}}(f) = \frac{1}{KC} \sum_{i=1}^{K} \hat{P}_i(f) \tag{12.13}$$

where C is the normalization factor for power in the window function given by

$$C = \frac{1}{K} \sum_{n=0}^{K-1} w^2(n)$$

Welch has shown that the variance of the estimator is

$$\text{Var}\left(\hat{P}_{\text{Wel}}(f)\right) \approx \frac{1}{K} P^2(f) \quad \text{for no overlapping} \tag{12.14a}$$

$$\approx \frac{9}{8K} P^2(f) \quad \text{for 50\% overlapping and Bartlett window.} \tag{12.14b}$$

By allowing overlap of subsequences, more number of subsequences can be formed than in the case of Bartlett's method. Consequently, the periodogram evaluated using the Welch's method will have less variance than the periodogram evaluated using the Bartlett method.

Example 12.3 Consider the random signal of Example 12.1 and evaluate its power spectrum using Welch's method with 50% overlapping and Hamming window.

Solution The following MATLAB program can be used to evaluate the power spectrum of the considered signal using Welch's method.

Fig. 12.3 Power spectrum
estimation using Welch's
method

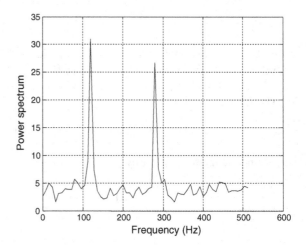

%**Program 12.3**

Power spectrum estimation using Welch's method
clear;clc;
N = 512;%total number of samples
k = 0 : N-1;
f1 = 120;
f2 = 280;
F_T = 1024;%sampling frequency in Hz
T = 1/F_T;
x = sin(2*pi*f1*k*T) + sin(2*pi*f2*k*T)+2*randn(size(k));%vector of length N
%containing input samples
L = 128;%length of subsequence
window = hamming(L);% window type
overlap = L/2;%number of overlapping samples(50%overlapping)
[pxx,f] = psd(x,L,F_T,window,overlap);
plot(f,pxx);grid;
xlabel('Frequency(Hz)');ylabel('Power spectrum');

The power spectrum obtained from the above MATLAB program is shown in
Fig. 12.3.

12.1.2.2 Blackman–Tukey Method

In this method, autocorrelation of the observed data sequence $x(n)$ is computed first.
Next, the autocorrelation is windowed and then the Fourier transform is applied on
it to obtain the power spectrum. Hence, the power spectrum using the Blackman–
Tukey method [4] is given by

$$\widehat{P}(f) = \sum_{k=-(N-1)}^{N-1} w(k)\widehat{r}(k)e^{-j2\pi fk} \qquad (12.15)$$

where the window $w(k)$ is real nonnegative, symmetric, and non-increasing with $|k|$, that is,

$$\text{(i)} \quad 0 \le w(k) \le w(0) = 1 \qquad (12.16a)$$

$$\text{(ii)} \quad w(-k) = w(k) \qquad (12.16b)$$

$$\text{(iii)} \quad w(k) = 0. M < |k|. \quad M \le N - 1. \qquad (12.16c)$$

It should be noted that the symmetry property of $w(k)$ ensures that the spectrum is real. It is obvious that the autocorrelation with smaller lags will be estimated more accurately than the ones with lags close to N because of the different number of terms that are used. Therefore, the large variance of the periodogram can be ascribed to the large weight given to the poor autocorrelation estimates used in its evaluation. Blackman and Tukey proposed to weight the autocorrelation sequence so that the autocorrelations with higher lags are weighted less. The bias, the variance, and the resolution of the Blackman–Tukey method depend on the applied window. For example, if the window is triangular (Bartlett),

$$w_B[k] = \begin{cases} \frac{M-|k|}{M}, & |k| \le M \\ 0, & \text{otherwise} \end{cases} \qquad (12.17)$$

and if $N \gg M \gg 1$, the variance of the Blackman–Tukey estimator is

$$\text{Var}(\widehat{P}_{\text{BT}}(f)) \cong \frac{2M}{3N} P^2(f) \qquad (12.18)$$

where $P(f)$ is the true spectrum of the process. Compared to Eqs. (12.6a) and (12.6b) it is clear that the variance of this estimator may be significantly smaller than the variance of the periodogram. However, as M decreases, so does the resolution of the Blackman–Tukey estimator.

Example 12.4 Consider the random signal of Example 12.1 and evaluate its power spectrum using Blackman–Tukey method.

Solution The following MATLAB program can be used to evaluate the power spectrum of the considered signal using Blackman–Tukey method.

Fig. 12.4 Power spectrum estimation using Blackman–Tukey method

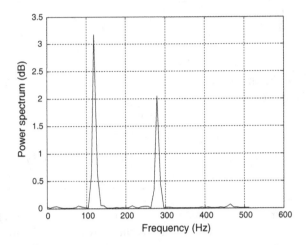

%**Program 12.4**

Power spectrum estimation Blackman–Tukey method
clear;clc;
N = 512;%total number of samples
k = 0 : N-1;
f1 = 120;
f2 = 280;
F_T = 1024;%sampling frequency in Hz
T = 1/F_T;
x = sin(2*pi*f1*k*T)+sin(2*pi*f2*k*T)+2*randn(size(k));%vector of length N
%containing input samples
r = f_corr(x,x,0,0);% evaluates correlation of input samples
L = 128;%length of window
window = Bartlett(L);% window type
[pxx,f] = psd(r,L,F_T,window);
plot(f,pxx);grid
xlabel('Frequency(Hz)');ylabel('Power spectrum(dB)');

The power spectrum obtained from the above MATLAB program is shown in Fig. 12.4.

12.1.3 Performance Comparison of the Nonparametric Methods

The performance of a PSD estimator is evaluated by quality factor. The quality factor is defined as the ratio of the squared mean of the PSD to the variance of the PSD given by

Table 12.1 Comparison of performance of classical methods

Classical method	Quality factor
Periodogram	1
Bartlett	1.11 N Δf
Welch (50% overlap)	1.39 N Δf
Blackman–Tukey	2.34 N Δf

$$Q_P = \frac{\text{var}(\hat{p}(f))}{E^2(\hat{p}(f))}. \tag{12.19}$$

Another important metric for comparison is the resolution of the PSD estimators. It corresponds to the ability of the estimator to provide the fine details of the PSD of the random process. For example, if the PSD of the random process has two peaks at frequencies f_1 and f_2, then the resolution of the estimator would be measured by the minimum separation of f_1 and f_2 for which the estimator still reproduces two peaks at f_1 and f_2. It has been shown in [5] for triangular window that the quality factors of the classical methods are as shown in Table 12.1.

From the above table, it can be observed that the quality factor is dependent on the product of the data length N and the frequency resolution Δf. For a desired quality factor, the frequency resolution can be increased or decreased by varying the data length N.

12.2   Parametric or Model-Based Methods for Power Spectrum Estimation

The classical methods require long data records to obtain the necessary frequency resolution. They suffer from spectral leakage effects, which often mask weak signals that are present in the data which occur due to windowing. For short data lengths, the spectral leakage limits frequency resolution.

In this section, we deal with power spectrum estimation methods in which extrapolation is possible if we have a priori information on how data is generated. In such a case, a model for the signal generation can be constructed with a number of parameters that can be estimated from the observed data. Then, from the estimated model parameters, we can compute the power density spectrum.

Due to modeling approach, we can eliminate the window function and the assumption that autocorrelation sequence is zero outside the window. Hence, these have better frequency resolutions and avoid problem of leakage. This is especially true in applications where short data records are available due to time variant or transient phenomena.

The parametric methods considered in this section are based on modeling the data sequence $y(n)$ as the output of a linear system characterized by a rational system function of the form

$$H(Z) = \frac{B(Z)}{A(Z)} = \frac{\sum_{k=0}^{q} b_k z^{-k}}{1 + \sum_{k=1}^{p} a_k z^{-k}}. \tag{12.20}$$

For the linear system with rational system function $H(Z)$, the output $y(n)$ is related to input $w(n)$ and the corresponding difference equation is

$$y(n) + \sum_{k=1}^{p} a_k y(n-k) = \sum_{k=0}^{q} b_k w(n-k) \tag{12.21}$$

where $\{b_k\}$ and $\{a_k\}$ are the filter coefficients that determine the location of the zeros and poles of $H(Z)$, respectively.

Parametric spectral estimation is a three-step process as follows

Step 1 Select the model
Step 2 Estimate the model parameters from the observed/measured data or the correlation sequence which is estimated from the data
Step 3 Obtain the spectral estimate with the help of the estimated model parameters.

In power spectrum estimation, the input sequence is not observable. However, if the observed data is considered as a stationary random process, then the input can also be assumed as a stationary random process.

Autoregressive Moving Average (ARMA) Model
An ARMA model of order (p, q) is described by Eq. (12.21). Let $P_w(f)$ be the power spectral density of the input sequence, $P_y(f)$ be the power spectral density of the output sequence, and $H(f)$ be the frequency response of the linear system, then

$$P_y(f) = |H(f)|^2 P_w(f) \tag{12.22}$$

where $H(f)$ is the frequency response of the model.

If the sequence $\omega(n)$ is a zero mean white noise process of variance σ_ω^2, the autocorrelation sequence is

$$r_{yy}(m) = \sigma_\omega^2 \delta(m). \tag{12.23}$$

The power spectral density of the input sequence $w(n)$ is

$$P_w(f) = \sigma_\omega^2. \tag{12.24}$$

Hence, the power spectral density of the output sequence $y(n)$ is

$$\begin{aligned}
P_y(f) &= |H(f)|^2 P_w(f) \\
&= \sigma_\omega^2 |H(f)|^2 \\
&= \sigma_\omega^2 \left| \frac{B(f)}{A(f)} \right|^2.
\end{aligned} \tag{12.25}$$

Autoregressive (AR) Model
If $q = 0$, $b_0 = 1$, and $b_k = 0$ for $1 \le k \le q$ in Eq. (12.21), then

$$H(Z) = \frac{1}{A(z)} = \frac{1}{1 + \sum_{k=1}^{p} a_k z^{-k}}$$
(12.26)

with the corresponding difference equation

$$y(n) + \sum_{k=1}^{p} a_k y(n-k) = w(n)$$
(12.27)

which characterizes an AR model of order p. It is represented as AR (p).

Moving Average (MA) Model
If $a_k = 0$ for $1 \le k \le p$ in Eq. (12.21), then

$$H(Z) = B(Z) = \sum_{k=0}^{q} b_k z^{-k}$$
(12.28)

with the corresponding difference equation

$$y(n) = \sum_{k=0}^{q} b_k w(n-k)$$
(12.29)

which characterizes a MA model of order q. It is represented as MA (q).

The AR model is the most widely used model in practice since the AR model is well suited to characterize spectrum with narrow peaks and also provides very simple linear equations for the AR model parameters. As the MA model requires more number of model parameters to represent a narrow spectrum, it is not often used for spectral estimation. The ARMA model with less number of parameters provides a more efficient representation.

12.2.1   Relationships Between the Autocorrelation and the Model Parameters

The parameters in AR(p), MA(q), and ARMA(p,q) models are related to the autocorrelation sequence $r_{yy}(m)$.

This relationship can be obtained by multiplying the difference Eq. (12.21) by $y^*(n-m)$ and taking the expected value on both sides. Then

$$E[y(n) \cdot y^*(n-m)] = -\sum_{k=1}^{p} a_k E[y(n-k) \cdot y^*(n-m)]$$

$$+ \sum_{k=0}^{q} b_k E[w(n-k) \cdot y^*(n-m)] \tag{12.30a}$$

$$r_{yy}(m) = -\sum_{k=1}^{p} a_k r_{yy}(m-k) + \sum_{k=0}^{q} b_k \gamma_{wy}(m-k). \tag{12.30b}$$

$\gamma_{\omega x}(m)$ is the cross-correlation between $w(n)$ and $y(n)$.
The cross-correlation $\gamma_{\omega x}(m)$ is related to the filter impulse response h as

$$
\begin{aligned}
\gamma_{\omega x}(m) &= E[x^*(n)w(n+m)] \\
&= E\left[\sum_{k=0}^{\infty} h(k)w^*(n-k)w(n+m)\right] \\
&= \sum_{k=0}^{\infty} h(k)E[w^*(n-k)w(n+m)] \\
&= \sigma_\omega^2 h(-m)
\end{aligned}
\tag{12.31}
$$

$$E[w^*(n)w(n+m)] = \sigma_w^2 \delta(m) \tag{12.32a}$$

$$\gamma_{wx}(m) = \begin{cases} 0, & m > 0 \\ \sigma_\omega^2 h(-m), & m \le 0 \end{cases}. \tag{12.32b}$$

By setting $q = 0$ in Eq. (12.30a), an AR model can be adopted. Then, the model parameters can be related to the autocorrelation sequence as

$$
r_{yy}(m) = \begin{cases}
-\sum_{k=1}^{p} a_k r_{yy}(m-k), & m > 0 \\
-\sum_{k=1}^{p} a_k r_{yy}(m-k) + \sigma_w^2, & m = 0 \\
r_{yy}^*(-m), & m < 0
\end{cases}
\tag{12.33}
$$

The above equation can be written in matrix form as

$$
\begin{bmatrix}
r_{yy}(0) & r_{yy}(-1) & \cdots & r_{yy}(-p+1) \\
r_{yy}(1) & r_{yy}(0) & \cdots & r_{yy}(-p+2) \\
\cdots & \cdots & \cdots & \cdots \\
r_{yy}(p-1) & r_{yy}(p-2) & \cdots & r_{yy}(0)
\end{bmatrix}
\begin{bmatrix}
a_1 \\ a_2 \\ \cdot \\ \vdots \\ a_p
\end{bmatrix}
= -
\begin{bmatrix}
r_{yy}(1) \\ r_{yy}(2) \\ \cdot \\ \cdot \\ r_{yy}(p)
\end{bmatrix}
\tag{12.34}
$$

From Eq. (12.33), we can obtain the variance

$$\sigma_w^2 = r_{yy}(0) + \sum_{k=1}^{p} a_k r_{yy}(-k) \tag{12.35}$$

Combining Eqs. (12.34) and (12.35), we get

$$\begin{bmatrix} r_{yy}(0) & r_{yy}(-1) & \cdots & r_{yy}(-p+1) \\ r_{yy}(1) & r_{yy}(0) & \cdots & r_{yy}(-p+2) \\ \cdots & \cdots & \cdots & \cdots \\ r_{yy}(p-1) & r_{yy}(p-2) & \cdots & r_{yy}(0) \end{bmatrix} \begin{bmatrix} a_1 \\ a_2 \\ \vdots \\ a_p \end{bmatrix} = - \begin{bmatrix} r_{yy}(1) \\ r_{yy}(2) \\ \vdots \\ r_{yy}(p) \end{bmatrix} \tag{12.36}$$

which is known as the Yule–Walker equation.

The correlation matrix is a Toeplitz non-singular and can be solved with Levinson–Durbin algorithm for obtaining the inverse matrix.

12.2.2 Power Spectrum Estimation Based on AR Model via Yule–Walker Method

Since the autocorrelation sequence actual values are not known a priori, their estimates are to be computed from the data sequence using

$$\widehat{r}_{yy}(m) = \frac{1}{N} \sum_{n=0}^{N-m-1} y^*(n) y(n+m) \quad m \geq 0 \tag{12.37}$$

These autocorrelation estimates and the AR model parameter estimates are used in Eq. (12.36) in place of their true values, and then the equation is solved using the Levinson–Durbin algorithm to estimate the AR model parameters. Then, the power density spectrum estimate is computed using

$$P_{\text{Yul}}(f) = \frac{\widehat{E}_p^2}{\left| 1 + \sum_{k=1}^{p} \hat{a}_k e^{-j2\pi fk} \right|^2} \tag{12.38}$$

where $\hat{a}_k$ are AR parameter estimates and

$$\widehat{E}_p = r_{yy}(0) \prod_{k=1}^{p} \left[1 - |\hat{a}_k|^2 \right] \tag{12.39}$$

is the estimated minimum mean squared value for the pth order predictor.

The following example illustrates the power spectrum estimation based on AR model via Yule–Walker method.

Example 12.5 Consider a fourth-order AR process characterized by

$$y(n) = 2.7607y(n-1) - 3.8106y(n-2) + 2.6535y(n-3) - 0.9238y(n-4)$$
$$= w(n)$$

where $w(n)$ is a zero mean, unit variance, white noise process.

Estimate the power spectrum of the AR process using the Yule–Walker method.

Solution The MATLAB function *pyulear(X,m,F_T)* gives the power spectrum of a discrete-time signal X using the Yule–Walker method. m being the order of the autoregressive (AR) model used to produce the PSD. F_T is the sampling frequency. A prediction filter with two zeros at $z_1 = 0.9804e^{j0.22\pi}$; $z_2 = 0.9804e^{j0.28\pi}$ gives the fourth-order AR model parameters. The two zeros are close to the unit circle; hence, the power spectrum will have two sharp peaks at the normalized frequencies 0.22π and 0.28π rad/sample.

The following MATLAB Program 11.5 is used to obtain the power spectrum using the Yule–Walker method.

%**Program 12.5**

Power spectrum estimation via Yule–Walker method

```
clear;clc;
F_T = 1024;
randn('state',1);
w = randn(200,1);
y = filter(1,[1-2.7607 3.8106 -2.6535 0.9238],w);
pyulear(y,4,F_T);
```

The power spectrum obtained from the above program based on 200 samples is shown in Fig. 12.5.

Due to lack of sufficient resolution, the two peaks corresponding to the frequencies 0.22π and 0.28π are not seen. The resolution can be improved by increasing the data length.

When the above program is run with 1000 data samples, the power spectrum estimate obtained is shown in Fig. 12.6 in which we can see clearly the two peaks corresponding to the frequencies 0.22π and 0.28π.

12.2.3   Power Spectrum Estimation Based on AR Model via Burg Method

The Burg method [6] can be used the estimation of the AR model parameters by minimizing the forward and backward errors in the linear predictors. Here we consider the problem of linearly predicting the value of a stationary random process either forward in time (or) backward in time.

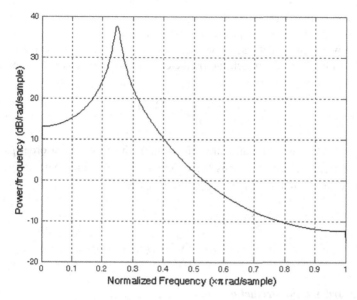

Fig. 12.5 Power spectral density estimate using Yule–Walker method based on 200 samples

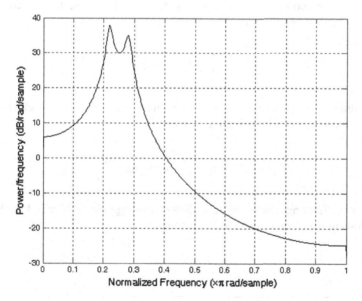

Fig. 12.6 Power spectral density estimate using Yule–Walker method based on 1000 samples

Forward Linear Prediction

Here in this case, from the past values of a random process, a future value of the process can be predicted. So, consider one-step forward linear prediction as depicted in Fig. 12.7, for which the predicted value of $y(n)$ can be written as

$$\hat{y}(n) = -\sum_{k=1}^{p} a_p(k)y(n-k) \tag{12.40}$$

where $\{-a_p(k)\}$ are the prediction coefficients of the predictor of order p.

The forward prediction error is the difference between the value $y(n)$ and the predicted value $(\hat{y}(n))$ of $y(n)$ and can be expressed as

$$
\begin{aligned}
e_p^f(n) &= y(n) - \hat{y}(n) \\
&= y(n) + \sum_{k=1}^{p} a_p(k)y(n-k).
\end{aligned} \tag{12.41}
$$

Backward Linear prediction

In the backward linear prediction, the value $y(n-p)$ of a stationary random process can be predicted from the data sequence $y(n), y(n-1), \ldots, y(n-p+1)$ of the process. For one-step backward linear prediction of order p, the predicted value of $y(n-p)$ can be written as

$$\hat{y}(n-p) = -\sum_{k=1}^{p} a_p^*(p-k)y(n+k-p). \tag{12.42}$$

The difference between $y(n-p)$ and estimate $\overset{...}{y}(n-p)$ is the backward prediction error which can be written as denoted as

$$e_p^b(n) = y(n-p) + \sum_{k=0}^{p-1} a_p^*(k)y(n+k-p). \tag{12.43}$$

For lattice filter realization of the predictor, a p-stage lattice filter is described by the following set of order-recursive equation

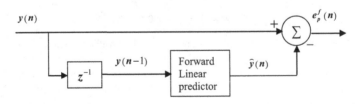

Fig. 12.7 One-step forward linear predictor

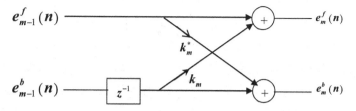

Fig. 12.8 A typical stage of a lattice filter

$$e_0^f(n) = e_0^b(n) = x(n) \tag{12.44a}$$

$$e_m^f(n) = e_{m-1}^f(n) + K_m e_{m-1}^b(n-1) \quad m = 1, 2, \ldots, p \tag{12.44b}$$

$$e_m^b = K_m^* e_{m-1}^f(n) + e_{m-1}^b(n-1) \quad m = 1, 2, \ldots, p \tag{12.44c}$$

where K_m is the mth reflection coefficient in the lattice filter.

A typical stage of a lattice filter is shown in Fig. 12.8.

From the forward and backward prediction errors, the least squares error for given data $y(n), n = 0, 1, \ldots, N-1$, can be expressed as

$$\varepsilon_m = \sum_{n=m}^{N-1} \left(\left| e_m^f(n) \right|^2 + \left| e_m^b(n) \right|^2 \right). \tag{12.45}$$

Now, the error is to be minimized with respect to predictor coefficients satisfying the following Levinson–Durbin recursion

$$a_m(k) = a_{m-1}(k) + K_m a_{m-1}^*(m-k), \quad 1 \le k \le m-1; 1 \le m \le p. \tag{12.46}$$

where $K_m = a_m(m)$ is the mth reflection coefficient in the lattice filter of the predictor.

Minimization of ε_m with respect to the reflection coefficient K_m yields

$$\widehat{K}_m = \frac{-\sum_{n=m}^{N-1} e_{m-1}^f(n) \left(e_m^b(n) \right)^*}{\frac{1}{2} \widehat{E}_m} \tag{12.47}$$

where $\widehat{E}_m$ is the total squared error which is an estimate of $\widehat{E}_{m-1}^f + \widehat{E}_{m-1}^b$, $\widehat{E}_{m-1}^f$ and $\widehat{E}_{m-1}^b$ being the least squares estimates of the forward and backward errors given by

$$\widehat{E}_{m-1}^f = \sum_{n=m}^{N-1} \left| e_{m-1}^f(n) \right|^2 \qquad (12.48)$$

$$\widehat{E}_{m-1}^b = \sum_{n=m}^{N-1} \left| e_{m-1}^b(n) \right|^2. \qquad (12.49)$$

The estimate $\widehat{E}_m$ can be computed by using the following recursion

$$\widehat{E}_m = \left(1 - \left|\widehat{K}_m\right|^2\right)\widehat{E}_{m-1} \left| e_m^f(m-2) \right|^2 - \left| e_m^b(m-2) \right|^2 \qquad (12.50)$$

The Burg method computes the reflection coefficients $\widehat{K}_m$ using Eqs. (12.47) and (12.50), and AR parameters are estimated by using Levinson–Durbin algorithm. Then, the power spectrum can be estimated as

$$P_{\text{BUR}}(f) = \frac{\widehat{E}_m^2}{\left|1 + \sum_{k=1}^{p} \widehat{a}_k e^{-j2\pi fk}\right|^2}. \qquad (12.51)$$

The following example illustrates the power spectrum estimation using the Burg method.

Example 12.6 Consider the AR process as given in Example 12.5. Evaluate its power spectrum based on 200 samples using the Burg method.

Solution The MATLAB Program 12.5 can be used by replacing pyulear(y,4,Fs) by pburg(y,4,Fs) to compute power spectrum using the Burg method. Thus, the PSD obtained based on 200 samples using the Burg method is shown in Fig. 12.9.

The two peaks corresponding to the frequencies 0.22π and 0.28π are clearly seen from Fig. 12.9. Using the Burg method based on 200 samples, whereas it is not as shown in Fig. 12.5 using Yule–Walker method for the same number of samples.

The main advantages of the Burg method are high-frequency resolution, stable AR model, and computational efficiency. The drawbacks of the Burg method are spectral line splitting at high SNRs and spurious spikes for high-order models.

12.2.4 Selection of Model Order

Generally, model order is unknown a priori. If the guess for the model order is too low, it will result in highly smoothed spectral estimate and the high-order model increases resolution but low-level spurious peaks will appear in the spectrum. The two methods suggested by Akaike [7, 8] for model order selection are:

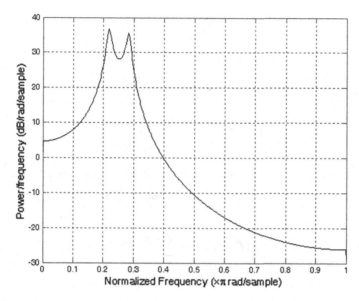

Fig. 12.9 Power spectral density estimate via Burg method based on 200 samples

1. Akaike forward prediction error (FPE) criterion states that

$$\mathbf{FPE}(p) = \sigma_{w_p}^2 \left(\frac{N + p + 1}{N - p - 1} \right) \qquad (12.52)$$

should be minimum. Here, N stands for the number of data samples, p is the order of the filter, and $\sigma_{w_p}^2$ is the white noise variance estimate.

2. Akaike forward prediction error (FPE) criterion states that

$$\mathbf{FPE}(p) = N \ln(\sigma_{w_p}^2) + 2p \qquad (12.53)$$

should be minimum.

12.2.5   Power Spectrum Estimation Based on MA Model

By setting $p = 0$ in Eq. (11.30a) and letting $h(k) = b(k), 1 \leq k \leq q$, a MA model can be adopted. Then, the model parameters can be related to the autocorrelation sequence as

$$r_{yy}(m) = \begin{cases} \sigma_\omega^2 \sum_{k=0}^{q} b_k b_{k+m}; & 0 \le m \le q \\ 0; & m > q \\ r_{yy}^*(-m); & m < 0 \end{cases} \qquad (12.54)$$

Then, the power spectrum estimate based on MA model is

$$P_{\mathbf{MA}}(f) = \sum_{m=-q}^{q} r_{yy}(m)\, e^{-j2\pi fm}. \qquad (12.55)$$

Equation (12.55) can be written in terms of MA model parameter estimates $\left(\widehat{b}_k\right)$, and the white noise variance estimate $(\widehat{\sigma}_w^2)$ can be written as

$$P_{MA}(f) = \widehat{\sigma}_w^2 \left| 1 + \sum_{k=1}^{q} \widehat{b}_k\, e^{-j2\pi fk} \right|^2. \qquad (12.56)$$

12.2.6 Power Spectrum Estimation Based on ARMA Model

ARMA model is used to estimate the spectrum with less parameters. This model is mostly used when data is corrupted by noise.

The AR parameters are estimated first, independent of the MA parameters, by using the Yule–Walker method or the Burg method. The MA parameters are estimated assuming that the AR parameters are known.

Then, the ARMA power spectral estimate is

$$P_{\mathbf{ARMA}}(f) = \widehat{\sigma}_w^2 \left| \frac{1 + \sum_{k=1}^{q} \widehat{b}_k e^{-j2\pi fk}}{1 + \sum_{k=1}^{p} \widehat{a}_k e^{-j2\pi fk}} \right|^2. \qquad (12.57)$$

12.3 Subspace Methods for Power Spectrum Estimation

The subspace methods do not assume any parametric model for power spectrum estimation. They are based solely on the estimate of the autocorrelation sequence of the random process from the observed data. In this section, we briefly discuss three subspace methods, namely, Pisarenko harmonic decomposition, MUSIC, and eigenvector method.

12.3.1   *Pisarenko Harmonic Decomposition Method*

Consider a process y consisting of m sinusoids with additive white noise. The autocorrelation values of the process y can be written in matrix form as

$$
\begin{bmatrix} r_{yy}(1) \\ r_{yy}(2) \\ \vdots \\ r_{yy}(m) \end{bmatrix} = \begin{bmatrix} e^{j2\pi f_1} & e^{j2\pi f_2} & e^{j2\pi f_3} & \dots & e^{j2\pi f_m} \\ e^{j4\pi f_1} & e^{j4\pi f_2} & e^{j4\pi f_3} & \dots & e^{j4\pi f_m} \\ \vdots & \vdots & \vdots & \dots & \vdots \\ e^{j2\pi m f_1} & e^{j2\pi m f_2} & e^{j2\pi m f_3} & \dots & e^{j2\pi m f_m} \end{bmatrix} \begin{bmatrix} P_1 \\ P_2 \\ \vdots \\ P_m \end{bmatrix} \tag{12.58}
$$

where P_i is the average power in the ith sinusoid.

If the frequencies f_i, $1 \le i \le m$, are known, then from the known autocorrelation values $r_{yy}(1)$ to $r_{yy}(m)$, the sinusoidal powers can be determined from the above equation.

The stepwise procedure for the Pisarenko harmonic decomposition method is as follows.

Step 1: Estimate the autocorrelation vector from the observed data.
Step 2: Find the minimum eigenvalue and the corresponding eigenvector (v_{m+1})
Step 3 : Find the roots of the following polynomial

$$
\sum_{k=0}^{M} v_{m+1}(k+1)z^{-k} \tag{12.59}
$$

where v_{m+1} is the eigenvector. The roots lie on the unit circle at angles $2\pi f_i$ for $1 \le i \le M$, M is dimension of eigenvector,
Step 4 : Solve Eq. (12.58.) for sinusoidal powers (P_i).

12.3.2   *Multiple Signal Classification (MUSIC) Method*

The MUSIC estimates the power spectrum from a signal or a correlation matrix using Schmidt's eigen space analysis method [9]. The method estimates the signal's frequency content by performing eigen space analysis of the signal's correlation matrix. In particular, this method is applicable to signals that are the sum of sinusoids with additive white Gaussian noise and more, in general, to narrowband signals. To develop this, first let us consider the 'weighted' spectral estimate

$$P(f) = S^H(f)\left(\sum_{k=m+1}^{M} c_k V_k V_k^H\right) S(f) = \sum_{k=m+1}^{M} c_k |S^H(f)V_k|^2 \qquad (12.60)$$

where m is the dimension of the signal subspace, $V_k, k = m+1, \ldots, M$ are the eigen vectors in the noise subspace, c_k are a set of positive weights, and

$$S(f) = \left[1, e^{j2\pi f}, e^{j4\pi f}, \ldots, e^{j2\pi(m-1)f}\right]$$ is complex sinusoidal vector.

It may be noted that at $f = f_i$, $S(f_i) = S_i$, such that at any one of the p sinusoidal frequency components of the signal we have,

$$P(f_i) = 0 \quad i = 1, 2, \ldots, m. \qquad (12.61)$$

This indicates that

$$\frac{1}{P(f)} = \frac{1}{\sum_{k=m+1}^{M} c_k |S^H(f)V_k|^2} \qquad (12.62)$$

is infinite at $f = f_i$. But, in practice due to the estimation errors, $\frac{1}{P(f)}$ is finite with very sharp peaks at all sinusoidal frequencies providing a way for estimating the sinusoidal frequencies.

Choosing $c_k = 1$ for all k, the MUSIC frequency estimator [10] is written as

$$P_{\text{MUSIC}}(f) = \frac{1}{\sum_{k=m+1}^{M} |S^H(f)V_k|^2} \qquad (12.63)$$

The peaks of $P_{\text{MUSIC}}(f)$ are the estimates of the sinusoidal frequencies, and the powers of the sinusoids can be estimated by solving Eq. (12.58). The following example illustrates the estimation of power spectrum using the MUSIC method.

Example 12.7 Consider a random signal generated by the following equation

$$x(n) = \sin\left(\frac{2\pi f_1 n}{F_s}\right) + 2\sin\left(\frac{2\pi f_2 n}{F_s}\right) + 0.1w(n)$$

where the frequencies f_1 and f_2 are 220 and 332 Hz, respectively, the sampling frequency F_s is 2048 Hz and $w(n)$ is a zero mean, unit variance, white noise process. Estimate power spectrum of the sequence $\{x(n), 0 \le n \le 1023\}$.

Solution The MATLAB function *pmusic(X,m, 'whole')* gives the power spectrum of a discrete-time signal X using the MUSIC method, m being the number of complex sinusoids in the signal X. If X is an autocorrelation data matrix of discrete-time signal x, the function *corrmtx* can be used to generate data matrices. The signal vector x consists of two real sinusoidal components. In this case, the dimension of the signal subspace is 4 because each real sinusoid is the sum of two complex exponentials.

The following MATLAB Program 12.6 is used to obtain the power spectrum using the MUSIC method.

Program 12.6

Power spectrum estimation using the MUSIC method
```
clear;clc;
randn('state',0);
N = 1024;%total number of samples
k = 0 : N-1;
f1 = 280;
f2 = 332;
FT = 2048;%sampling frequency in Hz
T = 1/FT;
x = sin(2*pi*f1*k*T) + 2*sin(2*pi*f2*k*T)+0.1*randn(size(k));%input vector of
length N
X = corrmtx(x,12);%estimates  (N+12)  by  (12+1)  rectangular autocorrelation
matrix
pmusic(X,4,'whole'); %estimates power spectrum of x containing two sinusoids
```

The power spectrum obtained from the above program is shown in Fig. 12.10.

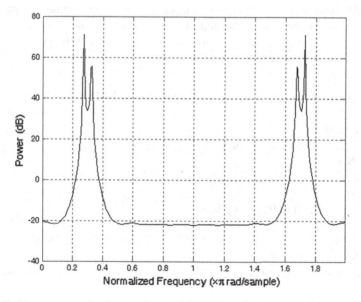

Fig. 12.10 Power spectral density estimate using MUSIC method

12.3.3 Eigenvector Method

The eigenvector method is a weighted version of the MUSIC method. Selecting $c_k = \frac{1}{\lambda_k}$ in Eq. (12.62) for all k, the eigenvector method produces a frequency estimator given by Johnson [11].

$$P_{\text{eig}}(f) = \frac{1}{\left(\sum_{k=m+1}^{M} \left(\frac{1}{\lambda_k} \right) \left| V_k^H S(f) \right|^2 \right)} \tag{12.64}$$

where M is the dimension of the eigenvectors and V_k is the kth eigenvector of the autocorrelation matrix of the observed data sequence. The integer m is the dimension of the signal subspace, so the eigenvectors V_k used in the sum correspond to the smallest eigenvalues λ_k of the autocorrelation matrix. The eigenvectors used in the PSD estimate span the noise subspace. The power spectrum estimation using the eigenvector method is illustrated through the following example.

Example 12.8 Consider the random signal generated in the Example 12.7 and estimate its power spectrum using the eigenvector method.

Solution The MATLAB function *peig(X,m, 'whole')* estimates the power spectrum of a discrete-time signal X, m being the number of complex sinusoids in the signal X. If X is an autocorrelation data matrix of discrete-time signal x, the function *corrmtx* can be used to generate data matrices. The signal vector x consists two real sinusoidal components. In this case, the dimension of the signal subspace is 4 because each real sinusoid is the sum of two complex exponentials.

The following MATLAB Program 12.7 is used to obtain the power spectrum using the eigenvector method.

Program 12.7
Power spectrum estimation using the eigenvector method

```
clear;clc;
randn('state',0);
N = 1024;%total number of samples
k = 0 : N-1;
f1 = 280;
f2 = 332;
FT = 2048;%sampling frequency in Hz
T = 1/FT;
x = sin(2*pi*f1*k*T)+2*sin(2*pi*f2*k*T)+0.1*randn(size(k));%input vector of length N
X = corrmtx(x,12);%estimates (N+12) by (12+1) rectangular autocorrelation matrix
peig(X,4,'whole');%estimates power spectrum of x containing two sinusoids
```

The power spectrum produced from the above program is shown in Fig. 12.11.

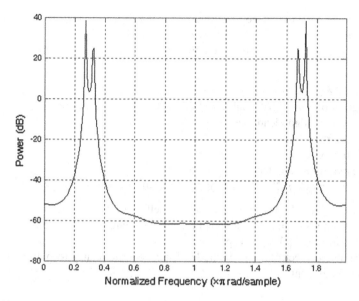

Fig. 12.11 Power spectral density estimate using eigenvector method

12.4 Spectral Analysis of Non-stationary Signals

A signal with time-varying parameters, for example, a speech signal, is called a
non-stationary signal. The spectrogram which shows how the spectral density of a
signal varies with time is a basic tool for spectral analysis of non-stationary signals.
Spectrograms are usually generated using the short-time Fourier transform (STFT)
using digitally sampled data. To compute the STFT, a sliding window which
usually is allowed to overlap in time is used to divide the signal into several blocks
of data. Then, an N-point FFT is applied to each block of data to obtain the
frequency contents of each block.

   The window length affects the time resolution and the frequency resolution of
the STFT. A narrow window results in a fine time resolution but a coarse frequency
resolution, whereas a wide window results in a fine frequency resolution but a
coarse time resolution. A narrow window is to be used to provide wideband
spectrogram for signals having widely varying spectral parameters. A wide window
is preferred to have narrowband spectrogram. The following example illustrates the
computation of the spectrogram of a speech signal.

Example 12.9 Consider a speech signal 'To take good care of yourself' from the
sound file 'goodcare.wav' (available in the CD accompanying the book). Compute
the spectrogram of the speech signal using Hamming window of lengths 256
samples and 512 samples with an overlap of 50 samples.

Solution The STFT of a non-stationary signal x can be computed by using the MATLAB file *specgram(x,wl,Fs, window,noverlap)*

where wl stands for window length, and noverlap is the number of overlap samples.

Program12.8
Spectrogram of a speech signal
[x,F_T] = wavread('goodcare.wav');
i = 1:length(x)
figure(1),plot(x)
xlabel('Time index i');ylabel('Amplitude');
figure(2), specgram(x,256, F_T,hamming(256),50)

The speech signal 'To take god care of yourself' is shown in Fig. 12.12.
The spectrograms of the speech signal for window lengths of 256 and 512 samples are shown in Fig. 12.13a, b respectively.
From the above spectrograms, the trade-off between frequency resolution and time resolution is evident.

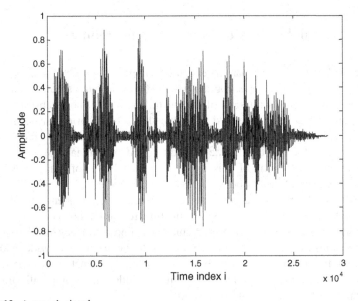

Fig. 12.12 A speech signal

Fig. 12.13 a Spectrogram
with window length 256,
overlap = 50 and
b spectrogram with window
length 512, overlap = 50

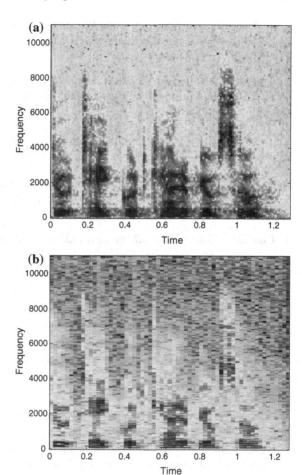

12.4.1 MATLAB Exercises

1. Consider a random signal of length 1024 composed of two sinusoidal components
   of frequencies 180 and 320 Hz with sampling frequency of $F_T = 2048$ Hz corrupted with zero mean, unit variance, white noise process. Evaluate its power
   spectrum using Bartlett's method with subsequence lengths of each 256 samples.
2. Consider the following signal of length $N = 1024$ with sampling frequency of
   $F_s = 2048$ Hz corrupted with zero mean, unit variance, white noise process.

$$x(i) = \sin\left(\frac{800\pi i}{F_T}\right)\cos\left(\frac{800\pi i}{F_T}\right) + w(i), \quad 0 \leq i < N$$

where $w(i)$ is zero mean Gaussian white noise.

Evaluate its power spectrum using Welch's method with subsequence lengths of each 256 samples using Blackman window for overlaps of 64 and 128 samples, respectively.

3. Consider the following signal of length $N = 1024$ with sampling frequency of $F_s = 2048$ Hz corrupted with zero mean, unit variance, white noise process.

$$x(i) = \sin^2\left(\frac{400\pi i}{F_T}\right)\cos^2\left(\frac{200\pi i}{F_T}\right) + w(i), \quad 0 \le i < N$$

where $w(i)$ is zero mean unit variance, white noise process.

Evaluate its power spectrum using Blackman–Tukey method with window length of 256 samples.

4. Consider a fourth-order AR process characterized by

$$y(n) + a_1 y(n-1) + a_2 y(n-2) + a_3 y(n-3) + a_4 y(n-4) = w(n)$$

where $w(n)$ is a zero mean, unit variance, white noise process. The parameters $\{a_1, a_2, a_3, a_4\}$ are chosen such that the prediction error filter

$$A(z) = 1 + a_1 z^{-1} + a_2 z^{-2} + a_3 z^{-3} + a_4 z^{-4}$$

has zeros at the locations.

$$0.98e^{\pm j0.15\pi} \text{ and } 0.98e^{\pm j0.35\pi}$$

Estimate the power spectrum of the AR process using the Yule–Walker method and Burg method based on 200 samples. Comment on the results.

5. Consider the following signal of length $N = 1024$ with sampling frequency of $F_s = 2048$ Hz corrupted with zero mean, unit variance, white noise process.

$$x(n) = \sum_{i=1}^{3} e^{j2\pi f_i n} + w(n), \quad 0 \le n < N$$

where $w(i)$ is zero mean unit variance, white noise process, and the frequencies are Hz $f_1 = 256$ Hz, $f_2 = 338$ Hz, and $f_3 = 338$ Hz.

Evaluate its power spectrum using the MUSIC method and the eigenvector method. Comment on the results.

6. Consider a speech signal from the sound file 'speech.wav' (available in the CD accompanying the book) and compute its spectrogram for different window lengths with and without overlap. Comment on the results.

References

1. A. Schuster, On the investigation of hidden periodicities with application to a supposed 26 day period of meteorological phenomena. Terr. Magn. Atmos. Electr. **3**, 13–41 (1898)
2. M.S. Bartlett, Smoothing *periodograms* from time series with continuous spectra. Nature **161**, 686–687 (1948)
3. P.D. Welch, The use of fast Fourier transform for the estimation of power spectra: A method based on time averaging over short, modified periodograms. IEEE Trans. Audio Electroacoust. **15**, 70–83 (1967)
4. R.B. Blackman, J.W. Tukey, *The Measurement of Power Spectra from the Point of View of Communication Engineering* (Dover Publications, USA, 1958)
5. J.G. Proakis, D.G. Manolakis, *Digital Signal Processing Principles, Algorithms and Applications* (Printice-Hall, India, 2004)
6. J.P. Burg, A new Analysis technique for time series data. Paper presented at the NATO Advanced Study Institute on Signal Processing, Enschede, 1968
7. H. Akaike, Power spectrum estimation through autoregressive model fitting. Ann. Inst. Stat. Math. **21**, 407–420 (1969)
8. H. Akaike, A new look at the statistical model identification. IEEE Trans. Autom. Control **19** (6), 716–723 (1974)
9. S.L. Marple, in *Digital Spectral Analysis*. (Prentice-Hall, Englewood Cliffs, NJ, 1987), pp. 373–378, 686–687
10. R.O. Schmidt, Multiple emitter location and signal parameter estimation. IEEE Trans. Antennas Propag. **AP-34**, 276–280 (1986)
11. D.H. Johnson, The application of spectral estimation methods to bearing estimation problems. Proc. IEEE **70**(9), 126–139 (1982)

Chapter 13
DSP Processors

DSP processors play a vital role in many consumer, communication, medical, and industrial products like mobile phones, codecs, radar analysis systems. DSP processors are specially designed microprocessors for DSP applications. In comparison with microprocessors, DSP processors are faster and energy efficient. This chapter deals with an evolution of DSP processors, key features of various DSP processors, internal architectures, addressing modes, and important instruction sets of Texas TMS320C54xx, and Analog Devices TigerSHARCTSxxx DSP processors, followed by implementation examples.

13.1 Evolution of DSP Processors

How DSPs are Different from Other Microprocessors?

The Von Neumann architecture and Harvard architecture are shown in Fig. 13.1a, b respectively. The program, i.e., instructions and data are stored in single memory, whereas in Harvard architecture, data and instructions are stored in two memory banks as shown in Fig. 13.1. The Von Neumann memory architecture is most commonly used in microcontrollers. The data operands and instructions cannot be accessed in parallel due to single data bus. Hence, the execution of DSP algorithms using microcontroller is slow. In Harvard architecture, a data operand and an instruction can be accessed in parallel in every cycle due to two data buses. Therefore, fast execution of DSP algorithms is possible.

13.1.1 Conventional DSP Processors

The first single-chip DSP processor was introduced by Bell Labs in 1979. Later, Texas Instruments produced the first DSP processor TMS32010 based on Harvard

© Springer Nature Singapore Pte Ltd. 2018
K. D. Rao and M. N. S. Swamy, *Digital Signal Processing*,
https://doi.org/10.1007/978-981-10-8081-4_13

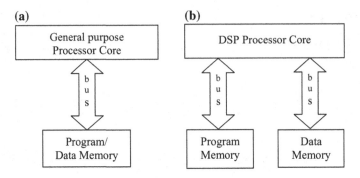

Fig. 13.1 a Von Neumann architecture, **b** Harvard architecture

architecture in 1982. The processors ADSP-21xx, TMS320C2xx, and DSP560xx operating at around 20–50 MHz yielded good performance with modest power consumption and memory usage. The next generation DSP processors operating at 100–150 Hz have exhibited better performance.

13.2   Modern DSP Processors and Its Architectures

As compared to the conventional DSP processors, modern DSP processors provide improved performance by incorporating dedicated hardware modules like multiple computational blocks, cache memories, barrel shifter, wider bus system. Some of the modern DSP processors support "multiprocessing," simultaneously more than one processor executing the same piece of code in parallel. TigerSHARC from Analog Devices is of this kind.

The very long instruction word (VLIW) and *superscalar* architectures having many execution units execute multiple instructions in parallel. For example, TMS320C62xx consists of eight independent execution units. VLIW DSP processors issue 4–8 instructions per clock cycle, whereas superscalar processors issue 2–4 instructions per cycle.

Some of the features of the DSP processor that accelerate the performance in DSP applications are described below.

13.2.1   Single-Cycle Multiply–Accumulate (MAC) Unit

A MAC operation in a single instruction cycle is achieved with built-in MAC hardware on the DSP processor. To avoid the possibility of arithmetic overflow, extra-bits are generally provided in the accumulator. High-performance DSP processors like TigerSHARC have two multipliers that enable more than one multiply–accumulate operation per instruction cycle.

13.2.2   Modified Bus Structures and Memory Access Schemes

A modified Harvard architecture with two separate banks of memory is shown in Fig. 13.2 in which each memory was accessed by its own bus during each cycle.

13.2.3   On-Chip Memory

A memory that physically exists on the processor itself is called on-chip memory. This memory is used to store program, data, or repeatedly used instructions (cache memory).

13.2.4   Pipelining

Pipelining is a technique used to increase the throughput (i.e., number of instructions that can be executed per unit time). Let T_n be the time per instruction on machine without pipelining and T_p the time per instruction on the machine with pipelining. Then,

$$T_p = T_n/\text{number of pipeline stages}.$$

As the number of pipeline stages increases, time per instruction decreases.

Fig. 13.2 Modified Harvard

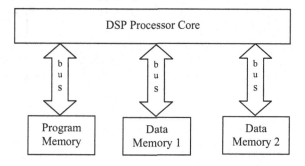

13.2.5 Parallelism

Data Parallelism—Single instruction, multiple data (SIMD) operations

SIMD is a technique employed to achieve data parallelism. As shown in Fig. 13.3, data parallelism is required when a large mass of data of a uniform type that needs the same instruction to be performed on it. An example of data parallelism is downconverting a signal sampled at F_T, with local oscillator ($F_T/4$) using a mixer. This operation involves multiplying the incoming signal with (**1 0 −1 0** 1 0 −1 0 ...,) sequence. This involves iteratively multiplying the incoming signal samples with 1 0 −1 0 sequence—multiple data points (input data samples), a single operation (multiply with 1 0 −1 0 pattern). TigerSHARC is capable of executing eight 16-bit multiplications per cycle using SIMD.

13.2.6 Special Addressing Modes

Bit-reversed addressing mode is an addressing mode that arises when a list of values has to be reordered by reversing the order of the address bits.

In applications like FIR filtering, filter coefficients are inverted and multiplied with the input to get the output sample. Such situations require repeated multiplications of reversed filter coefficients with input data shifted by one sample each time. For this, register-indirect addressing with post-increment is used to automatically increment the address of the pointer in the execution of these algorithms. Many DSP processors support "*circular addressing*," also called modulo-addressing, which is very helpful in implementing first-in, first-out buffers.

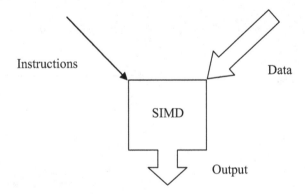

Fig. 13.3 Single instruction, multiple data

13.2.7 Specialized Execution Control

Most of the processors provide hardware support for efficient looping to perform repetitive computations. A for-next loop called *zero-overhead looping* is provided to implement without expending any instruction cycle for updating and testing the loop counter and branching to the top of the loop.

13.2.8 Streamlined I/O

Many DSP processors incorporate serial or parallel I/O interfaces. In direct memory access (DMA), processors read data from an I/O device and copy the data into memory or vice versa to provide improved performance for the input/output devices. In general, a separate DMA controller is used to handle such transfers more efficiently.

13.2.9 Specialized Instruction Sets

DSP processor instruction sets are designed to maximize the utilization of processors underlying hardware to increase the efficiency and minimize the amount of memory space required to store DSP programs to reduce the chip cost. To achieve maximum utilization of the processor hardware, DSP processor instruction sets allow many parallel operations in a single instruction. To minimize the amount of memory space, instructions are kept short.

13.2.10 On-Chip Peripherals

DSP processors communicate with the other systems through peripherals. These peripherals are the interfaces for the interrupts, DMA, and I/O transactions. DSPs also have one or more DMA controllers to perform data transfers without processor intervention.

13.3 Choice of DSP Processor

The following factors are to be considered in choosing a DSP processor

- Arithmetic Format: Fixed or Floating Point
- Size of Data Word

- Speed
- Organization of Memory
- Multiprocessor Support
- Power Consumption and Management
- Cost
- Ease of Development

13.4   TMS320C54xx Digital Signal Processor

13.4.1   Features of TMS320C54xx Digital Signal Processor

The features of the 54xx DSP processor are:

1. High performance, low power.
2. On-chip peripherals.
3. Power conservation features.
4. Emulation capability based on on-chip scan.
5. IEEE 1149.1 (JTAG) boundary-scan test capability.
6. 5.0, 3.3, 2.5, 1.8, 1.5 V power supply devices with speeds 40, 80, 100, 200, and 532 million instructions per second (MIPS).

13.4.2   The Architecture of TMS320C54xx Digital Signal Processor

The '54x DSPs are fixed-point processors based on an advanced, modified Harvard architecture with one program memory bus and three data memory buses. The functional block diagram of this processor is shown in Fig. 13.4.

Architectural Features of TMS320C54xx DSP Processors
(Courtesy of Texas Instruments Inc.)

• Cycle Performance
  – 25-ns single-cycle fixed-point instruction execution time [40 MIPS] for 5-V power supply ('C541 and 'C542 only)
  – 20-ns and 25-ns single-cycle fixed-point instruction execution time (50 MIPS and 40 MIPS) for 3.3-V power supply ('LC54x)
	– Dual-Access On-Chip RAM
	  – Single-Access On-Chip RAM ('548/ '549)
	  • Instruction Support
	  – Single-instruction repeat and block-repeat operations for program code
	  – Block-memory-move instructions for better program and data management

(continued)

(continued)

- 15-ns single-cycle fixed-point instruction execution time (66 MIPS) for 3.3-V power supply ('LC54xA, '548, 'LC549)
- 12.5-ns single-cycle fixed-point instruction execution time (80 MIPS) for 3.3-V power supply ('LC548, 'LC549)
- 10-ns and 8.3-ns single-cycle fixed-point instruction execution time (100 and 120 MIPS) for 3.3-V power supply (2.5-V Core) ('VC549)
- Core Performance
  - A 40-bit arithmetic logic unit (ALU)
  - Two 40-bit accumulators
  - A barrel shifter
  - A 17 × 17-bit multiplier/adder
  - A compare, select, and store unit (CSSU)
  - Exponent encoder to compute an exponent value of a 40-bit accumulator value in a single cycle
  - Two address generators with eight auxiliary registers and two auxiliary register arithmetic units (ARAUs)
  - Data bus with a bus holder feature
  - Address bus with a bus holder feature ('548 and '549 only)
- Memory Architecture
  - Extended addressing mode for 8 M × 16-Bit maximum addressable external program space ('548 and '549 Only)
  - 192 K × 16-bit maximum addressable memory space (64 K words program, 64 K words data, and 64 K words I/O)
  - On-chip ROM with some configurable to program/data memory

- Instructions with a 32-bit long word operand
- Instructions with two or three operand reads
- Arithmetic instructions with parallel store and parallel load
- Conditional store instructions
- Fast return from interrupt
- On-Chip Peripherals
  - Software-programmable wait-state generator and programmable bank switching
  - On-chip phase-locked loop (PLL) clock generator with Internal Oscillator or external clock source
  - Full-duplex serial port to support 8- or 16-bit transfers ('541, 'LC545, and 'LC546 only)
  - Time-division multiplexed (TDM) serial port ('542, '543, '548, and '549 only)
  - Buffered serial port (BSP) ('542, '543, 'LC545, 'LC546, '548, and '549 only)
  - 8-Bit Parallel Host Port Interface (HPI) ('542, 'LC545, '548, and '549)
  - One 16-Bit Timer
  - External input/output (XIO) off control to disable the external data bus, address bus and control signals
- On-chip scan-based emulation logic, IEEE Std 1149.1† (JTAG) boundary-scan logic
- Power consumption control with IDLE1, IDLE2, and IDLE3 instructions with power-down modes

13.4.3 Data Addressing Modes of TMS320C54xx Digital Signal Processors

An addressing mode specifies how to calculate the effective memory address of an operand. The TMS320C54xx devices offer seven basic addressing modes:

- Immediate addressing
- Absolute addressing

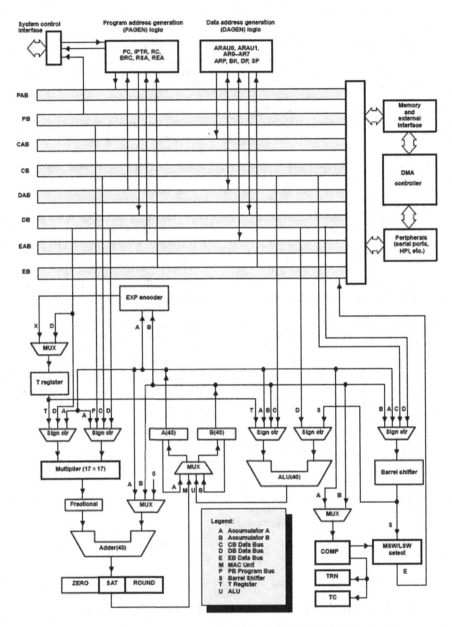

Fig. 13.4 Functional block diagram of **TMS320C54xx Digital Signal Processor** (Courtesy of Texas Instruments Inc.)

- Accumulator addressing
- Direct addressing
- Indirect addressing (Circular, Bit-reversed, and Dual Operand)
- Memory-mapped register addressing and
- Stack addressing.

13.5 Code Composer Studio

Code Composer Studio is an integrated development environment for developing DSP code for the TMS320 DSP processor family from Texas Instruments.

Steps required for creating a program

1. In project open a new project.
   Project → new
   Create a new project of type .mak in 'myprojects' or create in own directory.
2. In file option select new source file. Type your source code in that file
   File → new → source file
3. After writing source file save that file as .asm file.
4. Add source file to the project.
   Project → add files to the project → select .asm file → click open
5. Add command file to the project.
   Project → add files to the project → select c:\ti\c5400\tutorial\hello1
   In hello1 folder select helo.cmd, after selecting click to add.
6. If it is a 'C' program, then add libraries to the project.
   Project → add files to the project → select c:\ti\c5400\cgtools\lib\rts.lib
7. After adding required files, build the project. If there are any compilation errors, they are displayed on the build window.
8. If build is successful, .out file is created in directory or 'myprojects' directory.
9. Then load the program from the file menu.
   File → load program → select the .out file.
   Then a disassembly window is opened.
10. Change the program counter address to 0 × 1400.
    View → CPU registers → CPU register.
    In that double click on the program counter. Then change the address to 0 × 1400.
11. To view memory contents (to store the values at the specified memory locations)
    Select memory option.
    View → memory → select the required memory location.
    To see the contents of memory disassembly at the same time, select 'float in main window'.
    For that right click the mouse in memory window, and then select the option.

12. Put break point on the last instruction. For that keep cursor on the last instruction, click right button of the mouse and select toggle break point option.
13. After loading the contents of memory, run the program.
    Debug → run.
    To run the program step-by-step, select F8 key.

Example Programs on TMS320C54xx
Addition of two 16-bit numbers

```
ld #0,a ;              // clearing the accumulator contents
st #400 h, ar0;        // storing the address location in auxiliary register
st #402 h, ar1;
st #404 h, ar2;
ld *ar0,a;             // loading the value at 'ar0' register into accumulator
add *ar1,a;            // adding the contents of ar1 register to accumulator
stl a, *ar2 + ;        // storing the result in the memory locations
sth a, *ar2;
nop;                   // no operation
nop;
```

Addition of three 32-bit numbers

```
st #400 h, ar0;        // storing the address location in auxiliary register
st #402 h, ar1;
st #404 h, ar2;
st #406 h, ar3;
dld *ar0+,a;           // load long word into accumulator
dadd *ar1+,a;          //
adds the contents of accumulator to 32-bit long data memory operand
dadd *ar2+,a;
dst a, *ar3 + ;
nop;                   // no operation
nop;
```

Multiplication of two numbers using single memory operand

```
ld #0,a ;              // clearing the accumulator contents
st #400 h, ar0;        // storing the address location in auxiliary register
st #402 h, ar1;
st #404 h, ar2;
ld *ar0,t;             // loading the contents of auxiliary regis-
ter into temporary register
mpy *ar1,a;            // multiples the contents of 'ar1' with 't' and re-
sult is stored in accumulator
stl a,* ar2+;
```

```
sth a,*ar2;
nop;                    // no operation
nop;
```

Swapping of two numbers

```
mvdm #400 h, ar0;           // move data from data memory loca-
tion to 'ar0' register
mvdm #402 h, ar1;
mvdm #404 h, ar2;
mvmm ar0, ar2;
mvmm ar1, ar0;
mvmm ar2, ar1;
mvmd ar1, #402 h;
mvmd ar0, #400 h;
nop;                        // no operation
nop
```

Block transfer of 16 memory locations

```
st #400 h, ar3;
st #500 h, ar5;
rpt #0fh;               // counter is set to 16
mvdd *ar3+, *ar5+;      // move data from memory location to data memory lo-
cation
nop;
nop;
```

MAC instruction

```
.mmregs            // initialization of memory registers
st #400 h, ar0;
st #402 h, ar1;
ld #5 h,a;
st #4 h,t;
mac *ar0, a;       // multiply values at 'ar0' by 't' and re-
sult is added with accumulator
                   // and stored in accumulator
stl a, *ar1+;
sth a, *ar1;
nop;               // no operation
nop;
```

13.6 TMS320C67xx Digital Signal Processor

13.6.1 Features of TMS320C67xx Digital Signal Processor

The TMS320C67xx (including the TMS320C6713 device) is developed by Texas Instruments based on advanced VLIW architecture. It is an excellent choice for multichannel and multifunction applications. It operates at 225 MHz and delivers 1350 million floating-point operations per second (MFLOPS), 1800 million instructions per second (MIPS), and with dual fixed-/floating-point multipliers up to 450 million multiply–accumulate operations per second (MMACS). The features of the 67xx DSP processor are:

1. High-performance 'C67x CPU: The CPU consists of eight independent functional units: two ALUs (fixed point), four ALUs (floating and fixed point), and two multipliers (floating and fixed). It can process eight 32-bit instructions per cycle. The operating frequencies range from 167 to 225 MHz.
2. On-chip peripherals: It consists of a flexible phase-locked loop (PLL) clock generator with internal crystal oscillator or external clock source, full-duplex standard serial port.
3. On-chip scan-based emulation capability.
4. IEEE 1149.1 (JTAG) boundary-scan test capability.

13.6.2 The Architecture of TMS320C67xx Digital Signal Processor

TMS320C67xx processor architecture is shown in Fig. 13.5. The CPU consists of eight functional units L11, .S11, .M11, D11, D22, .M22, .S22, and .L22 divided into two sets. The first set contains L11, .S11, .M11, and D11 functional units, and the second set consists of D22, .M22, .S22, and .L22 functional units. It contains two register files A and B. The first set of functional units and the register file A compose side A of the CPU, and the second set of functional units and the register file B compose side B of the CPU. The C67x CPU executes fixed-point instructions, and in addition, the functional units .L11, .S11, .M11, .M2,2 .S22, and .L22 also execute floating-point instructions.

Features of TMS320C67xx Digital Signal Processor
(Courtesy of Texas Instruments Inc.)

Highest Performance Floating-Point Digital Signal Processor (DSP)
- Eight 32-bit instructions/cycle
- 32/64-bit data word
- 225, 200 MHz (GDP), and 200, 167 MHz (PYP) clock rates
- 4.4, 5, 6 instruction cycle times
- 1800/1350, 1600/1200, and 1336/1000 MIPS/ MFLOPS
- Rich peripheral set, optimized for audio
- Highly optimized C/C++ compiler
- Extended temperature devices available.

Advanced Very Long Instruction Word (VLIW) TMS320C67x DSP Core
- Eight independent functional units:
  - Two ALUs (fixed point)
  - Four ALUs (floating and fixed point)
  - Two multipliers (floating and fixed point)
- Load   –store architecture with 32, 32-bit. General-purpose registers
- Instruction packing reduces code size
- All instructions conditional

Instruction Set Features
- Native instructions for IEEE 754 (single and double precision)
- Byte addressable (8, 16, 32-bit data)
- 8-bit overflow protection
- Saturation; bit field extract, set, clear; Bit counting; normalization

L1/L2 Memory Architecture
- 4K Byte L1P program cache (direct mapped)
- 4K Byte L1D Data Cache (2-Way)
- 256K-Byte L2 Memory Total: 64K-Byte L2 Unified Cache/Mapped RAM, and 192K-Byte Additional L2 Mapped RAM

Device Configuration
- Boot Mode: HPI, 8-, 16-, 32-Bit ROM Boot
- Endianness: Little Endian, Big Endian

32-Bit External Memory Interface (EMIF)
- Glueless interface to SRAM, EPROM, Flash , SBSRAM, and SDRAM
- 512 M-Byte total addressable external memory space

Enhanced Direct Memory Access (EDMA) Controller (16 Independent Channels)
- 16-bit host port interface (HPI)

Two McASPs
- Two independent clock zones each (1 TX and 1 RX)
- Integrated digital audio interface transmitter (DIT) supports:
  - S/PDIF, IEC60958-1, AES-3, CP-430 Formats
  - Up to 16 transmit pins
  - Enhanced channel status/user data
- Extensive error checking and recovery
- Two inter-integrated circuit bus (I2C Bus) Multi-master and slave interfaces

Two Multi-channel Buffered Serial Ports:
- Serial Peripheral Interface (SPI)
- High-Speed TDM Interface
- AC97 Interface

Two 32-Bit General-Purpose Timers
IEEE-1149.1 (JTAG)
Boundary-Scan-Compatible
208-Pin PowerPAD Plastic (Low Profile)
Quad Flat pack (PYP)
272-BGA Packages (GDP)
0.13-μm/6-Level Copper Metal Process
CMOS Technology
3.3-V I/Os, 1.2-V‡ Internal (GDP & PYP)

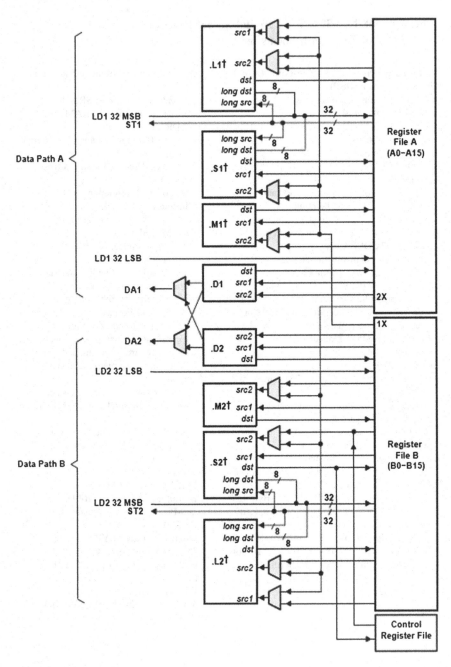

Fig. 13.5 Architecture of TMS320C67xx Digital Signal Processor(Courtesy of Texas Instruments Inc)

13.6.3 Design of a Digital Lowpass FIR Filter and Implementation Using TMS320C67xx DSP Processor

Problem statement

It is required to suppress unwanted high frequency noise from a corrupted speech signal. Toward this end, it is desired to design a digital lowpass FIR filter and to implement the filter on TMS320C67xx DSP processors.

Design of the digital FIR lowpass filter

(1) The input signal is a voice signal corrupted by a random noise as shown in Fig. 13.6.
(2) The desired filter is designed using MATLAB SPTOOL. The specifications and magnitude response of the lowpass filter designed are as shown in Fig. 13.7.
(3) The designed filter is of the order 114.

Implementation of 32-bit floating-point FIR lowpass filter using TMS320c67xx DSP Processor

Code Composer Studio is an integrated debugging and development environment (IDDE) for developing and debugging DSP code on processors of TI family.

Creating a Code Composer Studio Project

Create a folder to save project files.

- Open Code Composer Studio Environment.
- choose **New** → **Project** from the **File** menu …
- The **Project Wizard** dialog box appears. Set project options based on type of the board used (or simulator).

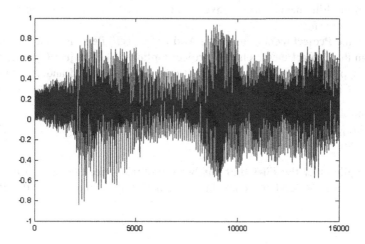

Fig. 13.6 Noisy speech signal

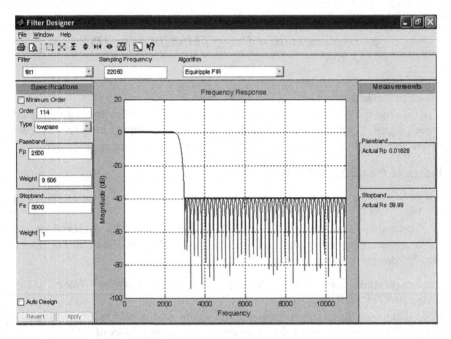

Fig. 13.7 Magnitude response of the FIR lowpass filter designed for suppressing the noise

- Fill in the project name in *Project Name*. Browse to the folder you created for the project in *Location*. Make sure you select *Executable (.out)* in *Project Type*, and select **TMS320c67xx** in *Target*, click *Finish* when you have done.
- Click on *File → Source File* to create a new source code file A window titled '*Untitled1*' will appear within the Composer environment:
- Copy the code from Appendix-A and paste it into the source file:
- From the **File** menu, choose **Save → File** to save this source file to c:\myprojects\filter.asm.
- From the **Project** menu, choose the **Add to Project/File(s)** and select filter.asm.
- From the **Project** menu, choose **Project Options**. The **Project Options** dialog box appears. From link menu, default cmd file is generated and select include files as required from the ti library.

To Compile Only the Current Source File
Select Project → Compile File, or click the Compile File button on the Project toolbar. No linking is performed.

To Compile Only the Files that has been modified since the Last Build
Select Project → Build, or click the Incremental Build button on the Project toolbar.

To Compile All Files
Select Project → Rebuild All, or click the Rebuild All button on the Project toolbar.
    The output file is relinked.
To Stop Building
Select Project → Stop Build, or click the Stop Build button on the Project toolbar. The build process stops only after the current file has finished compiling.
Build Clean
Select Project à Build Clean to remove all intermediary files in a CCS software project. That includes temp files, .obj files, and COFF files. It does not actually build your project.
Building the Project
From the **Project** menu, choose **Rebuild Project**.
Loading a Program
After code has been successfully compiled and built, load program onto the DSP. Select *File → Load Program...*
    Code Composer Studio is set at default to create a new folder in same directory called *Debug*. This is where the executable file is created. Double-click on the *Debug* folder to see **.out* file (executable).
   Running a Program
   Select *Debug → Run* or press the *F5* key to execute the program.
   Select Debug → Halt.to stop running the code.
   Press the *F5* button to resume running the code.
   To view Results:
   Right click on the variable and select memory to watch the results in a separate window. Or Go To View → memory; give starting address and length in hexadecimal.
   Plotting the results
   The graph menu contains many options. Use the View → Graph → Constellation command to view the Graph Property Dialog box. The resulted plot is shown in Fig. 13.8.

13.7   TigerSHARC TS201 DSP Processor

The ADSP–TSxxx processor is a 128-bit, high-performance TigerSHARC processor. The processors of TigerSHARC family operate from 250 to 600 MHz. TigerSHARC processors perform floating- and fixed-point operations by combining multiple computation units (Fig. 13.9).
    The TigerSHARC processor uses a variation of static superscalar architecture, which allows the programmer to specify the instructions to be executed in parallel in each cycle. To facilitate the high clock rate, the ADSP-TS201 processor uses a

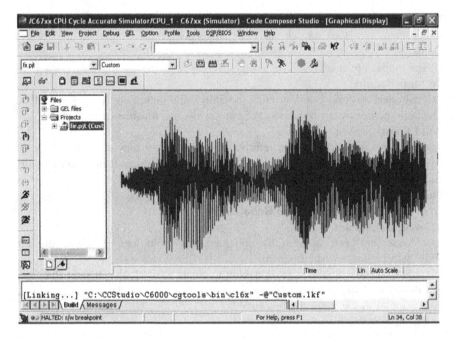

Fig. 13.8 Recovered speech signal

pipelined external bus with programmable pipeline depth for interprocessor communications. The ADSP-TS201 processor achieves its fast execution rate by means of a ten-cycle pipeline.

13.7.1 Architectural Features of TS201 Digital Signal Processor

(Courtesy of Analog Devices Inc.)

• **Cycle Performance**:	• **On-chip Peripherals**
– ADSP-TS101S operates at 250 MHz (4 ns cycle time) with 6 Mbits on-chip memory	– Dual computation blocks containing each an ALU, a multiplier, a shifter, a register file, and a communications logic unit (CLU)
– ADSP-TS203S operates at 500 MHz (2 ns cycle time) with 4 Mbits on-chip memory	– Integrated I/O includes 14-channel DMA controller, external port, four link ports, SDRAM controller, programmable flag pins, two timers, and timer expired pin for system integration
– ADSP-TS202S operates at 500 MHz (2 ns cycle time) with 12 Mbits on-chip memory	

(continued)

(continued)

– ADSP-TS201S operates at 500/ 600 MHz (2/1.67 ns cycle time) with 24 Mbits on-chip memory and executes 4.8 billion MACS with 3.6 GFLOPS of floating-point power and 14.4 BOPS of 16-bit processing	– Supports low overhead DMA transfers between internal memory, external memory, memory-mapped peripherals, link ports, host processors, and other (multiprocessor) DSPs

- **Parallelism and throughput**
  - Dual computation blocks
  - 10 cycle instruction pipeline
  - Dual integer ALUs, providing data addressing and pointer manipulation
  - 4 link ports @ up to 1 GByte/s each
  - Static superscalar architecture
  - Execution of up to four instructions per cycle
  - Access of up to eight words per cycle from memory
- **Multiprocessing Capabilities**
  - On-chip bus arbitration for glueless multiprocessing
  - Globally accessible internal memory and registers
  - Semaphore support
  - Powerful, in-circuit multiprocessing
  - emulation

- 1149.1 IEEE compliant JTAG test access port for on-chip emulation
- Single-precision IEEE 32-bit and extended-precision 40-bit floating-point data formats and 8-, 16-, 32-, and 64-bit fixed-point data formats
- Eases DSP programming through extremely flexible instruction set and high-level-language-friendly DSP architecture

Package
- 25 mm × 25 mm (576-ball) thermally enhanced ball grid array package
- Provides high performance static superscalar DSP operations, optimized for telecom, infrastructure and other large, demanding multiprocessor DSP applications

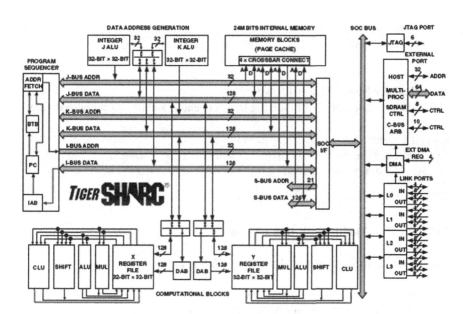

Fig. 13.9 Architecture of TigerSHARC TS201 processor (Courtesy of Analog Devices Inc.)

13.7.2   Addressing Modes

The following describes some of the addressing modes of TigerSHARC processor:

13.7.2.1   Data Addressing

Ureg Register Load Data Addressing: The address is not modified following transfers and maintains its initialized value.

```
e.g.: Ureg_s = {CB|BR} [Jm| < Imm32>] ;
/* Ureg suffix indicates: _s = single, _sd = double, _sq = quad */
```

Dreg Register Load Data Addressing:
The address is not modified following transfers and maintains its initialized value.

```
e.g.: { X|Y|XY}Rs = {CB|BR} [Jm | < Imm32>] ;
/* R suffix indicates: _s = single, _sd = double, _sq = quad */
/* m must be 0,1,2, or 3 for bit reverse or circular buffers */
```

Ureg Register Store Data Addressing:
The address is not modified following transfers and maintains its initialized value.

```
e.g.: [Jm | < Imm32>] = Ureg_s ;
/* Ureg suffix indicates: _s = single, _sd = double, _sq = quad */
```

Dreg Register Store Data Addressing:
The address is not modified following transfers and maintains its initialized value.

```
e.g.: { CB|BR} [Jm| < Imm32>] = {X|Y}Rs ;
/* R suffix indicates: _s = single, _sd = double, _sq = quad */
/* m must be 0,1,2, or 3 for bit reverse or circular buffers */
```

13.7.2.2   Post-Modify with Index Offsets

The address is incremented by 1, 2, 3, or 4 following each transfer.

```
Eg:   Ureg_s = {CB|BR} [Jm +|+= Jn| < Imm8> |< Imm32 >] ;
      L [Jm +|+= Jn| < Imm8 > |< Imm32 >] = Ureg_sd ;
```

```
{X|Y|XY} Rsd = {CB|BR} L [Jm += Jn| < Imm8> |< Imm32 >] ;
         {CB|BR} Q [Jm += Jn| < Imm8> |< Imm32 >] = {X|Y}Rsq ;
```

The address is incremented based on the value Jn

13.7.2.3 Circular Buffer Addressing

Circular buffer IALU load and store instructions generate the next address by adding the increment to the current address in a modulo-like fashion.

The address calculation formula is exactly the same for circular buffer addressing and linear addressing, assuming the LENGTH value equals zero and assuming the BASE value equals the base address of the buffer. Each circular buffer calculation has associated with its four separate values: a BASE value, a LENGTH value, an INDEX value, and a MODIFY value.

- The BASE value is the base address of the circular buffer and is stored in the associated base register.
- The LENGTH value is the length of the circular buffer (number of 32-bit words) and is stored in the associated length register.
- The INDEX value is the current address that the circular buffer is indexing and is stored in the associated IALU register.
- The MODIFY value is the post-modify value that updates the INDEX value.

The following pseudo-code uses these definitions and shows the address calculation:

```
INDEX = INDEX + MODIFY
if ( INDEX >= (BASE + LENGTH) )
  INDEX = INDEX - LENGTH
if ( INDEX < BASE)
  INDEX = INDEX + LENGTH
```

Circular buffer addressing may only use post-modify addressing. The post-modify addressing cannot be supported by the IALU for circular buffering, since circular buffering requires update of the index on each access.

Programs use the following steps to set up a circular buffer:

1. Load the starting address within the buffer into an index register in the selected J-IALU or K-IALU. In the J-IALU, J3–J0 can be index registers. In the K-IALU, K3–K0 can be index registers.
2. Load the buffer's base address into the base register that corresponds to the index register. For example, JB0 corresponds to J0.
3. Load the buffer's length into the length register that corresponds to the index register. For example, JL0 corresponds to J0.

4. Load the modify value (step size) into a register file register in the same IALU as the index register. The J-IALU register file is J30–J0, and the K-IALU register file is K30–K0. Alternatively, an immediate value can supply the modify value.

Circular Buffer Addressing Example

```
.section program ;
  JB0 = 0x100000 ;; /* Set base address */
  JL0 = 11 ;;      /* Set length of buffer */
  J0 = 0x100000 ;; /* Set location of first address */
  XR0 = CB [J0 += 4] ;; /* Loads from address 0x100000 */
  XR0 = CB [J0 += 4] ;; /* Loads from address 0x100004 */
  XR0 = CB [J0 += 4] ;; /* Loads from address 0x100008 */
  XR0 = CB [J0 += 4] ;; /* Loads from address 0x100001 */
  XR0 = CB [J0 += 4] ;; /* Loads from address 0x100005 */
  XR0 = CB [J0 += 4] ;; /* Loads from address 0x100009 */
  XR0 = CB [J0 += 4] ;; /* Loads from address 0x100002 */
  XR0 = CB [J0 += 4] ;; /* Loads from address 0x100006 */
  XR0 = CB [J0 += 4] ;; /* Loads from address 0x10000A */
  XR0 = CB [J0 += 4] ;; /* Loads from address 0x100003 */
  XR0 = CB [J0 += 4] ;; /* Loads from address 0x100007 */
  XR0 = CB [J0 += 4] ;; /* wrap to load from 0x100000 again */
```

The differences between the ADSP-TS20xS processors are highlighted in the table below.

Feature	ADSP-TS101	ADSP-TS201	ADSP-TS202	ADSP-TS203
Max. core clock (MHz)	250	500 /600	500	500
On-chip memory	6 Mbits SRAM	24 Mbits internal DRAM	12 Mbits internal DRAM	4 Mbits internal DRAM
Communications logic unit (CLU)	As a special block in ALU	YES	NO	NO
Link ports (GByte/s)	4 link ports total throughput of 1	4 link ports total throughput of 4	4 link ports total throughput of 4	2 link ports total throughput of 1
External port	64/32-bits	64/32-bits total throughput of 1 GByte/s	64/32-bits total throughput of 1 GByte/s	32-bits ONLY total throughput of 0.5 GByte/s

13.8   Implementation Example Using TigerSHARC (TS201) DSP Processor

Problem statement
It is required to suppress sinusoidal interference noise from a corrupted voice signal. Toward this end, it is desired to design a digital narrow bandstop filter and to implement the filter on TS201 DSP processor.

Design of a digital narrow bandstop filter

- The input signal is a voice signal corrupted by a sinusoidal interference of 900 Hz as shown in Fig. 13.10.
- The filter designed is a narrow bandstop filter with the following specifications as shown in Fig. 13.11.
- The designed filter is of the order 310.

13.8.1   Creating a VisualDSP++ Project

Visual DSP++ is a integrated debugging and development environment (IDDE) for developing and debugging DSP code on processors of Analog Devices family.

- Open VisualDSP++ Environment.
- From the **File** menu, choose **New** → **Project**. The **Project Wizard** dialog box appears.
- On the **Project : General** page, specify the following options

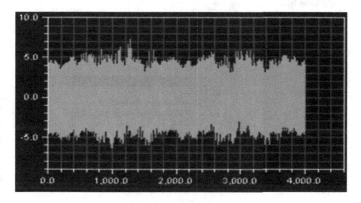

Fig. 13.10 Voice signal corrupted by sinusoidal interference

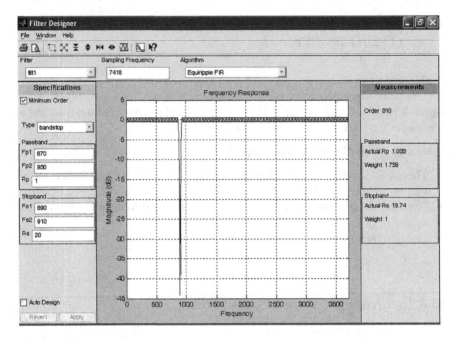

Fig. 13.11 Narrow bandstop filter

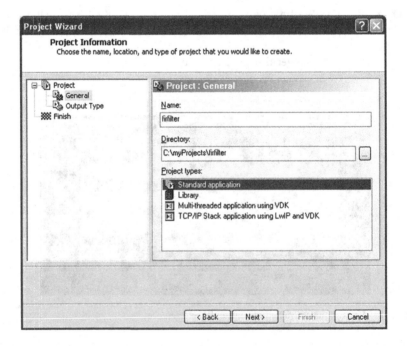

- Click **Next**.
- In Processor types, select ADSP-TS201, silicon version as Automatic, Project output type as Executable file and click **Finish**.

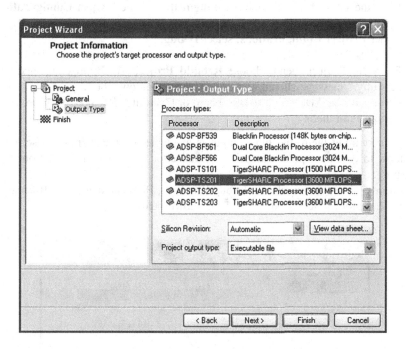

- From the **File** menu, choose the **New** → **File**.
- This creates a new source file.
- Copy the code from Appendix-B and paste it into the source file:
- From the **File** menu, choose **Save** → **File** to save this source file to c:\myprojects\filter.asm.
- From the **Project** menu, choose the **Add to Project/File(s)** and select filter.asm.
- From the **Project** menu, choose **Project Options**. The **Project Options** dialog box appears.
- From link menu, choose LDF Preprocessing and select "D:\Program Files \Analog Devices\VisualDSP 4.5\TS\include" from Additional include Directories or copy cache_macros.h file from visualDSP ++ installation folder to current project directory.

13.8.2   Building the Project

You must now build the project.

1. From the **Project** menu, choose **Configurations**. The **Project Configurations** dialog box appears.
2. Under **Project Configurations**, select **Debug**.
3. Click **OK**.
4. From the **Project** menu, choose **Rebuild Project**. A "Build completed successfully." message should appear on Output window's build page.
5. If you have an ADSP-TS201 DSP board or EZkit lite, download your programs to it and run them.

After running them, the filter output is available in DSP memory. Use a plot window to display a **plot**, which is a visualization of values obtained from DSP memory.

From the View menu, choose Debug Windows, Plot, and then New. Browse for variable "output" and set other values as shown below:

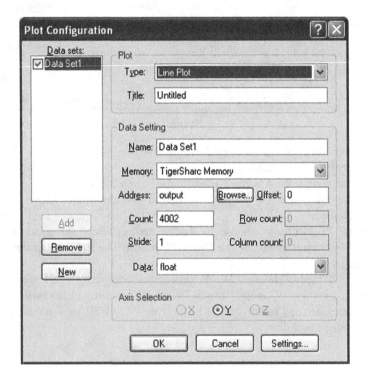

Click OK to visualize the graph as shown in Fig. 13.12.

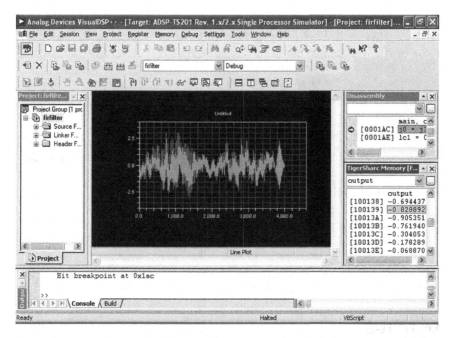

Fig. 13.12 Recovered voice signal

Appendix-A

C code for implementation of lowpass FIR filter on Texas Instruments TMS320C67xx DSP Processor

```c
#include < math.h>
#include < stdio.h>
#include < stdlib.h >
#include < time.h>
#include "inputs1.h"                    // this header file contains in-
put data and filter coefficients
#define   N                  14887      // length of the input data
#define   FILTER_LENGTH   114      // order of the filter
float output_filter[N];
void main()
{
unsigned int i;
unsigned int j, input_length;
double temp;
FILE *fp;
fp = fopen("dontworyrecccs.dat", "w");
```

```
        input_length = N+FILTER_LENGTH-1;
        for(i = (FILTER_LENGTH-1);i < input_length;i ++)
        {
                temp = 0.0;
                for(j = 0;j < FILTER_LENGTH;j++)
                {
                        temp += (filter_coefficients[j] * (*(input_fil-
ter + i - j)));
                }
                *(output_filter + i - FILTER_LENGTH + 1) = (float)temp;
        }
        input_length = input_length - FILTER_LENGTH;
        for(i = 0;i < 14886;i ++)
        fprintf(fp, "%f\n", output_filter[i]);
    fclose(fp);
```

Appendix-B

C code for implementation of narrowband FIR bandstop filter on TigerSHARC (TS201) Processor

```
#include < stdio.h>
#include < stdlib.h>
#include < defts201.h>
#define  N       4002 //  // number of data points in input
#define  FILTER_LENGTH    312 // coeffcients must be multiple of 4
float input_filter[N + FILTER_LENGTH-1] = {
            #include "input.dat"
            };
// input.dat and coefficients.dat files should be in the project folder
// In these files values should be seperated by comma,
float filter_coefficients[FILTER_LENGTH]; = {
            #include "coefficients.dat"
            };
float output_filter[N];
void main(void)
{
        unsigned int i,j,input_length;
        double temp;
        input_length = N+FILTER_LENGTH-1;
        for(i = (FILTER_LENGTH-1);i < input_length;i ++)
```

```
        {
                temp = 0.0;
                for(j = 0;j < FILTER_LENGTH;j ++)
                {
                        temp += ((double)filter_coefficients[j] * (*
(input_filter + i - j)));
                }
                *(output_filter + i - FILTER_LENGTH + 1) = (float)temp;
        }
        input_length = input_length - FILTER_LENGTH;
}
/
*********************End of the Program*********************************
*****/
```

Index

© Springer Nature Singapore Pte Ltd. 2018
K. D. Rao and M. N. S. Swamy, *Digital Signal Processing*,
https://doi.org/10.1007/978-981-10-8081-4

Printed in the United States
By Bookmasters